The Mammals of the Palaearctic Region

The Mammals of the Palaearctic Region: a taxonomic review

G. B. Corbet

British Museum (Natural History) 1978
Cornell University Press
LONDON AND ITHACA

Publication No 788

ISBN 0-8014-1171-8

Library of Congress Catalog Card Number 77-90899

Phototypeset in Great Britain by William Clowes & Sons Limited, Beccles, Suffolk
and printed by Richard Clay (The Chaucer Press) Ltd., Bungay, Suffolk.

Contents

Introduction 1
Zoological scope. 2
Geographical scope 2
The time scale 3
Composition of the Palaearctic mammal fauna 4
Taxonomic treatment 5
Nomenclature 7
New and recent names 8
Literature 9
Explanatory notes 11

The Mammals of the Palaearctic Region 13
Order Marsupialia: marsupials 13
Order Insectivora: insectivores 13
Family Erinaceidae: hedgehogs 13
Family Soricidae: shrews. 17
Family Talpidae: moles etc. 32
Order Macroscelidea: elephant-shrews 37
Order Chiroptera: bats 37
Family Pteropodidae: fruit bats 38
Family Rhinopomatidae: rat-tailed bats 39
Family Emballonuridae: sheath-tailed bats 40
Family Nycteridae: slit-faced bats 41
Family Rhinolophidae: horseshoe bats 42
Family Vespertilionidae: common bats 45
Family Molossidae: free-tailed bats 62
Order Primates: monkeys etc. 63
Order Lagomorpha: rabbits etc. 65
Family Ochotonidae: pikas 65
Family Leporidae: hares and rabbits 70
Order Rodentia: rodents 74
Family Sciuridae: squirrels 75
Family Castoridae: beavers 87
Family Cricetidae: hamsters, gerbils, voles etc 88
Subfamily Cricetinae: hamsters 88
Subfamily Myospalacinae: zokors 93
Subfamily Lophiomyinae: crested rat 94
Subfamily Microtinae: voles and lemmings 94
Subfamily Gerbillinae: gerbils and jirds 118
Family Spalacidae: mole-rats 129
Family Muridae: mice and rats 130
Family Gliridae: dormice. 143
Family Seleviniidae: desert dormice 147
Family Zapodidae: jumping mice 147
Family Dipodidae: jerboas 149
Family Hystricidae: porcupines 158
Family Capromyidae: coypus etc. 159
Family Ctenodactylidae: gundis 159

Order Carnivora: carnivores . . . 160
Family Canidae: dogs . . . 161
Family Ursidae: bears . . . 165
Family Procyonidae: raccoons etc. . . . 166
Family Mustelidae: weasels etc. . . . 167
Family Viverridae: mongooses etc. . . . 177
Family Hyaenidae: hyaenas . . . 179
Family Felidae: cats . . . 179
Order Pinnipedia . . . 185
Family Otariidae: sea-lions . . . 186
Family Odobenidae: walrus . . . 187
Family Phocidae: seals . . . 187
Order Proboscidea: elephants . . . 191
Order Hyracoidea: hyraxes . . . 192
Order Sirenia: sea-cows . . . 192
Order Perissodactyla: horses etc. . . . 193
Order Artiodactyla: even-toed ungulates . . . 195
Family Suidae: pigs . . . 196
Family Hippopotamidae: hippopotamuses . . . 196
Family Camelidae: camels . . . 197
Family Moschidae: musk deer . . . 198
Family Cervidae: deer . . . 198
Family Bovidae: oxen etc. . . . 204

Maps . . . 220
References . . . 272
Index . . . 298

Acknowledgements

This work could not have been undertaken without the solid foundation provided by the *Checklist of Palaearctic and Indian mammals* of Ellerman & Morrison-Scott and I am grateful to Sir Terence Morrison-Scott for his encouragement to embark upon the project. The number of occasions on which I have felt it necessary to deviate from their conclusions is no criticism of the *Checklist* but rather reflects the enormous influence it has had in stimulating and facilitating original taxonomic work on Palaearctic mammals.

All my colleagues in the Mammal Section of the British Museum (Natural History) have given freely of their assistance and advice. I am especially grateful to Miss Jean M. Ingles for her massive contribution in curatorial work and literature retrieval; to Miss Daphne M. Hills who prepared the final copies of the maps; and to Mr J. E. Hill for his critical comments on the typescript, especially the section on Chiroptera.

Useful information was gathered on visits to the following museums and I express my gratitude to all the curators concerned: U.S. National Museum of Natural History, Washington; American Museum of Natural History, New York; Field Museum, Chicago; Museum of Comparative Zoology, Harvard; Rijksmuseum van Natuurlijke Historie, Leiden; Museum Alexander Koenig, Bonn. I am also indebted to Professor Y. Imaizumi for providing information on material in the National Science Museum, Tokyo and to Dr O. L. Rossolimo for supplying details of some Russian bibliographic references unavailable in London.

Introduction

The Palaearctic Region can be defined approximately as the continent of Eurasia north of the Himalayas along with Africa north of the Sahara. It has been recognized and accepted as a natural zoogeographic region since it was first proposed by Sclater in 1858. The first comprehensive listing of the mammals of the region was the *Checklist of Palaearctic and Indian mammals, 1758 to 1946* by J. R. Ellerman and T. C. S. Morrison-Scott, published by the British Museum (Natural History) in 1951 and again, with corrections but no revision, in 1966. The *Checklist* has been used as a base-line for this volume and its authors will be referred to throughout the text as 'E & M-S'. The production of the *Checklist*, including all known names applied to the mammals of the region, was a complex task. Two difficulties that are more acute in the Palaearctic than in other parts of the world are the large number of descriptions, often very inadequate, from the very early period of Linnaean nomenclature in the late 18th century; and the very great diversity of countries and languages involved. The *Checklist* provided an invaluable foundation for any further work on the taxonomy of Palaearctic mammals by the thoroughness with which it summarized the taxonomic literature up to 1946. Since then a great deal of progress has been made but often in a very piecemeal fashion, resulting in a very scattered literature and very few comprehensive revisions of particular taxa.

My object in this work has been twofold: firstly to provide for the non-specialist a complete and independent list of Palaearctic mammals with keys for identification, concise statements of distribution and just sufficient taxonomic detail to allow the user to interpret intelligently and cautiously the results of the taxonomists' labours; and secondly to provide for the practicing taxonomist a supplement to the *Checklist* (but limited to the Palaearctic Region), by listing with full references all names proposed since 1946 and by taking account in the classification of revisionary work published since 1951.

This is a compilation, although I hope it is a critical one. The intention has been to present a statement of the current state of the classification of Palaearctic mammals. In certain groups and at certain levels this classification is approaching a definitive state, but there is still much to be done before even limited claims of completeness can be made.

There is a tendency amongst non-zoologists, and indeed amongst some zoologists, to believe that classifications can have some kind of official status. It must be emphasized that this is not so. Whether two forms are recognized as different species or are united in a single species, whether a species is included in genus A or genus B, these are subjective judgements that must be made by the individual taxonomist. The more evidence is considered, the more likely is it that different taxonomists will arrive at the same conclusion, but by its very nature classification cannot be absolute. Whether we work in the museum or 'in the field', we can only study samples, usually exceedingly small samples, of the populations that we envisage when we name and refer to species. When we say that tigers are striped we are not saying only that the ones we have examined are striped but we are making a generalization about tigers, including, wittingly or not, those still alive in their natural range and indeed those as yet unborn. We are in fact making a statistical statement whose probability of being true depends heavily upon the size of our sample.

Having decided that a number of animals belong to one species, its naming is indeed governed by a code prepared by the International Commission on Zoological Nomenclature (1961). But the official standing of a species' name should not be mistaken for any kind of official recognition of the species itself, and the same applies to the higher categories of classification.

Zoological scope

All groups of mammals are included except the order Cetacea (whales and dolphins). These are more appropriately listed by oceans rather than by continents, since even inshore species tend to occur on both sides of the Atlantic or Pacific oceans. There are no exclusively freshwater species in the Palaearctic Region although they do occur in the rivers Indus, Ganges and Yangtze Kiang, just beyond the southern limit of the region. The most recent comprehensive nomenclatural checklist of the Cetacea of the world is that of Hershkovitz (1966). For a concise list see Scheffer & Rice (1963).

All well-established introductions are included, a category not dealt with in the *Checklist*.

Geographical scope

A rigid boundary of the Palaearctic Region, illustrated in Map 1 and described below, has been used to the extent that all species that occur *within* the line, however marginally and even if predominantly non-Palaearctic, are listed. But the boundary has not been allowed to exclude full treatment of those segments of predominantly Palaearctic species that occur beyond it. Species that extend only a little way beyond the Palaearctic Region are treated in their entirety; but species that extend far into adjacent regions have had their extra-Palaearctic parts more briefly treated, without full listing of subspecies and synonyms.

The geographical scope employed in the *Checklist*, to 10°N in Malaya, was designed to complement that of Chasen's (1940) *Handlist of Malaysian mammals*. However, since that work, covering the southern half of the Oriental Region, is now out of print, out of date and scarce, it seemed unwise to attempt a revised listing of only the northern half of the Oriental Region. Hence the limitation of this work to the Palaearctic Region, in the hope that a revised list of the *entire* Oriental Region will be undertaken before too long.

In northern Africa this work overlaps extensively with *The mammals of Africa: an identification manual* (Meester & Setzer, 1971), but this seems quite fitting in view of the strongly transitional nature of the north African fauna.

The actual boundary used here can be described in more detail as follows, beginning in the west. The islands of Spitzbergen, Iceland, Azores, Madeira and Canaries are included, but the Cape Verde Islands are excluded. In Africa the boundary begins in the west at 21°30′N, i.e. between Rio de Oro and Mauritania, continues across Mauritania and Mali at the same latitude and thereafter follows the political boundaries such as to include in their entirety Algeria, Libya and Egypt and to exclude the whole of Niger, Chad and Sudan – the Hoggar Mountains are therefore included and the Tibesti excluded. The whole of the Arabian peninsula is included. The boundary in Asia begins between Pakistan and Iran and cuts across Afghanistan to exclude only those eastern parts draining into the Indus. From Afghanistan to central China it follows approximately the 3000 metre contour such that the forested and other lower slopes draining into Pakistan, India, Burma and China are excluded whilst the alpine zone of the mountains is included, along with the Tibetan Plateau. As a result Afghanistan is included except for the extreme eastern parts that drain to the Indus; Pakistan, India, Nepal, Burma and the Chinese provinces of Yunnan and Szechuan are excluded except for the alpine zones; Tibet is included except for the valley of the Tsangpo. In lowland China the boundary is taken very arbitrarily as latitude 35°N, corresponding in part with the Hwang-Ho. In Japan it follows latitude 30°N so that Yakushima is the southernmost island included. In the Bering region all territory of the USSR is included, all territory of the USA, e.g. St Lawrence Island and the Aleutians, is excluded.

Some predominantly Palaearctic genera are listed in their entirety by including species that fall just outside the boundary as defined above. Examples are *Apodemus* which has one species in the lower mountains of southern China and in Taiwan; and *Microtus* which has one species in the high alpine zone of Taiwan.

The sharpness of the faunal change across this boundary varies considerably from place to place and from group to group. In Africa the boundary is clearly meaningless for the desert fauna of the Sahara such as the gerbilline rodents. The Sahara does represent the southern limit of many Palaearctic groups that reach North Africa (e.g. *Microtus*, *Apodemus*, *Mustela*, *Cervus*), but a certain number of predominantly subsaharan African groups recur north of the Sahara (e.g. *Elephantulus*, *Praomys*, *Lemniscomys*, *Genetta*) or did before recent extinctions (*Alcelaphus*, *Loxodonta*). Southern Arabia has representatives of a number of African (*Papio*, *Ichneumia*, *Arvicanthis*, *Praomys*) although some of these can be suspected of introduction by human agency.

In Asia many dry-country species extend beyond my boundary into Pakistan and India (e.g. *Paraechinus*, *Gerbillus*) but the boundary serves to exclude such characteristically Oriental groups as the squirrels *Funambulus* and *Petaurista*. The Himalayan tree-line is probably the sharpest part of the boundary, although the coniferous forest belt, e.g. in Nepal, is a very transitional zone where Palaearctic groups like *Sorex*, *Apodemus*, *Pitymys* and *Ochotona* coexist with Oriental groups like *Petaurista* and *Presbytis*. There is also an endemic Himalayan/Chinese montane forest fauna (*Soriculus*, *Ailurus*, *Budorcas* etc.) that is excluded here because it tends to coexist with Oriental rather than Palaearctic species although some elements of it, e.g. *Soriculus*, have their closest relatives in the Palaearctic.

In lowland China there is gross overlap between the Palaearctic and Oriental fauna, with Palaearctic groups like *Apodemus* and the Microtinae reaching southern China and Oriental groups such as *Paguma* and *Callosciurus* extending north to Peking. In Japan the northern island of Hokkaido has a totally Palaearctic fauna; the more southern islands have a predominantly Palaearctic fauna but with several elements from the Oriental Region (e.g. *Macaca*, *Petaurista*, *Chimmarogale*).

The fauna of northeastern Siberia has a very close affinity with that of Alaska; in some groups a line in the region of Lake Baikal divides Palaearctic and Nearctic forms more clearly than do the Bering Straits (*Ovis*, *Spermophilus*).

The time scale

This account is limited to living species plus those that have become extinct in historic times. The latter category was not included in the *Checklist* – the additional species involved are detailed on p. 4. The synonymies however, like those in the *Checklist*, do not include names based on fossil or subfossil specimens. The inadvertent use of the same name for separate fossil and recent forms in the same genus is a frequent source of nomenclatorial confusion. Use of the *Checklist*, supplemented by this work, should provide a comprehensive list of names used for recent forms in the Palaearctic Region but there is no comparable checklist for fossil forms. It is therefore essential that a thorough search of the palaeontological literature be made before new names are proposed.

This work includes all names proposed up to the end of 1972 as indicated in the *Zoological Record*. No exhaustive search has been made for names subsequent to that date although many are included up to and including 1976. Compilation began in 1967 but important revisionary studies published up to 1972, and less comprehensively up to 1976, have been taken into account.

Composition of the Palaearctic mammal fauna

The composition of the fauna by orders is shown in Table 1. The differences between the figures in the two columns of Table 1 are accounted for by the following categories:

Table 1 Composition of the Palaearctic mammal fauna

	No. of extant, indigenous species occurring wholly or extensively within the Palaearctic Region	Total no. of species included in this work
Marsupialia	0	1
Macroscelidea	1	1
Insectivora	52	64
Chiroptera	62	82
Primates	2	5
Lagomorpha	22	23
Rodentia	217	241
Carnivora	45	57
Pinnipedia	14	15
Proboscidea	0	2
Hyracoidea	1	1
Sirenia	0	2
Perissodactyla	3	4
Artiodactyla	41	52
Total	460	550

Introduced species: *Macropus rufogriseus*, ? *Suncus murinus*, *Macaca cyclopis*, *Sciurus carolinensis*, *Callosciurus flavimanus*, *Castor canadensis*, *Ondatra zibethicus*, *Rattus rattus*, *Myocastor coypus*, *Procyon lotor*, *Mustela vison*, *Muntiacus reevesi*, *Cervus axis*, *Odocoileus virginianus*, *Ovibos moschatus*.

Totally extinct species: *Prolagus sardus*, *Hydrodamalis gigas*, *Gazella rufina*.

Species extinct in the wild but surviving in domestication or captivity: *Camelus dromedarius*, *Elaphurus davidianus*, *Bos primigenius*.

Species extinct in the Palaearctic Region but surviving elsewhere (all of these were never more than marginal in the Palaearctic Region during historic times): ?*Lophiomys imhausi*, *Panthera leo*, *Zalophus californianus*, *Elephas maximus*, *Loxodonta africana*, *Equus africanus*, *Hippopatamus amphibius*, *Alcelaphus buselaphus*, *Oryx dammah*.

Oriental Region species that occur marginally in the Palaearctic Region as here defined: *Chimmarogale himalayica*, *Pteropus dasymallus*, *Myotis formosus*, *M. ricketti*, *Macaca mulatta*, *Callosciurus swinhoei*, *Sciurotamias davidianus*, *Trogopterus xanthipes*, *Meriones hurrianae*, *Rattus niviventer*, *Martes flavigula*, *Arctonyx collaris*, *Lutra perspicillata*, *Paguma larvata*, *Herpestes auropunctatus*, *H. edwardsi*.

Oriental species that do not occur in the Palaearctic Region as here defined but are included because they are the sole Oriental members of Palaearctic groups: *Paraechinus micropus*, *Eothenomys melanogaster*, *Hyperacrius wynnei*, *H. fertilis*, *Gerbillus gleadowi*, *Apodemus draco*, *Salpingotus michaelis*.

Species of the Himalayan-Chinese montane forests that are included because they occur marginally on the Palaearctic side of the border or because they belong to Palaearctic genera: *Sorex cylindricauda*, *S. bedfordiae*, *Soriculus hypsibius*, *Blarinella quadraticauda*, *Scapanulus oweni*, *Apodemus gurka*, *Eozapus setchuanus*.

Ethiopian Region species that occur marginally in the Palaearctic Region as here defined: ?*Crocidura religiosa*, *C. lusitania*, *C. flavescens*, *C. sericea*, *Eidolon helvum*, *Rousettus aegyptiacus*, *Coleura afra*, *Taphozous perforatus*, *Nycteris thebaica*, *Rhinolophus clivosus*, *Hipposideros caffer*, *Myotis bocagei*, *Pipistrellus rueppelli*, *Nycticeius schlieffeni*, *Scotophilus leucogaster*, *Tadarida pumila*, *T. midas*, *Papio hamadryas*, *Xerus erythropus*, *Lemniscomys*

barbarus, *Praomys fumatus*, *P. erythroleucus*, *Lycaon pictus*, *Ichneumia albicauda*, *Felis serval*, *Tragelaphus imberbis*.
Oriental and Ethiopian Region species that occur only marginally in the Palaearctic Region: *Taphozous nudiventris*, *Tadarida aegyptaica*, *Dugong dugon*.
N. American species that occur marginally in N.E. Siberia: *Sorex cinereus*.
Vagrant bats from N. America: *Myotis lucifugus*, *Lasiurus cinereus*.

Taxonomic treatment

Care has been taken to give some indication of the degree of certainty or uncertainty with which the various taxonomic conclusions should be accepted. Since there is a tendency for the non-taxonomist to be either embarassingly credulous or totally incredulous of taxonomic work, it would seem useful to consider here in general terms the factors affecting the degree of definitiveness, dealing separately with the three very distinct operations involved in taxonomy. These are (1) the recognition of the species; (2) the description of variation within the species; (3) the classification of the species in a hierarchy of increasingly large groups.

The species

The recognition of species involves the recognition of those morphological (or other phenotypical) discontinuities that correspond to reproductive incompatibility between populations. Uncertainties at this level take three principal forms.

Firstly we have the problem of populations that we believe, on good grounds, to be completely isolated, e.g. the hedgehogs of the genus *Erinaceus* in the western Palaearctic (*europaeus*) and in eastern Asia (*amurensis*). In such a case we must recognize that there can be no *absolute* criterion for saying that the two forms are conspecific, although experimental work might demonstrate a sufficiently absolute form of reproductive barrier to enable us to say that they are *not* conspecific. But such experimental evidence is rarely available for Palaearctic mammals. Instead we must make a judgement on morphological grounds, occasionally assisted by evidence from reproductive behaviour in artificial conditions.

Secondly there is the situation where we consider it very likely, judging by continuity of suitable habitat, that two forms *are* in contact, but our available samples are from quite separate parts of the range. Here the question 'Are the two forms conspecific?' probably does have an objective meaning, and we must try to predict the actual relationship on the basis of our samples. In other words we must assess the probability that our two recognizably different forms will intergrade smoothly in the intervening region, or alternatively will meet at a boundary, or overlap, with morphological discontinuity caused by reproductive incompatibility. An example of this problem is provided by the hares (*Lepus*) where there is uncertainty of this kind as to the relationship between the forms *europaeus* of northern Europe and *capensis* of Africa.

Finally we have the problem of sibling species, i.e. sympatric species that are so similar morphologically that the differences have been overlooked or, more often, mistaken for individual variation. An example that has recently come to light is in the genus *Plecotus* (long-eared bats), of which the European representatives were treated as conspecific until recently but are now recognized as belonging to two species, *P. auritus* and *P. austriacus*.

A special category of sibling species that presents particular problems involves single morphological species that are found to consist of two or more groups that appear from cytological or experimental evidence to be reproductively isolated, but which cannot yet be recognized morphologically. The shrew *Sorex araneus* in Europe and the mole-rat *Spalax leucodon* in Palestine are well-studied examples but others are coming to light as karyological studies proceed. Where the karospecies are predominantly allopatric the policy that I have adopted here is to retain them as a single species as far as formal nomenclature is concerned until such time as satisfactory morphological differences are found. The alternative of treating them as separate named species results in the majority of records, past and present, having to remain unidentified and without a species name. In karyological studies a special nomenclature can be used, e.g. '*Sorex araneus* karyospecies A'. Where karyological studies have demonstrated widespread sympatry I have accepted the formal naming of each member of the pair, e.g. in the case of *Microtus arvalis* and *M. subarvalis* (p. 113).

In general considerably more confusion is caused by erroneously uniting under one specific name forms that are in fact distinct species, than by erroneously giving separate specific rank to two forms that are in fact conspecific. Therefore, in compiling this checklist lumping at the species level has been done cautiously, and several forms that are probably conspecific are tentatively given separate specific rank pending further data.

Specimens of all species listed have been examined unless it is stated otherwise under 'Remarks'. (The exceptions amount to 28 species many of which are doubtfully valid.) Where there is general agreement amongst recent authors, the current classification has been accepted with only a superficial examination of specimens to confirm that the classification seems reasonable. Where conflicting views have been expressed a more detailed examination has been undertaken to determine which classification is more likely to be correct or, more rarely, to produce a novel classification. All deviations at the species level from the arrangement in the *Checklist* are indicated by a dagger (†) and are explained under 'Remarks', either by referring to the alternative published classification that has been preferred or by providing reasons based on an original assessment of material. The most significant original contribution is probably in the genus *Eothenomys*.

In many groups the delimitation of the species is closely approaching a definitive state, whilst in others it is still very provisional. Such areas of uncertainty are indicated either in the introductory remarks to the genus or in the remarks under individual species. It helps to assess the degree of uncertainty if we remember that taxonomy, like all science, depends upon statistical predictions that the patterns of variation observed in our samples actually exist in the natural populations from which our samples were drawn, which in turn depends upon the size and randomness of our samples.

The subspecies

Whereas no alterations have been made to current classifications at the specific level without very careful assessment of published evidence and/or specimens, no such claim is made for the listing of subspecies, which must be treated as highly provisional. I begin with the conviction that there is no advantage to be gained by applying formal subspecific names to segments of a species that have no objective boundaries. Given adequate knowledge of a species' distribution, subspecific names can only be usefully applied to completely isolated segments whose members are recognizably different, or to contiguous segments sharing a relatively narrow zone of intergradation between two recognizably different forms, a rare phenomenon almost certainly indicating secondary contact. However, there are many cases where our knowledge of distribution is very incomplete and the available *samples* show discontinuous variation that lends itself to the recognition of subspecies. But it is important to recognize such subspecies as very provisional and liable to abandonment if further data should prove continuity of variation.

In most taxonomic works on mammals the presence of really objective subspecific boundaries, which are of great interest to the student of speciation, of historical zoogeography and of population genetics, is completely obscured by great numbers of so-called subspecies, based on subtle differences of colour, size or proportions, that are most unlikely to have any objective boundaries.

The authors of the *Checklist* took a conservative view and refrained from uniting subspecies if good evidence for continuity of variation was lacking. Whilst such caution is very desirable at the species level, the erroneous unification of subspecies is much less disastrous since a record accompanied by the species name *and the precise locality* is usually quite unambiguous. (There might be exceptions to this in the case of migratory species and species like the commensal rats and mice that are liable to be accidentally transferred from one part of their range to another.) The policy followed here has therefore been to unite forms that are certainly conspecific where one can reasonably predict continuity of range and variation, even if such continuity has not been demonstrated. Some mistakes are bound to have been made whilst in other cases subspecies that have been retained because of complete isolation may well prove not to be recognizably different when large samples are available. Only the study of subspecific variation accompanied by detailed mapping of distribution can resolve these difficulties. In many species no attempt has been made to list subspecies and all synonyms of the species name have been listed in a single sequence.

Finally it must be emphasized that the great majority of mammalian subspecies are based upon differences in a single character and that a subspecific difference, even if it stands up to the most rigorous criteria now applied, should never be taken to represent any particular level of genetic difference in the absence of genetic data relating to the particular characters involved. The adoption of the rather strict criteria detailed above for the recognition of named subspecies does not imply any denial of the importance of clinal variation but is based upon the belief that the application of formal subspecific names to parts of a continuous system of variation is an exceedingly inefficient and misleading way of describing this kind of subspecific variation. Whilst it can reasonably be expected that the delimitation of species will soon approach a definitive state, it must be recognized that the description of subspecific variation is, quite literally, limitless. Those who wish to pursue this study beyond the recognition of the major, *discrete* segments of the species should not attempt to apply the classical system of binominal and trinominal nomenclature below that level. For a more detailed discussion of subspecies see Corbet (1970).

Classification above the species

There are still many cases of instability in the higher classification of mammals, especially at the level of the genus. The classification used in the *Checklist* is followed here unless there have been specific reasons to deviate from it, and these reasons are stated in the appropriate part of the text. However, there have in recent years been very few attempts at revision of any group of mammals that have taken into account *all* the known species throughout the entire range of the group. Only such comprehensive revisions, taking account of all the available characters, will achieve a stable classification.

Subgenera have only rarely been employed in the list. Too often subgenera, or even genera, have been based on the discovery of a peculiar character in one species without careful study of the entire genus or related genera. If a genus contains more than two species, it is most unlikely that the differences between all pairs of species will be equally great. It is, therefore, always possible to recognize a finely graded hierarchy of small taxa. But this will serve little purpose unless the divisions are based upon the coincidence of a sufficient number of character differences to offer some prospect of stability.

The sequence of orders, families and genera is that of Simpson (1945) and is that used in the great majority of recent works on mammals. The sequence of orders in particular therefore differs from that used in the *Checklist*.

Nomenclature

Palaearctic mammals, having received considerable attention in the earlier years of the history of Linnaean nomenclature, suffer particularly from the chief weakness of the type system of nomenclature. The older a name the more important it is (by the law of priority), but at the same time the more likely it is to be based upon an inadequate description and/or a missing, undocumented or badly preserved type-specimen. In fact very few specific names of Palaearctic mammals are based upon adequate, useful and generally accessible original descriptions, and provided there is a good redescription, e.g. those of Miller (1912) for European species, and reasonably consistent usage in recent years, there is no case for allowing doubts about the original meaning and validity of the name to affect its current usage. The 'fifty-year rule', introduced in the 1961 edition of the *International code of zoological nomenclature*, was designed to avoid upsetting current nomenclature by zoologically sterile research into the early literature. According to this rule names that have not been used as senior synonyms in major works during the last fifty years should not be allowed to supplant the currently used names, but should be submitted to the International Commission to be considered for placing on a list of invalid names. Provided this rule is followed with a will to preserve stability rather than in a pedantic search for loopholes that will permit upsetting current usage, it should serve a useful purpose.

The *Checklist* embodies an enormous amount of painstaking work on the allocation of the earlier synonyms. The results, even when highly tentative, are accepted here without comment since their zoological importance is usually very small. Provided the 'fifty-year rule' is used to advantage, names more than fifty years old are irrelevant to current nomenclature unless they are the presently used names of taxa or have been deliberately used as such in the taxonomic literature during the last fifty years.

All synonyms are listed here in alphabetical order under the species or subspecies to which it is believed they apply. This will help to indicate the identity of records in the older literature provided it is realized that many of the older names in particular have at times been misapplied.

The nomenclature of domesticated animals presents a particular problem that affects also the names of their wild ancestors. The policy adopted here is to ignore names based upon domesticated forms and to use for the wild species the earliest valid name based upon a non-domesticated form, as advocated by Bohlken (1961) and Groves (1971). The naming of the domesticated forms themselves is outside the scope of this work but I would advocated a purely vernacular description following the name of the genus, e.g. *Capra* (domestic), *Equus* (domestic horse), depending upon whether one or more species of a genus has been domesticated.

Applications have been made to the International Commission on Zoological Nomenclature with respect to the names *Erinaceus sibiricus* (p. 15), *Sorex kinezumi* (p. 28), *Vulpes* (p. 162) and *Panthera* (p. 184) as explained on the pages quoted.

Vernacular names of species are given only when they are appropriate in a Palaearctic context or are very frequently and consistently used for the species as a whole.

New and recent names

No new names are proposed in this work. From 1946 (the latest date for inclusion in the *Checklist*) to 1972 (the latest volume of the *Zoological Record* available at the time of completion of the script) 60 new species have been named from the Palaearctic Region. Of these the following 26 are retained here as species (plus two marked *, described after 1972), although some of these must be considered of doubtful validity: *Sorex hosonoi*, *Talpa streeti*, *Myotis hosonoi*, *M. ozensis*, *M. pruinosus*, *Pipistrellus endoi*, *P. bodenheimeri*, *Vespertilio orientalis*, *Murina tenebrosa*, *Ochotona thomasi*, *O. kamensis*, *Clethrionomys rex*, *Pitymys tatricus*, *P. bavaricus*, *Microtus subarvalis*, *M. sachalinensis*, *Gerbillus perpallidus*, *G. hoogstraali**, *G. occiduus**, *Dipodillus kaiseri*, *D. maghrebi*, *Apodemus krkensis*, *A. microps*, *Dryomys laniger*, *Sicista pseudonapaea*, *Jaculus turcmenicus*, *Salpingotus heptneri*, *S. michaelis*.

In addition the following names, proposed since 1946 for subspecies, are used here for species: *Gerbillus mesopotamiae*, *G. aureus*.

Most of these additional species are very closely similar to previously known ones and can truly be said to have been 'newly recognized' rather than 'newly discovered'.

Ten new generic/subgeneric names have been proposed during the same period. Only two are used here as the valid names of genera, namely *Dinaromys* and *Sekeetamys*.

All subspecific names proposed since 1946 are listed, often without any attempt to comment upon their taxonomic validity. It is likely that very few would stand up to the criteria for subspecies laid down on p. 6.

Infrasubspecific names, *nomina nuda* and other names that were not validly proposed according to the International Code have been ignored, although it is possible that some subsequent validations of such names may have been overlooked.

All names that are additional to those in the first edition of the *Checklist* are marked * in the text, the original form of citation is given, with type-locality or type-species, and the bibliographic details are given in the list of references.

Literature

Literature on mammal taxonomy is very fragmentary and scattered. Many excellent regional works are available and these are listed below. What is lacking are comprehensive revisions of taxonomic groups such as genera or families, without regional limitations. For many, perhaps most, families of mammals there has been no comprehensive revision this century in spite of an enormous piecemeal accumulation of new taxonomic data. The major taxonomic works on particular groups are mentioned in the text introducing the groups concerned. Below are annotated lists of works on the whole class of mammals, on the separate parts of the Palaearctic Region, and on adjacent regions.

General works on mammals

Anderson & Jones (1967). A concise account of each family.

Butler *et al.* (1967). An outline classification of fossil and living mammals giving earliest and latest fossil records for each order or suborder and comments on the fossil history.

Grassé (1955). A comprehensive account in French, dealing with most genera and with much anatomical detail.

Hsu & Benirschke (1967). A continuing loose-leaf production of descriptions of karyotypes.

McKenna (1975). An outline of an original phylogenetic classification of mammals (fossil and recent).

Simpson (1945). A comprehensive classification of fossil and living mammals.

Walker (1975). An illustrated account of each genus of mammals.

Works on parts of the Palaearctic Region

Unless otherwise mentioned all the following works are comprehensive in dealing with all groups of mammals (other than Cetacea). In compiling the list priority has been given to modern works with illustrations, maps or keys that cover a large area, but some less ambitious works are included for areas lacking modern comprehensive coverage.

Afghanistan. Hassinger (1973): basically an expedition report but dealing with all species. Maps but no keys nor illustrations.

Arabian region. Harrison (1964a, 1968a, 1972): keys, maps and illustrations covering the Arabian peninsula north to and including Syria and Iraq (with maps extending to Asia Minor).

British Isles. Corbet & Southern (1977): maps, keys, illustrations, much biological data.

China. Allen (1938–40): keys, some maps, brief verbal descriptions; for the more difficult groups not an adequate substitute for the more detailed, scattered literature.

China (Chinghai). Chang & Wang (1963): a checklist with analysis of faunal regions (Chinese, English summary).

China (Manchuria). Zimmermann (1964): an expedition report in German but covering most species.

China (Tibet). Shen (1963): a checklist and analysis of distribution by vegetation zones (Chinese, Russian summary).

Europe. Brink (1967): a field guide with maps and coloured illustrations. Kurten (1968): a general account of the Pleistocene mammals of Europe. Miller (1912): keys, descriptions and illustrations of skulls and teeth, valuable if the obsolete taxonomy is allowed for. Niethammer (1963): a detailed account of introduced mammals in Europe. Niethammer & Krapp (in press): a detailed multivolume handbook.

Finland. Siivonen (1972). Detailed account in Finnish, with illustrations and maps which cover the whole of Scandinavia.

France. Saint Girons (1973): keys, illustrations, detailed maps except for ubiquitous species. No pinnipedes.

Himalayas. Mitchell (1975): a brief checklist of the mammals of Nepal. Weigel (1969), Gruber (1969): two consecutive papers dealing with the Insectivora and Rodentia of Nepal. Ellerman (1961): Rodentia only of India, Pakistan and Burma, keys and detailed descriptions.

Iran. Misonne (1960): maps of most species, no illustrations. Lay (1967): an expedition report but with notes on all recorded species. No keys, descriptions nor maps.

Italy. Toschi & Lanza (1959), Toschi (1965): illustrations and keys, no maps.

Japan. Imaizumi (1960): coloured illustrations, a few maps, text in Japanese; Imaizumi (1970a): Insectivora, Chiroptera, Primates, Lagomorpha only; keys, illustrations and maps for most species, text in Japanese and English.

Korea. Jones & Johnson (1960, 1965): Insectivora, Lagomorpha and Rodentia, no illustrations nor maps.

Mongolia. Bannikov (1954): detailed maps, some illustrations, in Russian; Bannikov (1953): keys, illustrations of diagnostic characters, in Russian.

Nepal. See Himalayas.

Scandinavia. Siivonen (1968): a field guide with coloured illustrations, keys and maps (in Swedish and Finnish editions).

Switzerland. Baumann (1949): detailed accounts, illustrations, keys (in German).

USSR. Bobrinskii *et al.* (1965): coloured illustrations, keys and maps for all species (in Russian); Formozov (1965): coloured illustrations, maps (in Russian); Gromov *et al.* (1963): monochrome illustrations, keys, no maps; Heptner & Naumov (1961–1972): the beginnings of a very detailed series dealing so far with Artiodactyla, Perissodactyla, Sirenia and Carnivora (in Russian; German translations available for first two parts); Ognev (1928–1950): a very detailed series, excluding ungulates; earlier volumes very out-dated (in Russian, English translations available).

USSR (Khirgizia). Gromov & Yanushchevich (1972): keys, some illustrations and maps.

USSR (Siberia). Stroganov (1957, 1962): Insectivora and Carnivora only, keys, illustrations, maps.

USSR (Ukraine). Abelentsev & Pidoplichko (1956), Abelentsev (1968): Insectivora, Chiroptera and Mustelidae only, detailed illustrated account, maps (in Ukrainian).

Yugoslavia. Dulic & Miric (1967): a checklist only; Petrov (1968): distribution maps of some Insectivora, Chiroptera and Rodentia.

Works on neighbouring regions

Africa. Meester & Setzer (1971–): a comprehensive account for the whole continent, with keys and a concise account of range for each species. The Carnivora and parts of the Insectivora and Rodentia have not yet appeared although provisional versions received a wide circulation. Dorst & Dandelot (1970): a field guide to the larger mammals with coloured illustrations and maps.

North America. Hall & Kelson (1959): keys, maps, few illustrations, covering the entire continent. Banfield (1974): Canada only, with coloured illustrations and maps.

Oriental Region. Nothing comprehensive. The checklists of Ellerman & Morrison-Scott (1951, 1966) and Chasen (1940) together cover most of the region. The *Fauna of India* series dealt in great detail with the Indian region but apart from the obsolete edition of Blanford (1888–91) only covers the Primates and Carnivora (Pocock, 1939, 1941) and the Rodentia (Ellerman, 1961). Lekagul & McNeely (1977): a comprehensive account of the mammals of Thailand. Van Peenen, Ryan & Light (1969): an identification manual for the mammals of southern Vietnam. Roberts (1977): a comprehensive account of the mammals of Pakistan.

Explanatory notes

Identification keys

The keys serve as a guide (but only as a guide) to identification and also provide some degree of justification and explanation of the classification adopted. Identifications made by using the keys should always be checked against more detailed descriptions. The keys to genera in particular are intended to be practical and may not apply to non-Palaearctic species.

The regional works listed above under 'Literature' will be found useful for the confirmation of determinations. Works on restricted taxonomic groups are mentioned in the remarks introducing the major taxa.

Synonyms

Generic synonyms that have been applied only to non-Palaearctic species are not listed. Objective synonyms are listed first, preceded by =, followed by subjective synonyms preceded by 'Incl.' meaning 'Including'. (Objective synonyms are those based upon the same type-species, in the case of generic names, or upon the same type-specimen, in the case of specific and subspecific names; subjective synonyms are those based upon different type-species or type-specimens.)

Type-localities

Detailed type-localities are given for all names not listed in the *Checklist*. For the valid names of other species and subspecies approximate type-localities are given; for synonyms the type-region is indicated only where the name concerned has been, or is likely to be, used for a possibly valid subspecies.

Abbreviations and symbols

AMNH	American Museum of Natural History, New York.
BM(NH)	British Museum (Natural History), London.
Checklist	The *Checklist of Palaearctic and Indian mammals* (Ellerman & Morrison-Scott, 1951, 1966).
E & M-S	Ellerman & Morrison-Scott (1951, 1966).
MCZ	Museum of Comparative Zoology, Harvard, USA.
USNM	United States National Museum, Washington.
*	A name not included in the *Checklist*. These are mostly names proposed after 1946 but also include: names overlooked by E & M-S (including those added to their second edition of 1966); names of African species that have only recently been recorded from the Palaearctic Region; the names of extinct and introduced species, categories that were not included in the *Checklist*.

† Names whose allocation to species differs from that in the *Checklist*. (Where a valid race has been transferred thus along with all its synonyms the latter have not been separately marked.)

The mammals of the Palaearctic Region

Order **MARSUPIALIA**

Marsupials

Represented in the Palaearctic Region only by one species introduced from Australia.

Family **MACROPODIDAE**

Genus ***MACROPUS****

Macropus Shaw, 1790. Type species *M. giganteus* Shaw. =*Kangurus* Geoffroy & Cuvier, 1795. Incl. *Protemnodon* Owen, 1873, *Wallabia* Trouessart, 1905.

This genus is frequently held to include the large wallabies as well as the large kangaroos, e.g. recently by Ride (1970), although the names *Protemnodon* and *Wallabia* also continue to be used.

Macropus rufogriseus* Red-necked wallaby

Kangurus rufogriseus Desmarest, 1817:36. King Island, Bass Strait, Australia.
Macropus (Halmaturus) fruticus Ogilby, 1838. Tasmania.
Macropus bennetti Waterhouse, 1838.

RANGE. Feral in two areas in England (Peak District and Sussex) and formerly in Germany. Native in eastern Australia and Tasmania.

REMARKS. For details of synonymy see Iredale & Troughton (1934).

Order **INSECTIVORA**

Insectivores

The families of the Insectivora are discrete and well-defined. The classification used here follows Butler (1956) in excluding the elephant-shrews, leaving six families of recent insectivores of which three are represented in the Palaearctic Region.

See Cabrera (1925) for a detailed account of the whole order; Abe (1967, 1968) for a detailed account, in English, of the Japanese species; Dolgov & Yudin (1975) for a comprehensive bibliography of the insectivores of the USSR, with a list of species; and Stroganov (1957) for a detailed account of Siberian species.

1. Dorsal pelage spiny; skull widest across zygomatic arches which are robust ***ERINACEIDAE*** (p. 13)
– Dorsal pelage not spiny; skull widest across brain-case, zygomatic arches slender or absent (2)
2. I^1 unicuspid; zygomatic arches present; tympanic bullae complete . ***TALPIDAE*** (p. 32)
– I^1 two-cusped; no zygomatic arches; tympanic bones annular . . ***SORICIDAE*** (p. 17)

Family **ERINACEIDAE**

Contains two subfamilies, the Erinaceinae (spiny hedgehogs) dealt with below, and the Echinosoricinae with five, spineless, species in the Oriental Region.

Subfamily **ERINACEINAE**

Hedgehogs

Contains three genera, all represented in the Palaearctic Region.

1. Crown of head with a median, spineless tract . . . . (2)
– Crown of head without a median, spineless tract; (post-glenoid processes larger than mastoid processes) . . . . ***HEMIECHINUS*** (p. 15)
2. Spines rugose; ears considerably longer than adjacent spines; auditory bones greatly inflated, especially the basisphenoid and squamosal elements; cavity of each bulla extending into pterygoid bones almost to posterior margin of palate . . ***PARAECHINUS*** (p. 16)
– Spines smooth; ears not longer than adjacent spines; auditory bones not greatly inflated; cavity of bulla scarcely penetrating pterygoid bone . . ***ERINACEUS*** (p. 14)

Genus *ERINACEUS*

Erinaceus Linnaeus, 1758. Type-species *E. europaeus* L. = *Herinaceus*. Incl. *Atelerix, Peroechinus, Aethechinus*.

In addition to the two species included here this genus contains probably three species in Africa south of the Sahara. It is arguable that some of the forms listed here as races of *E. europaeus* should be treated as distinct species.

Hallux larger, claw reaching base of 2nd digit; I^3 with a single root; spine on posterior margin of palate not flanked by flat shelves of bone; P_2 with a small lingual cusp . ***E. europaeus***

Hallux smaller, claw not reaching base of 2nd digit; I^3 with 2 roots; palatal spine flanked by flat shelves; P_2 lacking a small lingual cusp . . . . ***E. algirus***

Erinaceus europaeus Northern hedgehog

Erinaceus europaeus Linnaeus, 1758. Gothland Island, Sweden.

E. e. europaeus. Europe east to Oder/Adriatic. = *echinus, typicus*. Incl. *caniceps, caninus, erinaceus* (Germany), *hispanicus* (Spain), *italicus, meridionalis* (Italy), *occidentalis* (Britain), *suillus*.

E. e. concolor Martin, 1838. Asia Minor, Caucasus, Lebanon, Syria; type-locality in N.E. Asia Minor. Incl. *abasgicus* (W. Caucasus), *ponticus* (Georgia), *sacer* (Palestine), *transcaucasicus* (Transcaucasia); perhaps also *roumanicus* and *centralrossicus*, here listed separately.

E. e. roumanicus Barrett-Hamilton, 1900. Europe east of Oder/Adriatic; type-locality in Rumania. Incl. *bolkayi, danubicus, drozdovskii* (Macedonia), *kievensis*. Probably confluent with *E. e. concolor*.

E. e. centralrossicus Ognev, 1916. N. European Russia and W. Siberia; type-locality in Smolensk dist., Russia. Incl. *dissimilis* (E. Baltic coast), *pallidus* (W. Siberia). Probably confluent with *E. e. concolor*.

E. e. consolei Barrett-Hamilton, 1900. Sicily.

E. e. nesiotes Bate, 1906. Crete.

E. e. rhodius Festa, 1914. Rhodes.

E. e. amurensis Schrenk, 1859. E. Siberia, Manchuria and Korea; type-locality on Amur R., Manchuria. Incl. *chinensis* (Khingan Mts, Manchuria), *koreanus, koreensis* (Korea), *manchuricus* (S. Manchuria), *orientalis* (Vladivostok), *ussuriensis* (Ussuri). Doubtfully conspecific with *E. europaeus*.

E. e. dealbatus Swinhoe, 1870. China; type-locality Peking. Incl. *hanensis* (Hupeh), *kreyenbergi* (Shanghai), *tschifuensis* (Shantung). Perhaps confluent with *E. e. amurensis*.

RANGE (Map 2). The deciduous forest, Mediterranean and wooded steppe zones of Europe and W. Asia south to Palestine, Iraq and E. Iran, including the following isolated segments: S. Scandinavia; British Is. including Ireland; all the large islands of the Mediterranean except the Balearics. Also in the corresponding zones of E. Asia from the Amur to the Yangtze Kiang.

REMARKS. The forms *miodon* and *hughi*, included in this species by E & M-S, are referable to *Hemiechinus dauuricus*.

Recent studies of karyology (e.g. by Kral, 1967) have suggested that the eastern and western forms in Europe are specifically distinct (*E. europaeus* and *E. concolor*). However no comprehensive revision has been undertaken to show how the north Russian and eastern Asiatic forms relate to these.

Erinaceus algirus Algerian hedgehog

Erinaceus algirus Lereboullet, 1842. Oran, Algeria.
E. a. algirus. N. Africa, S. France and Spain (?introduced). Incl. *fallax* (Tunisia), ?*krugi* (Porto Rico – presumably an introduced specimen), *lavaudeni* (Morocco).
E. a. vagans Thomas, 1901. Majorca, Minorca, Ibiza; type-locality Minorca.
E. a. caniculus Thomas, 1915. Fuerteventura, Canary Is.

RANGE (Map 2). The Mediterranean and subdesert zones of N.W. Africa from S.W. Morocco to Libya. Also Fuerteventura and Tenerife (Canary Is.) (Niethammer, 1972a); Balearic Is.; Malta (Malec & Storch, 1972); and a few localities on the coasts of France and Spain (perhaps introduced to the islands and to Europe).

Genus *HEMIECHINUS*

Hemiechinus Fitzinger, 1866. Type-species *Erinaceus platyotis* Sundevall = *H. auritus* (Gmelin). Incl. *Ericius*, *Erinaceolus*.
*Mesechinus** Ognev, 1951. As a subgenus, type-species *Erinaceus dauuricus* Sundevall.

A genus of probably two species in the arid zone from the Sahara to central Asia, included in *Erinaceus* by some Russian authors. The Afghanistan form *megalotis*, given specific rank by E & M-S, is here included in *H. auritus* along with those names left *incertae sedis* by E & M-S.

1. Ears short, not reaching eyes when laid forwards . . . . . ***H. dauuricus***
– Ears long, reaching beyond eyes when laid forwards . . . . . ***H. auritus***

Hemiechinus auritus Long-eared hedgehog

Erinaceus auritus Gmelin, 1770. Astrakan, S.E. Russia.
H. a. auritus. Libya to W. Siberia. = *caspicus*. Incl. *aegyptius* (Egypt), *brachyotis*, *calligoni* (Transcaucasia), *chorassanicus*† (Kopet Dagh), *frontalis*, *insularis* (island of Barsa Kelmes, Aral Sea), ?*homalacanthus*†, *libycus*, *major*, *minor*, *persicus* (Iran), *platyotis*, *russowi*†, *syriacus* (Palestine), *turanicus*, *turkestanicus*. *Hemiechinus auritus metwallyi** Setzer, 1957a:6. Tell Basta, Zagazig, Sharquiya Prov., Egypt.
H. a. albulus Stoliczka, 1872. Western China; type-locality Yarkand, Sinkiang. Incl. *alaschanicus* (Ala Shan); *turfanicus* (Chami, Sinkiang); *holdereri* (Mongolia).
H. a. megalotis† (Blyth, 1845). Afghanistan.
H. a. collaris (Gray, 1830). Pakistan and N.W. India. Incl. *grayi*, *indicus*, *spatangus*.

RANGE (Map 2). Subdesert and dry steppe zones from the eastern Mediterranean (N. Libya and Egypt, Cyprus, Asia Minor) through S.W. Asia (not S. Arabia) to the Gobi Desert, Sinkiang and N.W. India.

REMARKS. The large, dark-bellied form in Afghanistan, given specific rank as *H. megalotis* by E & M-S, is included here following the demonstration by Niethammer (1973) that it intergrades with normal *auritus*.

Hemiechinus dauuricus† Daurian hedgehog

Erinaceus dauuricus Sundevall, 1842. Dauuria, Transbaikalia. Incl. *hughi*† (Shensi), *miodon*†, *przewalskii*† (N. China), *sibiricus*.†

RANGE (Map 2). Dry steppe round the eastern part of the Gobi Desert from near Lake Baikal to Shensi in China.

REMARKS. E & M-S queried the generic allocation of this species. Stroganov (1957), on the basis of topotypical material, treated it as a species of *Hemiechinus* and figured the skull. The types of *miodon* and *hughi* have been re-examined and show clearly all the characters of this species. The type of *sibiricus* Erxleben, 1777 has also been re-examined. It agrees in both external and cranial characters with *H. dauuricus*, a conclusion that was also reached by Sundevall (1842) when describing *H. dauuricus*. Sundevall, however, attributed the name *sibiricus* to Seba (1734) and was apparently unaware of its validation by Erxleben. Since the name *sibiricus* has not been used as a senior synonym for over 50 years whilst the name *dauuricus* is well established for this species in the Russian literature, an application has been made to the International Commission to have it placed on the Official Index of Rejected Names.

Genus *PARAECHINUS*

Desert hedgehogs

Paraechinus Trouessart, 1879. Type-species *Erinaceus micropus* Blyth. Incl. *Macroechinus*.

Contains two mainly allopatric species in the Palaearctic Region plus one (possibly more) in India (included here under *P. micropus*).

1. Spines with dark tips, giving overall blackish colour; ventral pelage (except on head) dark brown (*or* entire pelage, including spines, white); P^2 3-lobed, with 2 or 3 roots . . . ***P. hypomelas***
– Spines with white tips, except sometimes in the mid-dorsal line; ventral pelage predominantly white, with dark brown on groin and hind legs; P^2 rudimentary, with 1, occasionally 2, roots . . . (2)
2. Ventral pelage very soft and silky (Africa to Arabia) . . . ***P. aethiopicus***
– Ventral pelage coarse (India) . . . ***P. micropus***

Paraechinus aethiopicus Desert hedgehog

Erinaceus aethiopicus Ehrenberg, 1833. Dongola Desert, Sudan.
P.a. aethiopicus. Sudan and Egypt. =*brachydactylus*. Incl. *pallidus*, *sennaariensis*.
P.a. deserti (Loche, 1858). N. and W. Sahara; type-locality in N.E. Algeria.
P.a. pectoralis (Heuglin, 1861). Sinai, Jordan (type-locality).
P.a. dorsalis (Anderson & de Winton, 1901). Arabia and Iraq, type-locality Hadramaut. Incl. *albior*, *ludlowi* (Iraq), *oniscus* (S. Arabia).
P.a. blancalis Thomas, 1921. Djerba Island, Tunisia.
P.a. albatus Thomas, 1922. Tanb Island, Persian Gulf.
*P.a. wassifi** Setzer, 1957. N. Egypt.
Paraechinus deserti wassifi Setzer, 1957: 10·2 km W. of El Birigat, Kom Hamada, Tahreer Prov., Egypt. Probably a synonym of *deserti*.

RANGE (Map 3). Desert and dry steppe from Morocco, Mauritania, the Hoggar Mountains, Air and the middle Niger east to Egypt, Sudan, N. Ethiopia and N. Somalia; throughout Arabia and north-east to the Tigris. Probably on Gran Canaria, Canary Islands (Herter, 1972). Marginally sympatric with *Hemiechinus auritus* and *P. hypomelas*.

REMARKS. Setzer (1957) argued that *aethiopicus*, *deserti* and *dorsalis* represent three species, but if this is so they are completely allopatric and the differences are much fewer than supposed by Setzer. *P.a. dorsalis* differs from *pectoralis* and the African forms in having a dark dorsal stripe.

Paraechinus hypomelas Brandt's hedgehog

Erinaceus hypomelas Brandt, 1836. N. Iran.
P.h. hypomelas. Iran and neighbouring regions. Incl. *amir*, *blanfordi*, *eversmanni* (E. of Caspian Sea), *jerdoni* (Sind), *macracanthus*.
P.h. niger (Blanford, 1878). Muscat, Arabia.
P.h. seniculus Thomas, 1922. Tanb Island, Persian Gulf.
P.h. sabaeus Thomas, 1922. Near Aden.

RANGE (Map 3). The desert and dry steppe zones of S.W. Asia from Tashkent and the Aral Sea southeast to Punjab and southwest to S. Iran and S. Arabia. Marginally overlapping *P. aethiopicus* in the southwest and abutting with *P. micropus* in the southeast.

Paraechinus micropus Indian hedgehog

Erinaceus micropus Blyth, 1846. Bahawalpur, Pakistan.
P.m. micropus. Pakistan. Incl. *mentalis*, *pictus*.
P.m. nudiventris Horsfield, 1851. Southern India; type-locality Madras.

RANGE (Map 3). In two segments: from the Indus (including the west bank) through Rajasthan and south to Gujerat; and in Madras and Travancore (if *nudiventris* is conspecific).

REMARKS. Biswas & Ghose (1970) have argued that *micropus* and *nudiventris* are specifically distinct and have described a third species *P. intermedius* (with a new subspecies *P.i. kutchicus*) that is partly sympatric with *P. micropus* (*sensu strictu*). From their descriptions I would not rule out the possibility that these may all prove conspecific.

Family SORICIDAE
Shrews

The generic classification in this family is unsatisfactory. Ten genera occur in the Palaearctic, of which two are very marginal. Two further Oriental genera that come close to the southern border of the Palaearctic Region are *Anourosorex*, easily recognizable by the almost invisibly short tail; and *Nectogale*, easily recognizable by the complex, multiple fringes of hair on the tail. A comprehensive compilation dealing with all fossil and recent species worldwide has been produced, in Russian, by Gureev (1971).

1. Teeth with red or brown pigment at the tips (not conspicuous if the teeth are worn) (2)
– Teeth unpigmented . . . . (5)
2. First lower tooth with 3 lobes on the upper margin; 5 upper unicuspid teeth . . (3)
– First lower tooth with no more than one lobe on the upper margin; 3 or 4 unicuspid teeth (4)
3. Last two unicuspid teeth clearly visible in lateral view . . . . . ***SOREX*** (p. 17)
– Last two unicuspid teeth very small and crowded, scarcely visible in lateral view (E. Asia) ***BLARINELLA*** (p. 25)
4. Hind feet and at least part of ventral line of tail fringed with silvery hairs ***NEOMYS*** (p. 24)
– Hind feet and tail unfringed . . . . ***SORICULUS*** (p. 24)
5. Three upper unicuspid teeth . . . . (6)
– Two or 4 upper unicuspid teeth . . . . (7)
6. First unicuspid much larger than others; tail with scattered vibrissae projecting far beyond other hairs . . . . ***CROCIDURA*** (p. 26)
– Unicuspids subequal; tail shortly haired, with a ventral fringe (E. Asia) ***CHIMMAROGALE*** (p. 31)
7. Four upper unicuspids; dorsal pelage uniform . . . . . . ***SUNCUS*** (p. 30)
– Two upper unicuspids; dorsal pelage grey with a large white patch ***DIPLOMESODON*** (p. 31)

Genus *SOREX*

Sorex Linnaeus, 1758. Type species *S. araneus* L. Incl. *Amphisorex*, *Corsira*, *Homalurus*, *Oxyrhin*, *Soricidus*.
*Eurosorex** Stroganov, 1952. As subgenus, type-species *Sorex buchariensis* Ognev.
*Ognevia** Heptner & Dolgov, 1967 (cited therein as *Ognevia* Dolgov & Heptner). As a subgenus, type-species *Sorex mirabilis* Ognev.

A Holarctic genus including many species in both Eurasia and N. America. One species, *S. arcticus*, has a wide distribution in both continents whilst one American species, *S. cinereus*, occurs marginally in N.E. Siberia. The taxonomy of the Palaearctic species has been very confused and E & M-S were able to provide little more than a list of names. Recent work in USSR has done much to clarify the situation and by using characters of the genitalia and the karyotype has demonstrated the very subtle differences that distinguish species (Dolgov, 1967, 1968; Dolgov & Lukyanova, 1966; Yudin, 1971).

Complex chromosomal variation is present in some species, e.g. in *S. araneus*. Chromosome species are not listed separately below when no morphological differences are known. The key should be considered only as a guide to the characters distinguishing the species. The characters of the unicuspid teeth in particular can only be appreciated when they are reasonably unworn.

1. Tail over 90% of head and body; 5th unicuspid tooth usually well developed, almost as large as 4th, pigmented . . . (2)
– Tail under 90% of head and body; 5th unicuspid usually very small . . . (4)
2. Pelage uniformly dark grey; 2nd lower tooth clearly 2-cusped (Europe) . . ***S. alpinus***
– Pelage brown with a darker mid-dorsal stripe; 2nd lower tooth single-cusped or with a rudiment of a posterior cusp (China to Himalayas) . . . (3)
3. Head and body about 75 mm; hind foot about 15 mm; condylobasal length about 19–20 mm ***S. cylindricauda***
– Head and body about 55 mm; hind foot 11–13 mm; condylobasal length about 17 mm ***S. bedfordiae***
4. 2nd unicuspid smaller than 3rd which is equal to 1st . . . (5)
– 2nd unicuspid equal to or larger than 3rd . . . (6)
5. Unicuspids very long-cusped and crowded, height when unworn exceeding alveolar length; (head and body 40–50 mm; condylobasal length 15·0–17·5 mm) . . ***S. buchariensis***
– Unicuspids shorter and less crowded; (condylobasal length 14·0–16·4 mm; glans penis long and slender). . . . ***S. minutus***
6. Claws of front feet very large, over 3 mm; (condylobasal length 18·5–21·0 mm; glans penis long and very thick; E. Asia) . . . ***S. unguiculatus***
– Claws of front feet not over 3 mm . . . (7)
7. Very large: hind feet over 16 mm, condylobasal length over 22 mm; posterior cusp of I^1 very small, scarcely projecting beyond cingulum of 1st unicuspid; 4th unicuspid larger than 3rd; (eastern Siberia) . . . ***S. mirabilis***
– Smaller: hind feet under 16 mm, condylobasal length under 22 mm; posterior cusp of I^1 exceeding cingulum of 1st unicuspid; 4th unicuspid not larger than 3rd . . (8)
8. Pelage uniformly dark, ventral surface only slightly lighter than dorsal, without a clear line of demarcation on flank . . . (9)
– Ventral pelage distinctly lighter than dorsal . . . (10)
9. Unicuspids decreasing in size very gradually and evenly, 5th well developed . ***S. sinalis***
– 2nd unicuspid almost as large as 1st, 5th rather small; (Caucasus) . . . ***S. raddei***
10. Teeth heavily pigmented, pigment extending into basins of molariform teeth which are very broad and low-cusped; (unicuspids rather large and blunt; condylobasal length 17·1–18·8 mm; glans penis extremely long and slender; E. Asia). . . . ***S. daphaenodon***
– Teeth not so heavily pigmented . . . (11)
11. Larger: condylobasal length 18·3–20·3 mm . . . (12)
– Smaller: condylobasal length 12·0–18·3 mm . . . (15)
12. Upper molariform teeth with a very deep concavity in the posterior margin (as seen in palatal view); (rostrum rather slender and elongate; condylobasal length 19·0–21·6 mm) ***S. vir***
– Upper molariform teeth with a shallow posterior concavity . . . (13)
13. Unicuspid teeth tending to point backwards, tips forming a strongly convex line; (condylobasal length 18·0–19·4 mm; Tien Shan) . . . ***S. asper***
– Unicuspids not as above. . . . (14)
14. Pelage 3-coloured, with a distinct zone on flank intermediate in colour between dorsal and ventral pelage; dorsal pelage very dark brown in adults . . . ***S. araneus***
– Pelage 2-coloured; dorsal pelage pale brown in adults; (Caucasus). . ***S. caucasicus***
15. First 4 unicuspids in pairs, 1st equal in size to 2nd, 3rd smaller and equal to 4th, 5th very small; (rostrum slender, unicuspids elongate in axis of jaw and well spaced . (16)
– Unicuspids not grouped in this way . . . (17)
16. A canal linking the mandibular foramen with the coronoid fossa (as in most *Sorex*); condylobasal length 16·1–18·2 mm . . . ***S. caecutiens***
– No such canal on mandible; condylobasal length 14·6–16·7 mm; (N.E. Siberia) ***S. cinereus***
17. Unicuspids rather tall and pointed, 2nd equal to or larger than 1st, all inclined to point backwards with points forming a convex line; condylobasal length 15·7–18·4 mm ***S. arcticus***
– Unicuspids rather small and obtuse, decreasing rather evenly in size, not inclined backwards and with the points forming an almost straight line; condylobasal length 12·0–15·9 mm (18)
18. Extremely small: head and body usually under 50 mm, tail under 30 mm, hind feet 6–9 mm, condylobasal length 12–14 mm . . . ***S. minutissimus***

– Not so small: tail over 40 mm, condylobasal length over 15 mm . . . . (19)
19. Tail shorter: 40–45 mm; brain-case almost circular in outline, as seen from above; rostrum very narrow and slender; (condylobasal length 15·0–15·9 mm) . . ***S. gracillimus***
– Tail longer: 48–55 mm; brain-case elongate; rostrum less slender; (Honshu, Japan) ***S. hosonoi***

***Sorex minutissimus*†** Least Siberian shrew

Sorex minutissimus Zimmermann, 1780. R. Yenesei, Siberia.
S.m. minutissimus. European Russia, W. and central Siberia. Incl. *burneyi* (L. Baikal), *czekanovskii*, *exilis*, *neglectus* (near Moscow).
*Sorex minimus** Gmelin, 1793: 160. R. Yenesei.
*S.m. karelicus** Stroganov, ?1949b. Near Lake Suoyarvi, S.W. Karelia, USSR.
*S.m. barabensis** Stroganov, 1956. Upper Ob.
Sorex minutissimus barabensis Stroganov, 1956a. Near village of Egerbash, Barabinsk Region, Novosibirsk Oblast, W. Siberia.
*S.m. abnormis** Stroganov, 1949. Kazakhstan.
Sorex tscherskii abnormis Stroganov, 1949a. Near Chelkara, Kazakhstan.
*S.m. stroganovi** Yudin, 1964. S.E. Altai.
Sorex minutissimus stroganovi Yudin, 1964. Upper basin of River Yustid, S.E. Altai Mts, Siberia.
*S.m. caudata** Yudin, 1964. N.E. Altai.
Sorex minutissimus caudata Yudin, 1964. Near Kibizen, River Biya, N.E. Altai Mts, Siberia.
S.m. tscherskii Ognev, 1913. Ussuri Region, E. Siberia. Incl. *ussuriensis*.
*S.m. tschuktschorum** Stroganov, 1949. N.E. Siberia.
Sorex tscherskii tschuktschorum Stroganov, 1949a. Anadir, Markov, N.E. Siberia.
S.m. hawkeri Thomas 1906. ?Honshu, Japan.

RANGE (Map 6). The taiga zone from Norway, Sweden and Estonia to E. Siberia; the islands of Sakhalin, Hokkaido and perhaps Honshu.

REMARKS. The type of *S.m. hawkeri* is labelled Inukawa, Yedo, Japan. No subsequent specimens from Honshu have been reported. The species has only recently been detected in Scandinavia (see for example Østbye *et al.*, 1974).

Sorex minutus Pygmy shrew

Sorex minutus Linnaeus, 1766. Barnaul, W. Siberia.
S.m. minutus. Europe to central Siberia. =*pygmaeus*. Incl. *canaliculatus* (Sweden), *gmelini* Ognev, *gymnurus* (Greece), *hibernicus* (Ireland), *insulaebellae* (Belle Isle, France), ?*kastchenkoi* (W. Siberia), *lucanius* (S. Italy), *melanderi* (European Russia), *pumilio* (Germany), *pumilus*, *rusticus* (England), *volnuchini* (Caucasus).
*Sorex exiguus** Brink, 1952. Applescha, Frise, Netherlands. (Type examined; see also Leeuwen, 1954.)
*Sorex minutus becki** Lehmann, 1963. Silum, Liechtenstein.
*Sorex minutus carpatanus** Rey, 1971. Sierra de Guadarrama, Madrid, Spain.
S.m. thibetanus Kastschenko, 1905. Tsaidam, central China.
S.m. planiceps Miller, 1911. Kashmir.
*S.m. heptopotamicus** Stroganov, 1956. Tien Shan.
Sorex minutus heptopotamicus Stroganov, 1956a. Ala-Tau, Tien Shan, China.

RANGE (Map 4). Western Europe, from central Spain and the whole of Scandinavia, through western Siberia as far as the Yenesei and Lake Baikal, and south through the mountains of central Asia to Kashmir, Nepal (specimen in British Museum) and Tsaidam; Britain, Ireland and most of the Hebrides.

REMARKS. The eastern Asiatic form *gracillimus*, included in *S. minutus* by E & M-S, is here treated as specifically distinct following Dolgov & Lukyanova (1966) and other Russian authors.

***Sorex gracillimus*†** Slender shrew

Sorex minutus gracillimus Thomas, 1907. Darine, 25 miles N.W. of Korsakoff, Sakhalin Island. Incl. ?*hyojironis* (Manchuria), ?*leucogaster* (Amamushira, N. Kurile Is.), ?*yamashinai*.

RANGE (Map 4). S.E. Siberia from the southern shore of the Sea of Okhotsk to N. Korea and probably Manchuria; the islands of Sakhalin, Shantar, Hokkaido and some of the Kuriles.

REMARKS. This was treated as a subspecies of *S. minutus* by E & M-S but is given specific rank by all Russian authors. There are clear differences in the genitalia.

Sorex caecutiens Laxmann's shrew

Sorex caecutiens Laxmann, 1788. Lake Baikal, Siberia.

S.c. caecutiens. Main continental range. Incl. *altaicus*, *annexus* (Korea), *araneoides*† (L. Baikal), *buxtoni* (Sea of Okhotsk), *centralis* (Sayan Mts), *koreni* (Kolyma R.), *lapponicus* (N. Sweden), *macropygmaeus* (Kamchatka), *pleskei* (European Russia), *rozanovi*, *tasicus* (W. Siberia), *tungussensis*.
*Sorex macropygmaeus karpinskii** Dehnel, 1959. Bialowiesa Forest, Poland.

S.c. shinto Thomas, 1905. Honshu, Japan.
?*Sorex chouei** Imaizumi, 1954. Kitazawa-toge, Akaishi Mts, Nagano Pref., Honshu, Japan.

*S.c. shikokensis** Abe, 1967. Shikoku, Japan.
Sorex caecutiens shikokensis Abe, 1967. Mt Shimokabuto (900 m), Niihama City, Ehime Pref., Shikoku, Japan.

S.c. saevus Thomas, 1907. Sakhalin, Hokkaido and Kurile Islands. =*savenus*. Perhaps not distinct from *S.c. shinto*.

S.c. cansulus Thomas, 1912. Kansu, China.

RANGE (Map 4). The taiga and tundra zones from E. Europe (Sweden, Poland) to E. Siberia and south to central Mongolia, Kansu and Korea; the islands of Sakhalin, Hokkaido, N. Honshu and Shikoku (montane on the last two).

REMARKS. Imaizumi (1960, 1970a) treated the island form *shinto*, with *saevus* and *shikokensis*, as a distinct species and tentatively included *chouei*. Brink (1953) considered this species conspecific with *S. cinereus* of N. America, but Stroganov (1957) showed both species at three localities in N.E. Siberia. A record from Bulgaria (Markov, 1951) was later discredited (Peshev, 1964). Some of the forms here listed under *S. arcticus* (below) may be referable to *S. caecutiens*.

The form *buxtoni* Allen, 1903 was referred to *S. arcticus* by Stroganov. The type (USNM) seems clearly referable to *S. caecutiens*, but one of the two paratypes in the BM(NH) is *S. daphaenodon* and the other *S. caecutiens*.

Sorex araneus Common shrew

Sorex araneus Linnaeus, 1758. S. Sweden. Incl. *alticola*, *bergensis*, *bohemicus*, *bolkayi*, *carpathicus*, *castaneus* (England), *concinnus*, *coronatus*, *crassicaudatus*, *csikii*, *daubentonii*, *eleonorae*, *euronotus*, *fodiens* Duvernoy, *fretalis* (Jersey), *granti* (Islay, Inner Hebrides), *hermanni*, *ignotus*, *iochanseni*, *labiosus*, *macrotrichus*, *melanodon*, *mollis*, *monsvairani*, *nigra*, *nuda*, *pallidus*, *personatus*, *petrovi*, *peucinus*, *pulcher* (Terschelling Is., Holland), *pyrenaicus*, *quadricaudatus*, *rhinolophus*, *roboratus* (Altai), *samniticus* (S. Italy), *santonus*, *schnitnikovi* (S.E. Kazakhstan), *tetragonurus*, *tomensis*, *uralensis*, *vulgaris*.
*Blarina pyrrhonota** Jentink, 1910. Erroneously attributed to Surinam. For status see Husson (1962).
*Sorex araneus garganicus** Pasa, 1953. Monte Nero, Cagnano Varano, Gargano, Italy.
*Sorex araneus wettsteini** Bauer, 1960. Neusiedler See, Burgenland, Austria.
*Sorex araneus silanus** Lehmann, 1961. Camigliatello, Silano, Calabria, Italy.
*Sorex araneus hülleri** Lehmann, 1966a. Elmpter Schwalmbruch, Kr. Erkelenz, Niederrhein, W. Germany.
*Sorex gemellus** Ott, 1968. Rhone Valley, Switzerland (46°15′N, 7°03′E).

RANGE (Map 5). Western Europe (Arctic coast to mountains of Mediterranean zone) east to R. Yenesei and L. Baikal in all but the dry steppe and desert zones; Britain and most of the Inner Hebrides.

REMARKS. The following forms, here listed separately as possibly valid species, have been considered conspecific with *S. araneus* by other authors: *S. raddei* (Caucasus), *S. caucasicus* (Caucasus), *S. asper* (Kirgizia) and *S. vir* (N.E. Siberia) by Kuzyakin (1965); *S. isodon* (Siberia) by Stroganov (1957). Some of the forms here listed under *S. arcticus* (below) may be referable to *S. araneus*.

It has been demonstrated that this species consists of two basically allopatric sibling species differing in karyotype (Ott, 1968). Although small morphological differences have been detected in the area of contact in Switzerland (Hausser & Jammot, 1974) it is not yet possible to distinguish the species in other areas except by karyotype. The name *gemellus* Ott, 1968 applied to the southwestern member of the pair is almost certainly antedated, for example by *Sorex fretalis* Miller, 1909 from Jersey where the shrews have been demonstrated to belong to the 'non-*araneus*' chromosome species, and by other names applied to French and Spanish forms. Until satisfactory identification can be made on morphological grounds it seems better to refer to them as '*Sorex araneus*, NF 40 karyospecies' (which is probably *S. araneus* s.s.) and '*Sorex araneus*, NF 44 karyospecies' (which includes the forms *fretalis* and *gemellus*).

This species has been recorded from Nepal by Agrawal & Chakraborty (1971) but it seems best to treat this as possibly being one of the following species.

***Sorex raddei*†** Radde's shrew

Sorex raddei Satunin, 1895. Near Kutais, Georgia, USSR. Incl. *batis* (N.E. Asia Minor).

RANGE (Map 5). Caucasus and northeastern Asia Minor.

REMARKS. E & M-S and Kuzyakin (1965) listed this form under *S. araneus*. It differs from *S. araneus* in both colour and genitalia (Dolgov & Lukyanova, 1966) and there is also a very great difference in karyotype from the sympatric *S. caucasicus* (Kozlovsky, 1973).

***Sorex caucasicus*†** Caucasian shrew

Sorex caucasicus Satunin, 1913. Bakuryani, Tbilisi, Georgia, USSR.

RANGE (Map 5). Central Caucasus to Azov Sea; northern Asia Minor.

REMARKS. Listed under *S. araneus* by E & M-S but considered specifically distinct from both *S. araneus* and the sympatric *S. raddei* by Dolgov & Lukyanova (1966) on the basis especially of the genitalia. Reported from widely separated localities in Asia Minor by Spitzenberger (1968).

***Sorex granarius*†**

Sorex araneus granarius Miller, 1910. Segovia, Spain.

RANGE (not mapped). Central Spain.

REMARKS. This form was included in *S. caecutiens* by E & M-S, but Hausser *et al.* (1975) have demonstrated that its karyotype distinguishes it from both *S. araneus* and *S. caecutiens* and that it is also recognizable by its short skull.

***Sorex asper*†** Tien Shan shrew

Sorex asper Thomas, 1914. Tekes Valley, Tien Shan, central Asia.
S.a. asper. Tien Shan.
?*S.a. excelsus* Allen, 1923. Yunnan, China.

RANGE (Map 6). The western Tien Shan in Kirgizia and the adjacent part of Sinkiang. Yunnan, if *excelsus* is conspecific.

REMARKS. E & M-S listed this form under *S. araneus*, as did Kuzyakin (1965). It seems more closely related to *S. arcticus* than to *S. araneus*. The Chinese form *excelsus* (type examined in AMNH) more closely resembles *S. asper* than northern *S. araneus*.

***Sorex vir*†** Flat-skulled shrew

Sorex vir Allen, 1914. Nijni Kolymsk, near mouth of Kolyma R., N.E. Siberia.
S.v. vir. N.E. Siberia.
S.v. jacutensis Dukelski, 1933. W. Yakutia and Yenesei; type locality on R. Vilyui. Incl. *dukelskiae*, *turuchanensis*.
S.v. thomasi Ognev, 1921. L. Baikal.
S.v. platycranius Ognev, 1921. Ussuri, E. Siberia.

RANGE (Map 5). E. Siberia west as far as the R. Ob and south to the Altai, N. Mongolia, L. Baikal and Vladivostok (Dolgov, 1967).

REMARKS. E & M-S listed this form under *S. araneus*. Kuzyakin (1965) considered it a race of *S. araneus*, but confined to a more limited region of N.E. Siberia (excluding from it the forms *platycranius*, *thomasi* and *dukelskiae*). Stroganov (1957) considered

S. vir specifically distinct from *S. araneus* and recognized the four subspecies listed above, but if these are indeed conspecific they are likely to be confluent. Examination of the type of *vir* (MCZ) suggests that it is specifically distinct from *S. araneus*, and this is generally accepted by more recent Russian authors, e.g. Yudin (1971). Yudin used the name *roboratus* Hollister for this species but this is an error according to Dolgov (pers. comm.).

***Sorex sinalis*†** Dusky shrew

Sorex sinalis Thomas, 1912. 45 miles S.E. of Feng-hsiang-fu, Shensi, China (10 500 ft). Incl. *gravesi* (Maritime Province, S.E. Siberia), *isodon* (L. Baikal), *ruthenus* (European Russia).
*Sorex isodon princeps** Skalon & Rajevsky, 1940. Basin of Em-Engana, Kondo-Sosvinski reserve, W. Siberia (not seen; from Stroganov, 1957).

RANGE (Map 7). From S.E. Norway (Nilsson, 1971) and Finland to E. Siberia and south to Shensi in China; the island of Sakhalin.

REMARKS. This form was considered a race of *S. araneus* by E & M-S and by Stroganov (1957). The argument for giving it specific rank was given by Siivonen (1965). Dolgov (1964) erroneously considered *centralis* Thomas, 1911 to be the prior name for this species (see remarks under *S. caecutiens*). The type of *gravesi* has been examined along with a series from Sakhalin.

Since its clear recognition as a distinct species it has been known as *Sorex isodon* or, erroneously, as *S. centralis*. However if the Chinese form *sinalis* is correctly considered to be conspecific with *isodon* then the earlier name *sinalis* should be used. The name *isodon* is usually dated from Turov, 1924 but since it was then used in an infrasubspecific sense (*Sorex araneus tomensis isodon*) it is not available under the *Code* (Articles 1 and 45c). Its earliest valid use may be by Stroganov (1936), in which case it is predated by *gravesi* Goodwin, 1933. This is therefore an additional reason for using the prior name *sinalis* if the conspecificity is agreed.

Sorex arcticus* Arctic shrew

Sorex arcticus Kerr, 1792:206. Ontario, Canada.
S.a. borealis Kastschenko, 1905. Siberia; type locality Tomsk. Incl. *amasari* (Amur to L. Baikal), *baikalensis*, *irkutensis* (L. Baikal), *jenissejensis* (Sayan Mts), *midendorfi* (Lower Yenesei), *petschorae* (Petchora), *sibiriensis*, *ultimus* (Kolyma).
?*Sorex jenissejensis margarita** Fetisov, 1950a. Upper River Misovki, Khamar-Daban, E. Siberia.
*Sorex arcticus transrypheus** Stroganov, 1956a. Near Kustanaya, N. Kazakhstan.

RANGE (Map 7). The whole of Siberia (taiga and tundra) from the Urals to Vladivostok and Anadyr; N. America from Alaska to Nova Scotia, south to about 43°N in Wisconsin.

REMARKS. Neither E & M-S nor Kuzyakin (1965) recognized this species, and they included the above forms in *S. araneus* and *S. caecutiens*. More recent Russian workers are unanimous in recognizing it. Subspecific variation of cranial dimensions was described by Dolgov (1966). The possibility of the occurrence of this species in Germany was raised by Lehmann (1968).

Sorex cinereus* Masked shrew

Sorex arcticus cinereus Kerr, 1792:206. Fort Severn, Ontario, Canada.
*Sorex cinereus portenkoi** Stroganov, 1956b. Poberejhe Anadirsk estuary, near Anadyr, N.E. Siberia.
*Sorex cinereus caecutienoides** Stroganov, 1957. Yakutiya, Megino-Kangalask Region, Khaptagaisk, Siberia.
*Sorex beringianus** Yudin, 1967. Paramushir, Kurile Islands, USSR.
*Sorex cinereus camtschatica** Yudin, 1972. Kambalnaya Bay, Kamchatka, USSR.

RANGE (Map 4). In the Palaearctic known only from northeastern Siberia, from Anadyr to Kamchatka and the Kurile Is.; in North America almost ubiquitous north of central USA.

REMARKS. This species has been held to include also *S. caecutiens* (Brink, 1953), but

Stroganov (1957) recorded both species from three localities in northeastern Siberia. Kuzyakin (1965) treated *portenkoi* as a race of *S. minutus* and did not recognize *S. cinereus* in the Palaearctic. Yudin (1972) provided a detailed illustrated comparison of the two species in N.W. Siberia.

Sorex hosonoi*

Sorex hosonoi Imaizumi, 1954b:94. Tokiwa Mura Kita-Azumi Gun, Nagano Pref., Honshu, Japan.
S.h. hosonoi. Mountains of central Honshu.
*S.h. shiroumanus** Imaizumi, 1954. N. Japanese Alps, Honshu.
*Sorex hosonoi shiroumanus** Imaizumi, 1954b:97. Mt Shirouma (*c*. 2900 m), Japanese Alps, Honshu, Japan.

RANGE (Map 5). Montane forests of central Honshu, Japan.

REMARKS. Sympatric with *S. caecutiens shinto* according to Imaizumi (1970a).

***Sorex unguiculatus*†** Long-clawed shrew

Sorex unguiculatus Dobson, 1890. Sakhalin. Incl. *yesoensis*† (Hokkaido).

RANGE (Map 5). The Pacific coast of Siberia from Vladivostok to the Amur; the islands of Sakhalin and Hokkaido.

REMARKS. Listed under *S. araneus* by E & M-S. Imaizumi (1970a) considered *yesoensis* a synonym of *unguiculatus*. Finnish specimens identified as *S. unguiculatus* have subsequently been considered to represent *S. sinalis* (Siivonen, 1965). The representation of this species in Kamchatka on the map by Dolgov (1967) was an error (Dolgov, pers. comm.).

Sorex daphaenodon Large-toothed shrew

Sorex daphaenodon Thomas, 1907. Dariné, 25 miles N.W. of Korsakoff, Sakhalin.
S. d. daphaenodon. Sakhalin.
S. d. sanguinidens Allen, 1914. N.E. Siberia; type-locality near mouth of Kolyma R.
S. d. orii Koruda, 1933. Paramushiru, Kurile Is.
S. d. scaloni Ognev, 1933. W. Siberia; type-locality Turukhansk dist.

RANGE (Map 6). Siberia from just east of the Urals to the River Kolyma; the islands of Sakhalin and Paramushiru (N. Kuriles); N. Mongolia, Manchuria and Inner Mongolia (Wang, 1959).

REMARKS. The form *yesoensis* from Hokkaido, included in this species by E & M-S, is here placed in *S. unguiculatus* following Imaizumi (1960, 1970a).

Sorex buchariensis Pamir shrew

Sorex buchariensis Ognev, 1921. River Davan-su, N.W. Russian Pamir Mts.
*Sorex kozlovi** Stroganov, 1952. Kam, River Dze-Tchyou, Upper Mekong, E. Tibet.

RANGE (Map 7). Pamir Mountains, Tadzhikistan and E. Tibet.

REMARKS. A very distinct species (type-species of the subgenus *Eurosorex*). The Tibetan form *kozlovi* was included in *S. buchariensis* by Dolgov (1967).

***Sorex mirabilis*†** Giant shrew

Sorex mirabilis Ognev, 1937. Kishinka River, Ussuri region, E. Siberia.
*Sorex mirabilis kutscheruki** Stroganov 1956a. Bektan, North Korea.

RANGE (Map 7). North Korea and Ussuri region of E. Siberia.

REMARKS. Kuzyakin (1965) considered this form to be conspecific with *S. pacificus* of western USA, but Heptner & Dolgov (1967) showed convincingly that it is not. It is the type-species of *Ognevia* Heptner & Dolgov.

Sorex alpinus Alpine shrew

Sorex alpinus Schinz, 1837. St Gothard Pass, Switzerland. Incl. ?*intermedius*, ?*longobarda* (Italy), *hercynicus* (Harz Mts).
*Sorex alpinus tatricus** Kratochvil & Rosicky, 1952a. High Tatra Mountains, Czechoslovakia.

RANGE (Map 7). Montane forests of central Europe including the following isolated marginal areas: Pyrenees, Balkans, Carpathians, Tatra, Sudeten, Harz and Jura.

Sorex cylindricauda Greater striped shrew

Sorex cylindricauda Milne-Edwards, 1872. Moupin, W. Szechuan, China.

RANGE (Map 6). Montane forest of S. Szechuan, China (i.e. non-Palaearctic).

REMARKS. The difference in size between this and the next species is very considerable, and specimens of both species are available from precisely the same locality in S. Szechuan.

***Sorex bedfordiae*†** Lesser striped shrew

Sorex bedfordiae Thomas, 1911. Omisan (9500 ft), Szechuan, China.
S. b. bedfordiae. Szechuan. Incl. *fumeolus*.
S. b. wardi Thomas, 1911. Kansu.
S. b. gomphus Allen, 1923. Yunnan.
*S. b. nepalensis** Weigel, 1969. Nepal.
Sorex cylindricauda nepalensis Weigel, 1969. Ringmo, N.E. Nepal (3920 m).

RANGE (Map 6). Montane forest of China between 1800 and 3400 m from S. Kansu (35°N) to Yunnan, the adjacent part of Burma and Nepal.

Genus *SORICULUS*

Soriculus Blyth, 1854. Type-species *Corsira nigrescens* Gray. Incl. *Chodsigoa*, *Episoriculus*.

A genus of about eight species ranging from Kashmir to Burma and China. Only one species occurs (marginally) in the Palaearctic region as defined here, and it alone is dealt with below. It belongs to the subgenus *Chodsigoa*, characterized by the presence of only three unicuspid teeth.

Soriculus hypsibius De Winton's shrew

Soriculus hypsibius de Winton, 1899. Yangliupa, N.W. Szechuan, China.
S. h. hypsibius. Szechuan. Incl. *berezowskii*.
S. h. larvarum (Thomas, 1911). E. of Peking.
S. h. lamula (Thomas, 1912). Kansu.
S. h. parva (Allen, 1923). Yunnan.

RANGE (Map 8). Montane forest of China from Yunnan through Szechuan to S. Kansu and Shensi; also known from just east of Peking; and from Fukien in S.E. China (Lehmann, 1955).

REMARKS. A further species, *S. smithi* Thomas, (which is most unlikely to be conspecific with *S. salenskii* as claimed by E & M-S), has a similar range, but does not appear to have been recorded north of the Tsin Ling Shan in S. Shensi (about 34°N). It is larger than *S. hypsibius* with very large hind feet (18 mm).

Genus *NEOMYS*
Water-shrews

Neomys Kaup, 1829. Type species *Sorex fodiens* Pennant. Incl. *Amphisorex*, *Crossopus*, *Hydrogale*, *Hydrosorex*, *Leucorrhynchus*, *Pinalea*.

A clearly defined genus confined to the Palaearctic Region. Two widely distributed species are generally recognized and most Russian authors recognize a third, *N. schelkovnikovi*, in the Caucasus. I have not seen specimens clearly attributable to the last; the characters in the following key are taken from Gromov *et al.* (1963).

1. Tail with a mid-ventral keel of stiff hairs extending the entire length; condylobasal length usually 20·5–23·0 mm . . . . (2)
– Tail with a ventral keel confined to terminal third; condylobasal length usually 19–21 mm **N. anomalus**

2. Eyes almost hidden by long hair; upper surfaces of feet with sparse white hair and coarse black or brown scales; (Caucasus) . . . . ***N. schelkovnikovi***
– Eyes clearly visible; upper surfaces of feet with dense white hair and small brown scales ***N. fodiens***

Neomys fodiens Northern water-shrew

Sorex fodiens Pennant, 1771. Berlin, Germany.
N. f. fodiens. Mainland of Europe and W. Siberia. =?*canicularius, fluviatilis.* Incl. *albiventris, albus, ?alpestris, amphibius, aquaticus* Müller, *argenteus* (L. Baikal), *carinatus, collaris, constrictus, daubentonii, eremita, fimbriatus, griseogularis, hermanni, hydrophilus, ignotus, intermedius, leucotis, lineatus, linneana* (Sweden), *liricaudatus, longobarda, macrourus, minor, musculus, naias, nanus, natans, ?nigricans, nigripes, psilurus, remifer, rivalis, stagnatilis, stresemanni.*
N. f. bicolor (Shaw, 1791). Britain. Incl. *ciliatus, pennanti, sowerbyi.* Probably not distinct from *N. f. fodiens.*
N. f. teres† Miller, 1908. Asia Minor and Caucasus; type-locality 25 miles N. of Erzerum, N.E. Asia Minor. Incl. *balkaricus, dagestanicus, ?leptodactylus.*
N. f. orientis Thomas, 1914. S.E. Kazakhstan; type-locality near Djarkent. =*orientalis.* Incl. *brachyotis.*
N. f. watesei Kuroda, 1941. Sakhalin Is.
*N. f. niethammeri** Bühler, 1963. N. Spain.
Neomys fodiens niethammeri Bühler, 1963. Ramales de la Victoria, 30 km S. of Laredo, N. Spain.

RANGE (Map 8). Europe from arctic Scandinavia south to the montane parts of the Mediterranean zone, including Britain but not Ireland, eastwards through W. Siberia as far as the River Yenesei and Lake Baikal, and south to northern Asia Minor and the Tien Shan; E. Siberia from Vladivostok to the mouth of the Amur; the island of Sakhalin.

REMARKS. E & M-S placed *teres* (Asia Minor) in *N. anomalus.* Spitzenberger & Steiner (1962) considered it to be a form of *N. fodiens* and an examination of the type tends to support this.

Neomys anomalus Southern water-shrew

Neomys anomalus Cabrera, 1907. San Martin de la Vega, Madrid, Spain.
N. a. anomalus. Spain.
N. a. milleri Mottaz, 1907. Central and E. Europe; type-locality in Swiss Alps. Incl. *josti* (Macedonia), *soricoides* (Poland).
N. a. mokrzeckii Martino, 1917. Crimea.

RANGE (Map 8). Montane woodlands of western Europe, in Spain, Massif Central of France, Alps, Italy, Germany (west to the Eifel), Belgium, Balkans; lowlands of E. Europe east to Crimea and the River Don; perhaps W. Asia Minor (Osborn, 1965). A recent record from the eastern Elburz Mts, Iran (Lay, 1967) more probably refers to *N. fodiens.*

Neomys schelkovnikovi† Transcaucasian water-shrew

Neomys schelkovnikovi Satunin, 1913. Ushkul, Svanetiya, Transcaucasia.

RANGE (Map 8). Armenia and Georgia.

REMARKS. Left *incertae sedis* by E & M-S, this form is given specific rank by most recent Russian authors. I have not seen specimens.

Genus *BLARINELLA*

Blarinella Thomas, 1911. Type species *Sorex quadraticauda* Milne-Edwards.

Contains three named forms which are probably conspecific, although the differences are not negligible. The genus is only marginally distinct from *Sorex.* Its range falls just outside the Palaearctic Region as defined here.

Blarinella quadraticauda Chinese short-tailed shrew

Sorex quadraticauda Milne-Edwards, 1872. Moupin, Szechuan, China.

B. q. quadraticauda. Szechuan.
B. q. griselda Thomas, 1912. Kansu.
B. q. wardi Thomas, 1915. Upper Burma and Yunnan.

RANGE (Map 8). Known at present from three separate areas: S. Kansu, Szechuan and the Yunnan-Burma border, mostly between 2500 and 4000 m.

Genus *CROCIDURA*

Crocidura Wagler, 1832. Type-species *Sorex leucodon* Hermann. Incl. *Leucodon, Paurodus, Heliosorex*.

A large genus with species throughout the Oriental, Ethiopian and Palaearctic Regions except for the taiga and tundra zones of the Palaearctic. This is one of the most difficult genera of mammals taxonomically, and the taxonomy of even the limited number of Palaearctic species cannot yet be considered definitive at the species level. The differences between well-known sympatric species are so slight that allocation of poorly known allopatric forms is especially difficult.

The following key should be considered only as a guide to the specific differences amongst the Palaearctic species. The following wholly Oriental species are not included, although they closely approach the southern boundary of the Palaearctic Region used here: *C. horsfieldi* Tomes, *C. attenuata* Milne-Edwards, *C. dracula* Thomas. The relationship of Palaearctic and Oriental species has been reviewed by Jenkins (1976).

1. Tail longer than head and body (Egypt) . . . . ***C. floweri***
– Tail shorter than head and body . . . . (2)
2. Head and body over 100 mm; condylobasal length over 25 mm (Egypt) ***C. flavescens***
– Head and body under 100 mm; condylobasal length under 25 mm . . . (3)
3. Pale ventral pelage extending to flanks and forming a sharp boundary with the dark dorsal pelage . . . . (4)
– Ventral pelage merging gradually with dorsal pelage without a sharp boundary . . (5)
4. Larger: condylobasal length 20–22 mm; upper tooth-row usually over 9 mm (Asia Minor to Palestine) . . . . ***C. lasia***
– Smaller: condylobasal length 18–20 mm; upper tooth-row usually under 9 mm ***C. leucodon*** ***C. sibirica***
5. Dorsal colour pale grey, faintly washed with brown; ventral colour creamy white; tail almost entirely white; (condylobasal length 16·5–19·4 mm) . . . . . ***C. pergrisea*** ***C. zarudnyi***
– Dorsal and ventral colour darker (resembling *C. pergrisea* only in *C. suaveolens portali* (Palestine) which has the condylobasal length under 17 mm); tail bicoloured or uniformly dark brown . . . . (6)
6. Unicuspid teeth well spaced, their combined length at the cingulum over 130 % of the greatest (labial) length of the succeeding large premolar . . . . (7)
– Unicuspids more crowded, their combined length at the cingulum under 130 % of the length of the succeeding premolar . . . . (9)
7. Hind feet under 10 mm; condylobasal length about 15 mm (Egypt) . ***C. religiosa*** See also ***C. lusitania*** (Morocco)
– Hind feet over 11 mm; condylobasal length over 17 mm . . . . (8)
8. Condylobasal length usually over 21 mm (E. Siberia) . . . . . . ***C. lasiura*** See also ***C. sericea*** (Morocco)
– Condylobasal length usually under 21 mm (Europe and S.W. Asia) . . ***C. russula***
9. Condylobasal length over 17·8 mm; mandible (with first tooth) over 11 mm; hind feet over 12 mm; tail usually over 40 mm (S. Japan) . . . . ***C. dsinezumi***
– Condylobasal length under 17·8 mm; mandible usually under 11 mm; hind feet under 12 mm; tail usually under 40 mm . . . . ***C. suaveolens***

Crocidura floweri — Flower's shrew

Crocidura floweri Dollman, 1915. Giza, Egypt.

RANGE (Map 9). Known only from Giza and the Nile Delta, Egypt.

Crocidura religiosa Egyptian pygmy shrew

Sorex religiosus I. Geoffroy, 1827. Thebes, Egypt (based upon anciently mummified specimens).

RANGE (Map 10). Nile Valley, Egypt; possibly also further south in Africa.

REMARKS. This species was discovered alive in 1901 (de Winton, 1902) and has been consistently called *C. religiosa*. Heim de Balsac & Mein (1971) rejected the name *religiosa* as being unidentifiable and used the name *nana* Dobson, 1890 (Somalia) on the assumption that these forms were conspecific. Whilst they may well be conspecific there seems to be no reason to reject the familiar name *religiosa*. The account of the mummified specimens on which the name was based is perfectly consistent with the living species. There is no evidence that type-material of *religiosa* has survived. I therefore designate as neotype of *Sorex religiosus* I. Geoffroy, 1827 specimen number 10.6.18.4 in the British Museum (Natural History), an adult male in alcohol with the skull extracted, from Giza, Egypt, collected by Captain S. S. Flower and presented by the Giza Zoological Gardens.

Crocidura lusitania*

Crocidura lusitania Dollman, 1916:198. Trarza district, Mauritania.

RANGE (Map 9). W. Mauritania, S.W. Morocco (S. of Anti-Atlas).

REMARKS. The presence of this species in Morocco was reported by Heim de Balsac (1968).

Crocidura suaveolens Lesser white-toothed shrew

Sorex suaveolens Pallas, 1811. Khersones, Crimea.

C. s. suaveolens. Continental range from Europe to Mongolia. Incl. *antipae*, *astrabadensis*, *cantabra* (Spain), *de beauxi*, *dinniki* (Caucasus), ?*hyrcania* (N. Iran), *iculisma*, *ilensis* (Tien Shan), *italica*, *lar* (Mongolia), ?*lignicolor* (W. Sinkiang), *mimula* (Switzerland), *minuta*, *mordeni* (Kazakhstan).
*Crocidura suaveolens balcanica** Ondrias, 1970. 1 km W. of Kryonerion, Attica, Greece.
C. s. cassiteridum Hinton, 1924. Scilly Is., England.
C. s. oyaensis† Heim de Balsac, 1940. Yeu Is., W. France.
*C. s. uxantisi** Heim de Balsac, 1951. Is. of Ushant, France.
Crocidura uxantisi Heim de Balsac, 1951. Is. of Ushant, France.
*C. s. enezsizunensis** Heim de Balsac & Beaufort, 1966. Sein Is., N.W. France.
Crocidura suaveolens enez-sizunensis Heim de Balsac & Beaufort, 1966a. Sein Is., Finistère-Sud, France.
C. s. sarda Cavazza, 1912. Sardinia.
C. s. whitakeri de Winton, 1898. Morocco.
*C. s. matruhensis** Setzer, 1960. N.W. Egypt.
Crocidura suaveolens matruhensis Setzer, 1960. 3 miles W. of Mersa Matruh, Western Desert Gov., Egypt.
C. s. portali Thomas, 1920. Palestine. Incl. *tristrami*†.
C. s. shantungensis Miller, 1901. E. Asia; type-locality in Shantung. Incl. *coreae* (Korea), *longicauda*, *orientis* (Ussuri, E. Siberia), *phaeopus* (Szechuan).
C. s. utsuryoensis† Mori, 1937. Utsuryo Is. (=Ullung Do), Korea. Status *fide* Jones & Johnson (1960).

RANGE (Map 9). Temperate woodland and steppe zones of the entire Palaearctic from Spain to Korea: northern boundary through central France, S. Poland, Moscow region, Lake Balkash, Mongolia, S. Manchuria; southern boundary from Egypt through N. Iraq, N. Afghanistan, Sinkiang and the Yangtze-Kiang. Also in Morocco and Algeria; and the islands of Scilly, Jersey, Sark, Ushant, Yeu, Sardinia, Sicily in Europe; and Tsushima and Ullong Do between Korea and Japan.

REMARKS. According to Harrison (1964a) *tristrami* is a form of this species.

Crocidura russula Greater white-toothed shrew

Sorex russulus Hermann, 1780. Near Strasbourg, France.

C. r. russula. Mainland of Europe. Incl. ?*albiventris*, *araneus* Schreber (not of Linnaeus), *candidus*, *chrysothorax*, *cinereus*, ?*constrictus*, *fimbriatus*, ?*hydruntina*, ?*inodorus*, ?*leucurus*, *major*, *mimuloides*, *moschata*, *musaraneus*, *poliogastra*, *pulchra* (Portugal), *rufa*, *thoracicus*, *unicolor*.

?*C. r. gueldenstaedti* (Pallas, 1811). Caucasus and perhaps Asia Minor and Palestine; type-locality in Georgia. Incl. *aralychensis, bogdanowii, longicaudata,* ?*monacha* (Asia Minor).
?*C. r. caspica* Thomas, 1907. S. coast of Caspian Sea. Incl. ?*fumigatus.*
C. r. yebalensis Cabrera, 1913. N. Africa; type-locality in Morocco. Incl. ?*agilis, anthonyi* (Tunis), ?*foucauldi* (Spanish Morocco), ?*mauritanicus, pigmaea.*
*Crocidura heljanensis** Vesmanis, 1975. Heljani, Oran, Algeria.
C. r. sicula† Miller, 1901. Sicily. Incl. ?*caudata*† (see remarks below).
C. r. cypria Bate, 1904. Cyprus.
C. r. cyrnensis† Miller, 1907. Corsica. Incl. *corsicana.*
*C. r. ibicensis** Vericad & Balcells, 1965. Ibiza.
Crocidura russula ibicensis Vericad & Balcells, 1965. Ibiza, Spain.
C. r. balearica† Miller, 1907. Minorca.
C. r. caneae Miller, 1909. Crete.
*Crocidura russula zimmermanni** Wettstein, 1953. Nida Plateau, Ida Mts, Crete.
C. r. ichnusae Festa, 1912. Sardinia.
C. r. peta Montagu & Pickford, 1923. Guernsey, Channel Is.

RANGE (Map 10). Europe from Iberia and the Mediterranean northeast to northern Germany, southern Poland, Czechoslovakia, Rumania and Bulgaria; possibly conspecific forms from the Caucasus through Asia Minor to Israel, N. Iraq and N.E. Iran; north Africa from Tunisia to Morocco; perhaps Kashmir; almost all the large islands of the Mediterranean, the Atlantic islands of France, and Guernsey, Alderney and Herm in the Channel Isles.

REMARKS. E & M-S included the Japanese form *dsinezumi* in this species. It is here treated as a separate species since it is as different from *C. russula* as is the sympatric *C. suaveolens.* E & M-S treated most of the Mediterranean island forms as a separate species, *C. caudata,* but they have been considered races of *C. russula* by several recent authors. Two forms, *sicula* and *caudata,* have been described from Sicily. These differ greatly in length of tail but are otherwise very similar (types examined). Most available specimens from Sicily are short-tailed, agreeing with *sicula,* and this seems clearly referable to *C. russula* rather than to *C. leucodon* as listed by E & M-S. It is tentatively suggested that the type of *caudatus* is the same, with an aberrantly long tail (which is also abnormally flattened distally).

The form *judaica* from Palestine certainly does not represent *C. russula* and is here tentatively transferred to *C. leucodon,* following Harrison (1964a). The name *yebalensis* is used here for the N.W. African form following Vesmanis (1975) who considered *agilis* Levaillant, 1867 unidentifiable.

Crocidura dsinezumi† Japanese white-toothed shrew

Sorex dsi-nezumi Temminck, 1844. Kyushu, Japan.
C. d. dsinezumi. =*kinezumi* Temminck, 1843 (see Remarks), *kinczumi.* Kyushu, Shikoku and S. Honshu.
C. d. chisai Thomas, 1906. N. Honshu.
C. d. umbrina Temminck, 1844. Yakushima Is., Japan.
C. d. intermedia Kuroda, 1924. Tanegashima Is., Japan.
?*C. d. orii* Kuroda, 1924. Amamioshima Is., Japan.
C. d. quelpartis Kuroda, 1934. Quelpart Is., Korea.
*C. d. okinoshimae** Kuroda & Uchida, 1959. Okinoshima Is., Japan.
Crocidura suaveolens okinoshimae Kuroda & Uchida, 1959. Okinoshima Is., N. of Kyushu, Japan. Status from Imaizumi (1970a).

RANGE (Map 10). Honshu and the islands of Kyushu, Shikoku, Yakushima, Tanegashima, Amamioshima, Oki and Okinoshima, Japan; and Quelpart Island, Korea.

REMARKS. E & M-S listed this as a series of races of *C. russula.* Imaizumi (1960, 1970a) gave it specific rank and this seems justified. The inclusion of *chisai* in this species seems less certain.

According to Mazak (1967) Part 1 of Temminck's *Fauna Japonica,* including the illustrations of this species captioned *Sorex kinezumi* and *S. kinczumi,* was published in

1843 and Part 2, including the text under the name *Sorex dsinezumi*, in 1844. Since *kinczumi* and *kinezumi* have never been used as senior synonyms, an application has been made to the International Commission to have them placed on the Official Index of Rejected Names.

Imaizumi (1961) redescribed the form *orii* and considered that it might be specifically distinct from *C. dsinezumi*.

Crocidura leucodon Bicoloured white-toothed shrew

Sorex leucodon Hermann, 1780. Vicinity of Strasbourg, France.
C. l. leucodon. Europe. =*leucodus*, *microurus*. Incl. *albipes*, *narentae* (Yugoslavia).
?*C. l. persica* Thomas, 1907. Elburz Mts. =*caspica* Lydekker (not of Thomas). Doubtfully conspecific with *C. leucodon*.
?*C. l. judaica*† Thomas, 1919. Palestine and Lebanon, type-locality near Jerusalem.

RANGE (Map 11). From France through central Europe and southern Russia to the Volga, the Caucasus, and perhaps the Elburz Mts, Asia Minor and Palestine.

***Crocidura sibirica*†** Siberian white-toothed shrew

Crocidura leucodon sibirica Dukelski, 1930. Osnatschennoje, R. Yenesei, Siberia.
*Crocidura leucodon ognevi** Stroganov, 1956a. Near Berovlyank, Troitsk Region, Altai, Siberia.

RANGE (Map 10). Central Asia from L. Issyk Kul to the upper Ob, L. Baikal and Mongolia.

REMARKS. Dolgov (1974) has argued, mainly on the basis of genitalia, that this form cannot be considered conspecific with *C. leucodon*. I have not seen specimens.

Crocidura pergrisea

Crocidura pergrisea Miller, 1913. Skoro Loomba, Shigar, Baltistan, Kashmir (9500 ft). Incl. *serezkyensis* (Pamirs).
*Crocidura armenica** Gureev, 1963. Garni, Armenia, USSR.
*Crocidura pergrisea arispa** Spitzenberger, 1971a. *c*. 20 km E.S.E. of Ulikisla (*c*. 2000 m), Vil. Nigde, Turkey (Asiatic).

RANGE (Map 10). Mountains from S.W. Asia Minor, Armenia, Kopet Dag to Afghanistan, Kashmir and the Tien Shan (map in Spitzenberger, 1971a).

REMARKS. The above arrangement follows Spitzenberger (1971a) and Dolgov & Yudin (1975).

***Crocidura zarudnyi*†**

Crocidura zarudnyi Ognev, 1928. Baluchistan. =*tatianae*.
*Crocidura zarudnyi streetorum** Hassinger, 1970. 30 km N.W. of Ghazni, Afghanistan (*c*. 33°43′N, 68°15′E).

RANGE (Map 9). Baluchistan (S.E. Iran and W. Pakistan) and much of Afghanistan.

REMARKS. This form was considered conspecific with *C. pergrisea* by E & M-S but has been given specific rank by Hassinger (1970, 1973) and Spitzenberger (1971a). The relationship of both these to *C. suaveolens* remains uncertain.

Crocidura lasiura Ussuri white-toothed shrew

Crocidura lasiura Dobson, 1890. Ussuri R., Manchuria.
C. l. lasiura. Northern part of range. =*lasiura** Giglioli & Salvadori, 1887 (*nomen nudum*). Incl. *lizenkani*, ?*sodyi*†, *thomasi*, *yamashinai*.
?*C. l. campuslincolnensis* Sowerby, 1945. Shanghai, China.

RANGE (Map 9). The Ussuri region of Siberia and Manchuria south to Korea; also Shanghai if *campuslincolnensis* is conspecific.

REMARKS. E & M-S included *lasia* (Asia Minor) in this species but all Russian authors now treat these as distinct species, a view that is supported by the limited material in the British Museum.

***Crocidura lasia*†**

Crocidura leucodon lasia Thomas, 1906. Scalita, near Trebizond, Asia Minor.

RANGE (Map 11). N.E. Asia Minor; Lebanon; Lesbos Is., Àegean (as *C. lasiura:* Ondrias, 1969).

REMARKS. Clearly distinct from *judaica* with which it is sympatric in Lebanon, but it is uncertain whether one of these is conspecific with *C. leucodon.* This form is unlikely to be conspecific with *C. lasiura* of eastern Asia as indicated by E & M-S.

Crocidura flavescens* African giant shrew

Sorex flavescens Geoffroy, 1827. Southern Africa.
C. f. olivieri Lesson, 1827. Sakkara, Egypt.
*Crocidura flavescens deltae** Heim de Balsac & Barloy, 1966. Egypt. No type designated.
Also many other subspecies from south of the Sahara.

RANGE (Map 10). Nile Valley, Egypt; widespread in Africa south of the Sahara.

REMARKS. This large form is clearly distinct from any other Palaearctic species. The Egyptian form has been known consistently as *C. olivieri* since its discovery as a living species in 1901 and was listed as such by E & M-S. Heim de Balsac & Barloy (1966) included it in *C. flavescens* and at the same time proposed the name *deltae* on the grounds that *olivieri* Lesson, 1827 (based on anciently mummified specimens) is not identifiable. It seems quite unnecessary to change the name in this way. The mummified specimens as described by Olivier (1804) are very likely to be conspecific with the living animals and certainly all other living species in Egypt can be ruled out. In order to stabilize the nomenclature I therefore designate as neotype of *Sorex olivieri* Lesson, 1827 specimen number 9.7.1.13 in the British Museum (Natural History), an adult male skin and skull from near Giza, Egypt, dated 22 September 1908, collected by Mr M. J. Nicoll and presented by Capt. S. S. Flower and Mr J. C. Bonhote (cited by Bonhote, 1910:790). This same specimen can be designated as the holotype of *Crocidura flavescens deltae* Heim de Balsac & Barloy, 1966 since these authors did not designate a type nor locality but simply wrote 'Le type serait à choisir parmi les specimens du British Museum'.

Crocidura sericea*

Sorex sericeus Sundevall, 1843. Bahr-el-Abiad, Sudan. Incl. *bolivari* (Rio de Oro), *macrodon** Dobson, 1890.

RANGE (Map 10). S.W. Morocco from Tiznit to lower Dra; Rio de Oro; S. Sudan.

REMARKS. Heim de Balsac (1968) reported *C. sericea* in Morocco and considered *bolivari*, listed *incertae sedis* by E & M-S, to be a synonym. It is a medium-sized species, comparable to *C. russula* but with rather long tail and large ears.

Genus *SUNCUS*

Suncus Ehrenberg, 1833. Type-species *S. sacer* Ehrenberg = *S. murinus* (L.). Incl. *Pachyura, Paradoxodon, Plerodus, Sunkus.*

This group is only marginally distinguishable from *Crocidura* by the presence of fourth unicuspids, but it is convenient to retain it pending generic revision of the family. It is primarily Oriental and Ethiopian, but two species extend into the Palaearctic Region.

Large: head and body over 100 mm; condylobasal length over 25 mm ***S. murinus***
Small: head and body under 50 mm; condylobasal length under 14 mm ***S. etruscus***

Suncus murinus House shrew

Sorex murinus Linnaeus, 1766. Java. = *myosurus*. Incl. *caeruleus.*
S. m. sacer Ehrenberg, 1833. S.W. Asia: type-locality Suez, Egypt where it may have been a temporary introduction (Setzer, 1957a). Incl. *crassicaudus, duvernoyi.*
*S. m. temmincki** (Fitzinger, 1868). Kyushu and Goto Islands, Japan. Incl. ?*riukiuana.*
Pachyura temminckii Fitzinger, 1868:141. Kyushu, Japan. = *Sorex indicus* Temminck, 1844 (not of Geoffroy, 1811).

RANGE (Map 9). The entire Oriental Region, spread by man to many islands and isolated seaports, including, in the Palaearctic Region, ports from the Persian Gulf to the Red Sea (e.g. Bahrain, Basra, Aden, Suez), the Japanese island of Fukue (Goto Is.) and two localities on Kyushu (Nagasaki and Kagoshima) (also throughout the adjacent Ryukyu Islands).

REMARKS. *Suncus tristrami* Bodenheimer, tentatively referred to this species by E & M-S, is a synonym of *Crocidura suaveolens portali* according to Harrison (1964a).

Suncus etruscus Pygmy white-toothed shrew

Sorex etruscus Savi, 1822. Pisa, Italy. Incl. ?*nanula*† (Turkestan), *pachyurus* (Sardinia), *suaveolens* Blasius, not Pallas.

RANGE (Map 9). The Mediterranean zone of Europe; the Caucasus; Turkmenistan, north to the Aral Sea and east to Tashkent; W. Asia Minor; Afghanistan; Iraq; Aden; Palestine; the Nile Delta; Tunisia to Morocco; Ethiopia, northern Nigeria and Guinea; the islands of Crete, Corsica, Sardinia, Sicily, Majorca and Corfu. Perhaps also India and Ceylon (see below).

REMARKS. It seems doubtful if some, or any, of the Indian forms allocated to this species by E & M-S are in fact conspecific. Similar forms in Africa south of the Sahara are almost certainly not conspecific except for the records from Ethiopia (Corbet & Yalden, 1972), Nigeria (Morrison-Scott, 1946) and Guinea. The identity of the Nigerian specimen has been disputed by Petter & Chippaux (1962), but I would uphold Morrison-Scott's identification. Records from Asia Minor and Crete were given by Spitzenberger (1970).

Genus *DIPLOMESODON*

Diplomesodon Brandt, 1852. Type-species *Sorex pulchellus* Lichtenstein.

A distinct, monospecific genus characterized especially by the presence of only two upper unicuspid teeth on each side and by the uniquely patterned pelage.

Diplomesodon pulchellum Piebald shrew

Sorex pulchellus Lichtenstein, 1823. Kirghiz Steppe, Russian Turkestan. Incl. *pallidus* (S.E. Turkmenistan).

RANGE (Map 11). The dry steppe and subdesert zone of southern Asiatic USSR from the Volga to Lake Balkash.

Genus *CHIMMAROGALE*

Chimmarogale Anderson, 1877. Type-species *Crossopus himalayicus* Gray. Incl. *Crossogale*.

A predominantly Oriental genus of probably three species. In addition to *C. himalayica* dealt with below these are *C. phaeura* on Sumatra and Borneo, and *C. styani* in Szechuan.

Chimmarogale himalayica Himalayan water shrew

Crossopus himalayicus Gray, 1842. Chamba, Himalayas.
C. h. himalayica. Himalayas.
C. h. platycephala (Temminck, 1843). Japan; type-locality Kyushu.
C. h. leander Thomas, 1902. S. China.
C. h. varennei Thomas, 1927. Indo-China.
?*C. h. hantu** Harrison, 1958. Malaya.

RANGE (Map 11). Throughout Kyushu, Shikoku and Honshu, Japan; probably conspecific forms from Kashmir through Yunnan, Burma and Laos to Malaya, and on Taiwan.

REMARKS. It has recently been shown (Mazak, 1967) that the first part of Temminck's contribution to Siebold's *Fauna Japonica* was published in 1843, and therefore the name *platycephala* Temminck, used for this species by E & M-S and dated from 1842, must give way to the earlier *himalayica* Gray, 1842.

E & M-S included *styani* from Szechuan in this species but the very small size of *styani* seems to make this improbable. Medway (1965) considered the Bornean form *phaeura* (formerly in the genus *Crossogale)* conspecific with *C. himalayica*, but the differences seem too great to justify this step.

Family **TALPIDAE**
Moles etc.

A Holarctic family extending some way into the Oriental region. Four of the six subfamilies occur in the Palaearctic region. The others are Uropsilinae (genus *Uropsilus*) in southern China, and Condylurinae (genus *Condylura*) in North America.

1. Tail about equal in length to head and body; hind feet much wider than fore feet (subfamily Desmaninae). . . . . (2)
– Tail about half length of head and body or much less; fore feet much wider than hind feet (3)
2. Tail flattened laterally throughout; head and body up to 220 mm (Russia) ***DESMANA*** (p. 32)
– Tail flattened only at the tip; head and body not over 160 mm (Pyrenees and Iberia) ***GALEMYS*** (p. 32)
3. I^{1-3} small, followed by much larger C^1 (subfamily Talpinae) . . . ***TALPA*** (p. 32)
– I^1 larger than C^1 (subfamily Scalopinae) . . . . (4)
4. Ten upper teeth on each side (Japan) . . . . . . ***UROTRICHUS*** (p. 36)
– Nine upper teeth on each side (China) . . . . . . ***SCAPANULUS*** (p. 37)

Subfamily **DESMANINAE**
Genus ***DESMANA***

Desmana Güldenstaedt, 1777. Type-species *Castor moschatus* L. Incl. *Caprios, Desman, Desmanus, Mygale, Myogale, Myogalea.*

Contains one species.

Desmana moschata Russian desman

Castor moschatus Linnaeus, 1758. Russia. =*moscovitica.*

RANGE (Map 12). The rivers Don, Volga and lower Ural and their tributaries. Introduced to the Dnepr.

Genus ***GALEMYS***

Galemys Kaup, 1829. Type-species *Mygale pyrenaica* Geoffroy. =*Mygalina, Galomys.*

Contains one species.

Galemys pyrenaicus Pyrenean desman

Mygale pyrenaica Geoffroy, 1811. Near Tarbes, Haute-Pyrénées, France. Incl. *rufulus* (Segovia, Spain).

RANGE (Map 12). Rivers of the Pyrenees, N.W. Spain and N. Portugal.

REMARKS. For a recent account, with a detailed map, see G. Niethammer (1970). He considered that there were no grounds for subspecific separation of the two forms.

Subfamily **TALPINAE**
Genus ***TALPA***

Talpa Linnaeus, 1758. Type-species *Talpa europaea* L. Incl. *Asioscalops, Asioscaptor, Chiroscaptor, Eoscalops, Euroscaptor, Mogera, Parascaptor, Scaptochirus.*

E & M-S listed only three species in this genus whilst Stein (1960) recognised eight species in four genera. These genera are based upon differences in dental formula, but since other cranial differences cut across any such arrangement no subdivision of the genus *Talpa* is recognized here. E & M-S included all the eastern Asiatic talpine moles, Palaearctic and Oriental, in one species. I would place them in at least seven species, which seem very distinct. In addition to the species included below, at least three species are entirely Oriental: *T. micrura* (?including *longirostris* and *klossi*), *T. leucura* (=*Parascaptor leucurus*) and *T. latouchei* (including *hainana*).

In the western Palaearctic the moles of the *T. europaea* group present many problems. Along the southern fringe of the range of *T. europaea* s.s., from Iberia to Iran, many forms have been described that are very slightly differentiated although several cases of sympatry have been demonstrated. There is a very considerable literature on these but no comprehensive revision. I list the following members of this group as species: *T. europaea*, *T. romana*, *T. caucasica*, *T. caeca* and *T. streeti*.

1. Four upper premolars (11 upper teeth) on each side; 4 lower premolars, i.e. 2 small teeth, following the large, caniniform P_1; rostrum long and narrow, width across molars less than 70% (usually less than 60%) of upper tooth-row . . . . (2)
– Usually 3 upper premolars (10 upper teeth) on each side; usually 3 lower premolars; rostrum short and wide, width across upper molars over 70% of upper tooth-row (China) ***T. moschata***
2. Tympanic bullae rather swollen, projecting below surrounding bone on all sides, aperture elongate and more than half the length of the bulla (E. Asia) . . . . (3)
– Tympanic bullae very flat, not clearly demarcated from surrounding bone, aperture almost circular and less than half the length of the bulla (Europe to central Siberia) . (5)
3. Larger: head and body over 120 mm, condylobasal length over 30 mm; tail less than 20% of head and body; 3 lower incisiform teeth on each side; pelage brown . . . (4)
– Smaller: head and body under 105 mm, condylobasal length under 28 mm; tail over 20% of head and body; 4 lower incisiform teeth on each side; pelage black or greyish brown ***T. mizura***
4. Larger: condylobasal length over 36 mm; upper tooth-row over 15·2 mm; mandible over 24 mm; head and body usually over 150 mm . . . . . . ***T. robusta***[1]
– Smaller: condylobasal length under 36 mm; upper tooth-row under 15·2 mm; mandible under 24 mm; head and body usually under 150 mm . . . . . ***T. wogura***[1]
5. M^1, i.e. 3rd tooth from back, narrow, lacking hypocone, greatest width (perpendicular to labial margin) less than half length of labial margin (unique in genus); P^4 completely lacking a hypocone (central Siberia) . . . . ***T. altaica***
– M^1 wide, greatest width over half length of labial margin; P^4 with at least a small hypocone (6)
6. Rostrum wide, width across upper molars over 30% of condylobasal length . (7)
– Rostrum narrower, width across molars under 30% of condylobasal length . (8)
7. P2 and P3 in both jaws conspicuously smaller than any other teeth in their rows (Iran) ***T. streeti***
– P2 and P3 not conspicuously small . . . . ***T. romana***
8. Eyes capable of opening clearly; upper incisor row forming a smoothly rounded arc, I^1 usually less than twice the size of I^3; (condylobasal length usually over 32 mm; maxillary tooth-row over 12 mm). . . . . ***T. europaea***
– Eyes closed; upper incisor row slightly angled in front, I^1 usually more than twice the size of I^3 (9)
9. Larger: head and body usually over 110 mm, condylobasal length over 32 mm, middle claw of front foot over 2 mm wide . . . . ***T. caucasica***
– Smaller: head and body under 110 mm, condylobasal length under 32 mm, middle claw under 2 mm wide . . . . ***T. caeca***

Talpa europaea — European mole

Talpa europaea Linnaeus, 1758. S. Sweden. =*caudata, nigra, scalops, vulgaris*. Incl. *alba, albida, albomaculata, brauneri, cinerea, flavescens, friseus, grisea, lutea, maculata, pancici, rufa, uralensis, variegata*.
*Talpa europaea obensis** Skalon & Rajevsky, 1940. Near Shukhtunkurg, River Sosvye, River Ob, Siberia.
*Talpa europaea velessiensis** Petrov, 1941. Pepeliste, near Krivolak, 40 km S.E. of Veles, Yugoslavia.
*Talpa europaea transuralensis** Stroganov, 1956a. Tyumensk, Ubatsk region, Sorovaya Dept, Siberia.
Talpa europaea aberr. col. *argentea**, *atrimaculata**, *cremea**, *griseomaculata**, *innominabilis**, *murina**, *nebulosa**, *punctulata**, *sepiacea**, *sordida**, *subpunctulata** *ventromaculata**, Husson & Heurn 1959. Netherlands.
*Talpa europaea kratochvilli** Grulich, 1969. S. Slope Osobita Mt, western High Tatra, Czechoslavakia.

[1] These figures apply to Honshu where both species are present. Korean animals, probably referable to *T. robusta*, are rather smaller (upper toothrow 14·6–16·1 mm).

RANGE (Map 13). Europe, except for parts of the Mediterranean zone, and eastwards throughout Russia and as far as the rivers Ob and Irtish. Isolates in Britain and S. Sweden.

REMARKS. E & M-S included in this species *T. altaica* which is undoubtedly a distinct species, *T. romana* which is less distinctive but should probably be excluded, and several Caucasian forms which are also best excluded until their relationship is better known. The Tatra population has been described as distinct on the basis of small size (having formerly been referred to *T. caeca*) but no discontinuity of variation has been demonstrated.

***Talpa romana*†** Roman mole

Talpa romana Thomas, 1902. Near Rome, Italy. Incl. *major*, *montana*, ?*ognevi* (Caucasus), *stankovici* (Macedonia).

RANGE (Map 13). S. Italy, Macedonia and perhaps also the Caucasus.

REMARKS. There has been considerable uncertainty over the status of these forms. E & M-S included them as races of *T. europaea*. However, they appear to be clearly distinct (but predominantly allopatric) in Macedonia (Petrov, 1971, 1974). The Caucasian form *ognevi* was allocated to this species by Gromov *et al.* (1963).

***Talpa caucasica*†** Caucasian mole

Talpa caeca caucasica Satunin, 1908. Stavropol, Caucasus.

RANGE (Map 12). Northern side of the Caucasus from the Sea of Azov to the Caspian Sea.

REMARKS. Included in *T. europaea* by E & M-S and by Kuzyakin (1965), but considered a distinct species by Gromov *et al.* (1963) and by Kozlovsky *et al.* (1972), the latter authors basing their view on karyological data.

Talpa caeca Mediterranean mole

Talpa caeca Savi, 1822. Near Pisa, Italy. Incl. ?*davidianus* Milne-Edwards, *hercegovinensis* (Yugoslavia), *levantis* (Asia Minor), *occidentalis* (Spain), *olympica* (Greece), *orientalis* (W. Caucasus).
*Talpa orientalis talyschensis** Vereschchagin, 1945. Vilyazhchaya Gorge, Talysh, Azerbaijan, 1500–1600 m.
*Talpa minima** Deparma, 1960. Upper R. Beloi, near Khamishki, N.W. Caucasus (500 m).
*Talpa caeca dobyi** Grulich, 1971a. Peira Cava, Maritime Alps, France (1600–1700 m).
*Talpa caeca beaucournui** Grulich, 1971b. Pelister Mt., Macedonia, Yugoslavia.
*Talpa caeca steini** Grulich, 1971b. Lovcen Mt., Cetinje, Yugoslavia.

RANGE (Map 13). Parts of Iberia, the southern side of the Alps, Balkans, Asia Minor and Caucasus.

REMARKS. This species is undoubtedly distinct from *T. europaea* with which it is marginally sympatric, e.g. in the Alps. The existence of a yet smaller species in the Mediterranean region, variously called *T. hercegovinensis* or *T. mizura*, has been discredited by Niethammer (1969a) at least as far as Spain is concerned, and the allocation of these to the Japanese *T. mizura* has also been shown to be erroneous (Corbet, 1967).

Talpa streeti* Persian mole

Talpa streeti Lay, 1965. Hezer Darrak, Kurdistan Prov., Iran (35°25′N, 47°07′E).
*Talpa streetorum** Lay, 1967. Unjustified emendation of *streeti*.

RANGE (Map 13). Known only from the type locality.

REMARKS. Very similar to *T. romana* but the teeth appear distinctive and it seems wise to consider this as a potentially valid species. Only the type has been seen.

***Talpa altaica*†** Siberian mole

Talpa altaica Nikolsky, 1883. Altai Mts, Siberia. Incl. *irkutensis*, *saianensis*, *suschkini* (Sayan Mts), *salairici*, *sibirica* (W. Siberia), *tymensis*.
*Asioscalops altaica gusevi** Fetisov, 1956. Snejnoi, Kamar-Daban, Siberia.

RANGE (Map 13). The taiga zone of Siberia between the rivers Ob/Irtish and Lena, reaching north along the Yenesei to 70°N, and south to northern Mongolia.

REMARKS. E & M-S placed this form, with a query, in *T. europaea*; Stroganov (1957) placed it in a separate genus *Asioscalops*. This seems quite unjustified, but there seems no doubt that it is specifically distinct from *T. europaea*. Stein (1960) considered it conspecific with *T. longirostris* of Szechuan: this seems most unlikely judging by differences in P^4, M^1, auditory bullae and sacrum.

Talpa mizura† Japanese mountain mole

Talpa mizura Gunther, 1880. Near Yokohama, Honshu, Japan.
T. m. mizura. N.E. Honshu.
*T. m. hiwaensis** Imaizumi, 1955a. Hiwa, S.W. Honshu.
*Talpa mizura hiwaensis** Imaizumi, 1955a:33. Hiwa, Hiroshima Pref., Honshu, Japan (390 m).
*T. m. ohtai** Imaizumi, 1955a. N.W. Honshu.
*Talpa mizura ohtai** Imaizumi, 1955a:34. Mt Chogatake, Hida Mts, Honshu, Japan (2600 m).

RANGE (Map 13). Apparently confined to a few isolated montane areas on Honshu, Japan in forest and alpine grassland.

REMARKS. E & M-S considered this a synonym of *T. micrura wogura*. Niethammer & Niethammer (1964) considered it conspecific with *T. hercegovinensis*, but it can be distinguished by the tympanic bullae (Corbet, 1967). Imaizumi (1970a) placed it as the sole member of the genus *Euroscaptor*.

Talpa wogura† Lesser Japanese mole

Talpa wogura Temminck, 1843. Yokohama, Honshu, Japan (restricted by Thomas (1905): E & M-S gave Nagasaki, Kyushu without reason, whilst Imaizumi (1960) gave Yokohama). For the dating of Temminck's *Fauna Japonica* see Mazek (1967).
T. w. wogura. Mainland of Honshu, mainly N.E. of Tokyo district. =*moogura*. Incl. *minor* Kuroda (not of Freudenberg), *gracilis* Kishida (not of Kormos).
*Talpa micrura imaizumii** Kuroda, 1957. New name for *Mogera wogura minor* Kuroda, 1936 which is preoccupied.
T. w. kanai† Thomas, 1905. Islands of Yakushima (type-locality), Tanegashima, Goto, Tsushima and Amakusa.

RANGE (Map 13). N.E. Honshu; isolated montane areas of S.W. Honshu (according to Imaizumi, 1960); the southern Japanese islands listed above under *T. w. kanai*.

REMARKS. E & M-S treated this as a race of *T. micrura* from Nepal, but there seem to be no grounds for this. The absence of I_3 seems quite constant in this species and in the forms *kobeae*, *robusta*, *coreana*, *latouchei*, *hainana* and *insularis*. However, in view of the strong grounds for treating *wogura* and *kobeae* as specifically distinct on Honshu (see below under *T. robusta*) it would seem unwise at this stage to allocate any of these other forms to *T. wogura*. The one most likely to be conspecific with *T. wogura* is *insularis* on Taiwan.

When Thomas (1905) recognized two forms of *Talpa* on Honshu he assumed that the smaller one represented *T. wogura* Temminck and named the larger one *kobeae*. In fact the six syntypes of *wogura* in Leiden Museum seem to include both forms. One large albino skin, *a* in Jentink's (1887) catalogue, measures about 170 mm (head and body) and probably represents *kobeae*, whilst the remaining five, *b–f* (erroneously listed as *b–g* by Jentink) are smaller. To confirm Thomas's action I therefore select specimen *d*, also numbered 16249, as the lectotype of *Talpa wogura* Temminck. The skull of this specimen has been removed and has the following measurements: upper tooth-row 14·3 mm; length of mandible 22·5 mm.

Imaizumi (1970a) used the generic name *Mogera*. He recognized a second subspecies, *minor* on Honshu, found on higher ground but intergrading with *T. w. wogura*.

Talpa robusta† Greater Japanese mole

Mogera robusta Nehring, 1891. Vladivostok, E. Siberia.

T. r. robusta. Maritime Province of Siberia and Manchuria.
T. r. coreana (Thomas, 1907). Korea, except perhaps the extreme north. Probably confluent with *T. r. robusta.*
?*T. r. kobeae* (Thomas, 1905). S.W. Honshu and Shikoku, Japan, type-locality Kobe, Honshu.
Mogera wogura kiusiuana Kuroda, 1940: 199. Fukuoka, Kyushu, Japan.
?*T. r. tokudae** (Kuroda, 1940). Sado Is. and possibly the adjacent mainland of Honshu (Imaizumi, 1970a).
Mogera wogura tokudae Kuroda, 1940: 196. Sado Is., Japan.

RANGE (Map 13). From the Ussuri and Amur rivers south through Manchuria to Korea; if the Japanese races are conspecific, S.W. Honshu, Shikoku, Kyushu, Sado Is., Dogo Is.

REMARKS. Moles of the *wogura* type, i.e. lacking I_3, seem clearly referable to two species on Honshu. The length of the upper tooth-row in particular is clearly bimodal: 13·1–14·7 mm (n = 19) in *T. wogura* and 15·6–17·1 mm (n = 19) in *T. robusta kobeae.* These species were recognized as distinct by Imaizumi (1960, 1970a) and by Abe (1967). Abe also considered *tokudae* specifically distinct and closely related to *T. wogura* rather than to *T. robusta.*

Korean animals (*coreana*) are rather smaller than either typical *robusta* or *kobeae* but are closer to them than to *T. wogura* (upper tooth-row 14·7–16·1 mm (n = 15)). The types of *coreana* and *kobeae* seem scarcely distinguishable apart from a small difference in size.

One of the syntypes of *Talpa wogura* Temminck seems to represent this species (see remarks under *T. wogura* above). This specimen was listed, but not described, as var. *alba* by Jentinck (1887), but this name does not seem to have been validly published and in any case would be preoccupied by *Talpa europaea alba* Gmelin, 1788.

***Talpa moschata*†** Short-faced mole

Scaptochirus moschatus Milne-Edwards, 1867. Swanhwafu, 100 miles N.W. of Peking, China. = *davidianus* Swinhoe, *moschiferus.* Incl. *gilliesi, grandidens, leptura, sinensis.*

RANGE (Map 13). N.E. China; Hopei, Shantung, Shansi, Shensi.

REMARKS. Schwarz (1948), followed by E & M-S, considered this conspecific with *T. micrura* (including *T. wogura*) because the absence of one of the small premolars above and below is not absolutely constant. But in the few specimens having these teeth they are rudimentary and very crowded, so that these skulls are still quite different from those of *T. wogura, T. robusta* etc. The presence or absence of I_3 is also a constant difference between these species. The soft lustrous pelage of this species is also very characteristic.

Subfamily SCALOPINAE
Genus *UROTRICHUS*
Japanese shrew-moles

Urotrichus Temminck, 1841. Type-species *U. talpoides* Temminck. Incl. *Dymecodon.*

Two species, both confined to Japan.

Head and body about 90 mm; tail less than 40% of head and body; upper tooth-row over 10 mm; I^3 much smaller than C^1; 8 lower teeth on each side (I_3 missing) . ***U. talpoides***
Head and body about 70 mm; tail about 50% of head and body; upper tooth-row under 10 mm; I^3 about equal to C^1; 9 lower teeth (I_3 present but often very small). . ***U. pilirostris***

Urotrichus talpoides Greater Japanese shrew-mole

Urotrichus talpoides Temminck, 1841. Kyushu, Japan.
U. t. talpoides. Kyushu.
U. t. adversus Thomas, 1908. Tsuchima Is.
U. t. centralis Thomas, 1908. Shikoku.
U. t. hondonis Thomas, 1908. Honshu. Incl. *yokohamanis.*
U. t. minutus Tokuda, 1932. Dogo Is., Oki Islands.

RANGE (Map 12). Grassland and forest on Honshu, Shikoku, Kyushu; Dogo Is., and N. Tsushima Is., Japan.

Urotrichus pilirostris Lesser Japanese shrew-mole

Dymecodon pilirostris True, 1886. Enoshima (Yenosima), Bay of Yeddo, Honshu, Japan.

RANGE (Map 12). Montane, coniferous forests of Honshu, Shikoku and Kyushu, Japan.

Genus *SCAPANULUS*

Scapanulus Thomas, 1912. Type species *S. oweni* Thomas.

One species.

Scapanulus oweni Kansu Mole

Scapanulus oweni Thomas, 1912. 23 miles S.E. of Taochou, Kansu, China (9000 ft).

RANGE (Map 12). A limited area on the borders of Kansu, Shensi and Szechuan, central China (just south of the 35th parallel).

Order **MACROSCELIDEA**
Elephant-shrews

For the recognition of this group as a distinct order, containing only the family Macroscelididae, see Butler (1956).

Family **MACROSCELIDIDAE**

An entirely African family of about 15 species in four genera, of which one species occurs in North Africa. For a recent revision see Corbet & Hanks (1968).

Subfamily **MACROSCELIDINAE**
Genus *ELEPHANTULUS*

Elephantulus Thomas & Schwann, 1906. Type-species *Macroscelides rupestris* Smith (S. Africa). Incl. *Macroscelides* (in part).

Contains nine species south of the Sahara in addition to the following.

Elephantulus rozeti North African elephant-shrew

Macroscelides rozeti Duvernoy, 1833. Near Oran, Algeria.
E. r. rozeti. N. of Atlas Mts. Incl. *atlantis*, *moratus*.
E. r. deserti (Thomas, 1901). S. of Atlas Mts; type-locality Biskra, Algeria. Incl. *clivorum*.

RANGE (Map 3). The Mediterranean and subdesert zones of N.W. Africa from S.W. Morocco to W. Libya.

REMARKS. This species is *not* represented in E. Africa as suggested by E & M-S, the E. African *E. rufescens* and *E. revoili* being very distinct species (Corbet & Hanks, 1968).

Order **CHIROPTERA**
Bats

Bats constitute the second largest and by far the most distinctive order of mammals. Two clearly defined suborders are recognized of which one, the Megachiroptera, has no truly Palaearctic species although three tropical species extend into the southern extremities of the region. The suborder Microchiroptera is well represented with about 70 species in the Palaearctic Region. The families of Microchiroptera are moderately well defined but there is considerable instability of classification at the generic and species levels. There has been no comprehensive revision of the higher classification of bats since that of Miller (1907) in which all families and genera then known are diagnosed.

Second finger bearing a claw; margin of ear forming a complete ring; mandible with angular process broad and low or almost absent . . Suborder **MEGACHIROPTERA** (p. 38)
Second finger without a claw; margin of ear not forming a complete ring; mandible with angular process long and narrow . . . . . . Suborder **MICROCHIROPTERA** (p. 39)

Suborder MEGACHIROPTERA

Family PTEROPODIDAE

The only family of Megachiroptera, containing many species throughout the tropics of the Old World. Only three species reach the southern part of the Palaearctic Region. See Andersen (1912) for a detailed review.

1. No tail . . . . *PTEROPUS*
– A short tail present . . . . (2)
2. Wing membranes attached high on back, distance between them less than width of head; tympanic bone with a tubular meatus and a flat medial extension. . *EIDOLON*
– Wing membranes more widely separated on back, distance between them greater than width of head; tympanic bone annular . . . . *ROUSETTUS*

Genus *ROUSETTUS*

Rousettus Gray, 1821. Type-species *Pteropus aegyptiacus* Geoffroy. Incl. *Cercopteropus*.

Contains about eight species throughout the tropics of the Old World.

Rousettus aegyptiacus Egyptian fruit bat

Pteropus aegyptiacus Geoffroy, 1810. Giza, Egypt.
R. a. aegyptiacus. Egypt to Turkey; Cyprus; perhaps also south of the Sahara.
R. a. arabicus Anderson & de Winton, 1902. S. Arabia to Pakistan.

RANGE (Map 14). From Karachi through S. Iran to S. Arabia; the E. Mediterranean from Antioch to Egypt; Cyprus; throughout most of Africa south of the Sahara.

REMARKS. E & M-S gave *arabicus* specific rank. It is here considered conspecific with *R. aegyptiacus* following Harrison (1964a).

Genus *PTEROPUS*

Pteropus Brisson, 1762. Type-species *Pteropus niger* Kerr. Can be dated from Erxleben, 1777 if Brisson, 1762 is considered unavailable.

A genus of many species throughout the Oriental and Australasian Regions and on the islands of the Indian Ocean and western Pacific. One species reaches Kyushu, Japan from the Ryukyu Islands.

Pteropus dasymallus Ryukyu flying fox

Pteropus dasymallus Temminck, 1825. Kuchino Erabu, N. Ryukyu Islands.
P. d. dasymallus. N. Ryukyu Islands and S. Kyushu.
Also other subspecies from the southern Ryukyu Islands.

RANGE (Map 14). Some of the Ryukyu Islands and also at one locality on the south coast of Kyushu, Japan according to the map in Imaizumi (1970a).

Genus *EIDOLON*

Eidolon Rafinesque, 1815. Type-species *Vespertilio vampyrus helvus* Kerr. Incl. *Leiponyx*, *Liponyx*, *Pterocyon*, *Pteropus* (in part).

Contains a single species, *E. helvum*, if *dupreanum* of Madagascar is considered conspecific.

Eidolon helvum* Straw-coloured fruit bat

Vespertilio vampyrus helvus Kerr, 1792:91. Senegal, W. Africa.
E. h. sabaeum (Andersen, 1907). S.W. Arabia.

RANGE (Map 14). Extreme S.W. Arabia; also most of the forest and savanna zones of Africa south of the Sahara; Madagascar (assuming *dupreanum* to be conspecific). The mainland form reaches at least 15°N in Sudan.

REMARKS. E & M-S gave *sabaeum* specific rank, but examination of African and Arabian forms supports the view of Harrison (1964a) that *sabaeum* is conspecific with *E. helvum*.

Suborder MICROCHIROPTERA

Contains 17 families of which only six are represented in the Palaearctic. Only two of these, the Rhinolophidae and the Vespertilionidae, occur extensively in the region, the remainder being represented mainly by tropical species that extend into the southern fringes of the Palaearctic, for example in North Africa and the Arabian region. The arrangement of families follows that of Simpson (1945) and the *Checklist*, except that the Hipposideridae is reduced to a subfamily of Rhinolophidae as is done for example by Anderson & Jones (1967).

1. Tail completely enclosed in membrane or with only the terminal one or two vertebrae projecting beyond the margin (rather more in *Asellia* which has a large, complex nose-leaf) (2)
– Tail projecting far beyond the margin of the membrane or emerging from the dorsal surface of the membrane (never with a large, complex nose-leaf) . . . . (4)
2. Muzzle unspecialized; (tragus present) . . . . . **VESPERTILIONIDAE** (p. 45)
– Muzzle with a deep furrow or with complex nose-leaves . . . . (3)
3. Muzzle with a deep, median furrow; tragus present (projection of ventral margin of ear, in front of notch); last vertebra of tail bifid . . . . . **NYCTERIDAE** (p. 41)
– Muzzle with complex nose-leaves; tragus absent; last vertebra not bifid **RHINOLOPHIDAE** (p. 42)
4. Tail emerging from dorsal surface of membrane . . **EMBALLONURIDAE** (p. 40)
– Tail emerging from margin of membrane . . . . (5)
5. About half of tail projecting beyond the membrane; no nose-leaf; second finger with one rudimentary phalanx . . . . **MOLOSSIDAE** (p. 62)
– Much more than half of tail projecting beyond the membrane; a small triangular nose-leaf present; second finger with two phalanges . . . **RHINOPOMATIDAE** (p. 39)

Family RHINOPOMATIDAE

Rat-tailed bats

Contains a single genus, *Rhinopoma*.

Genus *RHINOPOMA*

Rhinopoma Geoffroy, 1818. Type-species *Vespertilio microphyllus* Brünnich.

Contains only the three species listed below, if the conspecificity of Oriental and Palaearctic forms within these groups is accepted.

1. Larger: forearm 57·5–75 mm, condylobasal length 17·3–20·6 mm; tail usually shorter than forearm; rostrum with ill-defined narial swellings . . . . ***R. microphyllum***
– Smaller: forearm 46–63 mm, condylobasal length 14·0–17·8 mm; tail usually longer than forearm; rostrum with prominent narial swellings . . . . (2)
2. Muzzle with well-developed transverse dermal ridge; narial swellings rounded, in profile extending to rear of upper canine; hind margin of palate rounded . ***R. hardwickei***
– Muzzle with a low dermal ridge; narial swellings more angular, in profile extending forward level with front face of upper canine; hind margin of palate angular . ***R. muscatellum***

Rhinopoma microphyllum Larger rat-tailed bat

Vespertilio microphyllus Brünnich, 1782. Arabia and Egypt.

R. m. microphyllum. The entire Palaearctic and African parts of the range. Including the following extralimital (African) synonyms: *cordofanicum, lepsianum.*

*Rhinopoma microphyllum harrisoni** Schlitter & DeBlase, 1974. 10 km S.E. of Kazerum, Fars Province, Iran (29°34′N, 51°46′E).

Also other races in the Oriental Region.

RANGE (Map 14). Probably very incompletely known. Recorded from isolated areas around the Sahara (Nigeria, Mauritania, Egypt, Sudan), Palestine, Jordan, Lebanon, Arabia, Iran and Afghanistan. Also in India and Sumatra if, as seems likely, the forms *kinneari* and *sumatrae* are conspecific.

Rhinopoma hardwickei Lesser rat-tailed bat

Rhinopoma hardwickii Gray, 1831. India.

R. h. hardwickei. India and extreme E. Afghanistan.

R. h. cystops Thomas, 1903. Northern Africa to Iraq; type-locality Luxor, Egypt. Incl. *arabium.*

Also other named forms from further south in Africa.

RANGE (Map 14). The periphery of the Sahara, including Somalia, Eritrea, Sudan, N.W. Kenya, Niger, Mauritania, Morocco, Tunisia, Egypt; Israel and Arabia to Afghanistan; the island of Socotra; the nominate race throughout India and as far as Thailand.

REMARKS. In the *Checklist* several small forms from Arabia and Iran were included in this species. These are here treated separately under *R. muscatellum* following DeBlase *et al.* (1973).

***Rhinopoma muscatellum*†**

Rhinopoma muscatellum Thomas, 1903. Muscat, Oman.

R. m. muscatellum. Oman and S. Iran. Incl. *pusillum* (S.E. Iran).

R. m. seianum Thomas, 1913. Seistan, Iran.

RANGE. Oman, W. and S. Iran; S. Afghanistan (detailed map in DeBlase *et al.*, 1973).

REMARKS. Included in *R. hardwickei* by E & M-S but demonstrated to be specifically distinct by DeBlase *et al.* (1973) who found the two species occurring together at several localities in Iran.

Family EMBALLONURIDAE

Sheath-tailed bats

Contains eight genera throughout the tropics except Australia, of which two reach the southern fringe of the Palaearctic.

Smaller: forearm under 50 mm, greatest length of skull under 18 mm; 3 pairs of lower incisors; frontal region of skull not conspicuously concave . . . . *COLEURA* (p. 40)

Larger: forearm over 60 mm, greatest length of skull over 19 mm; 2 pairs of lower incisors; frontal region conspicuously concave . . . . . *TAPHOZOUS* (p. 41)

Genus *COLEURA*

Coleura Peters, 1867. Type-species *Emballonura afra* Peters.

Besides *C. afra* listed below, contains only one other species, namely *C. seychellensis* from the Seychelle Islands.

Coleura afra*

Emballonura afra Peters, 1852:51. Tette, Mozambique.

C. a. gallarum† Thomas, 1915. S.W. Arabia and Somalia; type-locality in Somalia.

RANGE (Map 14). In the Palaearctic Region known only from the immediate vicinity of Aden. Throughout East and Central Africa from Eritrea and Sudan to Angola.

REMARKS. E & M-S treated *gallarum* as a species, but it is here given subspecific rank (which is marginally justified) following Harrison (1964a).

Genus *TAPHOZOUS*

Taphozous Geoffroy, 1818. Type-species *T. perforatus* Geoffroy. Incl. *Liponycteris*, *Saccolaimus* Temminck, 1841; *Taphonycteris*. (E & M-S dated *Saccolaimus* from Temminck, 1838 and, on p. 106 from Lesson, 1842. In fact Temminck, 1841 appears to be the first use.)

Contains about ten species throughout the tropics of the Old World, of which two reach the Palaearctic.

Abdomen completely furred; forearm under 66 mm; greatest length of skull under 22 mm . . . ***T. perforatus***

Abdomen naked for some distance in front of tail membrane; forearm over 66 mm; greatest length of skull over 24 mm . . . . ***T. nudiventris***

Subgenus *TAPHOZOUS*

Taphozous perforatus

Taphozous perforatus Geoffroy, 1818. Egypt. Including the following extra-limital (African) synonyms: *australis*, *haedinus*, *rhodesiae*, ?*senegalensis*, *sudani*, *swirae*.

RANGE (Map 15). South of the Sahara from Ghana through Sudan, Kenya, Tanzania, to Rhodesia; Egypt; S.W. Arabia; Oman (Harrison, 1968b); Iran (DeBlase, 1971); Kathiawar and Cutch in N.W. India.

RANGE. E & M-S and Harrison (1964a) used the name *haedinus* (type-locality in Kenya) for Arabian specimens. This depends upon the very slightly darker colour of these in comparison with Egyptian specimens and does not seem to justify subspecific naming. The presence of discrete races elsewhere in Africa seems equally dubious.

Subgenus *LIPONYCTERIS*

Taphozous nudiventris

Taphozous nudiventris Cretzschmar, 1830 or 1831. Giza, Egypt.
T. n. nudiventris. Africa to Palestine and Arabia. =*nudiventer*.
T. n. magnus† Wettstein, 1913. Iraq. Type-locality Basra. Incl. *babylonicus*.
*T. n. zayidi** Harrison, 1955. Oman.
Taphozous nudiventris zayidi Harrison, 1955. Al Ain, Buraimi Oasis, Oman, Arabia.
Also other forms in India and Burma that may be conspecific.

RANGE (Map 15). The savanna zone south of the Sahara from Senegal to Somalia; Mauritania (Panouse, 1951); Sudan, Egypt, Palestine, Arabia, Iraq, N.E. Iran (Lay, 1967), Afghanistan; perhaps also through most of India to Malaya.

REMARKS. E & M-S listed *kachhensis*, with *magnus* and *nudaster*, as a distinct species. They are here united, following Felten (1962) and Harrison (1964a), although it should be noted that Sinha (1970) has more recently recognized both *T. nudiventris* and *T. kachhensis* in the Indian region.

Family NYCTERIDAE

Slit-faced bats

Contains a single genus, *Nycteris*.

Genus *NYCTERIS*

Nycteris Cuvier & Geoffroy, 1795. Type-species *Vespertilio hispidus* Schreber. =*Nicteris*. Incl. *Nycterops*, *Petalia*.

Contains one species in S.E. Asia and about six in Africa of which one extends north of the Sahara.

Nycteris thebaica

Nycteris thebaicus Geoffroy, 1818. Egypt. Incl. *adana* (Aden), *albiventer* (Sudan) and several others from south of the Sahara.

RANGE (Map 15). W. Arabia, Palestine, Egypt, Libya and Morocco (Panouse, 1959); island of Corfu (Greece); most of the savanna zone of Africa south of the Sahara; probably Madagascar.

Family **RHINOLOPHIDAE**

This large family of mainly tropical, Old World bats contains two subfamilies although these are often given family rank. There are three small genera of Hipposiderinae in addition to those included here.

1. Noseleaf with a prominent, complex, median projection, the sella, immediately above the nostrils and with a *single*, dorsal, triangular process, the lancet; toes each with 3 phalanges (except the hallux which has 2) . . . . ***RHINOLOPHUS*** (p. 42)
– Noseleaf without, or with a simple, finger-like median projection, and without a single dorsal lancet; toes all with 2 phalanges . . . . (2)
2. Upper margin of noseleaf with 3 vertical projections . . . . (3)
– Upper margin of noseleaf without vertical projections ***HIPPOSIDEROS*** (p. 44)
3. Ear with a clear indentation on its inner margin; wing with a spur at the base of the terminal phalanx of the 4th finger; small 1st upper premolar present . . ***TRIAENOPS*** (p. 45)
– Ear without an indentation on its inner margin; wing without a spur; small upper premolar absent . . . . ***ASELLIA*** (p. 44)

Subfamily **RHINOLOPHINAE**

Horseshoe bats

Genus ***RHINOLOPHUS***

Rhinolophus Lacepède, 1799. Type-species *Vespertilio ferrumequinum* Schreber. =*Rhinocrepis*. Incl. *Euryalus*.

Contains a large number of species throughout the tropics of the Old World, of which seven occur in the Palaearctic Region.

1. Smaller: forearm under 43 mm, condylocanine length under 16 mm; (first upper premolar well developed, clearly separating C^1 and the succeeding premolar) . . . . (2)
– Larger: forearm over 43 mm, condylocanine length over 16 mm . . . . . (3)
2. Connecting process of noseleaf (the medial, laterally compressed flap connecting the central sella with the dorsal lancet) with the upper angle a rounded right angle (seen in lateral profile); (forearm 34–42 mm, condylocanine length 13–15 mm) (W. Palaearctic) ***R. hipposideros***
– Connecting process with the upper angle forming an acutely pointed process; (E. Asia) ***R. cornutus*** (also *R. lepidus* in Afghanistan)
3. Upper angle of connecting process (see dichotomy 2) short and bluntly rounded; (C^1 in contact with large premolar, small premolar displaced to outer side of tooth-row or missing) . . . . (4)
– Upper angle of connecting process long and acutely pointed . . . . . . (5)
4. Larger: forearm over 53 mm, $C–M^3$ over 7·5 mm . . . . . ***R. ferrumequinum*** ***R. clivosus***
5. Sella (in front view) suddenly narrowing above; connecting process with upper projection pointing dorsally, almost parallel to lancet; anterior and posterior lower premolars about equal in crown area; small upper premolar in tooth-row; (forearm 44–49 mm) ***R. blasii***
– Sella parallel-sided, rounded above; connecting process with upper projection pointing forwards; anterior and posterior lower premolars very unequal in crown area . (6)
6. Tip of lancet sharply acuminate, terminal process almost parallel-sided; zygomatic width greater than mastoid width; $C–M^3$ over 7·2 mm; (forearm 47–54 mm). . ***R. mehelyi***
– Tip of lancet evenly tapering; zygomatic width not greater than mastoid width; $C–M^3$ under 7·2 mm; (forearm 44–51 mm) . . . . ***R. euryale***

Rhinolophus ferrumequinum

Vespertilio ferrum-equinum Schreber, 1774. France.

R. f. ferrumequinum. Europe and N. Africa. =*hippocrepis*, *typicus*. Incl. *colchicus* (S. Russia), *equinus*, *germanicus*, *homodorensis*, *homorodalmasiensis*, *insulanus* (England), *italicus*, *major*, *obscurus* (Spain), *perspicillatus*, *solea*, *ungula*, ?*unifer*, *unihastatus*.

*Rhinolophus ferrumequinum martinoi** Petrov, 1941. Trifunovićevo, Brdo (Orl-Bajir), near Pepeliste, 40 km S.E. of Veles, Yugoslavia.

R. f. irani Cheesman, 1921. Shiraz, Iran.
R. f. proximus Andersen, 1905. Kashmir.
R. f. regulus Andersen, 1905. Kumaon, Himalayas.
R. f. tragatus Hodgson, 1835. Nepal (type-locality) to Yunnan. Incl. *brevitarsus*.
R. f. nippon Temminck, 1835. Japan (type-locality), Korea and China. Incl. *korai*, *mikadoi* (Honshu), *pachyodontus*, *quelpartis*.
*Rhinolophus fudisanus** Kuroda, 1940. Mt Fuji, Honshu, Japan.
*Rhinolophus kosidianus** Kuroda, 1940. Niigata Pref., Honshu, Japan.
*Rhinolophus norikuranus** Kuroda, 1940. Mt Norikura, Japanese Alps, Honshu, Japan.
*Rhinolophus ogasimanus** Kuroda, 1940. Akita Pref., Honshu, Japan.
(These four synonyms *fide* Imaizumi, 1970a.)

RANGE (Map 15). The entire southern Palaearctic from Britain to Japan. Northern boundary through S. England, central Germany, S. Poland, Crimea, Caucasus, Kopet Dag, Kirgizia, S. Korea, Hokkaido and Honshu. In the south in Morocco, Algeria and Tunisia, Palestine, Iran, Afghanistan and throughout the Himalayas to Yunnan.

REMARKS. In view of the apparent continuity of range and the mobility of this species, it seems probable that close study will render the recognition of subspecies invalid. Above is a provisional listing of the main races. Some southern African forms of *Rhinolophus*, e.g. *augur*, are scarcely distinguishable from *R. ferrumequinum* but appear to link up with *R. clivosus*.

Rhinolophus clivosus

Rhinolophus clivosus Cretzschmar, 1828. Mohila, W. Arabia.
R. c. clivosus. Arabia and N.E. Africa. Incl. *acrotis* (Eritrea), *andersoni*, *brachygnathus* (Egypt).
R. c. schwarzi Heim de Balsac, 1934. Algerian Sahara.
R. c. bocharicus† Kastschenko & Akimov, 1917. Turkestan.
*Rhinolophus bocharicus rubiginosus** Gubareff, 1939. Shusha, Azerbaijan, USSR.

RANGE (Map 16). W. and S.W. Arabia; Egypt, Sudan and most of the savanna zone of Africa south of the Sahara except the W. African region; the Algerian Sahara; Turkestan, N.E. Iran and N. Afghanistan if *bocharicus* is conspecific.

REMARKS. Aellen (1959) first suggested that *bocharicus* is conspecific with *R. clivosus*, but Kuzyakin (1965) did not accept this. However, two specimens of *bocharicus* from Samarkand, examined in the Museum Alexander Koenig, Bonn, seem indistinguishable from specimens of *R. clivosus* from Egypt. Hanak (1969) considered them specifically distinct and described the small differences involved.

The following forms from Africa south of the Sahara are probably conspecific: *augur*, *deckeni*, *keniensis*, *zuluensis*.

Rhinolophus hipposideros

Vespertilio hipposideros Bechstein, 1800. France.
R. h. hipposideros. Entire range except from Iraq eastwards. = *typicus*, *typus*. Incl. *alpinus*, ?*anomalus*, *bifer*, *bihastatus*, *eggenhoffner*, *escalerae* (Morocco), *helvetica*, ?*intermedius*, *kisnyresiensis*, *majori* (Corsica), *minimus* (Eritrea), *minutus* (England), *moravicus*, *pallidus*, *phasma* (Spain), *trogophilus*, *vespa*.
R. h. midas Andersen, 1905. Jask, Iran. From Iraq to Kashmir.

RANGE (Map 16). From Ireland, Iberia and Morocco through S. Europe and N. Africa to Turkestan and Kashmir.

Rhinolophus cornutus

Rhinolophus cornutus Temminck, 1835. Japan. Including the following extra-Palaearctic synonyms: *miyakonis*, *orii*, *perditus*, *pumilus* (all Ryukyu Islands); and doubtfully the following: *blythi* (Himalayas), *calidus*, *parcus*, *szechwanus* (S. China).

RANGE (Map 16). Most of Japan from Hokkaido south through most of the Ryukyu Islands; perhaps also most of S. China and west to Kumaon, if *blythi* and its allies are conspecific.

REMARKS. A very closely similar form, *lepidus*, occurs in N. India and the extreme east of Afghanistan (Gaisler, 1971). This has been treated as a distinct species by all previous authors, including E & M-S, but will appear as *R. cornutus* in the above key.

Rhinolophus euryale

Rhinolophus euryale Blasius, 1853. Milan, Italy. Incl. *atlanticus, cabrerae, toscanus* (all Europe); ?*algirus, barbarus, meridionalis* (N. Africa); *judaicus* (Palestine); *nordmanni* (Transcaucasia).

RANGE (Map 17). The Mediterranean zone of S. Europe and N. Africa and east to Turkestan and Iran (Lay, 1967).

Rhinolophus mehelyi

Rhinolophus mehelyi Matschie, 1901. Bucharest, Rumania. Incl. *carpetanus* (Spain).
*Rhinolophus euryale tuneti** Deleuil & Labbe, 1955. El Haouaria, Cap Bon, Tunisia.[1]

RANGE (Map 16). Known from rather few localities in S. Europe, the Mediterranean islands, N.W. Africa (Morocco to Cyrenaica – specimens from latter in B.M.), Asia Minor (Kahmann & Caglar, 1960), the Caucasus, and the Zagros Mts, Iran (specimens in B.M.). Western Mediterranean localities were mapped by Kahman (1958). See also DeBlase (1972).

REMARKS. This species was recorded from Egypt by Sanborn & Hoogstraal (1955), but Harrison (1964a) doubted this and identified Egyptian specimens as *R. euryale*. Strinati & Aellen (1958) have shown that the name *mehelyi* is quite probably based upon *R. euryale* in which case *R. carpetanus* Cabrera, 1904 would be the prior name of this species. But such a change would cause needless confusion since *mehelyi* has been used consistently for this species since the two names were synonymized by Miller (1912).

Rhinolophus blasii

Rhinolophus blasii Peters, 1866. Italy (restricted by Ellerman *et al.*, 1953). =*clivosus* Blasius (not of Cretzschmar), *blasiusi*. Incl. *andreinii* (Eritrea), *empusa* (Malawi).

RANGE (Map 17). From Italy through S.E. Europe and S.W. Asia to E. Afghanistan (Aellen, 1959), north to the Kopet Dag and Caucasus; Morocco and Tunisia; also in Eritrea and most of southern Africa if *andreinii* and *empusa* are conspecific.

Subfamily **HIPPOSIDERINAE**
Leaf-nosed bats
Genus ***HIPPOSIDEROS***

Hipposideros Gray, 1831. Type-species *Vespertilio speoris* Gray. =*Hipposiderus*. Incl. *Chrysonycteris, Ptychorhina*.

Contains about 43 species (according to a revision by Hill, 1963) throughout the tropics of the Old World. There are no truly Palaearctic species, but one African species reaches N. Africa and Arabia. An Oriental species, *H. fulvus*, reaches E. Afghanistan.

Hipposideros caffer

Rhinolophus caffer Sundevall, 1846. Near Durban, S. Africa.
H. c. caffer. S. and E. Africa and Arabia.
H. c. tephrus Cabrera, 1906. Morocco.

RANGE (Map 17). Extreme S.W. Arabia; Morocco; Hoggar (central Sahara); and most of Africa south of the Sahara.

REMARKS. There are many other named forms from south of the Sahara.

Genus ***ASELLIA***

Asellia Gray, 1838. Type-species *Rhinolophus tridens* Geoffroy.

A genus of two species: *A. tridens* (below) and the closely similar *A. patrizii* in Ethiopia.

[1] *Fide* Cockrum (1976)—Mammalia 40: 685–6.

Asellia tridens

Rhinolophus tridens Geoffroy, 1813. Egypt.
A. t. tridens. Range of species except N.W. Africa. Incl. *murraiana* (Pakistan).
A. t. diluta Andersen, 1918. Tunisia, Algeria and Morocco, type-locality in Algerian Sahara. Incl. *pallida*.

RANGE (Map 17). From Pakistan through Arabia to Egypt and south to Somalia, Ethiopia and Sudan; also in Tibesti, Tunisia, Algeria, Morocco and Gambia.

Genus *TRIAENOPS*

Triaenops Dobson, 1871. Type-species *T. persicus* Dobson.

Besides *T. persica*, listed below, this genus contains probably two species in Madagascar.

Triaenops persicus

Triaenops persicus Dobson, 1871. Shiraz, Iran.
T. p. persicus. S. W. Iran.
*T. p. afer** Peters, 1877. E. Africa and Aden; type-locality Mombasa, Kenya.
*T. p. macdonaldi** Harrison, 1955. Oman.
Triaenops persicus macdonaldi Harrison, 1955. Al Ain, Buraimi Oasis, Oman.

RANGE (Map 17). S.W. Iran, Oman, Aden, ?Egypt and E. Africa from Somalia to Mozambique.

REMARKS. The form *afer* was considered conspecific with *T. persicus* by Harrison (1964a). E & M-S included Egypt in the range, but Harrison could not trace any evidence of this.

Family VESPERTILIONIDAE

The largest family of bats with many species throughout the world. About 30 genera are currently recognized, but some are unsatisfactorily based upon single characters, and the classification is consequently unstable. Eleven genera are represented by Palaearctic species, whilst two others are represented respectively by a single specimen from Aden (*Scotophilus*) and a few vagrants from N. America (*Lasiurus*).

The genera *Vespertilio*, *Eptesicus*, *Pipistrellus* and *Nyctalus* in particular are very similar and the present diagnostic characters unsatisfactory. However, no satisfactory reclassification can be undertaken without considering also the extra-Palaearctic species of *Pipistrellus* and *Eptesicus* and several extra-Palaearctic genera that are clearly part of the same complex.

See Tate (1942) for a review covering much of this family.

1. Nostrils at the ends of conspicuous laterally protruding tubes (E. Asia) . . . *MURINA* (p. 62)
– Nostrils normal . . . . (2)
2. Second phalanx of 3rd finger about three times the length of the 1st, folded sharply back against the metacarpal at rest; upper canines preceded by a space equal to that occupied by an incisor . . . . *MINIOPTERUS* (p. 61)
– Second phalanx of 3rd finger less than twice the length of the 1st; no such gap in front of the upper canines . . . . (3)
3. Inner margins of ears close together (2–3 mm), united on top of head by a broad band; nostrils opening on upper surface of nose . . . . (4)
– Inner margins of ears widely separate; nostrils opening on anterior surface of nose (5)
4. Ears short, when laid forwards not exceeding muzzle by more than 3 mm . . . *BARBASTELLA* (p. 60)
– Ears very long, extending far beyond muzzle when laid forwards (about as long as forearm) . . . *PLECOTUS* (p. 61)
5. Two pairs of upper incisors . . . . (6)
– One pair of upper incisors . . . . (10)
6. Tragus long, slender and straight, its anterior length at least three times its greatest width; 6 upper cheek-teeth on each side, i.e. large premolar preceded by *two* small ones (except in the Japanese *M. ozensis* in which the two small premolars are represented by one bicuspid tooth) . . . . *MYOTIS* (p. 46)

– Tragus short, blunt and usually curved forwards, its anterior margin less than three times its greatest width; 4 or 5 upper cheek-teeth (i.e. large premolar preceded by no more than one small one) . . . . (7)
7. One small upper premolar on each side (absent only in some specimens of *Pipistrellus savii*) (8)
– No small upper premolars . . . . (9)
8. Fifth finger only a little longer than 4th metacarpal; forearm over 40 mm ***NYCTALUS*** (p. 54)
– Fifth finger longer than metacarpal and 1st phalanx of 4th finger; forearm under 40 mm ***PIPISTRELLUS*** (p. 51)
9. Dorsal pelage with a conspicuously frosted appearance due to long pale tips to the hairs; ear broader than long; anterior palatal emargination broader than deep ***VESPERTILIO*** (p. 58)
– Dorsal pelage not conspicuously frosted; ear longer than broad; anterior palatal emargination deeper than broad . . . . ***EPTESICUS*** (p. 56) (also some *Pipistrellus savii*)
10. Ear greatly exceeding muzzle when laid forwards (at least half length of forearm); (forearm 57–67 mm) . . . . ***OTONYCTERIS*** (p. 59)
– Ear falling short of muzzle . . . . (11)
11. Interfemoral membrane densely haired above . . . . . ***LASIURUS*** (p. 60)
– Interfemoral membrane not densely haired . . . . (12)
12. M^1 and M^2 with the w-pattern distorted or absent; (forearm over 45 mm in only species in Palaearctic) . . . . ***SCOTOPHILUS*** (p. 59)
– M^1 and M^2 with typical w-pattern; (forearm 28–33 mm in only Palaearctic species) ***NYCTICEIUS*** (p. 59)

Genus *MYOTIS*

Myotis Kaup, 1829. Type-species *Vespertilio myotis* Borkhausen. Incl. *Brachyotis*, *Capaccinius*, *Chrysopteron*, *Comastes*, *Euvespertilio*, *Isotus*, *Leuconoe*, *Nystactes* Kaup (not of Gloger), *Paramyotis*, *Rickettia*, *Selysius*.

A worldwide genus with a large number of species. Many of the characters that have been employed to recognize subgenera (or even distinct genera) do not provide a clear-cut division when the entire genus is considered, and therefore no subgenera are used here. Twenty-one species are listed from the Palaearctic Region, but there remain many uncertainties about the delimitation of certain species. The following key should be treated as only a provisional guide to identification.

1. Pelage bright reddish or yellowish brown . . . . (2)
– Pelage fawn, brown or grey, never with clear red or yellow tones (most yellowish in *M. emarginatus*) . . . . (3)
2. Wings boldly patterned black and yellow; forearm 45–53 mm (E. Asia) ***M. formosus***
– Wings not boldly patterned; forearm 36–40 mm (Arabia and Africa) ***M. bocagei***
3. Forearm over 52 mm; maxillary tooth-row (C–M^3) over 7·5 mm . . . . (4)
– Forearm under 52 mm; C–M^3 under 7·5 mm . . . . (6)
4. Hind foot, with claws, equalling or exceeding length of tibia (E. Asia) ***M. ricketti***
– Hind foot much shorter than tibia . . . . (5)
5. Smaller: forearm 53–63 mm, condylobasal length 18·6–21·4 mm, C–M^3 8·0–9·4 mm ***M. blythi***
– Larger: forearm 58–71 mm, condylobasal length 22·0–24·8 mm, C–M^3 9·8–10·8 mm ***M. myotis***
6. Ear more than half length of forearm, extending far beyond nose when laid forwards; (forearm 39–45 mm) . . . . ***M. bechsteini***
– Ear less than half length of forearm, not extending far beyond nose . . . . (7)
7. Margin of interfemoral membrane more densely haired between tail-tip and end of calcar than between end of calcar and foot . . . . (8)
– Margin of membrane less densely haired between tail and end of calcar than between there and foot, or entire margin almost or quite hairless . . . . (9)

8. Forearm over 45 mm (usually 48–51); C–M^3 over 6·5 mm (China) . **M. pequinius**
– Forearm under 45 mm (usually 36–41); C–M^3 under 6·5 mm . **M. nattereri**
9. Wing membrane meeting foot at the ankle or above . . . . (10)
– Wing membrane meeting foot below the ankle, usually at the base of the outer toe (but rather close to the ankle in *M. daubentoni*) . . . . (13)
10. Margin of membrane bordering calcar densely haired; (forearm 37–42 mm) (W. Palaearctic) **M. capaccinii**
– Margin of membrane bordering calcar with few or no hairs . . . . . . (11)
11. Forearm 43–48 mm; dorsal pelage light brown (W. Palaearctic) . **M. dasycneme**
– Forearm under 41 mm; dorsal pelage dark brown (E. Asia) . . . . . (12)
12. Forearm 36–40 mm . . . . **M. macrodactylus**
– Forearm about 34 mm (Sakhalin) . . . . **M. abei**
13. Foot about two-thirds length of tibia . . . . (14)
– Foot about half length of tibia . . . . (15)
14. Forearm 35–39 mm . . . . **M. daubentoni**
– Forearm 30–33 mm (Japan) . . . . **M. pruinosus**
15. Posterior margin of ear with a sharp indentation forming almost a right angle; pelage light yellowish brown; (forearm 37–43 mm) . . . . **M. emarginatus**
– Posterior margin of ear less sharply indented; pelage dark brown . . . . (16)
16. Tail about equal to or longer than head and body; tibia about 20 mm; (forearm about 39 mm; 2nd upper premolar displaced to medial side of tooth-row) (E. Asia) . **M. frater**
– Tail shorter than head and body; tibia under 18 mm . . . . (17)
17. Only one small upper premolar which is bicuspid; 2nd lower premolar displaced to the medial side of the tooth-row (Japan) . . . . **M. ozensis**
– Two small upper premolars which are unicuspid; 2nd lower premolar in the tooth-row (18)
18. Postcalcarial keel absent or if present then tapering evenly towards the tip of the calcar (19)
– Postcalcarial keel present, expanding in width along the course of the calcar . (21)
19. Hind foot (with claws) usually under 8 mm; (forearm 32–38 mm) . . . . . (20)
– Hind foot over 8 mm; (forearm 34–41 mm) (N. America, possible vagrant in Iceland and N.E. Siberia) . . . . **M. lucifugus**
20. Tail over 80% of head and body; posterior border of ear with distinct emargination; tail membrane extending to tip of tail . . . . **M. ikonnikovi**
– Tail under 80% of head and body; posterior border of ear without distinct emargination; tail membrane stopping 1–2 mm from tip of tail; (Japan) . . . . . **M. hosonoi**
21. Smaller (forearm 31·5–35 mm, condylobasal length 12·5–13·5 mm); dorsal pelage dark brown; penis uniformly thin; baculum with convex margin, under $0{\cdot}32 \times 0{\cdot}61$ mm **M. mystacinus**
– Larger (forearm 32–37·5 mm, condylobasal length 13–14 mm); dorsal pelage yellowish brown; penis club-shaped; baculum with basal notch, over $0{\cdot}36 \times 0{\cdot}73$ mm **M. brandti**[1]

Myotis mystacinus — Whiskered bat

Vespertilio mystacinus Kuhl, 1819. Germany.

M. m. mystacinus. Europe and W. Asia (perhaps also E. Siberia). Incl. *aurascens* (Caucasus), *bulgaricus*, *collaris*, *coluotus*† (*fide* Hanak, 1967), ?*gracilis* (E. Siberia), *hajastanicus* (Armenia), *humeralis*, *lugubris*, ?*nigricans*, *nigrofuscus*, *rufofuscus*, *schinzii*, *schrankii*, *sibiricus*.

M. m. przewalskii Bobrinskii, 1926. W. China and Mongolia; type-locality in Sinkiang. Incl. ?*kukunoriensis* (Tibet). A pallid race.

M. m. transcaspicus Ognev & Heptner, 1928. Kopet Dag, Turkestan.

M. m. sogdianus Kuzyakin, 1934. Tashkent, Turkestan.

M. m. meinertzhageni Thomas, 1926. Himalayas and Pamirs, type-locality in Kashmir. Incl. ?*pamirensis*. Perhaps *nipalensis* Dobson, 1871 (Nepal) is the prior name for this race.

*M. m. fujiensis** Imaizumi, 1954. Honshu, Japan.
Myotis mystacinus fujiensis Imaizumi, 1954a:42. Mt Fuji, Honshu, Japan.

?*M. m. davidi* Peters, 1869. China, type-locality Peking.

RANGE (Map 18). The entire Palaearctic Region from Ireland to Japan, north to about

[1] Characters, based on German specimens, from Gauckler & Kraus (1970).

65° in Europe and W. Siberia, south to Morocco (Brosset, 1960), N. Iran and the Himalayas.

REMARKS. The following forms, included in this species by E & M-S, should certainly be excluded: *blanfordi*, *caliginosus*, *montivagus*, *muricola* (types of these examined), and probably also *latirostris*, *moupinensis* and *orii*.

The above synonymy was compiled before the specific distinctness of *M. brandti* was demonstrated by Gauckler & Kraus (1970), but on the assumption that *muricola*, including *ikonnikovi*, was specifically distinct. Gauckler & Kraus did not consider Asiatic forms in detail, and therefore this account has not been altered except to remove those European forms allocated by Gauckler & Kraus to *M. brandti*. The status of all the Asiatic forms must be considered tentative pending a more detailed revision. Gauckler & Kraus indicated that *M. mystacinus* was probably confined to Europe, but Findley (1972) believed that *davidi* was 'the temperate Chinese representative of *mystacinus* (not *brandti*)'.

Myotis brandti

Vespertilio brandti Eversmann, 1845. Foothills of the Ural Mts, USSR. Incl. *aureus*.

RANGE (not mapped). Much of Europe from Britain to the Urals, south to Spain and Greece (Gauckler & Kraus, 1970; Hanak, 1970).

REMARKS. Gauckler & Kraus (1970) demonstrated that this form is specifically distinct from *M. mystacinus* in Bavaria, as had previously been shown in effect in Czechoslovakia by Hanak (1965), who however did not give them specific rank. It is probable that some of the Asiatic forms here allocated to *M. mystacinus* should be referred to *M. brandti*.

Myotis ikonnikovi

Myotis ikonnikovi Ognev, 1912. Ussuri Valley, S.E. Siberia.

RANGE (Map 18). Ussuri region and N. Korea to L. Baikal and the Altai; Sakhalin and Hokkaido.

REMARKS. Of very dubious status. This very small form seems to be sympatric with a larger form, *gracilis* (usually referred to *M. mystacinus*) in the Ussuri region. Kuzyakin (1965) considered *ikonnikovi* a race of the Himalayan *M. muricola*. This seems quite probably correct, but this whole group is so problematical that it seems best to keep it separate until the group can be revised as a whole. Records of *M. ikonnikovi* in E. Europe have been considered by Hanak (1965) to be based on *M. mystacinus* coexisting with the larger but previously unrecognized *M. brandti*.

Myotis abei

Myotis abei Yoshikura, 1944. Shirutoru, Sakhalin.

RANGE (Map 19). Known only from the type-locality.

REMARKS. Yoshikura (1956) contined to give this specific rank while recognizing *M. mystacinus*, *M. ikonnikovi* and *M. daubentoni ussuriensis* also on Sakhalin. It has not been seen.

Myotis emarginatus

Vespertilio emarginatus Geoffroy, 1806. France.

M. e. emarginatus. Europe and N. Africa. Incl. *budapestiensis*, *ciliatus*, *neglectus*, *rufescens* Crespon (not of Brehm), *schrankii* Kolenati (not of Wagner).

M. e. desertorum (Dobson, 1875). Iran and S. Turkestan, type-locality Jalk, S.E. Iran. Incl. *lanaceus*, *lanceus*, ?*turcomanicus* (S. Turkestan).

M. e. saturatus Kuzyakin, 1934. Tashkent, Uzbekistan. (Kuzyakin (1965) still considered this a distinct race.)

RANGE (Map 18). Southern Europe (north to Netherlands and S. Poland), Crimea, Caucasus, Kopet Dag, east to Tashkent and E. Iran; Palestine and Morocco (Brosset, 1960).

REMARKS. E & M-S listed two Indian forms, *peytoni* and *primula*, as 'allied to *emarginatus*'. These can confidently be excluded: *peytoni* as a race of *M. montivagus* (*fide* Hill, 1962) and *primula* as a synonym of *M. annectens* (*fide* Topal, 1970), both non-Palaearctic species distinct from *M. emarginatus*.

Myotis frater

Myotis frater Allen, 1923. Yenping, Fukien, S.E. China.
M. f. frater. China.
M. f. longicaudatus Ognev, 1927. Korea, Ussuri Region (type-locality) and the Altai Mts.
*M. f. bucharensis** Kuzyakin, 1950. Tadzhikistan.
Myotis longicaudatus bucharensis Kuzyakin, 1950:286. Aibadzh, S.W. Tadzhikistan.
*M. f. kaguyae** Imaizumi, 1956. Honshu, Hokkaido.
Myotis kaguyae Imaizumi, 1956. Kuzukawa, Takedate-Mura, Aomori Pref., Honshu, Japan.

RANGE (Map 19). Ussuri Region of E. Siberia, Korea, Manchuria (Wang, 1959), Sayan Mts, Altai Mts, Tadzhikistan, Japan (Honshu and Hokkaido), S.E. China (Fukien).

REMARKS. Wang (1959) doubted that *longicaudatus* was conspecific with *M. frater*. Imaizumi (1960, 1970a) allocated *kaguyae* to this species.

*Myotis hosonoi**

Myotis hosonoi Imaizumi, 1954a:44. Koumito, Tokiwa-mura (*c.* 30 km N. of Matsumoto City), Nagano Pref., Honshu, Japan (732 m).

RANGE (Map 19). Central and N. Honshu, Japan.

REMARKS. Imaizumi (1970a) still gave this form specific rank. It seems close to *M. muricola*.

*Myotis ozensis**

Myotis ozensis Imaizumi, 1954a:49. Ozegahara (*c.* 60 km N.E. of Maebashi City), Gunma Pref., Honshu, Japan (1400 m).

RANGE (Map 19). Central Honshu.

REMARKS. Imaizumi (1970a) continued to give this form specific rank. It has not been seen, and is known only from a single specimen.

*Myotis bocagei**

Vespertilio bocagii Peters, 1870. Duque de Braganca, Angola.
M. b. dogalensis† (Monticelli, 1887). Aden.

RANGE (Map 19). Throughout much of Africa south of the Sahara; known from Aden only by the type of *dogalensis* (Harrison, 1964a).

Myotis nattereri

Vespertilio nattereri Kuhl, 1818. Hessen, Germany.
M. n. nattereri. Europe. Incl. *escalerai* (Spain), *spelaeus*, *typicus*.
*M. n. araxenus** Dahl, 1947. Armenia and Iran.
Myotis nattereri araxenus Dahl, 1947. Vaikskoi Ridge, Araks River, Amgu, Mikoyansk Region, Armenia, USSR.
*M. n. hoveli** Harrison, 1964. Palestine.
Myotis nattereri hoveli Harrison, 1964b. Aqua Bella, near Jerusalem.
M. n. tschuliensis Kuzyakin, 1935. Kopet Dag, Turkestan.
M. n. amurensis Ognev, 1927. E. Siberia and Korea.
M. n. bombinus Thomas, 1905. Japan (type-locality on Kyushu).

RANGE (Map 19). Europe (except southeast and most of Scandinavia but including Britain and Ireland), Morocco (Brosset, 1963), Crimea, Caucasus, Palestine, Kopet Dag, Tadzhikistan, L. Baikal, S.E. Siberia, Korea, Kyushu and Honshu (Imaizumi, 1954a).

Myotis bechsteini

Vespertilio bechsteinii Kuhl, 1818. Hessen, Germany. Incl. *favonicus*, *ghidinii* (Spain).

RANGE (Map 19). Europe from Spain and France east to W. Russia; England; S. Sweden; the Caucasus.

Myotis myotis

Vespertilio myotis Borkhausen, 1797. Thuringia, Germany.
M. m. myotis. Europe. Incl. *alpinus, latipennis, murinus* of Schreber and many other early authors but not of Linnaeus, *myosotis, spelaea, submurinus, typus*.
*M. m. macrocephalicus** Harrison & Lewis, 1961. Lebanon and Palestine.
Myotis myotis macrocephalicus Harrison & Lewis, 1961. 2 km E. of Amchite, Lebanon.

RANGE (Map 20). Central and S. Europe, east to Ukraine; S. England; most of the Mediterranean islands; Asia Minor, Lebanon and Palestine.

REMARKS. The forms *omari, ancilla* and *risorius*, listed in *M. myotis* by E & M-S, are referable to *M. blythi* (types examined). The form *chinensis* Tomes (type examined) is unlikely to be conspecific with either *M. myotis* or *M. blythi*. Along with *luctuosus*, which is probably a race or synonym, *chinensis* is confined to southern China and is not listed here.

Myotis blythi

Vespertilio blythii Tomes, 1857. Nasirabad, Rajasthan, India.
M. b. blythi. India. Incl. *dobsoni* =*murinoides*.
M. b. ancilla† Thomas, 1910. China; type-locality in S.E. Shensi.
M. b. omari† Thomas, 1906. Iran, Turkestan, Palestine; type-locality in central Iran. Incl. *risorius*.
M. b. oxygnathus (Monticelli, 1885). Europe, N. Africa; type-locality in Italy.

RANGE (Map 20). Mediterranean zone of Europe and N.W. Africa; Crimea, Caucasus, Asia Minor, Palestine to Kirgizia, Afghanistan, Himalayas; Shensi and perhaps Inner Mongolia in China; an isolated record near the upper R. Ob, N.W. Altai according to Kuzyakin (1965).

REMARKS. The forms *ancilla, omari* and *risorius* are here transferred from *M. myotis* to *M. blythi*, a step already taken by Harrison (1964a) with respect to *omari* and *risorius*. The subspecies listed are unlikely to be valid, but are provisionally retained because of the very slight specific difference between this species and *M. myotis*. The species was reviewed in detail by Strelkov (1972).

Myotis formosus

Vespertilio formosa Hodgson, 1835. Nepal.
M. f. tsuensis Kuroda, 1922. Tsushima Is., Japan (type-locality) and Korea. Incl. *chofukusei* (Korea).

RANGE (Map 19). From E. Afghanistan (Meyer-Öhme, 1965), through N. India to S. China; Taiwan, Korea and Tsushima Is., Japan.

REMARKS. The following extra-Palaearctic named forms are probably conspecific; *andersoni, auratus, dobsoni, pallida* (all India); *rufoniger* (S. China); *watesei* (Taiwan).

Myotis daubentoni

Vespertilio daubentoni Kuhl, 1819. Hanau, Hessen-Nassau, Germany. Incl. *aedilis, albus* (England), *capucinellus, lanatus, loukashkini* (Manchuria), *minutellus, petax* (Altai), *staufferi, ussuriensis* (E. Siberia), *volgensis* (Urals); perhaps also *laniger* (S. China).

RANGE (Map 21). Europe and S. Siberia, east to Vladivostok, Korea (Kuroda, 1967) and Manchuria; Britain and Ireland; S. Scandinavia; Kurile Islands, Sakhalin and Hokkaido.

REMARKS. The form *macrodactylus* (Japan) has frequently been included in this species, most recently by Sanborn (1953), but this allocation seems improbable (see remarks under *M. macrodactylus* below).

Myotis capaccinii

Vespertilio capaccinii Bonaparte, 1837. Sicily. Incl. *blasii* Major, *bureschi, dasypus, majori, megapodius, pellucens*.

RANGE (Map 21). The Mediterranean zone of Europe and N.W. Africa, including most of the Mediterranean islands; S. Asia Minor and Palestine; S. Iraq; S. Iran (Lay, 1967); lower Amu-Darya, Uzbekistan.

REMARKS. Kuzyakin (1965) included *macrodactylus* in this species, but the type of *macrodactylus*, along with more recent Japanese specimens that apparently represent *macrodactylus*, seem clearly distinct. E & M-S suggested that *fimbriatus* (S. China) and *longipes* (Kashmir) might represent this species, but examination of a paratype of *longipes* and a probable paratype of *fimbriatus* does not suggest conspecificity. *M. longipes* has recently been recorded from eastern Afghanistan by Hanak & Gaisler (1969) who considered it to be specifically distinct.

Myotis pequinius

Myotis (Leuconoe) pequinius Thomas, 1908. 30 miles W. of Peking, China.

RANGE (Map 20). Hopei and Shantung, China.

Myotis macrodactylus

Vespertilio macrodactylus Temminck, 1840. Japan.

RANGE (Map 20). Maritime Prov. of E. Siberia; S. Kuriles; Hokkaido, Honshu, Shikoku, Kyushu and Tsushima, Japan.

REMARKS. This form has been variously treated as a race of *M. daubentoni*, e.g. by Sanborn (1953), or of *M. capaccinii*, e.g. by Kuzyakin (1965) and Wallin (1969). Neither the original description, the type-material (examined in Leiden), nor a series in the British Museum agreeing with the types support either of these allocations, and I follow E & M-S and Imaizumi (1960, 1970a) in giving it specific rank.

Myotis dasycneme

Vespertilio dasycneme Boie, 1825. Dagbieg, near Wiborg, Denmark. Incl. *limnophilus*, *major*.

RANGE (Map 21). N.E. Europe and W. Siberia; west to E. France and Switzerland, east to Yenesei, south to Austria and about 48°N in Russia; a single record from Manchuria (Lehmann, 1966b).

Myotis pruinosus*

Myotis (Leuconoe) pruinosus Yoshiyuki, 1971. Waga-Machi, Waga-Gun, Iwate Pref., Honshu, Japan.

RANGE (Map 21). N. Honshu, Japan.

REMARKS. This is a very small species that appears distinct from the most similar Japanese forms. It was described on the basis of 12 specimens. I have not seen it.

Myotis lucifugus*

Vespertilio lucifugus le Conte in McMurtie, 1831. Georgia, USA. Including many North American synonyms.

RANGE (not mapped). Most of USA and S. Canada, but extending north to Labrador in the east and central Alaska in the west. Single specimens have been recorded in Iceland (see Koopman & Gudmundsson, 1966) and Kamchatka (Hahn, 1905), but in both cases these authors considered it probable that they had been carried there by ship.

Myotis ricketti

Vespertilio (Leuconoe) ricketti Thomas, 1894. Foochow, Fukien, China. Incl. ?*pilosa*.

RANGE (Map 21). China: Fukien, Anhwei and Shantung.

REMARKS. Often placed in a separate subgenus or even genus, *Rickettia*.

Genus *PIPISTRELLUS*

Pipistrelles

Pipistrellus Kaup, 1829. Type species *Vespertilio pipistrellus* Schreber. Incl. *Alobus*, *Euvesperugo*, *Hypsugo*, *Nannugo*, *Romicia*, *Scotozous*, *Vansonia*.

A large genus, world-wide except for South America, but very dubiously distinct from the earlier named *Eptesicus* and *Nyctalus*. *Pipistrellus* differs from *Eptesicus* in possessing rudimentary small upper premolars, the presence of which is variable in two species, *P. savii* and the African *P. tenuis*. However, *Eptesicus* itself is doubtfully distinct from *Vespertilio*, and it therefore seems unwise to upset the present classification until a sufficiently comprehensive review can be made of the entire group to produce a stable classification based on a larger spectrum of characters.

About ten species occur in the Palaearctic region. Two other, Oriental, species have been recorded in eastern Afghanistan, namely *P. mimus* Wroughton and *P. babu* Thomas (Meyer-Öhme, 1965).

1. Inner upper incisor unicuspid . . . . (2)
– Inner upper incisor bicuspid . . . . (4)
2. Outer upper incisor very small, less than half the height of the inner; smaller upper premolars about equal in size to outer incisors; condylobasal length over 11 mm . . . . (3)
– Outer upper incisors larger, over half the height of the inner; small upper premolars rudimentary, much smaller than the outer inscisors; condylobasal length about 10·5 mm (Egypt and Sudan) . . . . **P. ariel**
3. Wing membrane with white margin, especially between foot and 5th finger; condylobasal length over 12 mm . . . . **P. kuhli**
– Wing membrane without white margin; condylobasal length under 12 mm (Madeira and Canary Is.) . . . . **P. maderensis**
4. Small upper premolars very rudimentary, occasionally absent . . . . . (5)
– Small upper premolars not rudimentary, usually exceeding cingulum of the canine and at least as large as the outer incisors. . . . . (6)
5. Condylobasal length over 12 mm; forearm over 32 mm; tail usually projecting 2 mm or more beyond membrane . . . . **P. savii**
– Condylobasal length 10·5–11·5 mm; forearm under 32 mm; tail scarcely projecting beyond membrane (Arabia) . . . . **P. bodenheimeri**
6. Ventral pelage white, hairs white to the roots; outer upper incisors scarcely exceeding cingulum of inners; (forearm 30–32·5 mm; condylobasal length 11·5–12·5 mm) (N. Africa and Iraq) . . . . **P. rueppelli**
– Ventral pelage at least partly dark; outer upper incisors more than half the height of the inners (7)
7. Upper canine clearly separated from large premolar so that small premolar is clearly visible from lateral aspect of skull; penis short, baculum rudimentary or absent (W. Palaearctic) (8)
– Upper canine almost in contact with large premolar so that only the tip of the small premolar is visible in lateral aspect; penis very long, baculum 9–12 mm; (forearm 30–35 mm) (E. Asia) . . . . (9)
8. Forearm usually under 33 mm; condylobasal length under 12 mm; C–M^3 under 4·5 mm; thumb short, its length equal to width of carpal joint; lower canine robust, length of cingulum about equal to length along anterior edge of shaft. . . . **P. pipistrellus**
– Forearm usually over 33 mm; condylobasal length over 12 mm; C–M^3 over 4·5 mm; thumb longer than carpal joint; lower canine slender, length of cingulum about half length of anterior margin . . . . **P. nathusii**
9. Baculum strongly curved; upper canine with rudimentary posterior cusp; pelage dark brown; C–M^3 usually over 4·6 mm; forearm about 33·5 mm; wing membrane light brown **P. javanicus**
– Baculum almost straight; upper canine with prominent posterior cusp; pelage tawny, especially below; C–M^3 usually under 4·6 mm; forearm about 32 mm; wing membrane very dark brown . . . . **P. endoi**

Pipistrellus pipistrellus — Common pipistrelle

Vespertilio pipistrellus Schreber, 1774. France.

P. p. pipistrellus. Europe, N. Africa, Asia Minor, Palestine. =*pipistrella, typus*. Incl. *brachyotos, flavescens, genei, ?griseus, limbatus, macropterus, ?mediterraneus, melanopterus, minutissimus, ?murinus* Gray,

?*nigra*, *nigricans* Bonaparte, *nigricans* Koch, *pusillus*, *pygmaeus* (England), ?*rufescens* de Selys Longchamps, *stenotus*.
P. p. aladdin† Thomas, 1905. Iran (type-locality), Turkestan and Afghanistan. Incl. *bactrianus*.

RANGE (Map 22). From W. Europe, including British Isles and S. Scandinavia, to the Volga and Caucasus; Morocco; Asia Minor and Palestine; Turkestan, Iran, Afghanistan and Kashmir; ?Korea.

REMARKS. E & M-S mentioned records from Japan, Taiwan and Korea but these seem doubtful, and Imaizumi (1970a) did not recognize this species in Japan. The Persian form *aladdin* was included in the '*coromandra* group' by Tate (1942) and tentatively listed as a race of *P. coromandra* by E & M-S, but an examination of the type shows that it is indistinguishable from *P. pipistrellus*. Lay (1967) allocated *alladin* to *P. kuhli* because the type has white margins to the wings, but he clearly did not examine the skull which shows no relationship to *P. kuhli*. See also Neuhauser & DeBlase (1971).

Pipistrellus nathusii

Vespertilio nathusii Keyserling & Blasius, 1839. Berlin, Germany. Incl. *unicolor*.

RANGE (Map 22). From W. Europe, where records are few and scattered, to the Urals, Caucasus, and W. Asia Minor. A single record from southern England (Stebbings, 1970).

REMARKS. There seems to be no good evidence for the presence of this species in Palestine.

*Pipistrellus javanicus** — Japanese pipistrelle

Scotophilus javanicus Gray, 1838. Java.
P. j. abramus (Temminck, 1840). Japan, Korea, China.
Incl. *akokomuli*, *irretitus* (Chekiang, China), *pomiloides*, *pumiloides*.
P. j. javanicus. S.E. Asia. Incl. *tralatitius* Thomas (not Horsfield) and other extra-Palaearctic names.

RANGE (Map 22). Japan (N. to Honshu), S. Ussuri Region, Korea, most of China and S.E. Asia as far as Timor.

REMARKS. The reasons for using the name *javanicus* for this species were given by Laurie & Hill (1954).

*Pipistrellus endoi**

Pipistrellus endoi Imaizumi, 1959. Horobe, *c.* 45 km N.W. of Morioka City, Ashiro-cho, Ninohe-gun, Iwate Pref., Honshu, Japan (550 m).

RANGE (Map 22). N. and central Honshu, Japan.

REMARKS. Very similar to *P. javanicus*, but apparently sympatric with it on Honshu.

Pipistrellus kuhli — Kuhl's pipistrelle

Vespertilio kuhlii Kuhl, 1819. Trieste. Incl. *aegyptius* (Egypt), *albicans*, *albolimbatus*, *alcythoe*, *calcarata*, *canus* (India), ?*deserti*† (Libya), *ikhwanius* (Arabia), *lepidus* (Afghanistan), *leucotis* (Punjab), ?*lobatus* (India), *marginatus* Cretzschmar (not of Aellen), *minuta* (Algeria), *pallidus*, *pullatus*, *vispistrellus*.
Some of the following African synonyms may have subspecific status: *fuscatus* (Kenya), *subtilis* (S. Africa), *broomi* (Natal).

RANGE (Map 23). Southern Europe, Crimea, Caucasus and Turkestan to Pakistan; throughout S.W. Asia and N. Africa; much of Africa south of the Sahara, especially in the east, but it has recently been found also in West Africa (Hayman & Hill, 1971).

REMARKS. The transition from the dark (nominate) form in Europe to the pale desert form of Arabia (*ikhwanius*) has been shown to be gradual (see Harrison, 1964a), and in view of the continuity of range it seems likely that most if not all of the other named forms do not represent discrete subspecies.

E & M-S listed *deserti* Thomas as a distinct species, but examination of the type suggests that it is a form of *P. kuhli*.

Pipistrellus maderensis

Vesperugo maderensis Dobson, 1878. Madeira.

RANGE (Map 23). Madeira and Las Palmas, Canary Is.

REMARKS. Perhaps conspecific with *P. kuhli*, but not close to *P. savii* as suggested by E & M-S.

Pipistrellus savii Savi's pipistrelle

Vespertilio savii Bonaparte, 1837. Pisa, Italy.

P. s. savii. Europe and N. Africa. =*agilis*. Incl. *aristippe, bonapartii, darwini* (Canary Is.), *leucippe, maurus, nigrans, ochromixtus*.

P. s. caucasicus (Satunin, 1901). Crimea to W. China. Incl. *pallescens* (Sinkiang), *tamerlani* (Uzbekistan), *tauricus* (Crimea).

P. s. alaschanicus (Bobrinskii, 1926). Mongolia.

P. s. velox (Ognev, 1927). Vladivostok, E. Siberia (type-locality) and Hokkaido, Japan (Imaizumi, 1955b).

*P. s. coreensis** Imaizumi, 1955. Korea and Tsushima, Japan.

Pipistrellus savii coreensis Imaizumi, 1955b. Taegu, S. Korea.

?*P. s. austenianus* Dobson, 1871. Assam.

RANGE (Map 22). From Iberia, Morocco, the Canary and Cape Verde Islands through the Crimea, Caucasus, Asia Minor, Turkestan and Mongolia to Korea and Japan; southeastwards through Iran and Afghanistan (Neuhauser & DeBlase, 1974) to Punjab (Neuhauser, 1970); perhaps also in Assam and Upper Burma if *austenianus* is conspecific.

REMARKS. The synonymy of the western Asiatic forms follows Kuzyakin (1950), of the eastern Asiatic forms, Imaizumi (1955b). The differences between the closely adjacent forms *velox* and *coreensis* would seem to make their conspecificity doubtful.

The Indian form *cadornae*, tentatively allocated to *P. savii* by E. & M-S, seems specifically distinct (type examined; see also Hill, 1962).

Pipistrellus bodenheimeri*

Pipistrellus bodenheimeri Harrison, 1960. Yotvata, Wadi Araba, 40 km N. of Eilat, Israel.

RANGE (Map 22). Known only from the type-locality, S.W. Arabia and perhaps the island of Socotra. (A single specimen from Socotra in the British Museum is possibly referable to this species but is very dark.)

Pipistrellus ariel

Pipistrellus ariel Thomas, 1904. E. Egyptian Desert (22°N, 35°E).

RANGE (Map 22). Egypt and Sudan.

Pipistrellus rueppelli Rüppell's bat

Vespertilio rüppellii Fischer, 1829. Dongola, Sudan.

P. r. rueppelli. Africa. Incl. *temminckii* Cretzschmar (not of Horsfield), and perhaps also the following from south of the Sahara: *fuscipes, leucomelas, pulcher, senegalensis, vernayi*.

P. r. coxi Thomas, 1919. Iraq.

RANGE (Map 23). Algeria, Egypt, Iraq; also much of Africa south of the Sahara.

Genus *NYCTALUS*

Noctules

Nyctalus Bowdich, 1825. Type-species *N. verrucosus* Bowdich. Incl. *Noctulinia* (in part), *Panugo, Pterygistes*.

In its restricted sense, i.e. excluding the very similar *Pipistrellus*, this genus contains only the four Palaearctic species listed here and *N. montanus* of N. India.

1. Forearm under 58 mm; condylobasal length under 20 mm; C–M^3 under 7·5 mm (2)
– Forearm over 58 mm; condylobasal length over 20 mm; C–M^3 over 8·0 mm . (3)
2. Forearm 42–46 mm; condylobasal length 13·0–16·1 mm; C–M^3 5·1–6·3 mm; small upper premolars not so reduced, visible from the labial side of the tooth-row; hairs of dorsal pelage usually with bases darker than tips . . . . ***N. leisleri***
– Forearm 48–57 mm; condylobasal length 17·0–19·4 mm; C–M^3 6·5–7·5 mm; small upper premolar very rudimentary, invisible from labial side of tooth-row; hairs of dorsal pelage quite or nearly unicoloured . . . . ***N. noctula***
3. Forearm 59–62 mm; C–M^3 8·1–8·5 mm; edge of wing membrane arising from the ankle (E. Asia) . . . . ***N. aviator***
– Forearm 63–69 mm; C–M^3 8·5–9·2 mm; edge of wing membrane arising from middle of metatarsus (Europe) . . . . ***N. lasiopterus***

Nyctalus leisleri Lesser noctule

Vespertilio leisleri Kuhl, 1818. Hanau, Hessen-Nassau, Germany.
N. l. leisleri. European mainland and British Isles. =*dasykarpos*. Incl. *pachygnathus*.
N. l. verrucosus Bowdich, 1825. Madeira. Incl. *madeirae*.
N. l. azoreum† (Thomas, 1901). Azores.

RANGE (Map 23). From W. Europe to the Urals and Caucasus; Britain, Ireland, Madeira and the Azores; the western Himalayas and E. Afghanistan.

REMARKS. E & M-S listed *azoreum* as a separate species and queried the allocation of *verrucosus* and *montanus* to this species. Examination of the types of *madeirae* and *azoreum* suggests that these are indeed likely to be conspecific with *N. leisleri*. The type of *N. montanus* Barrett-Hamilton (Kumaon) has been examined and is clearly not to be included in *N. leisleri*. It differs in having the pelage bicoloured, the teeth larger, the small upper premolars very rudimentary and the first lower premolars very small relative to the following ones. Both species were recorded from E. Afghanistan by Neuhauser & DeBlase (1974).

Nyctalus noctula Noctule

Vespertilio noctula Schreber, 1774. France.
N. n. noctula. Europe. =*proterus*. Incl. *altivolans*, *lardarius*, *magnus* (England), ?*major*, *minima*, *palustris*, *princeps* (Voronezh, Russia).
*N. n. lebanoticus** Harrison, 1962. Palestine and Lebanon.
Nyctalus noctula lebanoticus Harrison, 1962. Natural Bridge, Faraya, Lebanon.
N. n. meklenburzevi Kuzyakin, 1934. E. Turkestan.
N. n. labiatus (Hodgson, 1835). Himalayas, type-locality Nepal.
N. n. plancei (Gerbe, 1880). China, type-locality Peking. Incl. *sinensis*, ?*velutinus* (Fukien).
*N. n. furvus** Imaizumi & Yoshiyuki, 1968. Japan (N. Honshu only).
Nyctalus furvus Imaizumi & Yoshiyuki, 1968. Kado (300 m), Iwaizumi-Machi, Shimohei-Gun, Iwate Pref., N.E. Honshu, Japan.

RANGE (Map 23). From E. Europe, including Britain, S. Scandinavia and most of the Mediterranean islands, to the Urals and Caucasus; Morocco; S.E. Asia Minor to Palestine; W. Turkestan to the Himalayas and China; Taiwan (Jones, 1971); N. Honshu, Japan.

REMARKS. Imaizumi & Yoshiyuki (1968) considered *N. noctula*, *N. velutinus* (China) and *N. furvus* specifically distinct, but distinguished them mainly by average differences.

Nyctalus lasiopterus

Vespertilio lasiopterus Schreber, 1780. ?N. Italy. Incl. ?*ferrugineus*, *sicula*, *maxima*.

RANGE (Map 24). From W. Europe where records are sparse (Spain, France, Switzerland) to the Urals, Caucasus, Asia Minor, Iran (Etemad, 1970) and the Ust-Urt Plateau (Borovsky & Vorontsov, 1970).

Nyctalus aviator

Nyctalus aviator Thomas, 1911. Tokyo, Japan.

RANGE (Map 24). Hokkaido, Honshu, Shikoku and Kyushu, Japan and the small islands of Tsushima and Iki; Korea; and the island of Shaweishan at the mouth of the Yangtze Kiang, China.

REMARKS. E & M-S, following Tate (1942), listed this tentatively as a race of *N. lasiopterus*. However, the differences seem as great as between sympatric species, e.g. *N. lasiopterus* and *N. noctula*, and therefore I prefer to give it specific rank. Wallin (1969) and Imaizumi (1970a) both treated it as a race of *N. lasiopterus*.

Genus *EPTESICUS*

Serotines

Eptesicus Rafinesque, 1820. Type-species *E. melanops* Rafinesque = *Vespertilio fuscus* Beauvois (N. America). Incl. *Amblyotus, Cateorus, Cnephaeus, Noctula, Pachyomus, Rhyneptesicus*; and others relating to non-Palaearctic species.

A large, almost world-wide genus, doubtfully distinct from *Vespertilio* (which predates it) and *Pipistrellus* (see remarks under *Vespertilio*).

1. Tail membrane and proximal parts of wing membranes with warty projections; (forearm 36–41 mm) . . . . **E. walli**
– Membranes without warty projections . . . . (2)
2. Outer upper incisors large, more than half the length of the inner incisors which are deeply bicuspid; (forearm under 43 mm) . . . . (3)
– Outer upper incisors small, less than half the length of the inner incisors which are unicuspid or slightly bicuspid . . . . (4)
3. Forearm 38–43 mm; greatest width across upper canines exceeding interorbital width by at least 0·4 mm . . . . **E. nilssoni**
– Forearm 34–36 mm; canine width equal to interorbital width or up to 0·2 mm greater (Turkestan) . . . . **E. bobrinskoi**
4. Forearm 33–39 mm; condylobasal length under 13 mm; C–M^3 under 5 mm; (pelage pale, hairs unicoloured) . . . . **E. nasutus**
– Forearm over 39 mm; condylobasal length over 14·5 mm; C–M^3 over 5 mm . (5)
5. Forearm 42–46 mm; condylobasal length 14·5–17·0 mm; C–M^3 5·6–6·3 mm **E. bottae**
– Forearm 48–56 mm; condylobasal length 18·5–22·0 mm; C–M^3 6·5–8·2 mm **E. serotinus**

Eptesicus nasutus — Sind bat

Vesperugo (Vesperus) nasutus Dobson, 1877. Shikarpur, Sind, Pakistan.
E. n. nasutus. Pakistan.
E. n. pellucens (Thomas, 1906). Iran and Iraq.
E. n. matschiei (Thomas, 1905). S. W. Arabia.
*E. n. batinensis** Harrison, 1968. Oman, Arabia.
Eptesicus nasutus batinensis Harrison, 1968b. Harmul, 10 miles N. of Soher, Oman.

RANGE (Map 25). Pakistan, S.W. Iran and S.E. Iraq; Oman and S.W. Arabia.

REMARKS. The nominate form is very little known and the allocation of both *pellucens* and *matschiei* to *E. nasutus* is tentative.

Eptesicus bobrinskoi

Eptesicus bobrinskoi Kuzyakin, 1935. Aral Kara-Kum, Kazakhstan, USSR.

RANGE (Map 24). Kazakhstan and N.W. Iran (Harrison, 19 3a).

Eptesicus walli — Wall's serotine

Eptesicus walli Thomas, 1919. Basra, Iraq.

RANGE (Map 24). The Euphrates Valley, Iraq (Diyala to Basra) and the immediately adjacent part of Iran (Nader, 1971; Etemad, 1973).

Eptesicus nilssoni Northern bat

Vespertilio nilssoni Keyserling & Blasius, 1839. Sweden.
E. n. nilssoni. Europe, W. and central Siberia. Incl. *atratus* Kolenati, *borealis* Nilsson (not of Müller), *kuhlii* Nilsson (not of Kuhl).
*Vesperus propinquus** Peters, 1872. 'Guatemala'. Based upon a mislabelled *E. nilssoni* according to Davies (1965).
E. n. gobiensis Bobrinskii, 1926. Mongolia and adjacent parts of Siberia; type-locality in the Altai.
E. n. kashgaricus Bobrinskii, 1926. Chinese Turkestan.
E. n. centralasiaticus Bobrinskii, 1926. Chinghai, China.
?*E. n. parvus* Kishida, 1932. N. Korea and Hokkaido.
?*E. n. japonensis** Imaizumi, 1953. Honshu, Japan.
Eptesicus japonensis Imaizumi, 1953. Hokujo Mura (Shinden), Kita-Azumi Gun, Nagano Pref., Honshu, Japan (720 m).

RANGE (Map 24). Central and E. Europe to E. Siberia, north to beyond the Arctic Circle in Scandinavia, south to Iraq, the Elburz Mts, Pamirs and Tibet; also on Honshu and Hokkaido, Japan, if *japonensis* and *parvus* are conspecific.

REMARKS. Imaizumi (1970a) continued to give *japonensis* specific rank, but it seems to be scarcely distinguishable from European specimens of *E. nilssoni*. Wallin (1969) treated it as a subspecies of *E. nilssoni*.

***Eptesicus bottae*†** Botta's serotine

Vesperus bottae Peters, 1869. Yemen, Arabia.
E. b. bottae. Yemen.
E. b. innesi Lataste, 1887. Cairo (type-locality) and Sinai.
E. b. hingstoni† Thomas, 1919. Iraq (type-locality Baghdad) and Iran.
E. b. ognevi† Bobrinskii, 1918. Caucasus, Turkestan, N.E. Iran and N.E. Afghanistan.
*E. b. anatolicus** Felten, 1971. Asia Minor.
Eptesicus anatolicus Felten, 1971. Alanya, Antalya Prov., Turkey.
*E. b. omanensis** Harrison, 1976.
Eptesicus bottae omanensis Harrison, 1976. Mazjid Ma'illah, Jebel al Akhdar, Oman (1830 m).

RANGE (Map 25). S.W. Arabia and Oman; N.E. Egypt and Sinai; Iraq, S. Asia Minor and W. Iran; the Caucasus and Turkestan, east to L. Balkash.

REMARKS. The content of this species has been subject to a great deal of uncertainty. The above arrangement follows Hanak & Gaisler (1971) and Harrison (1976).

Eptesicus serotinus Serotine

Vespertilio serotinus Schreber, 1774. France.
E. s. serotinus. Europe to the Caucasus. Incl. *boscai*, *incisivus*, *insularis*, *intermedius* (Caucasus), *isabellinus* Cabrera (not of Temminck), *meridionalis* (Sardinia), *okenii*, *rufescens*, *serotine*, ?*sodalis*† (Rumania), *transylvanus*, *typus*, *wiedii*.
E. s. isabellinus (Temminck, 1840). N. Africa (for status see Harrison, 1963b).
E. s. turcomanus (Eversmann, 1840). Turkestan, N. Iran. Incl. *albescens*, ?*mirza* (Iran).
E. s. shiraziensis (Dobson, 1871). S.W. Iran.
*E. s. pashtomus** Gaisler, 1970. E. Afghanistan.
Eptesicus serotinus pashtomus Gaisler, 1970. Jalalabad, E. Afghanistan.
E. s. pachyomus (Tomes, 1857). N.W. India and Pakistan.
E. s. pallens Miller, 1911. N. China and Korea (type-locality in Kansu). Incl. ?*brachydigitus* (Korea), *pallidus*.
?*E. s. andersoni* (Dobson, 1871). S. China (type-locality in Yunnan). Possibly a distinct species.
?*E. s. horikawai* Kishida, 1924. Taiwan.

RANGE (Map 24). From W. Europe through southern Asiatic Russia to the Himalayas, Thailand and China, north to Korea; also in S. England, N. Africa and most of the islands of the Mediterranean.

REMARKS. The form *sodalis* from Rumania, known from a single specimen distinguishable from normal *E. serotinus* only by its small size, was treated as a species by Miller (1912), who included also a single specimen from Switzerland, and by E & M-S who included also the forms *ognevi* and *hingstoni*. Harrison (1964a) virtually followed E & M-S in suggesting that it was a race of *E. bottae* (including *hingstoni* and, tentatively, *ognevi*), whilst Kuzyakin (1965) rejected the association of *ognevi* and *sodalis*, suggesting that the latter was a variant of *E. serotinus*. An examination of the type of *sodalis* and the Swiss paratype suggests that *sodalis* is closer to *E. serotinus* than to any form here included in *E. bottae*, but its status remains uncertain.

Eptesicus, incertae sedis

E. kobayashii Mori, 1928. Korea.

Genus *VESPERTILIO*

Vespertilio Linnaeus, 1758. Type-species *V. murinus* L. Incl. *Aristippe*, *Meteorus*, *Vesperugo*, *Vesperus* Kaiserling & Blasius, *Marsipolaemus* (all but the last including species that are not here included in *Vespertilio*).

This genus is here limited to three species, i.e. as limited by Miller (1907). Several authors, e.g. Kuzyakin (1965), have included also *Eptesicus* and *Pipistrellus* in *Vespertilio*. The boundaries between these are indeed very dubious, but it seems unwise to alter the classification, however artificial, that has been used for most of this century without taking into account the entire range of these genera and the other, non-Palaearctic, genera that belong to the same complex.

In this restricted sense the genus is entirely Palaearctic. The classification and key below follow Wallin (1969).

1. Smaller: forearm 40–47 mm; maxillary tooth-row 5·0–6·1 mm . . . ***V. murinus***
— Larger: forearm 45–52 mm; maxillary tooth-row 6·1–6·6 mm (E. Asia) . . . . (2)
2. Underside, and especially throat, light; maximum width of tragus at base ***V. superans***
Underside, including throat, dark; maximum width of tragus in middle . ***V. orientalis***

Particoloured bat

Vespertilio murinus

Vespertilio murinus Linnaeus, 1758. Sweden.
V. m. murinus. Europe to Upper Amur. Incl. *albogularis*, *discolor*, *kraschennikovi*, *luteus* (Transbaikalia), *michnoi*, *siculus*.
*V. m. ussuriensis** Wallin, 1969. Ussuri region, S.E. Siberia.
Vespertilio murinus ussuriensis Wallin, 1969. Lake Chanka, River Odarki, Maritime Province, E. Siberia.

RANGE (Map 25). From central Europe through S. Siberia to the Ussuri Region of Siberia and perhaps Manchuria; S. Scandinavia and occasionally vagrant west to Britain; south to Iran and Afghanistan. A migrant species, having been recorded in the middle of the North Sea.

REMARKS. It should be noted that prior to the work of Miller (1907, 1912) the name *Vespertilio murinus* was widely used for the species now known as *Myotis myotis*.

Vespertilio superans

Vespertilio murinus superans Thomas, 1899. Sesalin, Ichang, Hupeh, China.
V. s. superans. S. China to Korea, Ussuri and Japan. Incl. *aurijunctus*† (Korea), *namiyei*† (Kyushu).
*Nyctalus noctula** Namie, 1889 (not of Schreber, 1774). Honshu.
Nyctalus montanus† Kishida, 1931 (not of Barrett-Hamilton, 1906). Substitute for *noctula* Namie, 1889.
Nyctalus noctula motoyoshii† Kuroda, 1934. Substitute for *montanus* Kishida, 1931.
*V. s. anderssoni** Wallin, 1963. Inner Mongolia.
Vespertilio namiyei anderssoni Wallin, 1963. Pang-Chiang, Inner Mongolia, China (*c.* 43°N, 113°30′E).

RANGE (Map 25). Central Yangtze north through Inner Mongolia to upper Amur, Ussuri region and Korea; Kyushu and Honshu, Japan. Mapped by Wallin (1969).

REMARKS. Wallin (1963) formerly gave *namiyei* specific rank, including *anderssoni*. The Japanese synonymy of *V.s. superans* follows Imaizumi (1968).

Vespertilio orientalis*

Vespertilio orientalis Wallin, 1969. Horobe, Ashiro, Ninoke district, Iwate pref., Honshu, Japan.

RANGE (Map 25). Honshu, Japan; China from Shansi south to Fukien; Taiwan (Jones, 1971). Mapped by Wallin (1969).

REMARKS. Wallin included in this species many specimens from China previously identified as *V. superans*.

Genus *OTONYCTERIS*

Otonycteris Peters, 1859. Type-species *O. hemprichii* Peters.

Contains only the following species.

Otonycteris hemprichi Hemprich's long-eared bat

Otonycteris hemprichii Peters, 1859. ?N.E. Africa.
O. h. hemprichii. Africa. Incl. *saharae* (Algerian Sahara), *ustus* (Egypt).
O. h. jin Cheesman & Hinton, 1924. Arabia and Israel (type-locality Hufuf, Hasa, Arabia).
O. h. petersi Anderson & de Winton, 1902. Iraq (type-locality Fao).
O. h. leucophaeus (Severtzov, 1873). Turkestan. Incl. *brevimanus*.
?*O. h. cinerea* Satunin, 1909. Iran (type-locality 'Zarakkuh country, near the Bamrud irrigation ditch in Khurasan' (N.E. Iran) according to Ognev (1928), not in Persian Baluchistan as stated in the original description).

RANGE (Map 26). The desert zone from Algeria through Egypt and N. Arabia to Turkestan and Afghanistan. A single, old record from Kashmir.

REMARKS. Harrison (1964a) recognized *jin* and *petersi* as subspecifically distinct, but it seems likely that variation will prove to be continuous throughout the range. Kuzyakin (1950) treated *cinerea* as specifically distinct on the basis of large size and unicoloured pelage.

Genus *NYCTICEIUS*

Nycticeius Rafinesque, 1819. Type-species *N. humeralis* Rafinesque. =*Nycticejus*, *Nycticeus*, *Nycticeyx*.

A genus of about ten species in North America, Africa, and Australia. One African species reaches the Palaearctic Region in Egypt and Arabia.

Nycticeius schlieffeni Schlieffen's bat

Nycticejus schlieffenii Peters, 1859. Cairo, Egypt. Incl. *adovanus*, *africanus*, *albiventer*, *australis*, *bedouin* (Aden), *fitzsimonsi* (all from Africa south of the Sahara except *bedouin*).

RANGE (Map 26). Most of the savanna zones of Africa south of the Sahara; Ethiopia; Egypt; S.W. Arabia.

REMARKS. Harrison (1964a) considered *bedouin* a synonym of *schlieffeni*.

Genus *SCOTOPHILUS*

Scotophilus Leach, 1821. Type-species *S. kuhlii* Leach. =*Pachyotus*.

A genus of about ten species in the Oriental and Ethiopian regions. One African species has been found near Aden.

Scotophilus leucogaster*

Nycticejus leucogaster Cretzschmar, 1831 (in Cretzschmar, 1826–1831). Kordofan, Sudan. Incl. *damarensis*, *nigritellus*, *viridis* – all from Africa south of the Sahara.

RANGE (Map 26). Most of the drier parts of Africa south of the Sahara, north to Ethiopia and Somalia; Aden.

REMARKS. The only record of a *Scotophilus* from Aden is that of Harrison (1964a) who recorded it as *S. nigrita* Schreber. Mr J. E. Hill (pers. comm.) has examined the single specimen from Aden and considers it to be referable to the smaller Africa species, *S. leucogaster*. (See also Hayman & Hill, 1971.)

Genus *LASIURUS**

Lasiurus Gray, 1831. Type-species *Vespertilio borealis* Müller. Incl. *Atalapha* (in part).

A genus of about ten species in N. and S. America, one of which has occurred as a vagrant in Iceland and Britain.

Lasiurus cinereus* Hoary bat

Vespertilio cinereus Palisot de Beauvois, 1796: 18. Philadelphia, Pennsylvania, USA.

RANGE. Most of N. and S. America, to British Columbia, Hudson Bay and Nova Scotia in the north, and to Buenos Aires (Argentina) and Valdivia (Chile) in the south. Vagrants have been reported in S.W. Iceland and Orkney (Scotland) (see Koopman & Gudmundsson, 1966).

Genus *BARBASTELLA*

Barbastelles

Barbastella Gray, 1821. Type-species *Vespertilio barbastellus* Schreber. =*Synotus*.

A genus of two species confined to the Palaearctic and the adjacent parts of the Oriental region.

Outer margin of ear with a prominent projecting lobe of skin; forearm 36–41 mm; condylobasal length 13·0–13·6 mm (Europe to Turkestan) . . . . ***B. barbastellus***

Outer margin of ear without a projecting lobe; forearm 38–45 mm; condylobasal length 14–15 mm (Asia) . . . . ***B. leucomelas***

Barbastella barbastellus Western barbastelle

Vespertilio barbastellus Schreber, 1774. Burgundy, France. =*communis*. Incl. *barbastelle, daubentonii*.

RANGE (Map 26). W. Europe, including England and S. Scandinavia, to the Volga; Crimea; Caucasus; Morocco (Panouse, 1956); the larger Mediterranean islands; perhaps Senegal (see remarks under *B. leucomelas* below).

Barbastella leucomelas Eastern barbastelle

Vespertilio leucomelas Cretzschmar, 1826. 'Arabia Petraea', =Sinai, Egypt.

B. l. leucomelas. Sinai, ?Eritrea (see remarks below), N. Iran.

B. l. darjelingensis (Hodgson, 1855). Rest of range. =*darjelinensis, blanfordi*. Incl. *caspica, walteri*.

RANGE (Map 26). From the Caucasus, where it is apparently sympatric with *B. barbastellus*, through the Elburz Mts, S. Turkestan and Afghanistan to the Pamirs, Himalayas and W. China; Honshu and Hokkaido, Japan; old records from Sinai and Eritrea.

REMARKS. The nomenclature of this species is made insecure by its dubious status in Sinai where it has not been found since the original description. The record from Eritrea (Heuglin, 1877) has likewise not been repeated. Another early record from Senegal (Rochebrune, 1883) could perhaps refer to *B. barbastellus* which is known from Morocco.

Neuhauser & DeBlase (1974) reported a pale animal from N. Iran as *B.l. leucomelas* and a dark one from Afghanistan as *B.l. dargelingensis*. If this interpretation is correct it may be that Caucasian animals (including *caspica*) should also be considered as *B.l. leucomelas*.

Genus *PLECOTUS*

Long-eared bats

Plecotus Geoffroy, 1818. Type-species *Vespertilio auritus* L. Incl. ?*Macrotus*.

A genus of two Palaearctic species and two Nearctic species, if *Corynorhinus* is included. Until the diagnostic characters are better known the baculum should be checked wherever possible before an identification is confirmed. The baculum is Y-shaped in both species.

Paired limbs of baculum about as long (on outer side) as they are wide; size smaller (C–M^3 usually under 5·6 mm, greatest width of tragus usually about 5 mm); proximal zone of dorsal pelage brown or brownish-grey . . . . ***P. auritus***

Paired limbs of baculum considerably longer than they are wide; larger (C–M^3 usually over 5·6 mm, greatest width of tragus usually about 6 mm); proximal zone of dorsal pelage dark grey or black . . . . ***P. austriacus***

Plecotus auritus Common long-eared bat

Vespertilio auritus Linnaeus, 1758. Sweden.

P. a. auritus. Europe, W. and central Siberia. Incl. *brevimanus* (England), ?*bonepartii*, *communis*, *cornutus*, *europaeus*, ?*megalotis*, *montanus*, *otus*, ?*peronii*, *typus*, ?*velatus*, ?*vulgaris*.

P. a. sacrimontis Allen, 1908. China and Japan, type-locality on Honshu. Incl. ?*ognevi* (Sakhalin).

?*P. a. homochrous* Hodgson, 1847. Himalayas. Incl. *puck*.

*P. a. uenoi** Imaizumi & Yoshiyuki, 1969. Korea.

Plecotus auritus uenoi Imaizumi & Yoshiyuki, 1969. Hwaamgul Cave, Cheonpo, Hwaam-ri, Dong-myeon, Jeongseon-gun, Kangweon-do, Korea.

RANGE (Map 27). W. Europe, including Britain, Ireland and S. Scandinavia, south to the Pyrenees, central Italy, Crimea and Caucasus and east to Mongolia, S.E. Siberia and N.E. China; Sakhalin, Hokkaido and N. Honshu; the Himalayas if *homochrous* is conspecific.

REMARKS. See under *P. austriacus* below.

***Plecotus austriacus*†** Grey long-eared bat

Vespertilio auritus austriacus Fischer, 1829. Vienna, Austria.

P. a. austriacus. Europe. Incl. *brevipes*, *kirschbaumii*, *meridionalis* (Yugoslavia), ?*teneriffae* (Canary Is.).

*Plecotus auritus hispanicus** Bauer, 1957. Lagunilla, Bejar, central Spain.

P. a. christiei Gray, 1838. N. Africa. Incl. *Vespertilio auritus aegyptius* Fischer (not *V. pipistrellus* var. *aegyptius* Fischer), *aegyptiacus*.

P. a. wardi Thomas, 1911. Himalayas (type-locality Ladak, Kashmir).

P. a. ariel Thomas, 1911. S.W. China (type-locality in Szechuan).

P. a. kozlovi Bobrinskii, 1926. W. China and Mongolia (type-locality in Tsaidam, W. China).

?*P. a. mordax* Thomas, 1926. Kashgar, Sinkiang, W. China.

?*P. a. macrobullaris*/ Kuzyakin, 1965. Caucasus.

Plecotus auritus macrobullaris Kuzyakin, 1965. Ordzhonikidze, N. Ossetian SSR., Caucasus, USSR.

RANGE (Map 27). S. Europe and N. Africa through the Caucasus and Palestine to the Himalayas, Mongolia and W. China; S. England, Jersey (Channel Is.), the Canary Is. and Cape Verde Islands (Dorst & Naurois, 1966); Senegal.

REMARKS. The genus *Plecotus* was for long considered monospecific in Europe but has recently been shown to contain two sibling species (see Hanak, 1966, for a summary). The allocation of the Asiatic forms is still rather doubtful, but if the above allocation to species is correct then the number of valid races is probably much less than indicated here (e.g. *ariel*, *kozlovi* and *mordax* may well prove to be consubspecific with *P.a. wardi*).

Genus *MINIOPTERUS*

Long-winged bats

Miniopterus Bonaparte, 1837. Type-species *Vespertilio ursinii* Bonaparte = *V. schreibersi* Kuhl.

A genus of about ten species found throughout the tropics of the Old World, with one species extending into the Palaearctic Region.

Miniopterus schreibersi Schreibers' bat

Vespertilio schreibersii Kuhl, 1819. Kulmbazer Cave, S. Bannat, Hungary.

M. s. schreibersi. Europe and N. Africa. Incl. *inexpectatus, italicus, oreinii, ursinii.*

*Miniopterus schreibersi baussencis** Laurent, 1944. Grotte des Fées, Les Baux (near Arles), Provence, France. (Status *fide* Bauer & Festetics, 1958.)

M. s. pallidus Thomas, 1907. Caucasus, Asia Minor, Kopet Dag, N. Iran, Palestine (type-locality on S. shore of Caspian Sea, Iran).

*Miniopterus schreibersi pulcher** Harrison, 1956. Ser' Amadia, Kurdistan, Iraq. (Status *fide* Harrison, 1964a.)

M. s. fulginosus (Hodgson, 1835). India to S. China and Japan (type-locality Nepal). Incl. ?*chinensis* (Hopei), *japoniae* (Kyushu, Japan), *parvipes* (Fukien).

Extralimital synonyms in Africa are *arenarius, breyeri, dasythrix, majori, natalensis, scotinus, vicinior* and *villiersi*; and in S.E. Asia to Australia *blepotis, eschscholzii, fuscus, magnater, orianae* and *yayeyamae.*

RANGE (Map 28). S. Europe and Morocco through the Caucasus and Iran to most of China and Japan; most of the Oriental Region; New Guinea the Solomon Is. and Australia; most of Africa south of the Sahara.

Genus *MURINA*

Tube-nosed bats

Murina Gray, 1842. Type-species *Vespertilio suillus* Temminck. Incl. *Ocypetes* Lesson (part).

A distinctive genus represented throughout the Oriental Region but extending northwards in E. Asia to Japan and the Altai, and south to New Guinea. There are probably about ten species of which two (perhaps three) extend into the Palaearctic Region.

Large: forearm 40–42·5 mm, condylobasal length 17·5–18·0 mm; 2nd upper incisor equal to 1st . . . ***M. leucogaster***

Small: forearm 29–32 mm, condylobasal length 14·0–14·5 mm; 2nd upper incisor smaller than 1st . . . ***M. aurata***

Murina aurata Little tube-nosed bat

Murina aurata Milne-Edwards, 1872. Moupin, Szechuan, China.

M. a. aurata. S. China and E. Himalayas. =*aurita.* Incl. *feae* (Burma).

M. a. ussuriensis Ognev, 1913. Ussuri region (type-locality) and Japan.

RANGE (Map 28). Ussuri Region of S.E. Siberia; Manchuria; Korea (Jones, 1960); Sakhalin; Japan; mountains of S.W. China to Burma and Nepal.

Murina leucogaster Great tube-nosed bat

Murina leucogaster Milne-Edwards, 1872. Moupin district, Szechuan, China.

M. l. leucogaster. S. China. =*leucogastra.*

M. l. rubex Thomas, 1906. Darjeeling, N.E. India.

M. l. hilgendorfi (Peters, 1880). Japan (type-locality Tokyo), Sakhalin, Korea to Inner Mongolia, Altai and upper Yenesei. Incl. *fuscus* (Manchuria), *intermedius* (Korea), *ognevi* (Vladivostock), *sibirica* (Tomsk, central Siberia).

RANGE (Map 28). Known from a few specimens from the E. Himalayas, Szechuan and Fukien; and more abundantly from the Upper Yenesei, Altai, Inner Mongolia (Wang, 1959), Manchuria, Korea, Ussuri region, Sakhalin and Japan (Honshu and Kyushu).

Murina tenebrosa*

Murina tenebrosa Yoshiyuki, 1970. Sago, Kamiagata-cho, Tsushima Is., Japan.

RANGE. Tsushima Is., Japan.

REMARKS. Known from a single specimen. Not seen. Forearm 34·4 mm.

Family MOLOSSIDAE

Free-tailed bats

A large family of about six genera found throughout the tropics and warm temperate regions of both hemispheres. Only one genus reaches the Palaearctic Region.

Genus *TADARIDA*

Tadarida Rafinesque, 1814. Type-species *Cephalotes teniotis* Rafinesque. Incl. *Nyctinoma, Nyctinomas, Nyctinomia, Nyctinomus*. Incl. *Chaerephon, Dinops, Dysopes, Mops* and others based on extra-Palaearctic species.

Contains a large number of species throughout the tropics and sub-tropics, four of which are listed here although three of these are very marginal.

1. Forearm 35–40 mm; condylocanine length under 15 mm . . . . . ***T. pumila***
– Forearm over 45 mm; condylocanine length over 16 mm . . . . . . (2)
2. Forearm under 54 mm; condylocanine length under 19 mm . . ***T. aegyptaica***
– Forearm over 54 mm; condylocanine length over 19 mm . . . . . . (3)
3. Anterior corner of tragus acutely pointed; premaxillae co-ossified; sagittal crest well developed; 2 pairs of lower incisors . . . . ***T. midas***
– Anterior corner of tragus bluntly rounded; premaxillae not co-ossified; sagittal crest absent; 3 pairs of lower incisors . . . . ***T. teniotis***

Tadarida teniotis — European free-tailed bat

Cephalotes teniotis Rafinesque, 1814. Sicily.

T. t. teniotis. W. Palaearctic. =*taeniotis*. Incl. *cestoni, midas* Schulze (not of Sundevall), *nigrogriseus, rüppellii* Temminck (Egypt), *savii*.

*Tadarida teniotis cinerea** Gubareff, 1939. Dashalty defile, Shusha, Azerbaijan, USSR.

T. t. insignis (Blyth, 1861). E. Asia, type-locality Fukien, China. Incl. *rueppelli* Swinhoe (not of Temminck), *latouchei* (N. China), *coecata* (Yunnan), *septentrionalis* (N. Korea).

RANGE (Map 18). The Mediterranean zone of Europe including most of the Mediterranean islands and Madeira; Morocco and Algeria; Egypt and Asia Minor east to Kirghizia and Afghanistan (Meyer-Öhme, 1965); also in eastern Asia from the eastern Himalayas through China to N. Korea; Hokkaido and Kyushu.

REMARKS. A detailed review with general and local distribution maps was given by Aellen (1966). Imaizumi (1970a) considered *insignis* to be specifically distinct.

Tadarida aegyptiaca

Nyctinomus aegyptiacus Geoffroy, 1818. Egypt.

T. a. aegyptiaca. N. and E. Africa and Arabia. =*geoffroyi*.

RANGE (Map 18). Algerian Sahara (Schlitter & Robbins, 1973), Egypt and S.W. Arabia; also throughout much of Africa south of the Sahara (including *bocagei* and other named forms) and in Iran (DeBlase, 1971), Pakistan, India and Sri Lanka (including *tragatus* and other named forms).

REMARKS. According to Chaturvedi (1964), who has examined the type, *tragatus*, listed as a species by E & M-S, is referable to *T. aegyptiaca*. It is therefore the earliest name amongst the Indian forms, with *gossei*, *sindiaca* and *thomasi* as possible synonyms.

Tadarida pumila

Dysopes pumila Cretzschmar, 1830 or 1831. Eritrea.

RANGE. (Map 18). S.W. Arabia; most of the savanna zone of Africa south of the Sahara.

*Tadarida midas**

Dysopes midas Sundevall, 1843: 207, p. 2, fig. 7. Bahr-el-Abiad, White Nile, Sudan.

RANGE (Map 18). Known from one locality in S.W. Arabia, just north of Yemen; throughout much of Africa south of the Sahara, in the savanna zone.

REMARKS. The two known Arabian specimens were collected in 1949. Harrison (1964a) gave a detailed account.

Order **PRIMATES**

The primates (monkeys, apes, lemurs etc.) are predominantly tropical, but two species of *Macaca* are Palaearctic, a third is present marginally in North China, whilst one species of African baboon occurs also in Arabia. See Napier & Napier (1967) for a summary of this order.

Family **CERCOPITHECIDAE**

Old-world monkeys

Contains about 16 genera of which two are represented in the Palaearctic Region. (In addition *Presbytis entellus*, the langur, a widespread Indian species, extends above the tree-line to about 4000 m in the Himalayas.)

Muzzle short; no mane . . . . ***MACACA***
Muzzle elongate; males with a mane of long hair on the shoulders (Arabia). . ***PAPIO***

Genus *MACACA*

Macaques

Macaca Lacepède, 1799. Type-species *Simia inuus* L. = *M. sylvanus* (L). = *Pithes*, *Salmacis*. Incl. *Inuus*, *Magotus*, *Rhesus*, *Simia* (suppressed), *Sylvanus*.

An Oriental genus of about ten species with one isolated species in North Africa and another in Japan. Two further species occur in N. China (probably introduced) and Japan (certainly introduced). For a review and key to the entire genus see Fooden (1976).

1. Tail completely absent (N.W. Africa and Gibraltar). . . . . . ***M. sylvanus***
– Tail present . . . . (2)
2. Tail very short, considerably shorter than hind feet; pelage long, hairs clearly banded (Japan) ***M. fuscata***
– Tail as long as hind feet, usually much longer . . . . (3)
3. Hind legs and feet darker than body; tail thickly haired, distal half predominantly black ***M. cyclopis***
– Hind legs and feet concolorous with body; tail less thickly haired, concolorous with body or slightly darker at tip . . . . ***M. mulatta***

Macaca mulatta Rhesus monkey

Cercopithecus mulatta Zimmermann, 1780. 'India'. Incl. *tcheliensis* (Hopei, N. China) and many other synonyms from the Oriental Region.

RANGE (Map 29). E. Afghanistan through the whole of N. India and S. China; an isolated colony, probably introduced, near Peking.

Macaca cyclopis Formosan macaque

Macacus cyclopis Swinhoe, 1862. Formosa. Incl. *affinis*.

RANGE (Map 29). Taiwan. Introduced on Oshima and another of the Izu Islands S. of Tokyo, Japan.

Macaca fuscata Japanese macaque

Innus fuscatus Blyth, 1875. Japan.
M. f. fuscata. Range except Yakushima Is.
*Macaca speciosa** Geoffroy, 1826. Japan. See Remarks below.
*Papio japonicus** Rennie, 1838. Japan. See Remarks below.
M. f. yakui Kuroda, 1941. Yakushima Is., S. Japan.

RANGE (Map 29). Honshu (to extreme north), Shikoku, Kyushu and Yakushima, Japan.

REMARKS. The specific name *fuscatus* Blyth, 1875 was validated and placed on the Official List of Specific Names in 1970 by Opinion 920 of the International Commission on Zoological Nomenclature and the names *speciosa* and *japonicus* were suppressed (*Bull. zool. Nom.* **27**: 77).

Macaca sylvanus Barbary ape

Simia sylvanus Linnaeus, 1758. Barbary coast. Incl. *ecaudatus*, *inuus*, *pithecus*, *pygmaeus*.

RANGE (Map 29). Morocco and Algeria; introduced in Gibraltar.

Genus *PAPIO*

Baboons

Papio Müller, 1773. Type species *Simia sphynx* L. (the mandrill). Incl. *Hamadryas*, *Comopithecus*.

A genus of about six species in Africa, one occurring also in Arabia and formerly in Egypt. The subgeneric name *Comopithecus* is available for this species.

Papio hamadryas Sacred baboon

Simia hamadryas Linnaeus, 1758. Egypt.
P. h. hamadryas, Mainland of Africa. =*aegyptiaca*. Incl. *chaeropithecus*, *cynamolgos*.
P. h. arabicus Thomas, 1900. Arabia.

RANGE (Map 29). S.W. Arabia, north to about Mecca; Ethiopia, N. Somalia and E. Sudan, formerly also in upper Egypt where now extinct.

Order **LAGOMORPHA**

Lagomorphs

In spite of their former inclusion in the order Rodentia, the lagomorphs constitute a very clearly defined group with little affinity to the rodents or any other order. They have a wide distribution in the northern hemisphere and in Africa. There are only two families, which are clearly differentiated and are both represented in the Palaearctic Region. The entire order, including the fossil members, was revised by Gureev (1964).

Hind legs and feet long (feet usually longer than skull); ears usually very elongate (always in Palaearctic forms); tail clearly visible; 6 upper cheek-teeth (last small); rostrum long, nasal bones over 40% of length of skull; zygomatic arches with short posterior processes; frontals wide in interorbital region, with supra-orbital processes . . **LEPORIDAE** (p. 70)

Hind legs and feet short (feet shorter than skull); ears short and rounded; no tail; 5 upper cheek-teeth; rostrum short, nasal bones under 40% of length of skull; zygomatic arches with long posterior processes; frontals with narrow interorbital constriction and no processes **OCHOTONIDAE** (p. 65)

Family **OCHOTONIDAE**

Pikas

A clearly defined family containing a single recent genus *Ochotona*, but with one very recently extinct genus which is also included here. Russian authors, e.g. Ognev (1940), Gureev (1964) and Kuznetsov (1965) all call this family Lagomyidae, but this is based on *Lagomys* Cuvier, 1800, which is a junior homonym of *Lagomys* Storr, 1780 (a synonym of *Marmota*), and the family name Lagomyidae is therefore clearly invalid according to Article 39 of the *International Code*.

First lower premolar with anterior part deeply separated by layer of cement from rest of tooth; 4 lower cheek-teeth, posterior molar consisting of 3 segments (extinct, Europe) ***PROLAGUS*** (p. 65)

First lower premolar without deeply separated anterior part; 5 lower cheek-teeth ***OCHOTONA*** (p. 66)

Genus *PROLAGUS**

Prolagus Pomel, 1853. Type-species *Anoema aeningensis* König, 1825, a Miocene species from Germany. Incl. *Myolagus* Hensel, 1856.

An extinct genus placed by Gureev (1964) in a separate subfamily Prolaginae. Gureev recognized five species, confined to Europe from Miocene to Recent. One species probably survived into historic times on Sardinia.

Prolagus sardus* Sardinian pika

Lagomys sardus Wagner, 1829: 1136. Sardinia.
Lagomys fossilis Wagner, 1829: 1139.
Lagomys corsicanus Wagner, 1829: 1139. Corsica.

RANGE. Extinct. Corsica and Sardinia, surviving perhaps as late as the 18th century: Kurten (1968) quoted a possible reference published in 1774 from the small island of Tavolara off Sardinia.

Genus *OCHOTONA*

Pikas

Ochotona Link, 1795. Type-species *Ochotona minor* Link = *Lepus dauuricus* Pallas. Incl. *Conothoa*, *Lagomys* G. Cuvier, *Ogotoma*, *Pika*, *Tibetholagus*.
*Abra** Gray, 1863: 11. Type-species *Lagomys curzoniae* Hodgson. Preoccupied by *Abra* Lamarck, 1818 (Mollusca).
*Abrana** Strand, 1928. New name for *Abra* Gray.

A genus of about 14 species confined to the mountains of central Asia except that one species extends north through N.E. Siberia and into western N. America, and another extends westwards across the steppes of Turkestan as far as the Volga. All species are included below.

It is a very difficult genus for the taxonomist, with rather small differences between the species and considerable geographical and seasonal differences within species, and some forms are very poorly known because of the inaccessability of their habitats. The classification used here is mainly that of Gureev (1964) who revised the genus, including four extinct species. It is certainly far from definitive, especially with regard to the forms found on and around the Tibetan plateau. Gureev's classification differs from that of E & M-S in the limitation of *O. pusilla* to forms in Turkestan; the inclusion of *macrotis* in *O. roylei*; the separation of *curzoniae* and other southern forms from *O. daurica*; the inclusion of the northern *hyperborea* and the N. American forms in *O. alpina*; and the separation of *nepalensis* from *O. roylei*. I accept most of Gureev's results primarily because they represent the latest comprehensive revision. Cases where I have deviated from them, or have supported them by personal observation, are noted under 'Remarks'.

There seem to me to be no grounds for recognizing subgenera. The character most used – the division of the palatal foramina by processes of the premaxillae – is a useful specific character but no more.

The following key is very imperfect and should be used only as a guide to the characters that distinguish the species. Skulls of all species were figured by Gureev (1964).

Data on the chromosomes of species in the USSR were given by Vorontsov & Ivanitskaya (1973).

1. Palatal foramina completely divided into small anterior and large posterior parts by meeting of processes of the premaxillae . . . . (2)
– Palatal foramina not completely divided as above . . . . (3)
2. Soles of feet brown (except in northern forms); interorbital constriction flat and wider (4–6 mm) . . . . ***O. alpina***
– Soles of feet white; interorbital constriction angular, with two longitudinal ridges, less than 5 mm wide . . . . ***O. pallasi***
3. Interorbital constriction very narrow, less than 11 % of condylobasal length (always less than 5 mm, and less than 4 mm in small species). . . . . (4)
– Interorbital constriction wider, over 11 % of condylobasal length (usually over 5 mm except in small species) . . . . (10)

4. A transverse band of pale fur on the neck behind the ears, bordered before and behind by more rufous hair; (condylobasal length 43–46 mm, hind feet over 30 mm) . . . ***O. rufescens***
– No pale transverse band on neck . . . . (5)
5. Palatal foramina narrow and parallel-sided in front, obscuring much of vomer, widening abruptly behind . . . . (6)
– Palatal foramina widening more evenly behind . . . . (7)
6. Larger: condylobasal length usually over 48 mm . . . . . . ***O. ladacensis***
– Smaller: condylobasal length under 48 mm . . . . ***O. koslowi***
7. Larger: condylobasal length usually over 40 mm, length of palate (front of incisors to back of palate) over 15 mm . . . . (8)
– Smaller: condylobasal length usually under 40 mm, length of palate usually under 15 mm (9)
8. Mid-dorsal line of skull very convex, nasals sloping down very abruptly from interorbital region . . . . ***O. curzoniae***
– Mid-dorsal line of skull less convex . . . . ***O. daurica***
9. Skull very narrow, zygomatic width under 15 mm . . . . . . ***O. thomasi***
– Skull wider, zygomatic width over 16 mm . . . . ***O. thibetana***
10. Small: condylobasal length under 41 mm, hind feet under 30 mm; ears short (under 17 mm); (palatal foramina very narrow in front) . . . . ***O. pusilla***
– Larger: condylobasal length over 41 mm, hind feet usually over 30 mm; ears long, over 20 mm . . . . (11)
11. Palatal foramina very narrow in front, partially obscuring vomer . . . . (12)
– Palatal foramina not so narrow in front, most of vomer visible . . . ***O. roylei***
12. Condylobasal length usually over 45 mm; (pelage rufous) . . . . ***O. rutila***
– Condylobasal length usually under 45 mm . . . . ***O. erythrotis***

Ochotona pusilla — Steppe pika

Lepus pusillus Pallas, 1769. Samara, S.E. European Russia.
O. p. pusilla. S.E. European Russia and N.W. Kazakhstan. Incl. *minutus*.
O. p. angustifrons Argyropulo, 1932. E. Kazakhstan.

RANGE (Map 30). Steppes from the Volga across N. Kazakhstan to upper Irtysh. Reached W. Europe and England in the late Pleistocene.

REMARKS. E & M-S tentatively included the forms *nubrica* (Kashmir), *forresti* (Yunnan) and *osgoodi* (Burma). Examination of the types of *nubrica* and *forresti* does not support their inclusion here, which is also improbable on ecological and zoogeographical grounds. Gureev (1964) also restricted this species to the forms listed above.

Ochotona thibetana — Moupin pika

Lagomys thibetanus Milne-Edwards, 1872. Moupin, Szechuan, China. =*tibetana*. Incl. *cansus* (Kansu), *hodgsoni* Bonhote, *huangensis* (S. Shensi), *huanghoensis*, *morosa*, ?*nubrica*† (Ladak), ?*osgoodi* (N.E. Burma), *sacraria*, *sikimaria* (Sikkim), *sorella* (Shansi), *stevensi*, *syrinx*, *zappeyi*.
*Ochotona tibetana cilanica** Bannikov, 1960. Tsilin, Tsinchilin, Koko Nor region, Nan Shan, Chinghai, China.
*Ochotona thibetana lhasaensis** Feng & Kao, 1974. Lhasa, Tibet (3950 m).

RANGE (Map 30). Mountains of W. China from S. Shansi and Nan Shan to Szechuan, Sikkim and S.E. Tibet; perhaps also Ladak and N. Burma.

REMARKS. The form *forresti* (Yunnan) was allocated here by Gureev (1964), but examination of the type does not support this. The form *nubrica* from Ladak is of doubtful status. E & M-S tentatively placed it in *O. pusilla*; Gureev (1963) placed it in *O. roylei*. My examination of the type suggests that it might refer to this species. Feng & Kao (1974) treated *cansus* as a distinct species.

Ochotona thomasi* — Thomas' pika

Ochotona thomasi Argyropulo, 1948. Lake Alyk Nor, N.E. Tibet (?=Alak Nor, Chinghai, 35°30′N, 97°20′E).

RANGE (Map 30). Vicinity of Lake Alyk Nor and Nan Shan range (*fide* Gureev, 1964).

REMARKS. Gureev (1964) placed this species in his '*tibetana* group'.

Ochotona roylei Royle's pika, Large-eared pika

Lagomys roylii Ogilby, 1839. Punjab. Incl. *auritus*, *baltina*, *chinensis* (Szechuan), *forresti*† (Yunnan), *griseus*, *hodgsoni*, *macrotis*† (Kuenlun Mts, S.W. Sinkiang), *nepalensis* (Nepal), *sacana* (N.E. Kirghizia), *sinensis*, *wardi* (Kashmir), *wollastoni* (Mt Everest).
?*Ochotona angdawai** Biswas & Khajuria, 1955. Khumbu Glacier (*c.* 18 000 ft), west of Mount Everest, Khumbu, Nepal.
?*Ochotona mitchelli** Agrawal & Chakraborty, 1971. Gosainkund (*c.* 4750 m), Nawakot Dist, Napal.
?*Ochotona himalayana** Feng, 1973. Qu-xiang, Bo-qu valley, Nei-la-mu Dist, Tibet (3500 m).
?*Ochotona lama** Mitchell & Punzo, 1975. Lupra, Mustang Dist, Napal (28°48′N, 83°47′E, 3640 m).

RANGE (Map 31). Mountains from Russian Tien Shan (Kirghizia) through Pamirs, Karakorum and Kunlun Mts (S.W. Sinkiang) to Kashmir; Nepal, Szechuan and N. Burma; up to 6000 m.

REMARKS. E & M-S treated *O. macrotis* as specifically distinct from *O. roylei*, but they are here united following Kuznetsov (1965) and Gureev (1964). E & M-S placed *forresti* from Yunnan tentatively in *O. pusilla*; Gureev placed it in *O. thibetana*. Examination of the type strongly suggests that it is a small race of *O. roylei*. There may well be distinctive subspecies in this species, but no attempt has here been made to separate them.

The four more recently described 'species' listed above all appear to be close to *O. roylei* and possibly conspecific, but none has been seen.

Ochotona rutila Turkestan red pika

Lagomys rutilus Severtzov, 1873. Vernoe Mts, S.E. Kazakhstan.

RANGE (Map 30). Mountains of S.E. Russian Turkestan and N. Afghanistan from Tien Shan to Pamirs, over 2000 m.

REMARKS. E & M-S included in this species *erythrotis*, *vulpina*, *brookei* and, tentatively, *gloveri*, all from E. Tibet or W. China. These are excluded here following Gureev (1964).

Ochotona erythrotis† Chinese red pika

Lagomys erythrotis Büchner, 1890. Burchan-Budda, E. Tibet. Incl. *brookei* (Kam, E. Tibet), ?*gloveri* (W. Szechuan), *vulpina* (Kansu).

RANGE (Map 30). E. Tibet, Chinghai and adjacent parts of Kansu and Szechuan.

REMARKS. E & M-S included this form in *O. rutila*, but it is here given specific rank following Gureev (1964).

Ochotona kamensis* Kam pika

Ochotona kamensis Argyropulo, 1948. Kam, W. Szechuan.

RANGE (Map 31). Known only from Kam (=Kham), W. Szechuan.

REMARKS. Gureev (1964) gave this form specific rank.

Ochotona daurica Daurian pika

Lepus dauuricus Pallas, 1776. Onon R., E. Siberia.
O. d. daurica. S.E. Siberia and Mongolia. Incl. *ogotona*.
O. d. altaina Thomas, 1911. Altai.
O. d. bedfordi Thomas, 1908. Shansi, China.
O. d. annectens Miller, 1911. Kansu, China.
*O. d. mursaevi** Bannikov, 1951. Khangai Mts, N.W. Mongolia.
Ochotona daurica mursaevi Bannikov, 1951. Dakhtin-Daba Pass, 80 km N. of Uliassutai, Khangai, Mongolia (*fide* Bannikov, 1954).

RANGE (Map 31). Steppes from Altai and Transbaikalia to Khingan Mts and south to Shensi.

REMARKS. E & M-S also tentatively included the forms *curzoniae* (S. Tibet) and *melanostoma* (Kansu) in this species. These were separated as *O. curzoniae* by Gureev (1964).

***Ochotona curzoniae*†** Black-lipped pika

Lagomys curzoniae Hodgson, 1858. Chumbi Valley, S. Tibet. Incl. *melanostoma* (Koko Nor and Kansu), ? *seiana* (Seistan, Iran).

RANGE (Map 31). Tibetan Plateau including adjacent parts of Kansu, Sikkim and Nepal; E. Iran if *seiana* is conspecific.

REMARKS. These forms were tentatively included in *O. daurica* by E & M-S, but Gureev (1964) recognized *O. curzoniae* as a distinct species, including *melanostoma*. He did not allocate *seiana*, but examination of the type suggests that it may also belong here.

Ochotona koslowi Kozlov's pika

Lagomys koslowi Buchner, 1894. Guldsha Valley, S.E. Sinkiang.

RANGE (Map 31). Known only from northern edge of Tibetan Plateau.

Ochotona ladacensis Ladak pika

Lagomys ladacensis Günther, 1875. Ladak, Kashmir.

RANGE (Map 30). S. Sinkiang and Kashmir at high altitude (type from 4000 m).

Ochotona pallasi Pallas's pika

Ogotoma pallasii Gray, 1867. 'Asiatic Russia-Kirgisen'.
O. p. pallasi. N.E. Kazakhstan. Incl. *ogotona* Waterhouse, *ogotona* Bonhote, *opaca*.
O. p. pricei Thomas, 1911. N.W. Mongolia.
O. p. sushkini Thomas, 1924. Chuya Alps, Russian Altai.
O. p. hamica Thomas, 1912. E. Tien Shan range, Sinkiang.

RANGE (Map 30). Mountains and high steppes from Altai to north of Lake Balkash and the eastern Tien Shan.

REMARKS. Most Russian authors call this species *O. pricei*, but Gureev (1964) included *pallasi* as a synonym, and there seems to be no reason to reject this name.

Ochotona rufescens Afghan pika

Lagomys rufescens Gray, 1842. Kabul, Afghanistan.
O. r. rufescens. Afghanistan and Pakistan.
O. r. vizier Thomas, 1911. Central Iran.
O. r. regina Thomas, 1911. Kopet Dag Mts, Turkmenia.
*O. r. shukurovi** Heptner, 1961. Great Balkhan Mts, S.W. Turkmenia.
Ochotona rufescens shukurovi Heptner, 1961. Great Balkhan Mts, S.W. Turkmenia, USSR.

RANGE (Map 31). Mountains of Afghanistan, Baluchistan, Iran and S.W. Turkmenia.

Ochotona alpina Northern Pika

Lepus alpinus Pallas, 1773. Tigeretski Range, Altai Mts.
O. a. alpina. Mountains from Altai to Transbaikalia and N. Korea; and Sakhalin. Incl. *ater*, *cinereofusca* (upper Amur), *coreana* (N. Korea), *mantchurica* (Manchuria), *nitida*, *scorodumovi* ('Zabaikalia', *fide* Gureev, 1964), *svatoshi* (Transbaikalia), *voshikurai* (Sakhalin).
O. a. argentata Howell, 1928. N. Kansu, China.
O.a. yesoensis Kishida, 1930. Hokkaido. =*yezoensis*. Incl. *convexa*, *inukaii*, *kinuta*, *kobayashii*, *ornata*, *rufa*, *sadakei*.
O. a. hyperborea (Pallas, 1811). N. Siberia, type-locality Chukotka region, extreme N.W. Siberia. Incl. *cinereoflava* (Okhotsk coast), *ferruginea* (R. Kolyma to R. Anadyr), *kamtschaticus*, *kolymensis*, *litoralis*, *normalis* (Kamchatka), *turuchanensis* (lower Tunguska R., central Siberia), *uralensis* (N. Urals).
O. a. collaris (Nelson, 1893). Alaska and Yukon.
O. a. princeps (Richardson, 1828). W. Canada and USA. Including many synonyms.

RANGE (Map 31). Mountains of central Asia from Altai through Mongolia to Manchuria and N. Korea; N. Kansu; mountains and tundra of N. Siberia from N. Urals to the Bering Strait; Sakhalin and Hokkaido; S.E. Alaska and Yukon; Rocky Mts from British Colombia to California and New Mexico.

REMARKS. E & M-S and Kuznetsov (1965) treated *O. alpina* and *O. hyperborea* as separate species. They are here united, along with the American forms listed, following Gureev (1964). The allocation of names to *O. a. alpina* and *O. a. hyperborea* also follows Gureev, who did not however comment on the status of *argentata* nor *yesoensis.*

Family **LEPORIDAE**

Rabbits and hares

The dominant family of lagomorphs, represented throughout the range of the order. Generic classification is rather unstable with varying views on the degree of generic separation amongst the rabbits and on the allocation of some rabbit-like species to the genus *Lepus*. The majority of species of *Lepus* are, however, immune from such instability. For recent treatments of the problem see Gureev (1964) and Angermann (1966a). *Lepus* is well represented in the Palaearctic Region. Of the other genera only *Oryctolagus* is included here. Gureev included several eastern Asiatic species in *Caprolagus*, but these are here referred to *Lepus* following Angermann, who retained *Caprolagus* only for the Oriental *C. hispidus*.

Width of mesopterygoid fossa considerably less than width of bony palatal bridge; hind feet usually less than 90 mm; (ear short, usually under 80 mm from crown, without black at tip) ***ORYCTOLAGUS*** (p. 74)

Width of mesopterygoid fossa slightly less, or greater than, width of palatal bridge; hind feet usually over 100 mm . . . . ***LEPUS*** (p. 70)

Genus ***LEPUS***

Hares

Lepus Linnaeus, 1758. Type-species *L. timidus* L. =*Chionobates*, *Eulepus*. Incl. *Allolagus*, *Eulagus*, *Tarimolagus*.

*Proeulagus** Gureev, 1964. As subgenus, type-species *Lepus oiostolus* Hodgson.

A large genus containing the hares, i.e. species adapted especially to fast running in open habitats throughout the Palaearctic, Oriental, Ethiopian and Nearctic regions. These adaptations include long hind legs, long ears, very wide nasal passage and the production of well developed young after a long pregnancy. This contrasts with the other genera of Leporidae whose members can be described as rabbits and are more fossorial. Seven species occur in the Palaearctic Region. Three others occur in the Oriental Region and about three in Africa south of the Sahara.

This is a very difficult group taxonomically with small differences between the species, some of which have large ranges and considerable intraspecific variation. Many of the characters that have been most often used to distinguish species show great individual and regional variation within single species. Length of ears and size of bullae both tend to be greater in arid habitats; width of mesopterygoid fossa (i.e. of the internal nares) also tends to be greatest in very open country, presumably correlated with respiratory and running efficiency. The last character is very conveniently expressed in comparison with the width (anterior–posterior) of the adjacent bony palatal bridge, but it would probably be more productive to relate these separately to the length of the skull.

The classification used here follows in general the work of Petter (1961a, 1971a) and Angermann (1966a, b; 1967a, b) rather than that of Gureev (1964) who placed *brachyurus*, *sinensis* and *yarkandensis* in *Caprolagus* and subgenerically separated *europaeus* from *capensis* etc. (here considered conspecific).

1. Ears considerably shorter than hind feet, not extending far beyond nose when laid forwards (2)
– Ears almost as long, or longer, than hind feet, extending well beyond nose when laid forwards; (pelage very soft; supraorbital processes with deep anterior notches) . . . (5)
2. Ears longer, reaching to about tip of nose when laid forwards; hind feet usually over 140 mm (shorter in Scotland); usually becoming white in winter; nasal bones short, falling short of incisors when seen in dorsal view; supraorbital processes usually with deep anterior notches; mesopterygoid fossa considerably wider than bony palatal bridge . ***L. timidus***
– Ears shorter, falling far short of nose when laid forwards; hind feet usually under 140 mm; nasals long, extending beyond incisors when seen in dorsal view; supraorbital processes with little or no anterior notch; width of mesopterygoid fossa about equal to or less than length of palatal bridge (E. Asia) . . . . (3)
3. Pelage rather coarse; tail brown above; bullae about equal in length to mid-line of basioccipital . . . . ***L. sinensis***
– Pelage soft; tail greyer than back, very short; bullae very small, shorter than basioccipital (4)
4. Dorsal pelage uniform reddish brown (may be white in winter); tail about 30–40 mm (Japan) ***L. brachyurus***
– Dorsal pelage speckled blackish and yellowish brown, neck contrasting in colour with back; tail about 60 mm (continental E. Asia) . . . . ***L. mandshuricus***
5. Tail with a black mid-dorsal line sharply defined from the white sides and ventral surface; (tips of ears black); nasals long, usually projecting beyond the incisors in dorsal view; mesopterygoid fossa usually much wider than palatal bridge; (auditory bullae usually under 13 mm in central Asian forms) . . . . ***L. capensis***
– Tail without a clearly defined black dorsal line; nasals usually shorter, falling short of the incisors in dorsal view . . . . (6)
6. Pelage very soft and silky; ears only very slightly darker at tips; bullae large (13–14·5 mm long); mesopterygoid fossa narrow, 6·5–7·5 mm . . . . . ***L. yarkandensis***
– Pelage very long and dense, but less silky than in *L. yarkandensis*; ears black at tips; bullae smaller (10–12 mm); mesopterygoid fossa wide, 8–10 mm . . . ***L. oiostolus***

Lepus capensis — Brown hare

Lepus capensis Linnaeus, 1758. Cape of Good Hope.

L. c. europaeus† Pallas, 1778. Europe, type-locality Burgundy, France. Incl. *alba*, *aguilonius*, *argenteogrisea*, *biarmicus*, *borealis* Kuznetzov, ?*campestris*, *campicola*, *caspicus* (Astrakhan, Russia), *caucasicus* (N. Caucasus), *cinereus*, ?*coronatus*, *cyrensis* (Azerbaijan), *flavus*, *gallaecius*, *hispanicus* (S.E. Spain), *hybridus* (Moscow region, USSR), *hyemalis*, *granatensis*, *iturissius*, *judeae* (Palestine), *kalmykorum*, *karpathorum*, *lilfordi*, ?*maculatus*, *medius*, *meridiei*, *meridionalis*, *niethammeri*, *niger*, ?*nigricans*, *occidentalis* (Britain), *parnassius* (Greece), *ponticus*, *pyrenaicus*, ?*rufus*, *syriacus* (Syria), *tesquorum* (Ukraine), *transsylvanicus*, *tumak*.

*Lepus cyanotis** Blanchard, 1957. Lanslebourg, Savoie, France, 2600 m.

L. c. schlumbergeri Saint-Loup, 1894. N. Africa, type-locality N.E. Morocco. Incl. *atlanticus*† (S.W. Morocco), *barcaeus*† (Cyrenaica), *harterti* (Rio de Oro), *innesi*, *kabylicus* (Algeria), *maroccanus*, *pallidior*, *pediaeus*, *rothschildi* (Egypt), *sefranus*, *sherif*, *tunetae* (Tunisia), *whitakeri*† (Tripolitania).

L. c. arabicus† Ehrenberg, 1833. Arabia, type-locality Qunfidha, S. of Mecca. Incl. *cheesmani* (S. of Bahrain), *omanensis* (Oman), ?*sinaiticus* (Sinai).

*L. c. atallahi** Harrison, 1972. Bahrain.

Lepus capensis atallahi Harrison, 1972. Isa, Bahrain Island.

L. c. tibetanus Waterhouse, 1841. Iran to Pakistan and Turkestan, type-locality in Kashmir. Incl. *aralensis*, *lehmanni* (lower Amu Darya, Turkestan), *pamirensis* (Pamirs), *craspedotis* (Baluchistan), ?*stoliczkanus* (W. Sinkaing), ?*biddulphi*, ?*kaschgaricus*, *zaisanicus* (E. of L. Balkash), *buchariensis* (Tadzhikistan), *desertorum* (Turkmenia).

L. c. tolai Pallas, 1778. Altai, Mongolia, W. China; type locality upper Amur basin S.E. of L. Baikal. Incl. *aurigineus* (China: see under 'Range' below), *brevinasus*, *butlerowi*, *cinnamomeus* (Szechuan), *filchneri* (S. Shensi), *gansuicus* (Kansu), *gobicus* (Mongolia), *kessleri*, *przewalski* (Tsaidam), *quercerus* (Altai Mts), *sowerbyae* (N. Shansi), *stegmanni*, *subluteus*, *swinhoei*.

L. c. mediterraneus Wagner, 1841. Sardinia. Incl. *typicus*.

L. c. corsicanus de Winton, 1898. Corsica.

L. c. creticus Barrett-Hamilton, 1903. Crete.

L. c. cyprius Barrett-Hamilton, 1903. Cyprus.

L. c. ghigii de Beaux, 1927. Stampalia Island, Aegian Sea.

?*L. c. aegyptius* Desmarest, 1882. Egypt. Petter (1963) considered this name unidentifiable.
Also many other synonyms from Africa south of the Sahara.

RANGE (Map 32). Open woodland, steppe and subdesert through the entire Palaearctic Region south of the coniferous forest zone, but excluding N.E. China and Japan; all the non-forested parts of Africa. In the north the range includes S. Sweden and Finland and overlaps widely with that of *L. timidus* in European Russia and S.W. Siberia. In China it extends south to the Yangtze Kiang, Szechuan and Sinkiang. (The type-locality of *aurigineus*, in Kiangsi south of the Yangtse, is erronous according to Sowerby (1943).) It occurs throughout S.W. Asia and eastwards to the Indus basin. It is present on Britain (but not Ireland, except by introduction) and on most of the islands of the Mediterranean. There has been much introduction for sport and this may have been the origin of some of the insular populations.

REMARKS. E & M-S recognized two wide-ranging species, *L. capensis* (southern) and *L. europaeus* (northern), almost entirely on the basis of supposed differences in size. Their allocation here to a single species follows Petter (1961a). Of the areas that are on the reputed boundary of the two forms Petter discussed the relevant data from Spain, and his views are supported by evidence of intergradation in Palestine (Yom-Tov, 1967) and Iran (Lay, 1967). The allocation to *L. capensis* of *atlanticus* (Morocco), retained as a species by E & M-S, and of *cyanotis* (Alps) follows Petter (1961a). The inclusion of *arabicus* follows Petter (1971a).

A very different classification is that of Gureev (1964) who placed *L. capensis* and *L. europaeus* in separate subgenera, but on the basis of minor cranial characters that seem to me to be variable even within the species in the most restricted sense. Other Russian authors likewise retain *L. europaeus*, e.g. Kuznetsov (1965), who recognized *L. europaeus*, *L. tolai* and *L. tibetanus*.

The above separation into subspecies is very arbitrary and is intended to demonstrate the principal segments of the species that have been recognized by various authors. In such a mobile animal occupying in the main the dominant and continuous habitats it is unlikely that even these few continental subspecies can be upheld. On the other hand some of the included synonyms may represent definable isolated populations, especially on the margins of the range, e.g. in central Asia. The insular populations listed above are unlikely to prove valid, being based mainly on small size, but they are retained provisionally pending a reassessment.

Kuznetsov (1965) gave specific rank to *europaeus*, *tibetanus* and *tolai* in the USSR. Although these are here given subspecific rank, the allocation of Russian names to these three forms follows Kuznetsov. Angermann (1967b) included *tolai* as a subspecies of *L. capensis* and considered *przewalskii* (from Tsaidam) a synonym of *tolai*. The above allocation of *stoliczkanus* and *kaschgaricus* (both from W. Sinkiang) to *L. c. tibetanus* rather than to *L. c. tolai* is arbitrary. Bannikov (1954) recognized *L. tolai* and *L. tibetanus* as specifically distinct in Mongolia.

Petter (1971a) listed 35 subspecies from the whole of Africa but remarked that only two (both non-Palaearctic), along with *L. c. arabicus*, were clearly distinguishable. It should be noted that whereas African forms of *L. capensis* as recognized by E & M-S are retained in that species, the African forms from south of the Sahara that E & M-S referred to *L. europaeus* are distinct from *L. capensis*. These are *L. saxatilis* in S. Africa, and East Africa species that Petter (1963, 1971a) referred to *L. crawshayi*, *L. whytei* and *L. habessinicus*. The relationship and nomenclature of these East African species are still in doubt, but they clearly do *not* represent *L. capensis* with which they coexist, *L. capensis* always occupying the more open habitats.

Lepus timidus — Arctic hare, Mountain hare, Varying hare

Lepus timidus Linnaeus, 1758. Sweden.

L. t. timidus. Continental Eurasia. =*variabilis, septentrionalis, typicus*. Incl. *abei* (Kurile Is), *algidus, alpinus* Erxleben, *altaicus, begitschevi* (Taimyr Peninsula), *borealis, canescens, collinus, gichiganus, kamtschaticus* (Kamchatka – *nom. nud.*), *kolymensis, kozhevnikovi* (European Russia), *lugubris* (Altai Mts), *mordeni* (S.E. Siberia), *orii* (Sakhalin), *rubustus, saghalinensis, sclavonius, sibiricorum* (W. Siberia), *sylvaticus, transbaicalicus* (L. Baikal), *tschuktschorum* (N.E. Siberia).
L. t. varronis Miller, 1901. Alps. Incl. *breviauritus*.
L. t. hibernicus Bell, 1837. Ireland. Incl. *lutescens*.
L. t. scoticus Hilzheimer, 1906. Scotland. Incl. ?*albus* Leach.
L. t. ainu Barrett-Hamilton, 1900. Hokkaido, Japan.
Other subspecies in N. America, including *L. t. arcticus* Ross, 1819 (N. Canada), *L. t. groenlandicus* Rhoads, 1896 (Greenland) and *L. t. othus* Merriam, 1900 (Alaska).

RANGE (Map 33). Tundra and coniferous forest zones from Scandinavia to E. Siberia, south to about 50° in European Russia and W. Siberia and including the mountains of central Asia south to the R. Ili in S.E. Kazakhstan, the Altai, and Sikhoto-Alin Mts. A widely isolated population in the Alps, and insular populations in Ireland, Scotland, Sakhalin and Hokkaido. Introduced to Spitzbergen in 1930's, but apparently died out about 1954 (Lønø, 1960b), and from Scotland to N. England. In N. America confined to the tundra zone from Alaska to Labrador and Greenland, and on Newfoundland. In the coniferous forest zone in N. America it is replaced by *L. americana* whereas in Eurasia the replacement by *L. capensis* occurs much further south.

REMARKS. The inclusion of the N. American forms in this species was suggested by E & M-S and supported by Gureev (1964). For a study of the variation in the Palaearctic part of the range see Angermann (1967a).

Lepus oiostolus — Woolly hare

Lepus oiostolus Hodgson, 1840. S. Tibet. Incl. ?*comus* (Yunnan), *grahami, hypsibius* (Ladak), *illuteus, kozlovi* (S.E. Tibet), ?*oemodias, pallipes, sechuanensis* (N.W. Szechuan), ?*tsaidamensis* (S.W. of Koko-Nor).

RANGE (Map 32). Tibetan Plateau between 3000 and 5000 m from Ladak to Yunnan (if *comus* is correctly allocated) and north through Chinghai and Szechuan to the Nan Shan range. The map is based on that of Angermann (1967b).

REMARKS. Reviewed by Angermann (1967b).

Lepus sinensis — Chinese hare

Lepus sinensis Gray, 1832. Canton, S. China.
L. s. sinensis. S. China. Incl. *flaviventris* (Fukien).
L. s. coreanus Thomas, 1892. Korea.
L. s. formosus Thomas, 1908. Taiwan.

RANGE (Map 33). Korea; lowland China south of the Yangtze Kiang; Taiwan.

REMARKS. Allen (1938) and Gureev (1964) allocated this species to *Caprolagus* which Angermann (1966a) limited to *C. hispidus* from India. The Korean form has been given specific rank, e.g. by Jones & Johnson (1965), but the differences are slight and there is no doubt of its close relationship with *L. sinensis*.

Lepus brachyurus — Japanese hare

Lepus brachyurus Temminck, 1844. Kyushu, Japan.
L. b. brachyurus. Kyushu, Shikoku and S.E. Honshu.
L. b. angustidens Hollister, 1912. N.W. Honshu. Incl. *etigo*.
L. b. okiensis Thomas, 1906. Oki Is.
L. b. lyoni Kishida, 1937. Sado Is.

RANGE (Map 32). Honshu, Shikoku, Kyushu, Oki Is. and Sado Is., Japan.

REMARKS. E & M-S tentatively included the continental form *mandshuricus* in this species, but it is here confined to the Japanese populations following Angermann (1966b). Imaizumi (1970a) diagnosed and mapped the subspecies listed above.

***Lepus mandshuricus*†** Manchurian hare

Lepus mandshuricus Radde, 1861. Bureja Mts, Amur basin. Incl. *melanonotus*.

RANGE (Map 32). Deciduous woodlands of Lower Amur, Manchuria and N. Korea.

REMARKS. E & M-S and Kuznetsov (1965) treated this as a race of the Japanese *L. brachyurus*, but it is here given separate specific rank following Angermann (1966b) who reviewed the differences in detail.

Lepus yarkandensis Yarkand hare

Lepus yarkandensis Günther, 1875. Yarkand, Sinkiang.

RANGE (Map 33). Steppes of S.W. Sinkiang (Chinese Turkestan).

REMARKS. Type-species of subgeneric name *Tarimolagus* Gureev. Angermann (1967b) reviewed the species, with description and map, and did not consider that its subgeneric separation was justified.

Genus *ORYCTOLAGUS*

Oryctolagus Lilljeborg, 1874. Type-species *Lepus cuniculus* L. =*Cuniculus*.

A monospecific genus, generally recognized as such although the differences between it and some other genera of rabbits, e.g. *Caprolagus*, are rather slight.

Oryctolagus cuniculus Rabbit

Lepus cuniculus Linnaeus, 1758. Germany. =*fodiens*. Incl. *algirus* (Algeria), *brachyotus*, *habetensis*, *huxleyi* (Puerto Santo, Madeira), *cnossius* (Crete), *oreas* (Spanish Morocco), ?*vermicula*, *vernicularis*.
*Oryctolagus cuniculus borkumensis** Harrison, 1952. Borkum Island, E. Friesian Islands, Germany.

RANGE (Map 33). Original range probably confined to N.W. Africa and Iberia. Now present in most of W. Europe east to E. Poland and Hungary, including Britain (since 12th century), Ireland, the islands of the western Mediterranean, the Azores, Madeira and Canary Is. More recently introduced and widely distributed in Australia, New Zealand, Chile and some oceanic islands.

REMARKS. The rabbits of the Mediterranean coast of Europe and the Mediterranean Islands have been allocated (by Miller, 1912) to *O. c. huxleyi* (type-locality Madeira) on the basis mainly of small size and paler colour. There seem to be no grounds for recognizing discrete subspecies on the mainland of Europe, although some of the insular forms may be distinctive. *O. c. borkumensis* was described on the basis of pale colour, but although paler than in mainland Germany they are scarcely separable from some Mediterranean forms, e.g. from Crete. I have not seen sufficient material to judge the status of North African forms.

An elaborate system of nomenclature for wild, domesticated and feral rabbits produced by Hochstrasser (1969) seems quite impractical in the absence of any demonstration that the names apply to discrete, definable populations. The early introductions of rabbits, e.g. to Britain, were of captive animals bred for meat and these may have already been modified by selective breeding.

Order **RODENTIA**

Rodents

The largest order of mammals, both worldwide and in the Palaearctic Region. The limits of the order are very clear-cut (as far as recent forms are concerned), but there is considerable controversy over higher classification within the order. However this concerns mainly the extent to which families can be grouped and the relationship of extinct groups. Comprehensive classifications were produced by Wood (1955) and Schaub (1958).

Twelve families are recognized here as being represented in the Palaearctic. This follows the arrangement used by Ellerman (in E & M-S) except that Zapodidae, Seleviniidae and Cricetidae are given family rank, instead of subfamily rank within the Dipodidae, Gliridae and Muridae respectively. At the family level this is the classification of Simpson (1945), Wood (1955) and Schaub (1958) except that the last author subdivided the Cricetidae, recognizing the Gerbillidae and Microtidae as distinct families.

There are too many examples of convergence in external form to compile a practical key to families based only on external characters. Instead the twelve families are listed below with some of their more prominent characteristics.

Sciuridae (squirrels). Three main bodily forms: tree squirrels with long bushy tail (cf. smaller dormice–Gliridae); flying squirrels, similar but with gliding membranes on flanks; ground squirrels with short, moderately hairy tail and very short ears (cf. similar but smaller hamsters and voles which all have less hairy tails); cheek-teeth 4–5/4. p. 75

Castoridae (beavers). Large; tail very broad, flat (horizontal) and scaly; cheek-teeth 4/4, hypsodont, rooted; aquatic. p. 87

Cricetidae (gerbils, voles etc). Tail long and moderately hairy, especially towards tip (Gerbillinae) or short and thinly haired; cheek-teeth 3/3, brachyodont to hypsodont, rooted or uprooted, cusps when clear in 2 longitudinal rows. p. 88

Spalacidae (blind mole-rats). Eyes invisible externally; no ear pinnae; no tail; incisors greatly enlarged; cheek-teeth 3/3. p. 129

Muridae (mice and rats). Tail long and thinly haired; cheek-teeth 3/3, usually brachyodont with cusps in 3 longitudinal rows. p. 130

Gliridae (dormice). Small; long bushy tail (except *Myomimus*); cheek-teeth 4/4, transversly ridged; mainly arboreal. p. 143

Seleviniidae (desert dormouse). Mouse-like, with most of the characters of the Gliridae but cheek-teeth only 3/3 and very small; tail long and thinly haired. p. 147

Zapodidae (jumping mice). Small, mouse-like; tail longer than head and body; hind feet normal or slightly enlarged but metatarsals always separate; cheek-teeth 3–4/3, brachyodont or slightly hypsodont; infraorbital canal large. p. 147

Dipodidae (jerboas). Hind feet greatly elongate (usually much longer than head and at least five times as long as front feet), metatarsals usually fused; locomotion bipedal; tail usually much longer than head and body; cheek-teeth 3–4/3, hypsodont. p. 149

Hystricidae (porcupines). Large; dorsal pelage of long black or banded spines; cheek-teeth 4/4, hypsodont, rooted; infraorbital canal very large. p. 158

Capromyidae (coypu). Large; hind feet with four toes webbed, fifth free; tail long, thinly haired, terete; cheek-teeth 4/4, hypsodont, increase in size and diverge posteriorly. p. 159

Ctenodactylidae (gundis). Medium sized; tail short (10–20% of head and body); legs short, front feet with combs of bristles on inner two toes; cheek-teeth 4/4 or 5/5, rootless, of simple shape. p. 159

Family SCIURIDAE
Squirrels

A large and clearly defined family including all the squirrels and represented in all continents except Australia. There are two well defined subfamilies, the Petauristinae (flying squirrels) and Sciurinae (tree and ground squirrels). The Petauristinae have frequently been given family rank, e.g. by Kuznetsov (1965). They are predominantly Oriental, but one genus, *Pteromys*, is entirely Palaearctic, another, *Aeretes*, is confined to northern China and representatives of two other genera reach northern China and Japan.

Amongst the Sciurinae, tree squirrels are poorly represented in the Palaearctic Region by three indigenous species of *Sciurus* with American affinities. Ground squirrels are better represented, mainly by the Holarctic genera *Spermophilus* and *Marmota*. Classification within the Sciurinae is rather unstable at both the specific and generic level, but this does not greatly affect the delimitation of genera in the Palaearctic Region.

Moore (1959) has revised the classification of the Sciurinae using especially cranial characters to delimit tribes. He appeared to give excessive weight to one character, the number of septae in the auditory bullae, that, by his own demonstration, shows considerable variation within individual genera.

1. Gliding membranes present (subfamily Petauristinae) . . . . (2)
– Gliding membranes absent (subfamily Sciurinae) . . . . (5)
2. Smaller (head and body under 200 mm, condylobasal length under 40 mm) ***PTEROMYS*** (p. 86)
– Larger (head and body of adults over 250 mm, condylobasal length over 50 mm) (3)
3. Ears tufted; P^4 considerably larger than M^1; (hind feet usually under 60 mm, head and body usually under 300 mm) . . . . ***TROGOPTERUS*** (p. 87)
– Ears not tufted; P^4 subequal to M^1; (hind feet usually over 60 mm, head and body over 300 mm) . . . . (4)
4. Very large (head and body over 350 mm); upper incisors narrow and ungrooved ***PETAURISTA*** (p. 86)
– Smaller (head and body under 350 mm); upper incisors broad and grooved ***AERETES*** (p. 87)
5. Tail less than half length of head and body; cheek-teeth more hypsodont . (6)
– Tail more than half length of head and body; cheek-teeth less hypsodont . (8)
6. Larger (head and body over 400 mm, condylobasal length over 70 mm); skull very angular with well developed ridges and sagittal crest . . . . . ***MARMOTA*** (p. 80)
– Smaller (head and body under 400 mm, condylobasal length under 65 mm); skull not angular and without a strong sagittal crest . . . . (7)
7. Claws very long, well over half length of digits; distal half of tail black below; palate extending far behind teeth . . . . ***SPERMOPHILOPSIS*** (p. 79)
– Claws not exceptionally long, not more than half length of digits; distal half of tail not black below; palate not extending far behind teeth . . ***SPERMOPHILUS*** (p. 82)
8. Back with multiple stripes . . . . (9)
– A single pale stripe on each flank . . . . ***XERUS*** (p. 79)
– Not striped . . . . (11)
9. Larger (head and body up to 220 mm); pelage harsh; mid-dorsal stripe pale; palate extending far behind teeth (N.W. Africa) . . . ***ATLANTOXERUS*** (p. 79)
– Smaller (head and body under 150 mm); pelage soft; mid-dorsal stripe dark; palate normal (10)
10. Ears with black and white tufts; infraorbital foramen constricted to form a canal ***CALLOSCIURUS*** (*C. swinhoei*) (p. 78)
– Ears without tufts; infraorbital foramen large and open . . . ***TAMIAS*** (p. 85)
11. Middle 2 toes of front feet equal in length; skull very flat ***SCIUROTAMIAS*** (p. 79)
– Middle 2 toes unequal; skull not very flat . . . . (12)
12. Pelage usually lighter below than above; auditory bullae usually with 2 or 3 septa; baculum without an accessory blade . . . . ***SCIURUS*** (p. 76)
– Pelage usually as dark below as above; bullae usually with a single septum; baculum with an accessory blade on the dorsal surface . . . . . ***CALLOSCIURUS*** (p. 78)

Genus ***SCIURUS***

Sciurus Linnaeus, 1758: Type-species *S. vulgaris* L. =*Aphrontis*. Incl. *Tenes* =*Oreosciurus*.

A large genus of very doubtful limits, although certainly including the three indigenous

Palaearctic species. As normally employed it also contains numerous species in North and South America. Ellerman (1940) believed that *Callosciurus*, the dominant genus of tree squirrels in the Oriental Region, should also be included, although in practice he kept them separate on grounds of convenience. Simpson (1945) on the other hand placed *Sciurus* and *Callosciurus* in different tribes and this was supported, on the basis of cranial differences, by Moore (1959).

One well established introduction from North America is included here besides the three Palaearctic species. Of the latter, *S. anomalus* is certainly distinct and has been separated subgenerically (*Tenes*); the Japanese *S. lis* has been included in *S. vulgaris*, e.g. by E & M-S, but I believe there are grounds for treating it as specifically distinct.

1. Ears tufted in winter; no pads on posterior parts of hind feet; (ventral pelage white; 5 upper cheek-teeth in each jaw) . . . . (2)
– Ears never tufted; a plantar pad on the posterior part of the sole of each hind foot (3)
2. Dorsal pelage red, grey or black, never finely speckled; ear tufts (in winter pelage) long, usually contrasting in colour with rest of dorsal pelage; hairs of tail without white tips **S. vulgaris**
– Dorsal pelage yellowish olive, finely speckled (darker in summer); ear tufts (in winter pelage) short and same colour as rest of dorsal pelage; hairs of tail with white tips (Japan) **S. lis**
3. Ventral pelage buff; 4 upper cheek-teeth in each jaw (S.W. Asia) . **S. anomalus**
– Ventral pelage white; 5 upper cheek-teeth in each jaw . . . **S. carolinensis**

Sciurus vulgaris Red squirrel

Sciurus vulgaris Linnaeus, 1758. Upsala, Sweden.

S. v. vulgaris. Continental range, probably also Sakhalin and Hokkaido. =*typicus*. Incl. *albus*, *alpinus* (Pyrenees), *altaicus* (Altai), *ameliae*, *anadyrensis* (Anadyr, N.E. Siberia), *arcticus*, *argenteus*, *baeticus*, *balcanicus*, *bashkiricus*, *brunnea*, *calotus*, *carpathicus*, *chiliensis* (Hopei, China), *coreae* (Korea), *coreanus*, *croaticus*, *dulkeiti* (Gt Shantar Is., E. Siberia), *europaeus*, *exalbidus* (W. Siberia), *fedjushini*, *formosovi*, *fuscoater* (Germany), *fusconigricans* (Transbaikalia), *fuscorubens* (E. Siberia), *gotthardi*, *graeca*, *infuscatus* (central Spain), *istrandjae*, *italicus*, *jacutensis*, *jenissejensis*, *kalbinensis*, *kessleri*, *lilaeus* (Greece), *mantchuricus* (Manchuria), *martensi*, *meridionalis*, *nadymensis*, *niger*, *nigrescens*, *numantius*, *ognevi*, ?*orientis* (Hokkaido), *rhodopensis*, *rufus*, ?*rupestris* (Sakhalin), *russus*, *rutilans*, *segurae*, *silanus*, ?*subalpinus*, ?*talahutky*, *ukrainicus*, *uralensis*, *varius*, *vilensis*.

*Sciurus vulgaris golzmajeri** Smirnov, 1960. Ozerninsk, near Zveringolovsk, Kurgan region, Transural Territory USSR.

*Sciurus vulgaris hoffmanni** Valverde, 1968. La Perdiz, Sierra de Espuna, Alhama de Murcia, Spain.

S. v. leucourus Kerr, 1792. Britain and Ireland. =*leucurus*.

RANGE (Map 34). All forested parts of the Palaearctic Region south to the Mediterranean, the southern Urals, the Altai, central Mongolia, Manchuria and Korea; Britain and Ireland; Sakhalin and Hokkaido. Introduced in the Caucasus, in Crimea and in several parts of Kazakhstan and Kirgizia south of the original range (not shown on map).

REMARKS. This species is exceedingly variable in colour, with considerable regional variation superimposed on a striking polymorphism and equally striking seasonal differences. The dorsal colour ranges from red or black to dark or pale grey, and the tail, feet and ear tufts may be concolorous or may contrast with the back. On the mainland, however, the range is likely to be continuous, with the possible exception of some marginal southern isolates. Until the range and variation are mapped and illustrated in greater detail it seems meaningless to list subspecies. For a recent attempt see Sidorowicz (1971). The situation is further complicated by introductions and transplantations of animals from one part of the range to another.

The British form is moderately differentiated by the progressive bleaching of the tail after each moult. The other insular forms on Sakhalin and Hokkaido may deserve subspecific rank, but they are undoubtedly much more similar to the continental forms than they are to *Sciurus lis* of the remainder of Japan. The latter form was included as a race of *S. vulgaris* by E & M-S but is here given specific rank.

***Sciurus lis*†** Japanese squirrel

Sciurus lis Temminck, 1844. Honshu, Japan.

RANGE (Map 34). Honshu, Shikoku and Kyushu, Japan.

REMARKS. E & M-S treated this as a subspecies of *S. vulgaris*. I prefer to give it specific rank, as did Imaizumi (1960), since it shows several characters (see key) that are quite outwith the range of variation of *S. vulgaris* elsewhere, in spite of the enormous variation found in that species.

Sciurus anomalus Persian squirrel

Sciurus anomalus Schreber, 1785. Georgia, Caucasus.
S. a. anomalus. Caucasus and Asia Minor. Incl. *caucasicus*, *russatus*.
S. a. syriacus Ehrenberg, 1828. Syria. Incl. *historicus*.
S. a. pallescens (Gray, 1867). Iran. Incl. *fulvus*.

RANGE (Map 34). Caucasus and Asia Minor south to Palestine and through the Zagros Mts to Shiraz; the island of Lesbos, Greece (Ondrias, 1966).

REMARKS. For a detailed comparison between this species and *S. vulgaris* in the Caucasus see Polyakova (1962).

Sciurus carolinensis * American grey squirrel

Sciurus carolinensis Gmelin, 1788. Carolina, USA.

RANGE (Map 34). Eastern half of USA. Introduced in Britain and Ireland.

REMARKS. This species has been separated generically in *Neosciurus*, e.g. recently by Brink (1967), but it seems difficult to find characters that would justify this even at a subgeneric level.

Genus *CALLOSCIURUS*

Callosciurus Gray, 1867. Type-species *Sciurus rafflesi* Vigors & Horsfield = *C. prevosti* Desmarest.

The dominant genus of squirrels in the Oriental Region with about 20 species. One species extends far enough north in China to be included here and another has been introduced in Japan.

For a revision of the Indo-Chinese species see Moore & Tate (1965). Their treatment of species is accepted here, but I have not followed them in giving the Chinese striped squirrels separate generic rank (*Tamiops*).

Dorsal pelage striped . . . . ***C. swinhoei***
Dorsal pelage unstriped . . . . ***C. flavimanus***

Callosciurus flavimanus Belly-banded squirrel

Sciurus flavimanus Geoffroy, 1831. Annam, Indo-China.
C. f. thaiwanensis† (Bonhote, 1901). Taiwan; introduced to the island of Oshima, Japan.

RANGE. Thailand and E. Burma through Indo-China south of the Yangtze; Taiwan, and introduced from there to Oshima, Japan (Kuroda, 1955).

REMARKS. This form was allocated to *C. caniceps* by E & M-S but is here placed in *C. flavimanus* following Moore & Tate (1965).

Callosciurus swinhoei Swinhoe's striped squirrel

Sciurus macclellandi var. *swinhoei* Milne-Edwards, 1874. Moupin, Szechuan.
C. s. vestitus (Miller, 1915). Hopei, type-locality Hsinlungshan, N.E. of Peking.
Also several other named forms from S.W. China.

RANGE (Map 35). N. Hopei and much of S.W. China.

REMARKS. Moore & Tate (1965) placed this species in the genus *Tamiops* which was given only subgeneric rank by E & M-S. Forms in S.E. China and Indo-China that were included in this species by E & M-S were separated as *Tamiops maritimus* by Moore & Tate.

Genus *SCIUROTAMIAS*

Sciurotamias Miller, 1901. Type-species *Sciurus davidianus* Milne-Edwards.

A genus of two species in China, one of which extends far enough north to be included here.

Sciurotamias davidianus Père David's rock squirrel

Sciurus davidianus Milne-Edwards, 1867. Near Peking, China.
S. d. davidianus. Hopei and N. Shansi. Incl. *latro*.
S. d. consobrinus Milne-Edwards, 1868. Shansi and Szechuan. Incl. *collaris*, *owstoni*, *saltitans* (Hupeh), *theyeri*.

RANGE (Map 35). Rocky mountain sides from S. Jehol and Hopei through Shansi to W. Szechuan and Hupeh; also one record from S.W. Kweichow.

REMARKS. The above classification and the map are based on the detailed account by Moore & Tate (1965).

Genus *ATLANTOXERUS*

Atlantoxerus Major, 1893. Type-species *Sciurus getulus* L.

A monospecific genus closely related to the genus *Xerus*.

Atlantoxerus getulus Barbary ground squirrel

Sciurus getulus Linnaeus, 1758. Morocco. Incl. *trivittatus*.

RANGE (Map 35). Morocco and Algeria.

Genus *XERUS**

Xerus Ehrenberg, 1833: folio ee. Type-species *Sciurus (Xerus) brachyotis* Ehrenberg =*Sciurus rutilus* Cretzschmar. Incl. *Euxerus* Thomas, 1909.

A genus of subsaharan ground squirrels with about four species, one of which extends marginally north of the Sahara. *Euxerus* (type-species *Sciurus erythropus* Desmarest) has frequently been given separate generic rank.

Xerus erythropus* Geoffroy's ground squirrel

Sciurus erythropus Desmarest, 1817. Senegal. Including many synonyms from south of the Sahara.

RANGE (Map 36). Savanna zone from Senegal to Ethiopia and Kenya; also in S.W. Morocco inland from Agadir (Blanc & Petter, 1959).

REMARKS. Blanc & Petter (1959) considered the Moroccan animals indistinguishable from the nominate race from Senegal. This species has frequently been separated from *Xerus* as the sole member of the genus *Euxerus*.

Genus *SPERMOPHILOPSIS*

Spermophilopsis Blasius, 1884. Type-species *Arctomys leptodactylus* Lichtenstein.

A monospecific genus, related to the African *Xerus* but differing especially in the extreme development of the claws.

Spermophilopsis leptodactylus Long-clawed ground squirrel

Arctomys leptodactylus Lichtenstein, 1823. N.W. of Bokhara, Uzbekistan.
S. l. leptodactylus. Karakum and Kyzylkum deserts, Turkestan. Incl. *turcomanus*.
S. l. bactrianus Scully, 1888. N. Afghanistan (type-locality) and adjacent parts of Russian Turkestan. Incl. *schumakovi*.
*S. l. heptopotamicus** Heptner & Ismagilov, 1952.
Spermophilopsis leptodactylus heptopotamicus Heptner & Ismagilov, 1952. Bakamas, R. Ili, Taukum, Zabalkhash, Kazakhstan, USSR.

RANGE (Map 36). Russian Turkestan from south side of L. Balkhash to S.E. shore of Caspian Sea and south to N. Afghanistan.

Genus *MARMOTA*
Marmots

Marmota Blumenbach, 1779. Type-species *Mus marmota* L. = *Arctomys*. (*Marmota* was first used by Frisch (1775), but this work was rejected by opinion 258 (1954) of the International Commission.)

A clearly defined Holarctic genus with about five species in N. America and at least five in the Palaearctic Region. Classification within the genus is very confused and unstable. At one extreme E & M-S recognized three Palaearctic species and Kuznetsov (1965) recognized only three in the USSR. At the other extreme Gromov *et al.* (1965) recognized eight Palaearctic species.

There is general agreement that *M. caudata* (Himalayas etc.) is distinct, and all Russian authors recognize (usually implicitly) the European *M. marmota*, as a distinct species. *M. menzbieri*, with a small range in the western Tien Shan, is also given specific rank, although sometimes tentatively, by all Russian authors. The remainder form a series of allopatric forms from the steppes of European Russia through the mountains of central Asia to N.E. Siberia and the Himalayas that Kuznetsov (1965) treated as a single species, *M. bobak*. The group in N.E. Siberia, *camtschatica* etc., was excluded from *M. bobak* by E & M-S who considered it conspecific with the European *M. marmota*. In view of the uncertainty it is here tentatively kept separate. The case for treating the Mongolian and Altai forms (*siberica* and *baibacina*) as conspecific with *M. bobak* of the western steppes seems stronger, although these are likewise given specific rank by some Russian authors, e.g. Gromov *et al.* (1965). The Himalayan form *himalayana* is given specific rank by Gromov but is not treated by other Russian authors. Examination of specimens in the British Museum suggests that, if *M. bobak* is extended to include *baibacina*, *sibirica* and *centralis*, then *himalayana* should also be included.

The arrangement adopted here is provisional. It is that used by Bibikov (1968) in his review of all Palaearctic marmots, except that he kept *himalayana* as a distinct species, without, however, a clear diagnosis.

1. Tail long, about half length of head and body (with terminal hair); (pelage bright yellow-brown below, yellow or strongly marked with black above; pelage course) **M. caudata**
– Tail short, less than half head and body . . . . (2)
2. Top of head black or very dark brown, contrasting strongly with rest of dorsal and ventral pelage . . . . (3)
– Top of head not very dark (but may be speckled with black or uniformly brown or rufous) (4)
3. End of muzzle white; pelage harsh (Europe) . . . . **M. marmota**
– Muzzle dark, not contrasting with top of head; pelage soft (N.E. Siberia) **M. camtschatica**
4. Smaller, head and body under 450 mm; pale face and side of neck sharply delimited from darker crown and nape (western Tien Shan) . . . . **M. menzbieri**
– Larger, head and body of adults over 450 mm; face and side of neck not sharply delimited from crown and nape . . . . **M. bobak**

Marmota marmota Alpine marmot

Mus marmota Linnaeus, 1758. Alps.
M. m. marmota. Alps. = *alpina*, *marmotta*. Incl. *alba*, *nigra*, *tigrina*.
*M. m. latirostris** Kratochvil, 1961. Tatra Mts.
Marmota marmota latirostris Kratochvil, 1961. Krivansky Zleb, Tatra Mts, Czechoslovakia.

RANGE (Map 36). Alps, Carpathians and Tatra Mountains; introduced in the central Pyrenees (Couturier, 1955).

REMARKS. E & M-S also included in this species the Asiatic montane forms here separated in *M. camtschatica*, *M. bobak* and *M. menzbieri*. These are excluded from *M. marmota* by all Russian authors.

***Marmota camtschatica*†** Black-capped marmot

Arctomys baibak var. *camtschatica* Pallas, 1811.
M. c. camtschatica. Kamchatka.
M. c. bungei (Kastchenko, 1901). Verkhoyansk Mts. Incl. *cliftoni*.
M. c. doppelmayri Birula, 1922. E. of L. Baikal.
Perhaps other races in N. America, e.g. *olympus*, *vancouverensis*.

RANGE (Map 36). Mountains of N.E. Siberia from L. Baikal to Kamchatka.

REMARKS. Included in *M. marmota* by E & M-S, in *M. bobak* by Kuznetsov (1965), but given specific rank by other Russian authors. The suggestion that the American forms *olympus* and *vancouverensis* might be conspecific was made by Liapounova & Vorontsov (1969), mainly on the basis of chromosome data (2n = 40, N.F. = 68 in *M. c. camtschatica*). On serological grounds it was considered the most distinctive of the Siberian species by Baranov & Vorontsov (1973).

***Marmota menzbieri*†** Menzbier's marmot

Arctomys menzbieri Kashkarov, 1925. W. Tien-Shan Mts.
M. m. menzbieri. Talass Alatau.
*M. m. zachidovi** Petrov, 1963. Chatkal Ridge.
  Marmota menzbieri zachidovi Petrov, 1963. Chatkal Ridge, N. Tien-Shan.

RANGE (Map 36). Confined to the western Tien-Shan Mountains in S.E. Kazakhstan.

REMARKS. Not seen. Tentatively included in *M. marmota* by E & M-S but treated as a separate species by all Russian authors including Kuznetsov (1965) who in other respects tended to minimize the number of species.

Marmota bobak Bobak marmot

Mus bobak Müller, 1776. Poland.
M. b. bobak. European range east to Volga. =*bobac*, *baibac*. Incl. *arctomys*.
  *Marmota bobac kozlovi** Fokanov, 1966. Volsk dist., Saratov region, European Russia.
M. b. tschaganensis Bazhanov, 1930. S. Urals to N. Kazakhstan. =*schaganensis*.
*M. b. aphanasievi** Kuznetsov, 1965.
  M[*armota*] *b*[*obak*] *aphanasievi* Kuznetsov, 1965. Kyzyl-Raya and neighbouring ranges, Kazakhstan.
M. b. baibacina† (Brandt, 1843). Altai Mts.
  '*M. b. ognevi** Scalon, 1950.' S.W. Altai. (Original citation not seen – *fide* Kuznetsov (1965).)
M. b. centralis† (Thomas, 1909). Tien Shan.
*M. b. caliginosus** Bannikov & Scalon, 1949. Mongolian Altai.
  Marmota sibirica caliginosus Bannikov & Scalon, 1949. Shar Us, Hangay Mts, Mongolia.
M. b. sibirica (Radde, 1862). Transbaikalia, N. Mongolia and Manchuria. Incl. *dahurica*.
M. b. himalayana (Hodgson, 1841). Himalayas and W. China; type-locality Nepal. Incl. *hemachalanus*, *hodgsoni*, *robustus* (Szechuan), *tataricus*, *tibetanus*.

RANGE (Map 36). Steppes of southern Russia and Kazakhstan, formerly west to Poland and Rumania but range now much reduced in the west; mountains of central Asia from Altai south to Himalayas and east to Szechuan and Manchuria. Introduced to the Caucasus from the Altai Mts (*M. b. baibacina*).

REMARKS. Kuznetsov (1965) also included in this species the forms here separated in *M. camtschatica*. Many Russian authors go to the other extreme and recognize *baibacina* (with *centralis*) and *sibirica* (with *caliginosus*) as separate species. The arrangement followed here agrees with those of Bannikov (1954) and Bibikov (1968), except that the latter excluded *himalayana*.

Of the recently described forms, *kozlovi* seems to be based on average differences and is unlikely to be a clearcut, discrete race. *M. b. aphanasievi* was based entirely on colour which is intermediate between that of *M. b. tschaganensis* and *M. b. baibacina* – it is retained tentatively in view of the possibility that *bobak* and *baibacina* may be specifically distinct as claimed for example by Sludskii (1969).

Marmota caudata Long-tailed marmot

Arctomys caudata Jacquemont, 1844. Kashmir.
M. c. caudata. Kashmir.
M. c. aurea (Blanford, 1875). Mountains of Turkestan, type-locality west of Yarkand, Sinkiang. Incl. *flavinus* (E. of Samarkand), *littledalei* (Altai Mts), *stirlingi* (Chitral, Pakistan).
M. c. dichrous (Anderson, 1875). Afghanistan (N. of Kabul).

RANGE (Map 36). Western Tien Shan south through Pamirs to Hindu Kush and Kashmir.

REMARKS. A clear-cut and universally recognized species.

Genus *SPERMOPHILUS*

Spermophilus F. Cuvier, 1825. Type-species *Mus citellus* L. = *Citellus*. Incl. *Anisonyx* Rafinesque, *Colobotis*, *Urocitellus*.

The dominant genus of ground squirrels in the Palaearctic and Nearctic Regions with about nine species in the Palaearctic and about twice as many in N. America.

This genus has more often been called *Citellus*. E & M-S, while recognizing that *Citellus* was invalid, continued to use it in the hope that it might be validated by the International Commission. However since then Oken's *Lehrbuch der Naturgeschichte* (1816), from which *Citellus* dates, has been rejected by the International Commission (opinion 417, 1956) and the correct but hitherto little-used name *Spermophilus* has been generally adopted by American authors (Anon., 1968). There therefore seems no alternative but to use it for the Palaearctic species also.

According to the classification of Moore (1959) the only genera very closely related to *Spermophilus* are *Cynomys* Rafinesque, 1817 and *Ammospermophilus* Merriam, 1892, both American and both generally recognized as distinct.

Classification of the Palaearctic species is very unstable. The most recent detailed revision is that of Gromov *et al.* (1965) who used a highly split classification recognizing 13 recent Palaearctic species. However, recent Russian work provides fairly strong grounds for supposing that the extreme reduction of species in the classification of E & M-S and of Kuznetzov (1965) represents an oversimplification. A summary of the abundant chromosome data for the genus was given by Vorontsov & Liapounova (1970).

The classification presented here is not original but is a compromise based on the various recent Russian classifications. Some aspects of it must be considered provisional and the following key is only an indication of the characters involved.

1. Tail long, length (without hairs) over one third that of head and body . . . (2)
– Tail short, less than one third head and body . . . . (3)
2. Back boldly spotted, tail usually under 40% of head and body . . . . ***S. parryi***
– Back finely spotted, tail usually over 40% of head and body . . ***S. undulatus***
3. Soles of hind feet hairy . . . . (4)
– Soles of hind feet naked except at sides and on extreme tip of heel . . . (6)
4. Back boldly spotted . . . . ***S. souslicus***
– Back unspotted or very faintly spotted . . . . (5)
5. Tail usually under 25% of head and body; tip of tail with boldly contrasting black subterminal and white terminal zones (Mongolia etc.) . . . . . ***S. dauricus***
– Tail usually over 25% head and body; black and white zones of tail rather obscure ***S. citellus***
6. Larger: hind feet usually over 42 mm; condylobasal length usually over 50 mm (7)
– Smaller: hind feet under 42 mm; condylobasal length under 50 mm . . . . (8)
7. Dorsal pelage yellowish, finely speckled due to black-tipped hairs, but without any light mottling . . . . ***S. fulvus***
– Dorsal pelage more greyish, with slight pale mottling . . . . . . ***S. major***
8. Larger: hind feet about 40 mm, condylobasal length about 45 mm; tail more than 25% of head and body; cheek-teeth large, row longer than diastema. . . . ***S. relictus***

– Smaller: hind feet about 32 mm, condylobasal length about 38 mm; tail less than 25 % of head and body; cheek-teeth small, row shorter than diastema . . . . ***S. pygmaeus***

Spermophilus citellus European souslik

Mus citellus Linnaeus, 1766. Austria.
S. c. citellus. Europe. =*citillus.* Incl. *gradojevici, istricus, karamani, laskarevi.*
*Citellus citellus martinoi** Peshev, 1955. Rhodopen Mts, Bulgaria.
*Citellus citellus balcanicus** Markov, 1957. Lukorsko, Sofia dist., Bulgaria.
*Citellus citellus thracius** Mursaloğlu, 1964. S.E. slope of Muraltepe, near Yenibedir, Lüleburgaz, Kirklareli, European Turkey.
S. c. xanthoprymnus (Bennett, 1835). Caucasus, Asia Minor and Palestine; type-locality Erzerum, Turkey. Incl. *concolor* Thomas, *schmidti.*

RANGE (Map 37). Grassland from S.E. Germany and S.W. Poland to Rumania and the adjacent part of Ukraine; from the south side of the Caucasus through Asia Minor to Palestine.

REMARKS. This species has been held, e.g. by E & M-S and by Kuznetzov (1965), to include also forms in central Asia, Mongolia and China, namely those included here in *S. dauricus* and *S. relictus.* Several other Russian authors, however, treat these as specifically distinct (e.g. Gromov *et al.*, 1965), and studies of chromosomes support this (Vorontsov & Liapounova, 1969, 1970).

These results also suggest that *citellus* and *xanthoprymnus* might be specifically distinct. They are morphologically distinguishable and are therefore at least very well differentiated subspecies in contrast to the forms listed above under *S. c. citellus* which are based on slight differences of proportions with no proof of discontinuity.

Spermophilus dauricus† Daurian souslik

Spermophilus dauricus Brandt, 1844. Tarei-Nor, *c.* 250 miles E. of L. Baikal.
S. d. dauricus. N.E. Mongolia, and adjacent parts of Siberia and China. Incl. *mongolicus* (Hopei), *ramosus* (Manchuria), *umbratus* (Mongolia), *yamashinai.*
S. d. alaschanicus Büchner, 1888. S. Alashan, Kansu, Incl. *dilutus* (Mongolian Altai), *obscurus, siccus* (Shansi).

RANGE (Map 37). Steppes of E. Mongolia and adjacent part of Siberia, and N.W. China from Manchuria to the Nan Shan.

REMARKS. These forms were included in *S. citellus* by E & M-S and by Kuznetzov (1965), but most Russian authors have given them specific rank and this is supported by data on chromosomes (see remarks under *S. citellus*). Ognev (1947), Bannikov (1954) and Gromov *et al.* (1965) all gave *alaschanicus* specific rank as well, but it is here tentatively treated as a subspecies of *S. dauricus.* The allocation of synonyms to these two subspecies follows these last three authors who, however, used most of them in a subspecific sense.

Spermophilus suslicus Spotted souslik

Mus suslica Güldenstaedt, 1770. Voronej Steppes, Russia. Incl. *averini, guttatus, guttulatus, leucopictus, meridioccidentalis, odessana, ognevi* (Rumania), *volhyensis* (E. Poland).
*Citellus suslicus boristhenicus** Pusanov, 1958. Between the rivers Dnestr and Bug, Ukraine.

RANGE (Map 37). Steppe zone of S. European Russia from the Volga and Don west to S.E. Poland and N.E. Rumania.

REMARKS. A clearly defined species. The range abuts with that of *S. pygmaeus* to the southeast and some evidence of hybridization has been reported from the boundary (Denisov, 1961).

Spermophilus pygmaeus Little souslik

Mus citellus var. *pygmaeus* Pallas, 1779. Between Emba and Ural Rivers, N.E. of Caspian Sea.
S. p. pygmaeus. Turkestan to N. of Caucasus and Crimea. Incl. *arenicola* Rall, *atricapilla, binominata, brauneri* (Crimea), *ellermani, ?flavescens, herbicola, kazakstanicus, kalabuchovi, musogaricus, nikolskii, orlovi* Ellerman, *pallidus* Orlov & Feniuk, *planicola* (Caucasus), *ralli* Heptner, *satunini, septentrionalis.*
S. p. musicus Ménétries, 1832. Caucasus, type-locality at foot of Mt Elbruz. Incl. *boehmi.*

RANGE (Map 37). Steppes from Crimea and the Caucasus east through most of Kazakhstan almost to L. Balkhash. The range abuts that of *S. suslicus* in the northwest and that of *Spermophilopsis leptodactylus* in the southeast.

REMARKS. This species is delimited as above by most Russian authors, but Gromov *et al.* (1965) gave *musicus* specific rank. All recent Russian authors agree in omitting from *S. pygmaeus* several forms included in it by E & M-S, namely *brevicauda*, *intermedius* and *carruthersi* (here allocated to *S. major*).

All Russian authors recognize most of the forms listed under *S. p. pygmaeus* as subspecies. Most of these, however, are based on slight differences of colour and size and extensive intergradation is admitted.

Natural hybridization with *S. major* has been claimed (Denisov, 1963) and also with *S. suslicus* (Denisov, 1961).

Spermophilus major Russet souslik

Mus citellus var. *major* Pallas, 1779. Samara, E. of R. Volga, European Russia.
S. m. major. Western part of range, east to the rivers Tobol and Ishim. Incl. *rufescens*.
*C(itellus) major argyropuloi** Bazhanov, 1947. River Uil, N.E. of Caspian Sea.
S. m. erythrogenys Brandt, 1841. N.E. Kazakhstan. Incl. *ungae* (Omsk).
*C(itellus) erythrogenys brunnescens** Belyaev, 1954. ?Locality.
S. m. brevicauda Brandt, 1843. S.E. Kazakhstan (type locality) and Mongolia. Incl. *carruthersi*† (Dzungaria), *intermedius*, *pallidicauda*† (Mongolia), ?*selevini*.
*Citellus major heptneri** Vasilieva, 1964b. ?Tien Shan.
*Citellus pygmaeus iliensis** Belyaev, 1945. River Ili.

RANGE (Map 38). From the Rivers Volga and Kama to Novosibirsk and the upper Yenesei, and through most of northern Kazakhstan east to L. Balkhash and the Mongolian Altai.

REMARKS. The taxonomy of this group has been very unstable. The northeastern form *erythrogenys* has frequently been given specific rank, e.g. by Gromov *et al.* (1965) who included *brevicauda* as a race but kept *ungae* in *S. major*, and by Sludskii (1969) who included *ungae* but excluded *intermedius* (i.e. *brevicauda)*.

E & M-S tentatively included *relictus* and gave *pallidicauda* specific rank. All Russian authors agree in excluding *relictus* whilst Bannikov (1954) and others include *pallidicauda* in *S. erythrogenys*. The arrangement shown above must therefore be considered very provisional.

Natural hybridization with *S. pygmaeus* has been claimed by Denisov (1963).

***Spermophilus relictus*†** Tien Shan souslik

Citellus musicus relictus Kashkarov, 1923. Tien Shan Mts, W. Kirghizia.
S. r. relictus. W. Tien Shan, between Tashkent and Issyk Kul.
*S. r. ralli** (Kuznetzov, 1948). Central Tien Shan, around east end of Issyk Kul.
Citellus relictus ralli Kuznetsov, 1948 (N.V.). Issyk Kul.

RANGE (Map 38). Mountains of Kirghizia from near Tashkent to a little way east of Issyk Kul (detailed map in Vasilieva, 1964a).

REMARKS. This form was tentatively listed under *S. major* by E & M-S. No Russian authors agree with this; it is given specific rank by most and was included as a race of *S. citellus* by Kuznetsov (1965). Chromosome data support its separation from *S. citellus* (2n = 36, N.F. = 68 according to Vorontsov & Liapounova, 1969). It has not been seen.

Spermophilus undulatus Long-tailed souslik

(Mus citellus) var. *undulatum* Pallas, 1779. R. Selenga (Lake Baikal), E. Siberia. Incl. *altaicus*, *eversmanni* (Altai), *intercedens* (E. Transbaikalia), *jacutensis* (Yakutsk), *menzbieri* (upper Amur), *stramineus* (N.W. Mongolia), *transbaicalicus*.

RANGE (Map 38). Mountains from Tien Shan through the Altai and Mongolia to the Amur region and the River Lena.

REMARKS. This species is clearly distinct from all other Palaearctic ones except *S. parryi*. It is frequently held to include also the forms (Asiatic and American) listed here under *S. parryi*. Several Russian authors, e.g. Gromov *et al.* (1965), treat these as specifically distinct and this is supported by differences in chromosomes (Liapounova, 1969) and voice (Nikolsky, 1969). This is tentatively accepted here although the two are undoubtedly very closely related.

Spermophilus parryi* Arctic souslik

Arctomys Parryii Richardson, 1825:316. Hudson Bay, Canada.
S. p. leucostictus Brandt, 1844. Extreme N.E. Siberia, type-locality Okhotsk River. Incl. *buxtoni*.
S. p. stejnegeri (Allen, 1903). Kamchatka.
S. p. janensis (Ognev, 1937). River Jana.
*S. p. coriakorum** (Portenko, 1963). Koryak Mts, N.E. Siberia.
Citellus undulatus coriakorum Portenko, 1963. Koryak Mts (Original description N.V.: *fide* Gromov *et al.*, 1965).
Also other races in North America.

RANGE (Map 38). Siberia northeast of the River Lena, including Kamchatka; also in the tundra zone of North America from Alaska east to Hudson Bay.

REMARKS. Frequently considered conspecific with *S. undulatus*– see remarks under that species.

Spermophilus fulvus Large-toothed souslik

Arctomys fulvus Lichtenstein, 1823. E. of Mugodshary Mts, Kazakhstan. Incl. *concolor* Fischer, *concolor* Geoffroy, *giganteus*, *hypoleucos* (N. Iran), *nanus*, *nigrimontanus* (Karatau Mts), *orlovi*, *oxianus* (S. Turkestan), *parthianus*.
[*Mus citellus*] *maximus** Pallas, 1779: 122. Lower River Ural.

RANGE (Map 38). Kazakhstan from the Caspian Sea and the River Volga east to Lake Balkhash and south through Turkestan to N. Iran and N. Afghanistan. Also in W. Sinkiang.

REMARKS. This is a clearly distinct species, which has for long been consistently known as *Citellus fulvus*. The name *maximus* Pallas has been quoted in synonymy by several Russian authors and was used as the name of the species by Kuznetsov (1965). This usage was quite invalid in terms of Article 23b of the International Code and the name should be considered a *nomen oblitum*. It is in any case very doubtful if its usage by Pallas could be interpreted as constituting a name.

Genus *TAMIAS*

Tamias Illiger, 1811. Type-species *Sciurus striatus* L. (a N. American species). Incl. *Eutamias*.

A predominantly Nearctic genus with about 16 species in North America, mostly in the west, and one in the eastern Palaearctic. The genus is frequently split, *Eutamias* being used for the Palaearctic and western Nearctic species, e.g. by Moore (1959) and Hall & Kelson (1959). However, this is based mainly on the presence in *Eutamias* of a rudimentary P^3 that is absent in *Tamias* s.s. This is not quite constant in *Eutamias* (e.g. Jones & Johnson (1965) found these teeth lacking in 2 out of 41 skulls of *T. sibiricus* from Korea) and does not seem to justify more than subgeneric rank. Studies of chromosomes support the concept of a single genus, with the Palaearctic species differing slightly from both of the Nearctic groups (Nadler *et al.*, 1969).

Tamias sibiricus Siberian chipmunk

Sciurus sibiricus Laxmann, 1769. Barnaul, W. Siberia.
T. s. sibiricus. Mainland except perhaps S. China. Incl. *asiaticus* (E. Siberia), *intercessor* (Shansi, China), *jacutensis*, *ordinalis* (Shensi, China), *orientalis* (Ussuri), *pallasi* (N. European Russia), *senescens* (Peking), *striatus* Pallas, *uthensis*.
T. s. albogularis (Allen, 1909). S. Shensi to Szechuan, China. Incl. *umbrosus*.

*T. s. barberi** (Johnson & Jones, 1955). S. Korea.
Eutamias sibiricus barberi Johnson & Jones, 1955b. Central National Forest, near Pup'yong-ni, Korea (37°44'N, 127°12'E).
T. s. lineatus (Siebold, 1824). Hokkaido, Sakhalin. S. Kurile Is., and perhaps the Amur region of the mainland. Incl. *okadae* (Kunashiri Is., S. Kuriles).

RANGE (Map 35). Entire Siberian taiga zone west to the White Sea; south to the Altai Mts, and in China south through Hopei to Shensi and N.W. Szechuan; Sakhalin, Hokkaido and the S. Kurile Islands.

REMARKS. There is considerable geographical variation especially in E. Asia, but most of it seems to be clinal and it is doubtful whether even the forms *albogularis* and *barberi* can be considered to be discrete subspecies.

Genus *PETAURISTA*

Petaurista Link, 1795. Type-species *Sciurus petaurista* Pallas.

A clearly defined Oriental genus of about five species, only one of them represented in the Palaearctic Region, in Japan.

Petaurista leucogenys — Japanese giant flying squirrel

Pteromys leucogenys Temminck, 1827. Kyushu, Japan.
P. l. leucogenys. Kyushu and Shikoku. Incl. *tosae* (Shikoku).
P. l. oreas Thomas, 1905. S. Honshu (type-locality Wakayama).
P. l. nikkonis Thomas, 1905. N. and central Honshu. Incl. *osiui*.
P. l. xanthotis (Milne-Edwards, 1872). Kansu to Szechuan (type-locality) and Yunnan. Incl. *filchnerinae* (Kansu).
P. l. hintoni Mori, 1923. ?Korea. =*thomasi* Kuroda & Mori, ?*watasei* (?Manchuria).

RANGE (Map 39). Honshu, Shikoku and Kyushu, Japan; W. China from Kansu to Yunnan; very doubtfully in Korea and Manchuria (see below).

REMARKS. According to Jones & Johnson (1965), the records of this species in Korea and Manchuria are based solely on the types of *thomasi* and *watasei*, both of which were skins purchased in markets and could therefore have been imported.

Genus *PTEROMYS*

Pteromys Cuvier, 1800. Type-species *Sciurus volans* L. =*Sciuropterus*.

A genus of two allopatric species, both Palaearctic. The following diagnostic characters are taken from Ellerman (1949).

Palatal foramina short, less than 10% of occipitonasal length; bullae small, less than 25% of occipitonasal length; frontals narrow, their width about 17% of occipitonasal length (Japan) . . . **P. momonga**
Palatal foramina long, over 10% of occipitonasal length; bullae large, over 25% of occipitonasal length; frontals wider, about 19–21% of occipitonasal length . . . . **P. volans**

Pteromys volans — Russian flying squirrel

Sciurus volans Linnaeus, 1758. Finland.
P. v. volans. Whole of Taiga forest. =*sibiricus*, *vulgaris*. Incl. *aluco* (Korea), *anadyrensis* (Anadyr), *arsenjevi* (S.E. Siberia), *betulinus* (W. Siberia), *gubari*, *incanus* (N.E. Siberia), *ognevi* (Volga estuary), *russicus*, *turovi* (L. Baikal), *wulungshanensis* (Manchuria).
P. v. buechneri Satunin, 1903. Kansu (type-locality) and Shansi, China.
P. v. athene (Thomas, 1907). Sakhalin.
P. v. orii (Kuroda, 1921). Hokkaido, Japan.

RANGE (Map 39). The entire coniferous forest zone from Finland and the Baltic coast of Russia to E. Siberia. South to S. Urals, Altai, N. Mongolia, Manchuria and Korea. Also in Kansu and Shansi, and on the islands of Sakhalin and Hokkaido.

Pteromys momonga — Small Japanese flying squirrel

Pteromys (Sciuropterus) momonga Temminck, 1844. Kyushu, Japan. Incl. *amygdali* (Honshu), *interventus*.

RANGE (Map 39). Honshu and Kyushu, Japan.

REMARKS. Imaizumi (1960) did not recognize a subspecific difference between the two populations nor within Honshu as indicated by E & M-S.

Genus *AERETES*

Aeretes G. Allen, 1940. Type-species *Pteromys melanopterus* Milne-Edwards.

A monosecific genus formerly included in *Petaurista* but differing in having the incisors broad and grooved.

Aeretes melanopterus

Pteromys melanopterus Milne-Edwards, 1867. N.E. Hopei, China. Incl. *sulcatus*.

RANGE (Map 39). Forests of N.E. Hopei, China.

Genus *TROGOPTERUS*

Trogopterus Heude, 1898. Type-species *Pteromys xanthipes* Milne-Edwards.

A monospecific genus closely related to *Belomys* from S.E. Asia.

Trogopterus xanthipes — Complex-toothed flying squirrel

Pteromys xanthipes Milne-Edwards, 1867. N.E. Hopei, China. Incl. *edithae* (Yunnan), *himalaicus* (S. Tibet), *minax* (Szechuan), *mordax* (Hupeh).

RANGE (Map 39). Montane forests of China from N.E. Hopei to Yunnan and adjacent parts of S.E. Tibet.

Family **CASTORIDAE**

Beavers

A clearly defined family with a single genus.

Genus *CASTOR*

Castor Linnaeus, 1758. Type-species *Castor fiber* L. =*Fiber* Duméril.

A genus of two species, formerly throughout the Holarctic. The relationship of the Palaearctic and American beavers is debatable. Most recent authors, including Hinze (1950) in his monograph, treat them as specifically distinct. Freye (1960) on the other hand treated them as conspecific. More recently Lavrov & Orlov (1973) have maintained their distinctness on the basis of chromosome number and skull form, and have shown that the Asiatic beavers are clearly *C. fiber* and not intermediate between European and American animals. Both species are included here since American beavers have been introduced in Europe.

Nasal bones rather parallel-sided, especially in anterior half, usually over 54 mm in length; interorbital width over 25 mm; 2n=48 . . . . ***C. fiber***

Nasal bones with lateral margins strongly convex, usually under 54 mm in length; interorbital width under 25 mm; 2n=40 . . . . ***C. canadensis***

Castor fiber — Eurasian beaver

Castor fiber Linnaeus, 1758. Sweden.

C. f. fiber. Europe. =*proprius*. Incl. *albicus* (R. Elbe, Germany), *albus*, *balticus*, *flavus*, *fulvus*, *gallicus*, *niger*, *solitarius*, *variegatus*, *vistulanus* (Poland).
*Castor galliae** Desmarest, 1822:278. France.

C. f. pohlei Serebrennikov, 1929. W. Siberia and Mongolia, type-locality N.E. Urals. Incl. *birulae* (W. Mongolia).
?*Castor fiber tuvinicus** Lavrov, 1969. Azas River, Upper Yenesei.

RANGE (Map 78). Formerly throughout the forested parts of the Palaearctic except for the Mediterranean zone and Japan. Greatly reduced by early 20th century. Indigenous populations survive on the Rhone and Elbe, in southern Norway, in parts of European Russia, N.W. Siberia and the Altai. It has been extensively reintroduced in W. Siberia, European Russia, Finland, Sweden, Switzerland and Germany.

REMARKS. It is probable that variation was once continuous. Hinze (1950) diagnosed six subspecies but Russian authors (Kuznetsov, 1965; Bannikov, 1954) tend to synonymize *birulae* with *pohlei* and *vistulanus* with *fiber*. Even the two subspecies retained here are not very clearly definable.

Castor canadensis* American beaver

Castor canadensis Kuhl, 1820. Hudson Bay.

RANGE. Formerly all forested parts of N. America south to N. Mexico; now absent from much of USA but reintroductions have taken place. Introduced and well established in Finland (details in Lahti & Helminen, 1974).

Family **CRICETIDAE**

As used here this family contains a number of groups of rodents, principally the hamsters, voles, gerbils and New World mice, that are generally agreed to be closely related, although these groups are variously given subfamily or family rank. Ellerman (1940–41, and in E & M-S) included all these in the Muridae but this course was opposed by Simpson (1945) and by most subsequent authors.

Palaearctic species fall into four clearly defined groups which are here treated as subfamilies.

1. Cheek-teeth extremely hypsodont, usually growing continuously without the development of roots (except *Clethrionomys*, *Ellobius*, *Dinaromys*), crown consisting of a series of connected prisms . . . . (2)
– Cheek-teeth not extremely hypsodont, rooted (except *Rhombomys*), cusps remaining independent or coalescing to form transverse laminae . . . . (3)
2. Highly modified for subterranean life: claws of front feet greatly enlarged (longer than digits), eyes very small, ears completely hidden in pelage; occiput sloping forwards, crest level with posterior ends of zygomata; infraorbital foramina large and triangular *MYOSPALACINAE* (p. 93)
– Less modified for subterranean life: claws not greatly enlarged, eyes and ears less reduced; occiput nearly vertical; infraorbital foramina small and narrow *MICROTINAE* (p. 94)
3. Upper incisors with longitudinal grooves on front surface (except in *Psammomys*); tail over 50 % of length of head and body; cusps of molar teeth usually combining to form unbroken transverse laminae . . . . *GERBILLINAE* (p. 118)
– Upper incisors without grooves; tail less than 50 % of head and body (except in *Calomyscus*); cusps of molar teeth not forming discrete transverse laminae *CRICETINAE* (p. 88)

Subfamily **CRICETINAE**
Hamsters

The hamsters form a group of about 14 species varying little in general form from the familiar golden hamster and confined to the Palaearctic Region, along with one aberrant mouse-like species, *Calomyscus bailwardi*. Ellerman (1941) included also in the subfamily Cricetinae the American mice (*Peromyscus*, *Sigmodon* and allies). More recently Reig (1972) has proposed separate subfamilies for the American genera (Peromiscinae, Sigmodontinae) and has limited the subfamily Cricetinae to the Palaearctic hamsters and *Mystromys* of southern Africa.

Apart from the distinctive *Calomyscus*, the generic divisions within the group are rather unstable and a fresh, comprehensive classification is required. Four genera are recognized here, following Ellerman, but *Mesocricetus* is frequently included in *Cricetus*, e.g. by Kuznetsov (1965).

1. Tail equal to or longer than head and body, with a terminal tuft; cheek-pouches rudimentary *CALOMYSCUS* (p. 89)
– Tail not more than half length of head and body, not tufted; cheek-pouches large (2)
2. All or part of ventral pelage black or dark grey; mammae 7 pairs or more; skull angular with prominent crests . . . . (3)

– Ventral pelage entirely pale; mammae 4 pairs; skull less angular with poorly developed crests (4)
3. Ventral pelage uniformly black; zygomatic plate broad and projecting forwards from arch *CRICETUS* (p. 89)
– Ventral pelage not uniformly black, black band on chest darker than belly; zygomatic plate narrow, not projecting in front of arch . . . . *MESOCRICETUS* (p. 92)
4. Soles of feet densely hairy; tail less than 14 mm . *PHODOPUS* (p. 89)
– Soles of feet naked or only thinly hairy; tail over 14 mm . *CRICETULUS* (p. 90)

Genus *CALOMYSCUS*

Calomyscus Thomas, 1905. Type-species *C. bailwardi* Thomas.

A distinctive monospecific genus, not very closely related to other Palaeartic cricetine genera but, according to Ellerman (1941), closer to the American genus *Peromyscus*. Its aberrant position is also demonstrated by the chromosomes, the autosomes resembling those of Palaearctic cricetines, the sex chromosomes resembling those of *Peromyscus* (Matthey, 1961).

Calomyscus bailwardi — Mouse-like hamster

Calomyscus bailwardi Thomas, 1905. S.W. Iran. Incl. *baluchi*, *elburzensis* (Elburz Mts), *hotsoni* (Baluchistan), *mystax* (Kopet Dag).
*Calomyscus bailwardi mustersi** Ellerman, 1948. Paghman, 17 miles W. of Kabul, Afghanistan.
*Calomyscus bailwardi grandis** Schlitter & Setzer, 1973. 11 km E.N.E. of Fashan, Teheran Prov., Iran.

RANGE (Map 40). Iran, Baluchistan, the more arid parts of Afghanistan and S. Turkmenistan.

REMARKS. There is considerable variation in colour, size and proportions but this seems to form a mosaic pattern and it seems unlikely that any major, discrete regional groups can be recognized.

Genus *PHODOPUS*

Phodopus Miller, 1910. Type-species *Cricetulus bedfordiae* Thomas. Incl. *Cricetiscus*.

A genus of two clearly distinct but closely related species, characterized by small size, very short tail and short, broad feet with hairy soles. The genus must be considered only marginally distinct from *Cricetulus* and *Cricetus*.

A dark, mid-dorsal stripe; dividing line on flank between dorsal and ventral pelage very sinuous *P. sungorus*
No mid-dorsal stripe; dividing line on flank almost straight . . . *P. roborovskii*

Phodopus sungorus — Striped hairy-footed hamster

Mus sungorus Pallas, 1773. Gratschefskoi, 100 km W. of Semipalatinsk, W. Siberia.
P. s. sungorus. Western part of range in Kazakhstan. =*songarus*, *songorus*.
P. s. campbelli (Thomas, 1905). Mongolia. Incl. *crepidatus*.

RANGE (Map 42). Most of Mongolia and the adjacent parts of USSR; west through Kazakhstan as far as the River Ishim and east to Manchuria.

Phodopus roborovskii — Desert hamster

Cricetulus roborovskii Satunin, 1903. Nanshan, China. Incl. *bedfordiae* (Shensi), *praedilectus* (Manchuria).
?*Phodopus przhewalskii** Vorontsov & Kriukova, 1969. Eastern desert of Zaissan Basin between Ulken-Karatal and Akzhon on left bank of the Black Irtish.

RANGE (Map 42). Western and southern Mongolia and adjacent parts of Manchuria and Kansu, south to Shensi.

REMARKS. *P. przhewalskii* is tentatively included here since it was diagnosed entirely on the basis of a distinctive karyotype.

Genus *CRICETUS*

Cricetus Leske, 1779. Type-species *Mus cricetus* L. Incl. *Hamster*, *Heliomys*.

A monospecific genus, as used here, but the species here listed in *Mesocricetus* are sometimes included, e.g. by Kuznetsov (1965).

Cricetus cricetus Common hamster

Mus cricetus Linnaeus, 1758. Germany. =*frumentarius*. Incl. *albus*, *babylonicus* (N. Caucasus), *canescens* (Belgium), *fulvus*, *fuscodorsis* (Kirgizia), *germanicus*, *jeudii*, *latycranius*, *nehringi* (Rumania), *niger*, *nigricans*, *polychroma*, *stavropolicus*, *tauricus*, *tomensis* (W. Siberia), *varius*, *vulgaris*.

RANGE (Map 41). From Belgium across central Europe, W. Siberia and N. Kazakhstan as far as the upper Yenesei and the Altai.

Genus *CRICETULUS*

Cricetulus Milne-Edwards, 1867. Type-species *C. griseus* Milne-Edwards. Incl. *Urocricetus*, *Tscherskia*, *Cansumys*, *Asiocricetus*, *Allocricetulus*.

The largest genus of hamsters with (as understood here) eight species, occupying the steppe zone from S.E. Europe to N.E. China. Three species are frequently excluded from the genus, *eversmanni* and *curtatus* in *Allocricetulus* and *triton* in *Tscherskia*. Of these the latter is the more distinctive. However pending a review of generic classification in the subfamily as a whole they are here retained in *Cricetulus*.

There are several uncertainties about the classification at the species level. Study of karyotypes suggests that there are distinct species that differ very little in morphology. For a general account of the entire genus see Flint (1966).

1. Tail relatively long, over 50 % of head and body, 60–100 mm; large: head and body up to 170 mm, hind feet 21–26 mm (E. Asia) . . . . ***C. triton***
- Tail of medium length, 25–55 % of head and body; small: head and body usually under 120 mm, hind feet under 20 mm . . . . (3)
- Tail very short, under 20 % of head and body; medium size: head and body 110–150 mm, hind feet 16–20 mm . . . . (2)
2. Greyish or yellowish brown mark on chest, between fore legs; dorsal pelage dark greyish or reddish brown . . . . ***C. eversmanni***
- No mark on chest; dorsal pelage yellowish grey . . . . . . ***C. curtatus***
3. A narrow, dark, dorsal stripe . . . . ***C. barabensis***
- No dorsal stripe . . . . (4)
4. Bullae very small, their length less than the distance between bulla and last molar (Tibet and Himalayas) . . . . (5)
- Bullae larger . . . . (6)
5. Tail bicoloured, longer (35–45 mm) . . . . ***C. kamensis***
- Tail entirely white, shorter (25–35 mm) . . . . ***C. alticola***
6. Ears dark with white margins; tail over 30 % length of head and body (31–50 mm); antero-external angles of parietal bones produced forwards into sharp points ***C. longicaudatus***
- Ears without white margins; tail under 30 % of head and body (18–35 mm); antero-external angles of parietals rounded . . . . ***C. migratorius***

Cricetulus migratorius Grey hamster

Mus migratorius Pallas, 1773. Lower R. Ural. Incl. *accedula*, *arenarius*, *atticus* (Greece), *bellicosus*, *caesius*, *cinerascens* (Syria), *cinereus*, *caerulescens* (Pamirs), *falzfeini*, *fulvus* (Sinkiang), *griseiventris* Thomas, *griseus* Kashkarov, *isabellinus*, *murinus*, *neglectus*, *ognevi*, *pamirensis*, *phaeus*, *pulcher*, *sviridenkoi*, *vernula*, *zvieresombi*.

*Cricetulus migratorius elisarjewi** Afanasiev, 1953:237. Near Shemonaikh, Altai. (Perhaps a *nomen nudum*).

RANGE (Map 40). Steppes from southern European Russia (and marginally in Greece, Rumania and Bulgaria) through Kazakhstan to southern Mongolia and Sinkiang. North almost to Moscow; south to Palestine, most of Iran and central Afghanistan.

REMARKS. Size and colour are very variable and many races have been described. Some of the more marginal ones may be well differentiated but most are likely to represent local responses to habitat and until the variation can be mapped it does not seem helpful to recognize subspecies.

It has been suggested by Flint (1966) that the forms *lama* and *alticola* might be conspecific but the magnitude of the difference in the auditory bullae makes this rather unlikely.

Cricetulus barabensis Striped hamster

Mus barabensis Pallas, 1773. Pavlovsk, Barnaul, Siberia.
C. b. barabensis. W. Siberia. Incl. *furunculus*.
C. b. griseus Milne-Edwards, 1867. Mongolia and N. China. Incl. *mongolicus*, *obscurus*.
C. b. fumatus Thomas, 1909. Manchuria. Incl. *manchuricus*.
C. b. ferrugineus Argyropulo, 1941. Ussuri Region.
*C. b. pseudogriseus** Orlov & Iskhakova, 1975. N.E. Mongolia.
Cricetulus pseudogriseus Orlov & Iskhakova, 1975. Kyakhta Region, S. Buryatskaya ASSR, USSR.

RANGE (Map 41). Steppes of S. Siberia from R. Irtish to the Ussuri, south to Mongolia, N. Shansi and Manchuria.

REMARKS. The Chinese form *griseus* was treated as specifically distinct by Flint (1966) because of the difference in chromosomes demonstrated by Matthey (1960): *barabensis* – 2n = 20, *griseus* – 2n = 22. However Matthey himself considered the difference to warrant only subspecific rank. *C. pseudogriseus* was also diagnosed by karyotype. Orlov & Iskhakova (1975) recognized these as three allopatric species and presented a detailed distribution map.

Cricetulus longicaudatus Lesser long-tailed hamster

Cricetus (Cricetulus) longicaudatus Milne-Edwards, 1867. N. Shansi, China. Incl. *andersoni*, *dichrootis* (Nan Shan), *griseiventris* Satunin (Mongolia), *kozhantscikovi* (Sayan Mts), *nigrescens* (Hopei).
*Cricetulus longicaudatus chiumalaiensis** Wang & Cheng, 1973. Chiumali Dist. (Sewukou Valley), Chinghai Province, China.

RANGE (Map 40). Mongolia and adjacent parts of USSR (Sayan Mts) and China (Hopei, Shansi, Kansu). Perhaps also Chinghai and parts of Tibet.

REMARKS. Flint (1966) suggested that the forms *kamensis* and *kozlovi* might be conspecific with this species but they are here included under *C. kamensis* following Wang & Cheng (1973).

Cricetulus kamensis Tibetan hamster

Urocricetus kamensis Satunin, 1903. R. Moktschjun, Mekong Dist, Tibet. Incl. *kozlovi* (Nan Shan), *lama* (Lhasa, Tibet), *tibetanus*.

RANGE (Map 41). Tibetan Plateau.

REMARKS. These forms have been of very uncertain status: *lama* was given specific rank by E & M-S whilst *kamensis* and *kozlovi* were tentatively included in *C. longicaudatus* by Flint (1966). However on the basis of much more abundant material Wang & Cheng (1973) have considered *C. kamensis*, with the above synonyms, as a good species distinct from *C. longicaudatus*.

Cricetulus alticola Ladak hamster

Cricetulus alticola Thomas, 1917. Shushul, Ladak (13 500 ft).

RANGE (Map 41). Ladak and Kashmir.

REMARKS. E & M-S included *tibetanus* in this species but examination of the types and other limited material available suggests that *tibetanus* should be associated with *lama*, here allocated to *C. kamensis*. This course was also taken by Wang & Cheng (1973) who did not however comment on the position of *alticola*.

Cricetulus eversmanni Eversmann's hamster

Cricetus eversmanni Brandt, 1859. N. Kazakhstan.
C. e. eversmanni. W. of range. Incl. *microdon*.
C. e. beljawi Argyropulo, 1933. Zaysan region of E. Kazakhstan. Incl. *belajevi*, *beljaevi*.

RANGE (Map 41). Steppes of N. Kazakhstan from R. Volga to the upper Irtysh at Zaysan.

REMARKS. The Mongolian form *curtatus* was included in this species by E & M-S and other authors. It was excluded by Flint (1966) in view of the considerable difference in karyotype. Subspecific variation was described by Mitina (1959).

***Cricetulus curtatus*†** Mongolian hamster

Cricetulus migratorius curtatus Allen, 1925. Iren Dabasu, Inner Mongolia, China.

RANGE (Map 41). Steppes of Mongolia north of the Altai and eastwards to Inner Mongolia.

REMARKS. Formerly considered conspecific with *C. eversmanni* – see remarks under that species.

Cricetulus triton Greater long-tailed hamster

Cricetus (Cricetulus) triton de Winton, 1899. N. Shantung, China. Incl. *albipes* (Ussuri Region), *arenosus*, *bampensis*, *canus* (S. Kansus), *collinus* (Shensi), *fuscipes* (Hopei), *incanus* (Shansi), *meihsienensis*, *nestor* (Korea), *yamashinai*.

RANGE (Map 40). N.E. China from Shensi to Manchuria, Korea and north to the upper Ussuri.

REMARKS. Several races were described by Allen (1940) but the very slight differences involved, the amount of intergradation and the apparent continuity of range make it unlikely that really discrete races exist.

Genus *MESOCRICETUS*

Mesocricetus Nehring, 1898. Type-species *Cricetus nigricans* Brandt =*Mesocricetus nigriculus* Nehring. =*Semicricetus* =*Mesocricetus*.

These hamsters are very similar to *Cricetus* and are included in that genus by many authors. However since they are also very close to some species of *Cricetulus* it seems best not to alter the most frequently used generic classification until a revision of the entire group can be undertaken. The number of species in *Mesocricetus* is very debatable. Ellerman (1940–41) recognized three but later (in E & M-S) considered the genus monospecific. Recent work on chromosomes and genetics suggests that there are at least three species, and possibly four if *auratus* and *brandti* are considered specifically distinct (Hamar & Schutowa, 1966). They are here given subspecific rank.

1. Ventral pelage entirely dark; skull ridges strongly developed . . . ***M. raddei***
– Ventral pelage pale except for a dark spot or band on chest; skull ridges poorly developed (2)
2. Longitudinal dark stripe on nape; a dark patch around and especially in front of eye (Europe) ***M. newtoni***
– No dark stripe on nape; no dark mark near eye (Asia) . . . . ***M. auratus***

Mesocricetus auratus Golden hamster

Cricetus auratus Waterhouse, 1839. Aleppo, Syria.
M. a. auratus. Syria.
M. a. brandti (Nehring, 1898). Palestine to Caucasus (type-locality in Georgia). Incl. *koenigi*.

RANGE (Map 42). Asia Minor (except extreme west) south to Syria and doubtfully to Israel; east to the Caucasus and Kurdistan.

REMARKS. These two subspecies differ rather sharply in pelage and slightly in karyotype. They may prove to be specifically distinct. The laboratory golden hamster was derived from *M. a. auratus*. A detailed bibliography has been published by Kittel (1969).

***Mesocricetus raddei*†** Ciscaucasian hamster

Cricetus nigricans raddei Nehring, 1894. R. Samur, Dagestan, Caucasus.

M. r. raddei. N.E. Caucasus. Incl. *avaricus*.
M. r. nigriculus Nehring, 1898. Plains N. of Caucasus.

RANGE (Map 42). From N. slopes of Caucasus to the R. Don and Sea of Azov.

REMARKS. All recent authors seem agreed that this form is specifically distinct from the *M. auratus/brandti* group, e.g. Kuznetsov (1965), Hamar & Schutowa (1966).

***Mesocricetus newtoni*†** Rumanian hamster

Cricetus newtoni Nehring, 1898. Schumla, E. Bulgaria.

RANGE (Map 42). E. Rumania and Bulgaria.

REMARKS. This species can with difficulty be crossed with *M. a. auratus* but the offspring are sterile (Raicu & Bratosin, 1968).

Subfamily **MYOSPALACINAE**
Genus *MYOSPALAX*
Zokors

Myospalax Laxmann, 1769. Type-species *Mus myospalax* Laxmann. Incl. *Myotalpa* = *Siphneus*, *Eospalax* = *Zokor*.

A very distinctive genus of about five species of mole-rats confined to N.E. Asia. Superficially they resemble the mole-rats of the genus *Spalax* and of the African genus *Tachyoryctes* and like these groups they show confusing variability. Two subgenera are usually recognized. The members of *Eospalax* are confined to China. No recent data appear to be available and therefore the classification used by E & M-S (following Allen, 1938–40) is retained, although it must be considered provisional. In the subgenus *Myospalax*, in Siberia, Mongolia and N. China, Ognev (1947) retained three species but Kuznetsov (1965), on the basis of more abundant material, has reduced these to races of a single species and this course is followed here.

1. Occipital surface of skull forming almost a flat plane between the transverse occipital crest and the foramen magnum; anterior palatal foramina confined to premaxillae **M. myospalax**
– Occipital surface strongly convex; anterior palatal foramina equally divided between premaxillae and maxillae . . . . (2)
2. Temporal ridges separate and parallel; posterior edge of nasals usually deeply notched (3)
– Temporal ridges converging in median line; posterior edge of nasals not notched; (pelage very grey) . . . . **M. smithi**
3. Smaller: greatest length of skull not over 40 mm, head and body not over 170 mm; claws rather delicate . . . . **M. rothschildi**
– Larger: skull up to 58 mm, head and body up to 200 mm; claws strongly developed **M. fontanieri**

Subgenus *EOSPALAX*

Myospalax fontanieri Common Chinese zokor

Siphneus fontanierii Milne-Edwards, 1867. Near Peking, China.
M. f. fontanieri. Hopei, Shensi and Shansi. Incl. *fontanus*.
M. f. cansus (Lyon, 1907). Shensi and Kansu. Incl. *rufescens*, *shenseius*.
M. f. baileyi Thomas, 1911. Szechuan.
M. f. kukunoriensis Lönnberg, 1926. Nan Shan.

RANGE (Map 43). Dry grasslands of northern and western China, from Hopei to Kansu, E. Chinghai and Szechuan.

Myospalax rothschildi Rothschild's zokor

Myospalax rothschildi Thomas, 1911:722. 40 miles S.E. of Taochow, Kansu. Incl. *minor*.

RANGE (Map 43). Kansu and Hupeh, China.

Myospalax smithi Smith's zokor

Myospalax smithi Thomas, 1911:720. 30 miles S.E. of Taochow, Kansu.

RANGE (Map 43). Kansu.

Subgenus *MYOSPALAX*

Myospalax myospalax Siberian zokor

Mus myospalax Laxmann, 1773. Near Barnaul, W. Siberia.

M. m. myospalax. Altai Mts and adjacent parts of W. Siberia. Incl. *incertus*, *laxmanni*, *tarbagataicus* (Tarbagatai Mts).

M. m. aspalax (Pallas, 1776). Upper Amur basin and N. Mongolia. Incl. ?*armandii*, *dybowskii*, *talpinus*.

M. m. psilurus (Milne-Edwards, 1874). Chinese part of range and adjacent parts of USSR and Mongolia. Incl. *epsilanus*, *spilurus*.

RANGE (Map 43). The Altai Mts and adjacent parts of the valleys of the Ob and Irtish; steppes east of the Gobi Desert from the upper Amur through extreme eastern Mongolia to Manchuria, Hopei and adjacent parts of Shansi, Honan and Shantung.

REMARKS. These three races were given specific rank by Ognev (1947) but were considered conspecific by Kuznetsov (1965). The first, in the Altai, seems to be widely separated from the remainder, but it is likely that the range of *aspalax* and *psilurus* is continuous.

Subfamily **LOPHIOMYINAE**

Genus *LOPHIOMYS**

Lophiomys Milne-Edwards, 1867. Type-species *L. imhausi* Milne-Edwards.

A monospecific genus.

Lophiomys imhausi* Crested rat

Lophiomys imhausi Milne-Edwards, 1867. Somalia.

RANGE. Known as a living species only from the mountains of East Africa, north to the northern edge of the Ethiopian Plateau. A record of a subfossil find from Israel has been dated to the 2nd century AD but the published report (Dor, 1966) does not provide enough information to permit a judgement as to whether the skull, as distinct from the archaeological context in which it was found, can be attributed conclusively to that date. The skull of this species is very distinctive but the taxonomic position of the genus is debatable.

Subfamily **MICROTINAE**

Voles and lemmings

The microtine rodents constitute a rather discrete group, sometimes given family rank (Microtidae or Arvicolidae). They are the dominant small herbivorous rodents throughout the Holarctic, being equally well represented in North America and Eurasia. All the Old World species are Palaearctic, the southernmost representatives being in N. Africa and the high alpine zones in the Himalayas and S.W. China.

Classification within the subfamily presents many problems both in delimiting the species and in defining genera. Much of the interest in microtine classification has arisen from the abundance of Pleistocene fossils, with a consequent overemphasis on dental characters at the expense of other available characters. The genera used here are mainly those used by Ellerman, the only difference being my inclusion of *Blanfordimys* in *Pitymys*. The most unstable areas of classification at the generic level are the *Clethrionomys*/*Eothenomys*/*Alticola* group, and the *Microtus*/*Pitymys*/*Arvicola* group. Only a very comprehensive revision using all the available characters is likely to result in a stable arrangement. There appear to be good grounds for considering *Ellobius*,

Prometheomys and *Lemmus*/*Myopus* as three distinctive groups; the remainder form a very closely interrelated group with *Dicrostonyx* and *Lagurus* as the most distinctive members.

The status of *Aschizomys*, based upon *A. lemminus* from N.E. Siberia, remains enigmatic and is likely to remain so until the genera in the *Clethrionomys* group can be redefined without using the very unsatisfactory character of the rooting of the teeth, which shows all shades of development. It is here included in *Eothenomys*.

1. Tail laterally compressed; very large: head and body up to 400 mm, condylobasal length of adults over 45 mm . . . . ***ONDATRA*** (p. 106)
– Tail terete; smaller: head and body under 220 mm, condylobasal length under 45 mm (2)
2. Pinna of ear rudimentary, consisting of a slightly raised rim around ear opening; (tail shorter than hind feet) . . . . (3)
– Pinna normally developed (but may be hidden in pelage) . . . . . . (4)
3. Upper incisors projecting forwards very prominently; incisors entirely white; cheek-teeth rather simple, becoming rooted, M^3 shorter than M^2; no dorsal stripe ***ELLOBIUS*** (p. 117)
– Upper incisors projecting downwards; incisors yellow or orange on anterior faces; cheek-teeth complex, remaining unrooted, M^3 as long as M^2; a black dorsal stripe; (pelage white in winter) . . . . ***DICROSTONYX*** (p. 96)
4. Upper molar tooth-rows diverging behind; M^3 very broad, showing 4 transversely elongate areas on wearing surface, very different from M^2 . . . . (5)
– Upper tooth-rows almost parallel; M^3 not unusually broad, showing triangular areas on wearing surface as in M^2 . . . . (6)
5. Pelage mottled or with a dark dorsal stripe; tail equal to or shorter than head and body; feet broad, soles hairy concealing pads . . . . ***LEMMUS*** (p. 96)
– Pelage dark grey; tail longer than hind feet; hind feet normal, soles naked in front revealing plantar pads . . . . ***MYOPUS*** (p. 96)
6. Claws of fore feet very long, central one longer than toe; (molar pattern rather simple, molars becoming rooted; Caucasus) . . . . ***PROMETHEOMYS*** (p. 117)
– Claws on fore feet not greatly enlarged, always shorter than toes . . . (7)
7. Soles of hind feet densely hairy; tail very short, scarcely exceeding length of hind feet; M_3 with dentine areas forming separated triangles . . . . ***LAGURUS*** (p. 116)
– Soles of hind feet not densely hairy; tail usually considerably longer than hind feet; M_3 usually with confluent transverse areas not divided into triangles. . . . . (8)
8. Posterior edge of bony palate consisting of a thin transverse shelf in the centre, extending on either side to form the floor of a lateral pit . . . . (9)
– Posterior edge of palate consisting of a thick median projection bordered on either side by deep, open pits . . . . (12)
9. Molars becoming rooted (sometimes only late in life); pelage usually rufous above ***CLETHRIONOMYS*** (p. 97)
– Molars remaining unrooted; pelage usually greyish brown . . . . . . (10)
10. Lateral grooves of molars wide and deep, giving wearing surface a very attenuated appearance . . . . (11)
– Grooves of molars narrow, teeth appearing more robust; (interorbital crest absent or weak) ***EOTHENOMYS*** (p. 100)
11. No interorbital crest; M_1 with several closed triangles; bullae well developed ***ALTICOLA*** (p. 103)
– Interorbital crest well developed; M_1 usually with only one closed triangle, remaining dentine areas forming confluent transverse loops; bullae rather small ***HYPERACRIUS*** (p. 104)
12. Molars becoming rooted; (surface of M_1 with 5 closed triangles; Balkans) ***DINAROMYS*** (p. 104)
– Molars remaining unrooted . . . . (13)
13. Wearing surface of M_1 with 3 closed triangles in front of posterior loop, 4th, if present, widely confluent with 5th (counting from rear) . . . . (14)
– Wearing surface of M_1 with at least 4 closed triangles . . ***MICROTUS*** (p. 110)

14. Very large: head and body up to 210 mm, hind feet over 20 mm, condylobasal length up to 45 mm . . . . *ARVICOLA* (p. 105)
– Small: head and body not over 120 mm, hind feet under 20 mm, condylobasal length under 28 mm; (4th and 5th triangles of M_1 usually widely confluent with each other but often constricted from area in front) . . . . *PITYMYS* (p. 106)

Genus *DICROSTONYX*

Dicrostonyx Gloger, 1841. Type-species *Mus hudsonius* Pallas (an American species). Incl. *Cuniculus* Wagler, *Misothermus* = *Borioikon*.

A distinctive genus with a circumpolar distribution in the tundra zone of Eurasia and N. America. All recent authors agree that there is only one species in Eurasia. Two additional species are generally recognized in N. America but more recently one of these, *groenlandicus*, has been considered conspecific with the Palaearctic *D. torquatus*.

Dicrostonyx torquatus Arctic lemming

Mus torquatus Pallas, 1779. Mouth of the R. Ob, Siberia.
D. t. torquatus. Mainland range west of Taimyr Peninsula.
D. t. lenae (Kerr, 1792). Mainland east of Taimyr Peninsula. Incl. *chionopaes*, *lenensis*.
D. t. ungulatus (von Baer, 1841). Island of Novaya Zemblya. Incl. *pallida*.
*D. t. vinogradovi** Ognev, 1948. Wrangel Island.
Dicrostonyx torquatus vinogradovi Ognev, 1948a. Wrangel Island.

Also other races in N. America of which *groenlandicus* Traill, 1823 is the earliest name.

RANGE (Map 43). Tundra of Siberia from the White Sea eastwards. In E. Siberia it has only recently been reported as far south as Kamchatka (Lazarev & Paramonov, 1973). The islands of Novaya Zemblya, Novosibirsk and Wrangel. The tundra of N. America from Alaska to Greenland (but replaced on the mainland east of Hudson Bay by *D. hudsonius*).

Genus *MYOPUS*

Myopus Miller, 1910. Type-species *Myodes schisticolor* Lilljeborg.

A distinctive, monospecific genus closely related to *Lemmus* but confined to the Palaearctic.

Myopus schisticolor Wood lemming

Myodes schisticolor Lilljeborg, 1844. Norway. Incl. *middendorfi*, *morulus* (Altai), *saianicus* (Sayan Mts), *thayeri* (N.E. Siberia), *vinogradovi*.

RANGE (Map 44). Coniferous forest zone from Norway and Sweden through most of Siberia east to the R. Kolyma and Kamchatka, south to the Altai, N. Mongolia and the Sikhote Alin Range. This is an elusive species but a spot-map by Gorbunov & Kulik (1974) shows a greatly increased number of locality records compared with previous maps. E & M-S wrote 'has been recorded from Sakhalin' but no records from the island of Sakhalin are given in any of the recent Russian works.

Genus *LEMMUS*

Lemmus Link, 1795. Type-species *Mus lemmus* L. Incl. *Myodes*, *Hypudeus*.

A clearly defined genus, closely related only to *Myopus*, represented in the tundra zones of the Palaearctic and Nearctic regions by a number of allopatric forms. The number of species involved is debatable. E & M-S recognized two, *L. lemmus* in Scandinavia and *L. sibiricus* in Siberia and N. America. Most Russian authors, e.g. Kuznetsov (1965), also recognize *L. amurensis* as distinct from *L. sibiricus* and this course is tentatively followed here. On the other hand Sidorowicz (1964) considered these and the American forms as races of a single species, mainly on the basis of absence of variation in cranial characters. The genus was reviewed by Rausch & Rausch (1975).

1. Pelage boldly patterned: dorsal surface of head and shoulders black with a large yellow patch on each side of occiput; (west of White Sea). . . . . . . *L. lemmus*
– Pelage uniform brown above except for a narrow mid-dorsal black stripe; (east of White Sea) (2)
2. Larger: hind feet over 15·5 mm, condylobasal length over 28 mm . . *L. sibiricus*
– Smaller: hind feet under 15·5 mm, condylobasal length under 28 mm . *L. amurensis*

Lemmus lemmus Norway lemming

Mus lemmus Linnaeus, 1758. Lappmark, Sweden. Incl. *borealis*, *norvegicus*.

RANGE (Map 44). Mountains of Scandinavia and the tundra from Lappland east to the White Sea.

Lemmus sibiricus Siberian lemming

Mus lemmus sibiricus Kerr, 1792. N. Urals.

L. s. sibiricus. Siberia, east to the R. Lena. Incl. *bungei* (R. Lena), *iterator*, *migratorius*, *novosibiricus* (Novosibirsk Islands), *obensis*.

L. s. chrysogaster Allen, 1903. N.E. Siberia and Kamchatka. Incl. *flavescens*, *kittlitzi (nomen nudum)*, *ognevi*, *paulus*.

*L. s. portenkoi** Tchernyavsky, 1967. Wrangel Island.
Lemmus sibiricus portenkoi Tchernyavsky, 1967. Wrangel Island.

Also races in N. America for which *trimucronatus* Richardson, 1825, is the earliest name.

RANGE (Map 44). Siberian tundra from White Sea to N.E. Siberia and Kamchatka; islands of Novaya Zemblya, Novosibirsk and Wrangel.

REMARKS. Variation in this species has been described by Krivosheev & Rossolimo (1966) who recognized only two subspecies, *sibiricus* and *chrysogaster*. The relationship of Siberian and American forms was discussed in detail by Rausch & Rausch (1975).

Lemmus amurensis Amur lemming

Lemmus amurensis Vinogradov, 1924. R. Zeya, Amur.

RANGE (Map 44). Mountains of the upper Amur basin; Verkhoyansk Mts.

REMARKS. I have not seen this species. It is very little known but is recognized as distinct from *L. sibiricus* by all recent Russian authors.

Genus *CLETHRIONOMYS*

Red-backed voles

Clethrionomys Tilesius, 1850. Type-species *Mus rutilus* Pallas. =*Evotomys*, *Euotomys*. Incl. *Craseomys*, *Neoaschizomys*.

*Glareomys** Razorenova, 1952. As subgenus of *Clethrionomys*, type-species *C. glareolus* (Schreber).

A genus with rather uncertain limits, represented throughout the northern Holarctic Region and characterized especially by the development of roots on the molar teeth coupled with the type of palate shared by *Alticola*, *Eothenomys* and *Hyperacrius*. Besides the type-species *C. rutilus*, the Palaearctic *C. glareolus* and Nearctic *C. gapperi* clearly belong here. In all of these the cheek-teeth become rooted at an early age. *C. rufocanus*, in which the formation of roots occurs later in life, is also included in most modern classifications but was formerly separated as the genus *Craseomys*. Great confusion has been caused by Japanese forms (included here in *C. andersoni*) in which rooting takes place so late in life that a majority of individuals in most samples do not show it. These have recently been referred to the genus *Aschizomys*, e.g. by Imaizumi (1960), but since there is no evidence that the type-species of *Aschizomys (A. lemminus)* has rooted molars the Japanese species in which rooting does occur, even late in life, are here retained in *Clethrionomys*. '*Aschizomys lemminus*' is difficult to place. Russian authors refer it to *Alticola* but this course would seem to make *Alticola* indefinable. It seems more likely to be a derivative of *Clethrionomys* (perhaps *C. rutilus*) that has developed rootless molars, in which case it is more appropriate to allocate it to *Eothenomys*.

Many eastern Asiatic forms were believed by E & M-S to be based upon young *C. rufocanus* (*smithi*, *regulus*, *shanseius*, *inez*, *nux*, *eva*, *alcinous* and *aquilus*), but there seem to be no grounds for retaining these in *Clethrionomys* and they are here included in *Eothenomys*.

As understood here the genus contains three widespread Palaearctic species along with two species restricted to Japan; the three Nearctic species are conspecific with or vicariants of the widespread Palaearctic ones.

It has been claimed by Kretzoi (1964) that the name *Clethrionomys* Tilesius, 1850 is a synonym of *Myodes* Pallas, 1811, but since the latter was based on ten species including *Lemmus lemmus* and has never been used for species of *Clethrionomys* for over a century there are no grounds for using *Myodes* instead of *Clethrionomys*.

1. Cheek-teeth becoming rooted very late in life so that in most samples the majority of adults show no sign of roots; tail usually over 50% of head and body; (Japan south of Hokkaido) . . . ***C. andersoni***
– Cheek-teeth rooting earlier in life so that most adults show at least the first signs of rooting; tail usually under 50% . . . . (2)
2. Cheek-teeth rooting very early, within 2 months, development of roots visible in all but conspicuously juvenile animals; smaller: condylobasal length usually under 26 mm, maxillary tooth-row usually under 6 mm . . . . (3)
– Cheek-teeth rooting later in life, roots rarely becoming long; larger: condylobasal length usually over 26 mm, maxillary tooth-row usually over 6 mm . . . . . (4)
3. Tail densely hairy, terminal pencil usually over 10 mm long; posterior margin of palate usually interrupted at each side; M^3 with 4 ridges on inner side . . . ***C. rutilus***
– Tail less hairy, terminal pencil under 10 mm; posterior margin of palate usually continuous; M^3 variable, with 3 or 4 ridges on inner side . . . . . . ***C. glareolus***
4. M^3 with 3 ridges on outer and inner sides; anterior loop of M_3 asymmetrical; posterior margin of palate usually entire; condylobasal length usually under 28 mm; (widespread) ***C. rufocanus***
– M^3 with 4 ridges on outer and inner sides; anterior loop of M_3 symmetrical; posterior margin of palate broken at each side; very large; condylobasal length 26—32 mm; (Hokkaido, Japan) . . . . ***C. rex***

Clethrionomys rutilus — Northern red-backed vole

Mus rutilus Pallas, 1779. East of the Ob, W. Siberia. Incl. *amurensis* (E. Siberia), *baikalensis*, *dorogostaiskii*, *hintoni*, *jacutensis*, *jochelsoni*, *latipes*, *latigriseus*, *lenaensis*, *mikado* (Hokkaido), *mollessonae*, *narymensis*, *otus*, *parvidens*, *rossicus*, *russatus*, *salairicus*, *tugarinovi*, *uralensis*, *vinogradovi*.

*Evotomys rutilus volgensis** Kaplanov & Raevskii, 1928. Kulakovo, Tver Province, Bezhetsk region, European Russia.

*Clethrionomys rjabovi** Belyaeva, 1953. Zorin, Kaganovich region, valley of R. Budund, 100 km from junction with R. Zeya, Amur Province, E. Siberia. (Considered to be a colour mutation of *C. rutilus* by Rossolimo, 1971b).

*Clethrionomys rutilus tundrensis** Bolshakov & Shvartz, 1965. River Khadit, Yamal Peninsula, N.W. Siberia. Also other forms in N. America of which the earliest named is *dawsoni* Merrian, 1888 from the Yukon, Canada.

RANGE (Map 46). The entire tundra and taiga zone of the Palaearctic from N. Scandinavia to N.E. Siberia; south to Moscow, N. Kazakhstan, the Altai, Manchuria and Korea; the islands of Sakhalin and Hokkaido; also in the corresponding zones in North America.

REMARKS. This species appears to be precisely allopatric with the more southern *C. glareolus* in Scandinavia but they are widely sympatric in western USSR.

Clethrionomys glareolus — Bank vole

Mus glareolus Schreber, 1780. Island of Lolland, Denmark. Incl. *alstoni* (Is. of Mull, Scotland), *britannicus* (England), *caesarius*† (Jersey, Channel Is), *centralis*† (Tien Shan), *erica*† (Is. of Raasay, Scotland), *frater* (Tien Shan), *fulvus*, *gorka*, *hallucalis* (S. Italy), *helveticus*, *hercynicus*, *insulaebellae* (Belle Ile, France),

intermedius Burg, *istericus* (Rumania), *italicus*, *jurassicus*, ?*minor*, *nageri* (Alps), *norvegicus* (Norway), *ognevi*, *ponticus*, *pratensis*, *reinwaldti*, *riparia* Yarrell, *rubidus*, *rufescens*, *ruttneri*, *saianicus* (Sayan Mts, Siberia), *skomerensis* (Skomer Is., Wales), *sobrus*, *suecicus*, *tomensis*, *vasconiae*, *vesanus*, *wasjuganensis* (W. Siberia).
*Clethrionomys glareolus bernisi** Rey, 1972. Valle de Valvanera, Sierra de la Demanda, Logrono, N. Spain.
Clethrionomys glareolus natio *bosniensis** Martino, 1945. Lah, Sarajevo, Bosnia, Yugoslavia.
*Clethrionomys glareolus cantueli** Saint Girons, 1969. Besse-en-Chandesse, Puy-de-Dôme, Auvergne, Massif Central, France.
*Clethrionomys glareolus curcio** Lehmann, 1961. Camigliatello Silano, Calabria, Italy.
*Clethrionomys glareolus devius** Stroganov & Turyeva, 1948. R. Petchora, W. Siberia.
*Clethrionomys glareolus garganicus** Hagen, 1958. Umbra Forest, Monte Gargano, Italy (800 m).
*Clethrionomys glareolus makedonicus** Felten & Storch, 1965. Pelister Mt (1600 m), near Bitola, Macedonia, Yugoslavia.
Clethrionomys glareolus natio *patrovi** Martino, 1945. Lisic, Rugova, between Metohija and Montenegro.
*Clethrionomys glareolus pirinus** Wolf, 1940. Bandariza Hut (between 1150 and 1800 m), Pirin Mts, Bulgaria.
*Clethrionomys glareolus variscicus** Wettstein, 1954. Klein Perthenschlag, Waldviertel, Lower Austria.

RANGE (Map 45). Forested zones of the western Palaearctic from France and Scandinavia to Lake Baikal; south to N. Spain, N. Italy (with isolated montane populations further south in Italy), the Balkans (but absent from most of Greece), N. Kazakhstan and the Altai Mts; isolates in N. Asia Minor and the Tien Shan; Britain and S.W. Ireland (probably recently introduced to the latter).

REMARKS. Two insular forms allocated by E & M-S to *C. rufocanus* are now universally recognized as representing this species (*caesarius* on Jersey, *erica* on Raasay). E & M-S considered *Evotomys centralis* Miller, 1906 from the Tien Shan Mts to represent *C. rutilus* but an examination of the type suggests that it belongs here or should perhaps be considered a distinct species (with *frater* a synonym).

Some of the isolated, marginal populations are moderately distinctive and merit recognition as subspecies (e.g. on the islands of Raasay, Jersey and Skomer, and in southern Italy, Asia Minor and Tien Shan). But in the absence of any descriptive revision of subspecific variation covering the entire range of the species most of the above names have little meaning.

Clethrionomys rufocanus — Grey red-backed vole

Hypudaeus rufocanus Sundevall, 1846. Lappmark, Sweden. Incl. *arsenjevi*, *bargusinensis*, *bedfordiae* (Hokkaido), *irkutensis* (Irkutsk), *kamtschaticus* (Kamchatka), *kolymensis* (N.E. Siberia), *kurilensis* (Paramushir Is., Kuriles), *latastei*, *sibirica*, ?*sikotanensis* (Sikotan Is., Kuriles), *wosnessenskii*, ?*yesomontanus*.

RANGE (Map 45). Northern Palaearctic from Scandinavia through the whole of Siberia to Kamchatka; south to the S. Urals, Altai, Manchuria and Korea; isolates on Hokkaido and some of the Kurile Islands.

REMARKS. E & M-S included a great many forms in this species that should be excluded according to all subsequent students of the group. In the west these are the insular forms on Raasay and Jersey which are undoubtedly *C. glareolus*. In eastern Asia the situation is more complex. All Japanese forms south of Hokkaido should be excluded, namely *Eothenomys smithi* and *Clethrionomys andersoni*, including *niigatae* (Jameson, 1961; Imaizumi, 1960). Imaizumi (1960) gave *sikotanensis* specific rank (in *Clethrionomys*) but later (1971) suggested that it might be considered a race of *C. rufocanus*. There seems to be no doubt that the Chinese and Korean forms allocated to *C. rufocanus* by E & M-S should be excluded. Although the skulls of some of these approach the condition of young *C. rufocanus*, externally they are very different and there is no indication that the molars become rooted.

*Clethrionomys rex**

Clethrionomys rex Imaizumi, 1971. Kanrosen, Mt Rishiri, Rishiri Is., Hokkaido, Japan.
*Clethrionomys montanus** Imaizumi, 1972. Mt Poroshiri, Hidaka Mts, Hokkaido, Japan (970 m).

RANGE (Map 45). Known only from Rushiri Island, at the northwestern extremity of Hokkaido, Japan, and from the mainland of Hokkaido, if *montanus* is conspecific.

REMARKS. According to Imaizumi (1971) *C. rex* is related to *C. rufocanus* but is sympatric on Rushiri Is. with another *Clethrionomys* which he called *C. sikotanensis* but considered might be a race of *C. rufocanus*. I have not seen *rex* nor *montanus*.

Clethrionomys andersoni†

Evotomys andersoni Thomas, 1905. Tsunagi, near Morioka, Iwate Ken, N. Honshu, Japan.
C. a. andersoni. N. Honshu.
C. a. niigatae (Anderson, 1909). Central Honshu.
*C. a. imaizumii** Jameson, 1961. S. Honshu.
*Clethrionomys imaizumii** Jameson, 1961. Nachi Falls (300 ft), Wakayama-ken S. Honshu, Japan.

RANGE (Map 45). Confined to Honshu, Japan.

REMARKS. This form was considered conspecific with *C. rufocanus* by E & M-S but was transferred to the genus *Aschizomys* by Imaizumi (1960). Although the type-species of *Aschizomys* (*A. lemminus* of N.E. Siberia) is of very doubtful affinities, it seems very reasonable to retain this species in *Clethrionomys* as was done by Jameson (1961).

Imaizumi (1960) considered *niigatae* specifically distinct from *andersoni* as did Jameson (1961) who also considered *imaizumii* to be specifically distinct but closer to *niigatae* than to *andersoni*. Aimi (1967) considered *imaizumii* conspecific with *andersoni*. The key given by Jameson to distinguish *niigatae* and *andersoni* does not serve to separate the specimens in the British Museum (including the types of both) and I am therefore inclined to consider them all conspecific.

Genus *EOTHENOMYS*

Eothenomys Miller, 1896. Type-species *Arvicola melanogaster* Milne-Edwards. Incl. *Anteliomys*, *Aschizomys*†, *Phaulomys*†, *Caryomys*†.

A genus of small voles found mainly in China and Japan. They closely resemble *Clethrionomys*, being distinguished mainly by the absence of rooting of the cheek-teeth even in old age. However the species of *Clethrionomys* show such a gradation in the age at which rooting takes place that the difference does not constitute a very satisfactory generic character. Only the Japanese, northern Chinese and Siberian species fall clearly into the Palaearctic Region as defined here but since this and *Hyperacrius* are the only Old-World genera of Microtinae falling mainly outside the region they are both included here in their entirety. The southern Chinese forms however have not been revised in detail.

Eleven species are recognized here. Five of these, from southern China, were included in *Eothenomys* by E & M-S; five more were included, more or less tentatively, in *Clethrionomys rufocanus*, the types having been interpreted as young individuals. All these ten species are represented by substantial series in the collection of the British Museum which also includes the holotypes of all but *E. melanogaster*. The five northern species are clearly distinguishable from each other and from species of *Clethrionomys*, and none of the specimens examined, including apparently old individuals, showed any sign of the development of roots on the molar teeth.

The inclusion of *lemminus* is more tentative since I have seen only a single individual. Russian authors place it in *Alticola* but the molar pattern is much closer to *Eothenomys* than to *Alticola*.

1. M_1 with adjacent triangles confluent and more or less opposite to each other . (2)
– M_1 with adjacent triangles discrete and alternating in position . . . . . (7)
2. M^1 with four ridges on inner side; (ventral pelage usually dark grey) ***E. melanogaster***
– M_1 with three ridges on inner side (as in *Clethrionomys*) . . . . (3)

3. Tail long, over 50% of head and body . . . . (4)
– Tail short, under 50% of head and body . . . . (5)
4. Large: hind feet over 20 mm; M^3 complex, with 4 or 5 inner and outer ridges, first outer groove shallow, second and third dentine spaces separated; (S.W. China) ***E. chinensis***
– Small: hind feet under 20 mm; M^3 simple, with usually 3 inner and outer ridges, second and third dentine spaces confluent; (Japan) . . . . ***E. smithi***
5. Hairs with pale tips giving a slightly frosted appearance to otherwise dark dorsal and ventral pelage; (M^2 with a prominent medial projection on the posterior lobe; M^3 complex with usually 4 ridges on each side) . . . . ***E. olitor***
– Pelage lacking frosted appearance, dorsal pelage browner . . . . . . (6)
6. M^3 complex, with 4 or 5 ridges on inner side; tail usually over one third of head and body; hind feet usually under 17 mm . . . . ***E. custos***
– M^3 simpler, with usually only 3 inner ridges; tail less than one third of head and body; hind feet usually over 17 mm . . . . ***E. proditor***
7. Tail longer, about 60% of head and body; (M^3 rather short and simple) ***E. eva***
– Tail shorter, under 40% of head and body . . . . (8)
8. Pelage very long, soft and shaggy at all seasons; molars robust, M^3 usually short and simple (10)
– Pelage shorter and neater; molars smaller, M^3 rather elongate . . . . (9)
9. Dorsal pelage reddish brown; larger (hind feet over 18 mm); rostrum normal (diastema distinctly longer than molar tooth-row) . . . . ***E. regulus***
– Dorsal pelage not distinctly reddish; smaller (hind feet under 18 mm); rostrum very short (diastema about equal to molar tooth-row) . . . . ***E. inez***
10. Pelage yellowish brown; tail over 30% of head and body; M^3 with first and second outer grooves much shallower than corresponding inner ones . . . . ***E. shanseius***
– Pelage greyish brown in summer, white in winter; tail about 20% of head and body, sharply bicoloured and with a long terminal pencil of hairs; M^3 with outer grooves as deep as inner ones . . . . ***E. lemminus***

Eothenomys melanogaster Père David's vole

Arvicola melanogaster Milne-Edwards, 1872. Moupin, Szechuan, China. Incl. *aurora, bonzo, cachinus, colurnus, confinii, eleusis, fidelis, kanoi* (Taiwan), *libonotus, miletus, mucronatus*.

RANGE (Map 45). Western and southern China, north to extreme south of Kansu, south to N. Burma; Taiwan.

REMARKS. The forms listed above appear to constitute a natural group which E & M-S considered a single species. However the variation is considerable and there might be justification for recognizing more than one species as was done for example by Allen (1938–40).

Eothenomys olitor

Microtus (Eothenomys) olitor Thomas, 1911. Yunnan, China.

RANGE (Map 46). Known only from Yunnan, around 2000 m.

Eothenomys proditor

Eothenomys proditor Hinton, 1923. Likiang Range, Yunnan, China.

RANGE (Map 45). Mountains of Yunnan and Szechuan, China, reaching 4200 m.

Eothenomys chinensis

Microtus chinensis Thomas, 1891. Szechuan, China. Incl. *tarquinus, wardi* (Yunnan).

RANGE (Map 46). Szechuan and Yunnan, China, 2000–4000 m.

Eothenomys custos

Microtus (Anteliomys) custos Thomas, 1912. N.W. Yunnan, China. Incl. *hintoni* (Szechuan), *rubellus, rubelius*.

RANGE (Map 46). Mountains of Szechuan and Yunnan, China, reaching at least 4200 m.

Eothenomys smithi†

Evotomys (Phaulomys) smithii Thomas, 1905. Kobe, Honshu, Japan. Incl. *okiensis* (Dogo Island). *Eothenomys kageus** Imaizumi, 1957. Yamura-Machi, Minamitsuru-Gun, Yamanashi Pref., Honshu, Japan.

RANGE (Map 46). Japan: Honshu, Shikoku and Kyushu; the small island of Dogo.

REMARKS. This form was treated as a race of *Clethrionomys rufocanus* by E & M-S. All subsequent authors are agreed that it is not a *Clethrionomys*. Imaizumi (1960) allocated it to *Eothenomys* and this course is followed here. Tanaka (1971) made a detailed study of its relationships and concluded that it held a rather indeterminate position between *Eothenomys* and *Clethrionomys*. He provisionally used the generic name *Phaulomys* of which *smithi* is the type.

According to Tanaka (1971) the specific status of *kageus* cannot be upheld.

Eothenomys regulus†

Craseomys regulus Thomas, 1907. Mingyong, 110 miles S.E. of Seoul, S. Korea.

RANGE (Map 46). Korea and Hopei.

REMARKS. Examination of the type and 54 other specimens of this form shows that it is quite distinct from *Clethrionomys rufocanus* in which it was placed by E & M-S. In old animals the molar teeth appear to stop growing but do not form multiple roots as in *Clethrionomys*. True *Clethrionomys rufocanus* also occurs in Korea. *E. regulus* is easily distinguished by its short pelage, brown flanks, narrow molar teeth and usually elongate M^3.

Eothenomys shanseius†

Craseomys shanseius Thomas, 1908. 100 miles N.W. of Taiyuenfu, Shansi, China.

RANGE (Map 45). Shansi and perhaps Hopei, China.

REMARKS. This form was treated as a race of *Clethrionomys rufocanus* by Allen (1940) as well as by E & M-S. However examination of the type and a considerable series in the British Museum indicates that, like the following forms, it is not a *Clethrionomys* but should be allocated to *Eothenomys*. Its long shaggy pelage (in both summer and winter) contrasts with the other species of *Eothenomys* which tend to have the pelage short and 'tidy'.

Eothenomys inez†

Microtus (Eothenomys) inez Thomas, 1908. 12 miles N.W. of Kolanchow, Shansi, China. Incl. ?*jeholicus* (Jehol), *nux* (Shensi).

RANGE (Map 46). Shansi, Shensi and perhaps east to Jehol, China.

REMARKS. These forms were tentatively placed in *Clethrionomys rufocanus* by E & M-S, following the conclusion of Hinton (1926) that the types were based on young animals of that species. However the types agree with considerable series in the British Museum which show no sign of rooting of the molar teeth and are externally very different from *Clethrionomys rufocanus*. I therefore follow Allen (1940) in giving *inez* specific rank in *Eothenomys*. The teeth and pelage suggest a close relationship to *E. regulus*, but the very short rostrum is distinctive.

Eothenomys eva†

Microtus (*Caryomys*) *eva* Thomas, 1911. S.E. of Taochow, Kansu, China. Incl. *alcinous* (Szechuan), *aquilus* (Hupeh).

RANGE (Map 46). Mountains of Southern Kansu and adjacent parts of Szechuan and Hupeh, China.

REMARKS. The remarks given under *E. inez* apply equally to this species. Its long tail and short M^3 appear to distinguish it very clearly from *E. inez*.

***Eothenomys lemminus*†**

Aschizomys lemminus Miller, 1898. Kelsey Station, Plover Bay, Bering Strait, N.E. Siberia.

RANGE (Map 47). N.E. Siberia from Anadyr region west to the mouth of the R. Lena and south throughout the Lena basin.

REMARKS. An enigmatic species, allocated to *Alticola* by Ognev (1950), who treated it as the sole member of a subgenus *Aschizomys*, and by Kuznetsov (1965) who treated it as a race of *Alticola macrotis*. Both these courses would seem to make the genus *Alticola* indefinable. Accepting the Russian view that the molars remain unrooted, I allocate it to *Eothenomys* on the basis of dental pattern. It seems very close to the northern forms of *Eothenomys* such as *E. shanseius*.

Genus *ALTICOLA*

Alticola Blanford, 1881. Type-species *Arvicola stoliczkanus* Blanford. Incl. *Platycranius*.

A genus of small voles confined to the mountains of central Asia. They are related to *Clethrionomys* but are characterized especially by the cheek-teeth (which are rootless) being very slender and elongate.

Three species seem clearly attributable to *Alticola*; two other forms, *macrotis* and *lemminus*, are of less certain status. The latter, and sometimes also *macrotis*, has been separated generically as *Aschizomys*. Here *macrotis* is retained in *Alticola* but *lemminus* is allocated to *Eothenomys*.

1. Head very flat, height of skull only half width of brain-case; (tail over 30 % of head and body) . . . ***A. strelzovi***
– Head normal, height of skull about two-thirds width of brain-case . . . (2)
2. Tail over 25 % of head and body, white or bicoloured; (M^3 with anterior dentine area widely confluent with area behind and with usually 3 angles on medial side) . . ***A. roylei***
– Tail under 25 % of head and body . . . . (3)
3. Tail bicoloured; M^3 with anterior dentine area separate from one behind and with 3 angles on medial side . . . . ***A. macrotis***
– Tail white; M^3 with anterior dentine area confluent with area behind and with only 2 angles on medial side . . . . ***A. stoliczkanus***

Alticola roylei Royle's mountain vole

Arvicola roylei Gray, 1842. Kumaon, N.W. India. Incl. *acmaeus, albicauda, altaicus, argentata* (Pamirs), *argurus, blanfordi* (Kashmir), *cautus, glacialis, gracilis, imitator, lahulius, longicauda, montosa, phasma* (Karakorum), *rosanovi, semicanus* (Mongolia), *severtzovi, shnitnikovi, subluteus, villosa, worthingtoni* (Tien Shan).

*Arvicola leucurus** Severtzov, 1873:82. Summit of Mt Masat, Tien Shan, USSR (preoccupied by *Arvicola leucurus* Gerbe, 1852 = *Microtus nivalis*; *severtzovi* Tichomirov & Kortchagin, 1889 was proposed as a replacement according to Ognev, 1950).

*Alticola argentata saurica** Afanasiev & Bazhanov, 1948. Saur Mts, Tarbagatai Range, E. Kazakhstan (*fide* Kuznetsov, 1965). Perhaps a *nomen nudum*.

*Alticola tuvinicus** Ognev, 1950. Near Kyzyr, Tuva Region (Upper Yenesei).

*Alticola argentatus olchonensis** Litvinov, 1960. Olkhon Island, Lake Baikal, USSR.

*Alticola roylei parvidens** Schlitter & Setzer, 1973. 20·5 m N. of Dir, 10 400 ft, Dir State, Pakistan.

RANGE (Map 47). Mountains of central Asia from the western Himalayas and Afghanistan through the Pamirs, Tien Shan and Altai to Mongolia and L. Baikal.

REMARKS. This species was for long known as *A. argentata* in the Russian literature but Kuznetsov (1965) has followed E & M-S in combining the more northern forms with the Himalayan *roylei*. He recognized six races in USSR but there has been no review of subspecific variation in the species as a whole.

Alticola stoliczkanus Stoliczka's mountain vole

Arvicola stoliczkanus Blanford, 1875. N. Ladak, Himalayas. Incl. *acrophilus, cricetulus, lama* (W. Tibet), *nanschanicus* (Nan Shan), *stracheyi*.

*Alticola barakshin** Bannikov, 1947b. Dzun Saihan, S.E. Gurban Saihan Range, Gobi Altai, Mongolia.

RANGE (Map 47). Himalayas, Tibet, Nan Shan, Altai.

Alticola strelzowi Flat-headed vole

Microtus strelzowi Kastschenko, 1900. Altai Mts. Incl. *depressus, desertorum*.

RANGE (Map 47). Altai Mts in N.W. Mongolia west across Kazakhstan as far as Karaganda.

Alticola macrotis Large-eared vole

Arvicola macrotis Radde, 1862. Eastern Sayan Mts, Siberia.
A. m. macrotis. Sayan Mts.
A. m. vinogradowi† Rasorenova, 1933. Altai Mts.

RANGE (Map 47). Altai and Sayan Mts, east to Lake Baikal.

REMARKS. I have not seen this species. Kuznetsov (1965) included '*Aschizomys lemminus*' in this species but this seems quite unjustified. Judging by all descriptions and by the single specimen available the teeth of *lemminus* are quite unlike those of *Alticola* and the genus would be made indefinable by its inclusion. On the other hand *A. macrotis* appears from descriptions, e.g. those of Ognev (1950), to be a true *Alticola*.

Genus *HYPERACRIUS*

Hyperacrius Miller, 1896. Type-species *Arvicola fertilis* True.

Contains two Himalayan species, both on the fringe of the Palaearctic Region as defined here but included to complete the Microtinae. They are related to *Alticola* but are more fossorial. The genus was reviewed by Phillips (1969).

Larger (greatest length of skull usually over 24·5 mm, upper tooth-row over 6·2 mm); tail usually more than one third length of head and body; pelage long and lax . . . ***H. wynnei***
Smaller (skull under 25 mm, upper tooth-row under 6·2 mm); tail usually under one third of head and body; pelage short, stiff and dense . . . . ***H. fertilis***

Hyperacrius wynnei Murree vole

Arvicola wynnei Blanford, 1881. Murree, Punjab.
H. w. wynnei. Murree Hills, E. of R. Indus.
*H. w. traubi** Phillips, 1969. W. of R. Indus.
Hyperacrius wynnei traubi Phillips, 1969. Yakh Tangai, 6600 ft, Swat, Pakistan.

RANGE (Map 48). Extreme north of Pakistan, in coniferous forest between 2000 and 2500 m.

Hyperacrius fertilis True's vole

Arvicola fertilis True, 1894. Kashmir. Incl. *aitchisoni, brachelix*.
*Hyperacrius fertilis zygomaticus** Phillips, 1969. 6 miles S.W. of Utror, 8900 ft, Swat, Pakistan.

RANGE (Map 48). Kashmir and upper Punjab, on open ground between 2500 and 3800 m.

Genus *DINAROMYS**

Dinaromys Kretzoi, 1955. Type-species *Microtus (Chionomys) marakovici* Bolkay, 1924.

A monospecific genus confined to S.E. Europe. This species has usually been referred to the genus *Dolomys*, based on a Pleistocene species from Hungary, but Kretzoi (1955) concluded that the various forms allocated to *Dolomys* represented three distinct genera of which only one, *Dinaromys*, is represented by a recent species. He allocated *Dinaromys* (and *Dolomys*) to the tribe Ondatrinae, along with the Nearctic *Ondatra* and *Neofiber*, but later (Kretzoi, 1969) placed it as the sole recent member of the tribe Pliomyini.

Dinaromys bogdanovi Martino's snow vole

Microtus (Chionomys) bogdanovi Martino, 1922. Cetinje, Montenegro, Yugoslavia. Incl. *grebenscikovi, korabensis, marakovici, preniensis*.

*Dolomys bogdanovi coeruleus** Mirić, 1960. Trebevic Mts, S. of Sarajevo, Yugoslavia (1600 m). (Name from *D. b. marakovici* natio *coeruleus* Martino, 1948.)
*Dolomys bogdanovi trebevicensis** Gligić, 1959. Mt Trebević, Bosnia, Yugoslavia.
*Dolomys bogdanovi longepedis** Dulic & Vidimic, 1967. Rupe-Kolibe Petra Cetnika, Catrna koda, Mt Dinara, N. Dalmatia, Yugoslavia (1420 m).

RANGE (Map 48). Mountains of Yugoslavia and presumably N. Albania (map in Mirić & Dulic, 1962).

REMARKS. This species has been called *Dolomys milleri* Nehring, e.g. by Brink (1967) on the grounds that the living form is conspecific with the fossil form of that name. Kretzoi (1955) has argued that these are not only specifically, but generically distinct. The subspecies that have been described have mostly been based on average differences in size and proportions or slight differences in colour.

Genus *ARVICOLA*

Arvicola Lacepède, 1799. Type-species *Mus amphibius* L. Incl. *Hemiotomys* (part), *Paludicola, Praticola, Ochetomys.*

A moderately distinctive genus of large voles, closely related to *Microtus*. Besides the Palaearctic forms one North American species, *A. richardsoni*, has frequently been included (e.g. lately by Jannett & Jannett, 1974) although it has more often been allocated to *Microtus*. E & M-S recognized only one (Palaearctic) species but more recently the presence of two species in Europe has been demonstrated (Corbet *et al.*, 1970).

Smaller (especially in area of overlap): maxillary tooth-row usually under 10 mm, condylobasal length usually under 40 mm; nasals relatively narrow, usually under 4·75 mm . . . . ***A. terrestris***
Larger: maxillary tooth-row usually over 10 mm, condylobasal length usually over 40 mm; nasals relatively wide, usually over 4·75 mm (S.W. Europe) . . . . ***A. sapidus***

Arvicola terrestris — Water vole

Mus terrestris Linnaeus, 1758. Upsala, Sweden. Incl. *abrukensis* (Estonia), *albus, americana, amphibius* (England), *aquaticus, argentoratensis, argyropus, armenius, ater, barabensis, brigantium, buffonii, canus, castaneus, caucasicus, cubanensis, destructor, djukovi, exitus* (Switzerland), *ferrugineus, fuliginosus, hintoni* (Syria), *hyperryphaeus, illyricus* (Yugoslavia), *italicus* (Italy), *jacutensis* (Yakutia), *jenissijensis* (Sayan Mts), *karatshaicus, korabensis, kuruschi, kuznetzovi, littoralis, meridionalis* (S. Urals), *minor, monticola* (Pyrenees), *musignani, niger, nigricans, obensis, ognevi* (Caucasus), *paludosus, persicus* (Iran), *pertinax, reta* (N. Scotland), *rufescens, scherman* (E. France), *schermous, scythius* (S.E. Kazakhstan), *tanaiticus, tataricus* (Ukraine), *tauricus, turovi, uralensis, variabilis, volgensis.*
*Arvicola terrestris martinoi** Petrov, 1949: 186. Near Belgrade, S. of rivers Sava and Danube, Yugoslavia.
*Arvicola terrestris stankovici** Petrov, 1949: 189. Suvo Rudiste, Kapaonik Mts, Yugoslavia (1700 m).
*Arvicola terrestris cernjavskii** Petrov, 1949: 190. Ponot, Dojkinci, Stara Planina Mts, Yugoslavia (1350 m).

RANGE (Map 49). Most of Europe (but absent most of Iberia, W. France, S. Italy) from the mountains of the Mediterranean zone to the Arctic Sea and east through most of Siberia almost to the Pacific coast; south to Palestine, Zagros Mts and north side of Tien Shan; Britain.

REMARKS. This is a variable species but the variation cannot be usefully described by orthodox subspecific taxonomy. In some areas local adaptation to more or less aquatic habitats leads to the proximity of dissimilar forms.

It has been maintained by Brink (1967) that if the forms *terrestris* and *amphibius* of Linnaeus (1758) are considered conspecific, as here, then *amphibius* should have priority (presumably following Blasius (1857) as first reviser). Although strictly correct this is contrary to long-established usage and would cause considerable confusion and ambiguity.

Arvicola sapidus

Arvicola sapidus Miller, 1908. Burgos, Spain. Incl. *musiniani* Lataste, *tenebricus.*

RANGE (Map 49). Iberia and western France.

REMARKS. This form was shown to be specifically distinct from *A. terrestris* by Reichstein (1963). Brink (1967) included also the lowland British form in this species (which he therefore called *A. amphibius*) but a detailed analysis of British animals indicates that only *A. terrestris* is represented in Britain (Corbet *et al.*, 1970).

Genus *ONDATRA**

Ondatra Link, 1795. Type-species *Castor zibethicus* L. Incl. *Fiber.*

A monospecific genus in N. America. As a result of escapes from fur farms it has become established in many parts of the Palaearctic.

Ondatra zibethicus — Muskrat

Castor zibethicus Linnaeus, 1766. E. Canada.

RANGE (Map 48). Almost the whole of N. America. Introduced in much of central and northern Europe, most of the USSR, adjacent parts of Mongolia and China and in Japan (central Honshu).

REMARKS. A monograph, with emphasis on the feral populations, was produced by Hoffmann (1958).

Genus *PITYMYS*

Pitmys McMurtie, 1831. Type-species *Psammomys pinetorum* Le Conte (USA). Incl. *Arbusticola, Blanfordimys*†, *Micrurus, Neodon, Phaiomys, Terricola* Fatio.

A large genus of very ill-defined limits, represented in the Palaearctic and Nearctic Regions. It is closely related to *Microtus* and is included in that genus by some authors, e.g. Kuznetsov (1965). Pending a comprehensive revision of this entire group I follow the generic classification of E & M-S except to include *Blanfordimys* in *Pitymys*.

Delimitation of species is very provisional, especially in the European part of the genus. Recent work has demonstrated considerable differences in karyotype between forms that are morphologically very similar, as in other rodents of subterranean habit. A great deal of work is being undertaken on the genus but as yet no comprehensive revision is available. The following account is based mainly upon the work of Kratochvil (1970) who has illustrated the skulls and teeth of most species, supplemented by many other works which are mentioned under the individual species. The key below is based upon that of Kratochvil (1970) but is no more than an indication of the characters that have been used to distinguish the species. The character of the coronoid process (dichotomy 7) seems particularly unsatisfactory.

1. Auditory bullae greatly enlarged: greatest length (excluding posterior mastoid part) greater than diastema; (M_1 with dentine fields of 3rd outer and 4th inner ridges widely confluent but rather constricted from simple rounded anterior field; temporal ridges obscure) . . . ***P. afghanus***
– Auditory bullae normal: greatest length less than diastema . . . . . . (2)
2. M_1 with dentine fields of 3rd outer and 4th inner ridges (counting from posterior end) confluent with each other and with the field in front; temporal ridges usually prominent and often uniting to form an interorbital crest . . . . (3)
– M_1 with these dentine fields sharply separated from the one in front; temporal ridges very obscure, interorbital region usually remaining broad and flat . . . . . (5)
3. M_1 with anterior loop simple and rather circular; skull broad and angular; upper incisors slightly pro-odont . . . . ***P. leucurus***
– M_1 with anterior loop complex; upper incisors orthodont . . . . . . (4)
4. Pelage dark brown above, brownish grey below; M_1 usually with 6 inner ridges . . . ***P. sikimensis***
– Pelage pale yellowish brown above, pale grey below; M_1 usually with 5 inner ridges . . . ***P. juldaschi***
5. M^3 complex, with 4 ridges on inner side . . . . (6)
– M^3 simple, with 3 ridges on inner side . . . . (11)

6. M^3 very complex, with 4 or 5 ridges on the outer side; M^2 with a well developed third inner ridge (as in *Microtus agrestis*); 4 pairs of mammae . . . ***P. schelkovnikovi***
– M^3 less complex, with 3 ridges on the outer side; M^2 with a third inner ridge poorly developed or absent; 2 or 3 pairs of mammae . . . . (7)
7. Coronoid process of mandible appearing medial to or concealed by articular process in posterior view; bullae large but flat, not reaching plane of molar teeth; M^3 with second inner and outer ridges almost opposite, their dentine fields confluent ***P. subterraneus***
– Coronoid process appearing lateral to articular process in posterior view . . (8)
8. M^3 with dentine spaces of second inner and outer ridges confluent; 3 pairs of mammae ***P. majori***
– M^3 with dentine spaces of second inner and outer ridges separated; 2 pairs of mammae (i.e. inguinal only) . . . . (9)
9. Bullae flat, not projecting ventrally beyond plane of molars . . . ***P. multiplex*** ***P. liechtensteini***
– Bullar larger, projecting ventrally beyond plane of molars . . . . . . (10)
10. Posterior edge of nasals about level with that of premaxillae; third outer ridge of M^3 usually associated with a closed dentine field . . . . ***P. tatricus***
– Posterior edge of nasals anterior to that of premaxillae; dentine field of third outer ridge of M^3 usually confluent with posterior field . . . . ***P. bavaricus***
11. M^3 with second outer ridge large, level with first and third; posterior edge of nasals usually broad and truncate . . . . ***P. savii***
– M^3 with second outer ridge small, not reaching plane of first and third; posterior edge of nasals narrow . . . . (12)
12. Skull deeper at level of M^2 than at level of bullae . . . . . ***P. thomasi***
– Skull not deeper at M^3 than at bullae . . . . (13)
13. Larger: zygomatic width 13·4–16·0 mm, upper diastema 7·0–10·1 mm ***P. duodecimcostatus***
– Smaller: zygomatic width 12·2–14·3 mm, upper diastema 6·3–7·7 mm ***P. lusitanicus***

Pitymys leucurus Blyth's vole

Phaiomys leucurus Blyth, 1863. Ladak. Incl. *blythi*, *everesti* (Mt Everest), *fuscus* (E. Tibet), *petulans*, *strauchi* (Sinkiang), *waltoni* (Tibet).

RANGE (Map 49). Tibetan Plateau and Himalayas at high altitude.

REMARKS. If this species is allocated to the genus *Microtus* the name *M. blythi* (Blanford, 1875) would have to be used since *leucurus* is preoccupied.

Pitymys sikimensis Sikkim vole

Neodon sikimensis Hodgson, 1849. Sikkim. Incl. *forresti*† (Yunnan), *irene*† (Szechuan), *oniscus*† (Kansu), *thricolis*.

RANGE (Map 49). Himalayas from W. Nepal eastwards, to Upper Burma, Szechuan and S. Kansu.

REMARKS. E & M-S treated *irene* (including *forresti* and *oniscus*) as specifically separate from *P. sikimensis*. These have been united by most recent authors, e.g. by Gruber (1969) who made a detailed study of the species.

Pitymys juldaschi Juniper vole

Arvicola juldaschi Severtzov, 1879. Lake Karakul, Pamir Mts. Incl. *carruthersi*† (Gissar Mts, E. of Samarkand), *pamirensis*.

RANGE (Map 50). Tien Shan and Pamirs, west to Samarkand.

REMARKS. E & M-S gave *carruthersi* specific rank; I include it in *P. juldaschi* following Kuznetsov (1965) and Bolshakov *et al.* (1969).

***Pitymys afghanus*†** Afghan vole

Microtus (Phaiomys) afghanus Thomas, 1912. Gulran, N.W. Afghanistan.
P. a. afghanus. Afghanistan and S.E. Turkmenistan.
P. a. bucharicus (Vinogradov, 1928). W. Pamirs.
*P. a. balchanensis** (Heptner & Shukurov, 1950). S.W. Turkmenistan.
Microtus afghanus balchanensis Heptner & Shukurov, 1950. Great Balkhan Mts, W. Turkmenistan.

RANGE (Map 50). Mountains of Afghanistan and S.W. Turkmenistan; apparently isolated populations in the western Pamirs (Tadzhikistan) and in the Great Balkhan Mts on the east coast of the Caspian Sea.

REMARKS. This species was placed in a monospecific genus, *Blandfordimys*, by E & M-S. J. Niethammer (1970) considered it most closely related to *juldaschi*, *leucurus*, and *sikimensis* which he placed in *Microtus*. These seem to form a group (subgenus *Neodon*) with some affinity to European *Pitymys*.

Pitymys subterraneus European pine vole

Arvicola subterraneus de Sélys Longchamps, 1836. Liège, Belgium. Incl. *atratus*, *capucinus*, *dacius* (Rumania), *ehiki*, *hungaricus*, *incertoides*, *klozeli*, *kupelwieseri*, *martinoi*, *matrensis*, *mustersi*, *nyirensis*, *transsylvanicus* (Hungary), *ukrainicus* (Ukraine), *wettsteini*, *zimmermanni*.
*Microtus (Pitymys) dinaricus** Kretzoi, 1959. New name for *Pitymys mustersi* Martino 1937, preoccupied by *Microtus mustersi* Hinton, 1926 if these genera are united.

RANGE (Map 50). From France through central Europe to Ukraine and the R. Don, with apparently isolated populations northeast in Russia almost to L. Onega. Absent from Baltic and Mediterranean coasts but range in southern Europe rather uncertain.

REMARKS. This species was reviewed in detail by Niethammer (1972b) who included the synonyms listed above. Other synonyms listed by E & M-S are referred to the next five species except for the following which are best treated as '*Pitymys*, *incertae sedis*' (all are from the Alpine region except for *brauneri* from Serbia): *brauneri* Martino, *fusca*, *orientalis*, *rufofuscus*, *selysii* Gerbe. (*Microtus* (*Pitymys*) *subterraneus serbicus** Kretzoi, 1958 has been proposed as a new name for *P. multiplex brauneri* Martino, 1926, preoccupied by *Microtus arvalis brauneri* Martino, 1926.)

Pitymys multiplex Alpine pine vole

Arvicola multiplex Fatio, 1905. Lugano, Ticino, Switzerland. Incl. *druentius* (French Alps), *fatioi*, *leponticus*.

RANGE (Map 50). Southern side of Alps, in Switzerland, Austria, Italy and France.

REMARKS. This form could with some justification be treated as a rather discrete subspecies of *P. subterraneus* with which it is interfertile in captivity (Meylan, 1972). However the very close proximity, perhaps even sympatry of the two forms, differing clearly in chromosomes, provides some grounds for giving them specific rank. The use of the earlier name *incertus* de Sélys-Longchamps, 1841 for this species by Dottrens (1962) was considered to be incorrect by Meylan (1970). The French form *druentius* is included following Saint Girons (1973). The holotype of *P. m. hercegoviniensis* Martino, 1940 appears to be a mismatched skin of a *Pitymys* and a skull of *Microtus*.

Pitymys tatricus* Tatra pine vole

Pitymys tatricus Kratochvil, 1952b. High Tatra Mts, Czechoslovakia.

RANGE (Map 50). Tatra Mts, between Czechoslovakia and Poland. Apparently restricted to alpine grassland above the treeline.

REMARKS. This species has been considered distinct from *P. subterraneus* by most subsequent authors, mainly on the basis of its distinctive karyotype.

Pitymys bavaricus* Bavarian pine vole

Pitymys bavaricus König, 1962. Garmisch-Partenkirchen, Bavarian Alps, Germany (730 m).

RANGE (Map 50). Bavaria.

REMARKS. This form was accepted as a species by Kratochvil (1970) who compared it in some detail with *P. tatricus*.

Pitymys liechtensteini†

Pitymys liechtensteini Wettstein, 1927. Croatia, Yugoslavia.

RANGE (Map 51). N.W. Yugoslavia.

REMARKS. This form was listed as a race of *P. subterraneus* by E & M-S and of *P. multiplex* by Dulic & Miric (1967). Petrov & Zivkovic (1971, 1974) have considered it a distinct species.

Pitymys schelkovnikovi†

Microtus schelkovnikovi Satunin, 1907. Near village of Dzhi, Talysh Mts, S.W. of Caspian Sea.
*Pitymys subterraneus dorothea** Ellerman, 1948. South of Kuramabad, Elburz Mts, Iran.

RANGE (Map 51). Talysh and Elburz Mts, S. of Caspian Sea.

REMARKS. This form was included in *P. subterraneus* by E & M-S, left *incertae sedis* by Ognev (1950) and Kuznetsov (1965) and treated as a distinctive species by Kratochvil (1970). The type and paratypes of *dorothea* show the characters used by Kratochvil in separating *P. schelkovnikovi* from *P. majori*.

Pitymys majori†

Microtus (Pitymys) majori Thomas, 1906. Sumela, S. of Trebizond, Asia Minor. Incl. ?*ciscaucasicus* (N. Caucasus), *colchicus*, ?*daghestanicus* (E. Caucasus), *fingeri* (N.W. Asia Minor), *intermedius* Shidlovsky, *nasarovi*, *rubelianus*.
*Arbusticola rubelianus vinogradovi** Sviridenko, 1936. N. Caucasus (nomen nudum).
*Microtus majori labensis** Heptner, 1948b. New name for *A. r. vinogradovi* Sviridenko, 1936, preoccupied by *Microtus (Lasiopodomys) vinogradovi* Fetisov, 1936.
*Microtus majori suramensis** Heptner, 1948b. New name for *Microtus (Arbusticola) rubelianus intermedius* Shidlovsky, 1919 preoccupied by *Microtus (Lagurus)* intermedius Taylor, 1911.

RANGE (Map 50). Caucasian region, northern and western Asia Minor.

REMARKS. This species was treated as such by Ognev (1948a). It was included in *P. subterraneus* by E & M-S and by Kuznetsov (1965) although the latter gave it specific rank in the caption to the maps. Kratochvil (1970) treated *P. majori* and *P. daghestanicus* as separate species but included them both in the same group. The species was dealt with in some detail by Felten *et al.* (1971b). More recently (Kratochvil & Kral, 1974) *ciscaucasicus* has also been given specific rank, along with *P. majori* and *P. daghestanicus*, on karyological grounds.

Pitymys savii

Arvicola savii de Sélys Longchamps, 1838. Near Pisa, Italy.
P. s. savii. Italy. Incl. *selysii*.
P. s. pyrenaicus (de Sélys Longchamps, 1838). France and N. Spain. Incl. *brunneus*, *gerbii*, *planiceps*.
P. s. nebrodensis Mina-Palumbo, 1868. Sicily.
*P. s. brachycercus** Lehmann, 1961. Calabria, S. Italy.
  Pitymys savii brachycercus Lehmann, 1961. Camigliatello Silano, Calabria, Italy.
?*P. s. felteni** Malec & Storch, 1963. Macedonia, Yugoslavia.
  Pitymys savii felteni Malec & Storch, 1963. Trnovo, Macedonia, Yugoslavia.

RANGE (Map 51). Italy, S.W. France, Pyrenees, N.W. Spain, Sicily; perhaps also in Macedonia.

REMARKS. The above names are arranged in subspecies primarily to indicate the geographical groups. The allocation of all these to *P. savii* is very tentative. The French form was given specific rank as *P. pyrenaicus* by Saint Girons (1973). Recent work on the chromosomes of Balkan *Pitymys* suggests that *felteni* may be specifically distinct (Petrov *et al.*, 1976).

***Pitymys lusitanicus*†** Lusitanian pine vole

Arvicola (Microtus) lusitanicus Gerbe, 1879. Portugal. Incl. ?*depressus*, *hurdanensis* (Salamanca), *mariae* (Galicia, Spain), ?*pelandonius* (Burgos).
*Microtus (Pitymys) savii gerritmilleri** Kretzoi, 1958. New name for *Pitymys depressus* Miller, 1908 (not *Arvicola arvalis depressa* Rörig & Börner, 1905).

RANGE (Map 51). The northwestern half of the Iberian Peninsula and extreme S.W. France.

REMARKS. For a detailed comparison with other Iberian species see Almaca (1973). All the above forms were allocated to *P. savii* by E & M-S.

Pitymys duodecimcostatus

Arvicola duodecimcostatus de Sélys Longchamps, 1839. Montpelier, Gard, France. Incl. *centralis* (Burgos), *flavescens* (Catalonia), *fuscus*, *ibericus* (Murcia), *pascuus* (Valencia), *provincialis*, *regulus* (Granada).

RANGE (Map 51). S.E. France (Provence and Rhone Valley), eastern and southern Spain.

REMARKS. The Balkan forms (*thomasi* and *atticus*), included in this species by E & M-S, are now generally considered to represent a distinct species, *P. thomasi.*

***Pitymys thomasi*†**

Microtus (Pitymys) thomasi Barrett-Hamilton, 1903. Vranici, Montenegro, Yugoslavia. Incl. *atticus* (Greece), *byroni*.

RANGE (Map 51). Southern coastal Yugoslavia, Greece (widespread except in east, and including the island of Evvoia). Presumably also in Albania.

REMARKS. Considered specifically distinct from *P. duodecimcostatus* (to which it was allocated by E & M-S) by Petrov & Zivkovic (1972) and others, primarily on the basis of its very different karyotype.

Genus *MICROTUS*
Grass voles

Microtus Schrank, 1798. Type-species *M. terrestris* Schrank = *Mus arvalis* Pallas. Incl. *Agricola*, *Alexandromys*, *Campicola*, *Chionomys*, *Euarvicola*, *Lasiopodomys*, *Lemmimicrotus*, *Proedromys*, *Stenocranius*, *Sumeriomys*, *Sylvicola* Fatio.
*Pallasiinus** Kretzoi, 1964. Type-species *Mus oeconomus* Pallas.

A very large genus represented throughout the Holarctic Region. Sometimes extended to include also all the species here placed in *Arvicola* and *Pitymys*. The characters distinguishing these three genera are somewhat trivial but they are definable and it seems best to retain them until a comprehensive examination of generic classification can be undertaken.

Problems remain in delimiting the species. *M. arvalis* has recently been found to comprise two widely sympatric sibling species and other such cases may emerge. Most species occur in the USSR where considerable attention is being paid to the taxonomic problems in this genus. The following account is in accordance with most recent Russian views. Subspecies have not generally been listed as such since most names are based on differences of size and colour which tend to vary clinally or upon dental pattern which shows great intrapopulation variability.

All Eurasian species are included below. Those on the southeastern fringe of the Palaearctic (Taiwan, S. China) are all confined to high altitude. Twenty-four species are listed, compared with 25 by E & M-S. However five of the species accepted by E & M-S have been considered conspecific with others (*guentheri*, *irani*, *hyperboreus*, *igmanensis* and *orcadensis*) whilst two of their species have been split (*M. arvalis* and *M. transcaspicus*). The following key does not allow critical identification of single specimens in every case but serves to indicate the principal characters used to separate the species.

1. Upper incisors each with a distinct groove on the anterior face; M^3 very reduced with only 2 inner ridges; (tail about 40% of head and body; China) . . . ***M. bedfordi***
– Upper incisors ungrooved . . . (2)
2. Tail long, over 50% of head and body (if near 50% see also dichotomy 15) . (3)
– Tail short, under 50% of head and body . . . (8)
3. M^2 with a well-developed postero-internal ridge; (E. Asia) . . . (4)
– M^2 without such a ridge. . . . (5)
4. Larger: hind foot 20–21 mm, head and body up to 135 mm; skull angular, temporal ridges fusing in interorbital region; M^3 with usually 4 inner ridges; M_1 with 6 inner and 5 outer ridges . . . ***M. clarkei***
– Smaller: hind foot about 18 mm; skull less angular, broad and flat in the interorbital region which is concave in profile; M^3 with 3 or 4 inner ridges; M_1 with 5 inner and 4 outer ridges ***M. millicens***
5. Tail 60–80% of head and body . . . (6)
– Tail 50–60% of head and body . . . (7)
6. M^3 complex with 4 or 5 ridges on each side; M_1 complex with usually 5 ridges on each side; (W. Asia) . . . ***M. roberti***
– M^3 less complex with 4 inner and 3 or 4 outer ridges; M_1 with usually 5 inner and 4 outer ridges; (Taiwan) . . . ***M. kikuchii***
7. Tail distinctly darker above than below; M^3 usually with 4 inner and outer ridges ***M. gud***
– Tail scarcely darker above than below; M^3 with only 3 inner and outer ridges ***M. nivalis***
8. M^3 with only 3 ridges on inner side; (tail very short, about 25% of head and body; E. Asia) (9)
– M^3 with 4 or more ridges on inner side . . . (10)
9. Pelage very yellowish, above and below; interorbital constriction narrow, about 3·0–3·5 mm; M^3 prolonged a little behind the 3rd outer ridge . . . ***M. brandti***
– Pelage more normal, brown above and grey below; interorbital constriction wider, about 4 mm; M^3 very short, not prolonged behind 3rd outer ridge which may be poorly developed . . . ***M. mandarinus***
10. Skull very narrow, zygomatic width about half greatest length or less, interorbital constriction under 3·3 mm . . . ***M. gregalis***
– Skull less narrow, width considerably more than half length, interorbital constriction over 3·3 mm . . . (11)
11. Auditory bullae rather large, length (without mastoid part) greater than diastema; tail very short, usually about 25% of head and body, uniformly pale or slightly darker above than below; mastoid parts of bullae inflated; (teeth fairly normal – M^2 may have a 3rd inner ridge, M^3 and M_1 moderately complex, all dentine triangles rather small and deeply separated) . . . ***M. socialis***
– Bullae normal; tail usually about 30–40% of head and body . . . (12)
12. M_1 rather simple, with usually only 3 sharp ridges on outer side; (pelage rather dark, hind feet rather dark brown above; 6 plantar pads; tail rather long – *c.* 40% of head and body) ***M. oeconomus***
– M_1 more complex, with at least 4 external ridges . . . (13)
13. M^2 with 3 inner ridges; (6 plantar pads; pelage rather long and shaggy) ***M. agrestis***
– M^2 with only 2 inner ridges . . . (14)
14. Tail usually considerably more than $\frac{1}{3}$ head and body; (hind feet dark brown above; 5 or 6 plantar pads) . . . (15)
– Tail $\frac{1}{3}$ head and body or less; (hind feet usually pale grey above; usually 6 plantar pads (except *M. mongolicus*) . . . (18)
15. M^3 with only 2 or 3 outer ridges . . . (16)
– M^3 with 4 outer and 5 inner ridges; (island of Sakhalin) . ***M. sachalinensis***
16. Large: head and body up to 160 mm; interorbital region angular, under 4·5 mm (17)
– Small: head and body up to 120 mm; interorbital region wide (over 4·5 mm) and smooth; (Japan) . . . ***M. montebelli***
17. Skull angular, interorbital ridge usually well developed; anterior lobe of M_1 grooved on outer side . . . ***M. maximowiczii***

– Skull less angular; anterior lobe of M_1 without groove on outer side . . . *M. fortis*
18. Tail very sharply bicoloured, rather short (*c.* 30% of head and body); (Siberia) . . . *M. middendorffi*
– Tail less sharply bicoloured, without a distinct line separating dark upper and light under surface, generally over 30% of head and body . . . (19)
19. M_1 with 4 outer ridges . . . (20)
– M_1 with 5 outer ridges . . . *M. arvalis*
M. subarvalis
M. transcaspicus
M. ilaeus
20. Six plantar pads; upper surface of skull very convex (Iberia) . . . *M. cabrerae*
– Five plantar pads; upper surface of skull flatter (hind feet dark brown above; Mongolia) . . . *M. mongolicus*

Microtus roberti — Robert's vole

Microtus roberti Thomas, 1906. Sumela, S. of Trebizond, N.E. Asia Minor. Incl. *circassicus, occidentalis* Turov, *personatus, pshavus* (Caucasus).
*Microtus roberti turovi** Hoffmeister, 1949. New name for *M. r. occidentalis* Turov.

RANGE (Map 54). Woodlands of N.E. Asia Minor and W. Caucasus.

Microtus gud

Microtus gud Satunin, 1909. Gudaur, Caucasus. Incl. *lasistanius* (N. Asia Minor), *lucidus, neujukovi, oseticus.*

RANGE (Map 54). Caucasus and N.E. Asia Minor.

REMARKS. Kuznetsov (1965) listed *gud* as a race of *M. nivalis* but Spitzenberger (1971b) found the two species to be sympatric in N.E. Asia Minor and provided a detailed comparison.

Microtus nivalis — Snow vole

Arvicola nivalis Martins, 1842. Bernese Oberland, Switzerland. Incl. *abulensis* (central Spain), *alpinus, aquitanius* (Pyrenees), *dementievi* (Kopet Dag Mts), *gotshobi, hermanis* (Palestine), *ighesicus* (Caucasus), *lebrunii* (Massif Central, France), *leucurus, malyi* (Yugoslavia), *mirhanreini*† (Tatra Mts), *nivicola, olympius, petrophilus, pontius* (Asia Minor), *radnensis, satunini, trialeticus, ulpius* (Rumania), *wagneri.*
*Microtus (Chionomys) nivalis loginovi** Ognev, 1950: 448. Mt Atcheshbok, Caucasus, USSR.
*Microtus (Chionomys) nivalis cedrorum** Spitzenberger (in Felten *et al.*), 1973. Ciglikara, *c.* 25 km SSW of Elmali, Kohu Dag, Antalya, Turkey (1750 m).

RANGE (Map 52). Mountains of Europe from Spain to Tatra, Carpathians and Balkans; W. Caucasus, Asia Minor, Palestine, Kopet Dag and Iran (Zagros Mts – Lay, 1967).

REMARKS. The form *mirhanreini* from the Tatra Mts, left *incertae sedis* by E & M-S, was allocated to *M. nivalis* by Kratochvil (1956). Kuznetsov (1965) included *gud* in this species but they appear to be distinct and sympatric in some areas (Spitzenberger, 1971b). Because of the fragmentation of the montane habitat many of the above names probably apply to moderately discrete subspecies but in the absence of any comprehensive revision it does not seem practicable to provide a meaningful list of subspecies.

Microtus socialis — Social vole

Mus socialis Pallas, 1773. Lower Ural River. Incl. *astrachanensis, binominatus, colchicus* Argyropulo, *goriensis, gravesi, guentheri*† (Asia Minor), *hartingi*† (Greece), *hyrcania* (N.E. Iran), *irani*† (S.W. Iran), *lydius*†, *mustersi*† (Libya), *mystacinus*† (Iran), *paradoxus* (Kopet Dag), *parvus* (Caucasus), *philistinus*† (Palestine), *satunini* Ognev, *schidlovskii, shevketi*†, ?*syriacus*†.
*Sumeriomys guentheri martinoi** Petrov, 1939. Pepeliste, near Krivolak, 40 km S.E. of Veles, Yugoslavia.
*Microtus guentheri macedonicus** Kretzoi, 1964. New name for *Sumeriomys guentheri martinoi* Petrov, 1939 (preoccupied by *Pitymys nyirensis martinoi* Ehik, 1935).
*Microtus (Chilotus) socialis nikolajevi** Ognev, 1950: 387. Kuyuk-Tuk Island, Crimea.
*Microtus guentheri strandzensis** Markov, 1960. Near village of Gramatikovo, Bezirk Burgas, Strndza Mts, E. Bulgaria.

RANGE (Map 52). Steppes from R. Dnepr and Crimea east to L. Balkash and

Dzungaria, south through Asia Minor to Palestine and S.W. Iran; also in S.E. Europe (S. Yugoslavia, Bulgaria, Greece, Thrace) and N.E. Libya.

REMARKS. E & M-S treated *M. socialis*, *M. guentheri* and *M. irani* as separate species. They have been considered conspecific by most recent authors (Kuznetsov, 1965; Lay, 1967; Harrison, 1972). Several Iberian forms, namely *cabrerae*, *dentatus* and, doubtfully, *asturianus*, were included in *M. guentheri* by Brink (1967) but these are excluded here following Niethammer *et al.* (1964). The Libyan form *mustersi* was treated as specifically distinct from, but closely related to *M. guentheri* by Ranck (1968). The name *syriacus* Brants, 1827, left *incertae sedis* by E & M-S, was tentatively considered a synonym of *M. socialis* by Harrison, 1972. The Iranian form *mystacinus*, allocated to *M. arvalis* by E & M-S and by Lay (1967), is included here following examination of the syntypes in the British Museum.

Microtus arvalis Common vole

Mus arvalis Pallas, 1779. Germany (see Remarks below). = *arvensis*. Incl. *albus*, *angularis*, *assimilis*, *asturianus* (Spain), *brauneri*, *brevirostris*, *calypsus*, *cimbricus*, *contigua*, ?*cunicularius*, *depressa*, *duplicatus*, *flava*, *fulva*, *galliardi*, *ghalgai*, *gudauricus*, *howelkae*, ?*igmanensis*† (Yugoslavia), *incertus*, *incognitus*, *iphigeniae* (Crimea), *levis* (Rumania), *macrocranius* (Caucasus), *meridianus* (Pyrenees), *muhlisi* (Asia Minor), *obscurus* Eversmann (Altai Mts), *orcadensis*† (Orkney Is), *principalis*, *relictus*, *rhodopensis*, *ronaldshaiensis*†, *rossiaemeridionalis*, *rousiensis*†, *rufescentefuscus*†, *sandayensis*†, *sarnius* (Guernsey, Channel Isles), *simplex*, *terrestris* Schrank, *transcaucasicus*, *transuralensis*, *variabilis*, *vulgaris*, *westrae*†.
*Microtus arvalis oyaensis** Heim de Balsac, 1940. Ile de Yeu, France.
*Microtus arvalis grandis** Martino & Martino, 1948. Gatacko Polje, near Gacko, Hercegovina, Yugoslavia.
*Microtus arvalis caspicus** Ognev, 1950: 215. Tchanyushkima, S. Astrakhan Dist, USSR.
*Microtus arvalis ruthenus** Ognev, 1950: 210. Omutninsk Dist, Kirov Region, N. European Russia.
*Microtus arvalis meldensis** Delost, 1955. Mareuil-les-Meaux, Seine et Marne, France.
*Microtus arvalis epiroticus** Ondrias, 1966. Perama, near Ioannina, Epirus, Greece.[1]

RANGE (Map 53). Europe from N. Spain and Denmark east through Russia and Siberia as far as upper Yenesei; south to Caucasus, Altai and L. Balkash; isolates on several of the Orkney Islands, Guernsey (Channel Isles), Yeu (France) and perhaps Spitzbergen (for last see Nyholm, 1966). A map of localities in which critical distinction from *M. subarvalis* has been made by analysis of karyotype was produced by Malygin & Orlov (1974).

REMARKS. *M. arvalis* as recognized by E & M-S has recently been found to comprise two sibling species over much of its range (see *M. subarvalis* below). All synonyms falling within the range of *M. arvalis* have been listed above although some may properly refer to *M. subarvalis*. The form *mongolicus*, included in *M. arvalis* by E & M-S, is here considered specifically distinct following recent Russian work (e.g. Malygin & Orlov, 1974). Other forms given specific rank by E & M-S are here included in *M. arvalis*, namely *orcadensis* and other forms from Orkney (Stein, 1958) and *igmanensis* from Yugoslavia, although the status of the last is unclear.

A neotype for *M. arvalis* Pallas was designated by Meyer *et al.* (1972) using a specimen from Leningrad Oblast, USSR.

Microtus subarvalis*

Microtus subarvalis Meyer, Orlov & Skholl, 1972. Leningrad Oblast, USSR.

RANGE (Map 53). European Russia from Finnish border to Caucasus (Armenia), east to Urals and west to Ukraine (map in Malygin & Orlov, 1974); also in S. Yugoslavia (Zivkovic & Petrov, 1974) and Bulgaria (Kral, 1975).

REMARKS. This species was diagnosed as distinct from *M. arvalis* primarily on the basis of its very distinctive karyotype (2n = 54, mostly acrocentric, compared with 2n = 46, mostly metacentric, in *M. arvalis*). The shapes of the spermatozoa and baculum are also distinctive (Aksenova & Tarasov, 1974). It is listed here as a distinct species in view of its wide sympatry with *M. arvalis* s.s. I have not seen authenticated specimens.

[1] Considered an independent species by Ruzic *et al.*, 1975 (*Arh. Poljoprivredne Nauk* 28: 153–160).

Microtus transcaspicus Transcaspian vole

Microtus transcaspicus Satunin, 1905. Ashabad, Turkmenistan. Incl. ?*khorkoutensis* (N.E. Iran).

RANGE (Map 53). S. Turkmenistan, N. Afghanistan and probably N. Iran.

REMARKS. This member of the *M. arvalis* group was given specific rank by E & M-S, included in *M. arvalis* by Kuznetsov (1965) and by Niethammer (1970) but again given specific rank by Malygin & Orlov (1974). Its status as a separate species is supported by the sterility of hybrids between it and both *M. arvalis* and *M. ilaeus* (Meyer, 1976).

Microtus ilaeus† Tien Shan vole

Microtus ilaeus Thomas, 1912. R. Ussek, S.E. Kazakhstan.
*Microtus arvalis kirgisorum** Ognev, 1950:219. Tuyouk, Khirgiski Mts, Khirgizia.
*Microtus arvalis innae** Ognev, 1950:221. Lake Issik-Kul, Khirgizia.

RANGE (Map 53). Tien Shan Mts, Khirgizia.

REMARKS. This form was included in *M. transcaspicus* by E & M-S but treated as a distinct species by Malygin & Orlov (1974) mainly on karyological evidence. Hybrids between it and both *M. arvalis* and *M. transcaspicus* are sterile (Meyer, 1976).

Microtus mongolicus† Mongolian vole

Arvicola mongolicus Radde, 1862. Transbaikalia. Incl. *baicalensis*, *poljakovi*, *xerophilus*.

RANGE (Map 53). N.E. Mongolia and adjacent parts of USSR south of Lake Baikal.

REMARKS. Included in *M. arvalis* by E & M-S, Kuznetsov (1965) and Bannikov (1954), but given specific rank by recent Russian authors, e.g. by Malygin & Orlov (1974).

Microtus cabrerae

Microtus cabrerae Thomas, 1906. Sierra de Guadarrama, Madrid Prov., Spain. Incl. *dentatus*.

RANGE (Map 53). Central and S.E. Spain.

REMARKS. The status of this form was reviewed by Niethammer *et al.* (1964) who concluded that it was a good species and refuted its association with *M. guentheri* (= *M. socialis*) by Brink (1956).

Microtus maximowiczi†

Arvicola maximowiczii Schrenk, 1859. Mouth of Omutnaya River, upper Amur, E. Siberia. Incl. *ungurensis*.

RANGE (Map 54). E. shore of Lake Baikal to the upper Amur Basin, in 'open spaces in the southern taiga' (Orlov *et al.*, 1974).

REMARKS. The name *maximowieczi* is in general use for this species in the USSR. E & M-S used the name *ungurensis* and listed *M. maximowieczi* as of doubtful status. Several Russian authors have considered *M. fortis* conspecific with this species but Orlov *et al.* (1974) have demonstrated that they are sympatric and distinct in parts of Transbaikalia and Meyer (1976) was unable to produce hybrids.

Microtus fortis Reed vole

Microtus fortis Büchner, 1889. Valley of Huang Ho, Ordos Desert, Inner Mongolia, China. Incl. *calamorum* (lower Yangtze Kiang), *dolichocephalus* (Manchuria), *michnoi* (Transbaikalia), *pelliceus* (Ussuri).
*Microtus fortis uliginosus** Jones & Johnson, 1955. Chip'o-ri (38°08'N, 127°19'E), central Korea.

RANGE (Map 54). China from Chekiang and the lower Yangtze Valley north through Manchuria and Korea to the Amur Valley, west to Lake Baikal. Sympatric with *M. maximowieczi* in parts of Transbaikalia. Records from Sakhalin may properly refer to *M. sachalinensis*.

Microtus sachalinensis* Sakhalin vole

Microtus sachalinensis Vasin, 1955. Olen River, Poronaisk region, Sakhalin, USSR.

RANGE (Map 55). Only known from the island of Sakhalin.

REMARKS. I have not seen this species. The baculum was illustrated along with those of related species by Aksenova & Tarasov (1974). It was considered closely similar to *M. fortis* and *M. maximowiczi* by Meyer (1976) but with a distinctive karyotype and apparently unable to hybridize with these species.

Microtus kikuchii

Microtus kikuchii Kuroda, 1920. Taiwan.

RANGE (Map 52). Taiwan (2800 m).

REMARKS. Considered a member of the *M. maximowiczi* group by Zimmermann (1964).

Microtus clarkei — Clarke's vole

Microtus clarkei Hinton, 1923. Yunnan, China.

RANGE (Map 55). Yunnan and N. Burma (3400–4000 m).

Microtus millicens — Szechuan vole

Microtus millicens Thomas, 1911. Weichoe, W. Szechuan, China (12 000 ft).

RANGE (Map 52). Szechuan, China.

Microtus montebelli

Arvicola montebelli Milne-Edwards, 1872. Honshu, Japan. Incl. *brevicorpus* (Sado Is.), *hatanezumi*.

RANGE (Map 55). Honshu, Kyushu and Sado Island, Japan; Sikotan Island, Kuriles (Sokolov, 1954).

REMARKS. Although consistently treated as distinct, this species was considered by Zimmermann (1964) to be an insular vicariant of *M. maximowiczi* (*sensu latu*, i.e. including *fortis*).

Microtus agrestis — Field vole

Mus agrestis Linnaeus, 1761. =*nigricans*. Incl. *angustifrons*, *arcturus* (Zungaria), *argyropuli*, *bailloni* (France), *campestris*†, *britannicus*, *estiae*, *exsul* (Outer Hebrides, Scotland), *fiona*, *gregarius* (Germany), *hirta* (England), *insularis*, *intermedia*, *latifrons*, *levernedii*, *luch*, *macgillivrayi* (Island of Islay, Scotland), *mial*, *mongol* (Mongolia), *neglectus*, *nigra*, *ognevi* Scalon (N.W. Siberia), *orioecus*, *pannonicus*, *punctus*, *rozianus* (Portugal), *rufa*, *tridentinus*, *wettsteini* Ehik.
*Microtus agrestis scaloni** Heptner, 1948. New name for *M. a. ognevi* Scalon, 1935 (not Turov, 1926).
*Microtus agrestis carinthiacus** Kretzoi, 1958. New name for *M. a. wettsteini* Ehik, 1928, preoccupied by *Pitymys subterraneus wettsteini* Ehik, 1926.
*Microtus agrestis heptneri** Hamar, 1963. Bucegi Mts, S. Carpathians, Rumania.
*Microtus agrestis enez-groezi** Heim de Balsac & Beaufort, 1966b. Groix Island, Morbihan, France.
*Microtus agrestis armoricanicus** Heim de Balsac & Beaufort, 1966b. Quimper, Finisterre, France.
Also other synonyms in N. America of which the earliest is *Mus pennsylvanica* Ord, 1815.

RANGE (Map 54). From Atlantic coast through N. and C. Europe and Siberia as far as the R. Lena. Montane in south: N. Portugal, Pyrenees, N. Yugoslavia, S. Urals, Altai, L. Baikal. Britain, including many small islands, e.g. in Hebrides.

REMARKS. The relationship between Palaearctic *M. agrestis* and Nearctic '*M. pennsylvanicus*' was studied in detail by Klimkiewicz (1970) who concluded that there were no morphological reasons for treating them as separate species.

Microtus oeconomus — Root vole

Mus oeconomus Pallas, 1776. Ishim Valley, S.W. Siberia. Incl. *altaicus* (Altai), *anikini*, *arenicola* (Netherlands), *dauricus* (Transbaikalia), *flaviventris* (Kansu), *hahlovi*, *kamtschaticus* (Kamchatka), *kjusjurensis*, *koreni* (N.E. Siberia), *limnophilus* (Tsaidam, Chinghai, China), *malcolmi*, *medius*, *mehelyi*, *montiumcaelestinum* (Dzungar Alatau, S.E. Kazakhstan), *naumovi*, *ouralensis* (S. Urals), *petshorae*, *ratticeps* (N. European Russia), *shantaricus* (Gt Shantar Is., E. Siberia), *stimmingi*, *suntaricus*, *tschuktschorum* (E. Siberia), *uchidae* (Kurile Is.).
Also other forms in N. America of which the earliest name is *Arvicola operarius* Nelson, 1893 from Alaska.

RANGE (Map 55). The entire tundra and taiga zones of the northern Palaearctic from

Scandinavia to N.E. Siberia, south to N. Germany, Ukraine, Semirechyia, and Shensi. Southern isolates in Netherlands, Hungary, Kirghiz Mts and Tsaidam. Also in Alaska and adjacent territories in N. America.

Microtus middendorffi Middendorff's vole

Arvicola middendorffi Poliakov, 1881. Taimyr Peninsula, N. Siberia. Incl. *hyperboreus*† (Verkhoyansk Mts), *obscurus* Middendorff, *ryphaeus* (Urals), *swerevi*†, *tasensis*, *uralensis* Skalon.

RANGE (Map 54). Tundra of N. Siberia from mouth of R. Kolyma in east to the Yamal Peninsula and south in the Urals to about 62°N. Also in the Lena Valley near Yakutsk.

REMARKS. Recent studies have shown that the form *hyperboreus* is fully interfertile with *M. middendorffi* s.s. and suggest that these should be considered conspecific (Gileva, 1972).

Microtus bedfordi Duke of Bedford's vole

Proedromys bedfordi Thomas, 1911. 60 miles S.E. of Minchow, Kansu, China.

RANGE (Map 55). Kansu, China. Only known from the type-locality.

Microtus brandti Brandt's vole

Arvicola (Hypudaeus) brandtii Radde, 1861. N.E. Mongolia. Incl. *aga*, *warringtoni*.
*Microtus (Lasiopodomys) brandtii hangaicus** Bannikov, 1948. River Dzak, near Dzak, S.W. Khangai, Mongolia.

RANGE (Map 52). Mongolia and adjacent parts of USSR in Transbaikalia.

Microtus mandarinus Mandarin vole

Arvicola mandarinus Milne-Edwards, 1871. N. Shansi, China. Incl. *faeceus* (Hopei), *jeholensis*, *johannes*, *kishidae* (Korea), *pullus*, *vinogradovi*†.

RANGE (Map 52). China from S.E. Shensi through Shansi to N.E. of Peking; also in N. Mongolia and the adjacent part of USSR to the south of Lake Baikal.

REMARKS. E & M-S left *vinogradovi* as *incertae sedis* but recent Russian authors agree in treating it as a synonym of *M. mandarinus*.

Microtus gregalis Narrow-skulled vole

Mus gregalis Pallas, 1779. East of R. Chulym, central Siberia. Incl. *angustus* (Inner Mongolia), *brevicauda* (Yakutia), *buturlini*, *castaneus* Kashkarov, *dolguschini*, *eversmanni* (Altai), *kossogolicus*, *major*, *montosus* (Pamirs), *nordenskioldi* (Taimyr Peninsula), *raddei* (Transbaikalia), *ravidulus* (Sinkiang), *slowzovi*, *tarbagataicus*, *tianschanicus* (Tien Shan), *tundrae*, *unguiculatus*, *zachvatkini* (Aral Sea).
*Microtus (Stenocranius) gregalis dukelskiae** Ognev, 1950:477. Vostochnoye, 35 km S.E. of Minusinsk, Upper Yenesei, Siberia.
*Microtus gregalis talassicus** Heptner, 1948b. New name for *M. g. castaneus* Kashkarov, 1923 (not de Sélys Longchamps).
*Microtus gregalis sirtalensis** Yung, 1966. Bargan Plateau, Inner Mongolia, China.

RANGE (Map 55). The tundra zone of Siberia from the White Sea to the far north-east; the wooded steppe zone from the southern Urals east to the Amur and Manchuria, south to the Aral Sea, Pamirs and Sinkiang.

REMARKS. Closely related vicariant species occur on the St Matthews Islands in the Bering Sea (*M. abbreviatus* Miller, 1899) and in Alaska (*M. miurus* Osgood, 1901).

Genus *LAGURUS*
Steppe lemmings

Lagurus Gloger, 1841. Type-species *L. migratorius* Gloger = *Georychus luteus* Eversmann. Incl. *Eremiomys*, *Lemmiscus*.
*Eolagurus** Argyropulo, 1946. Subgenus for type-species *Lagurus luteus* Eversmann (presumably in the belief, following Ellerman (1940–41), that *L. lagurus* was the type of *Lagurus*).

A genus of three distinctive species, two Palaearctic and one Nearctic. Kretzoi (1969) placed it in a tribe Lagurini, with *Hyperacrius* as another possible member.

Black mid-dorsal stripe; rest of dorsal pelage greyish brown; condylobasal length under 27 mm; M3 more complex, with 4 external and 3 internal ridges . . . . . . ***L. lagurus***

No mid-dorsal stripe; dorsal pelage sandy yellow; condylobasal length over 27 mm; M3 simpler, usually with 3 external and 2 internal ridges . . . . ***L. luteus***

Lagurus lagurus Steppe lemming

Mus lagurus Pallas, 1773. Mouth of R. Ural, W. Kazakhstan. Incl. *abacanicus, aggressus, altorum* (Zungaria), *occidentalis* (Ukraine).

RANGE (Map 48). Steppes from Ukraine through N. Kazakhstan to western parts of Mongolia and Sinkiang.

Lagurus luteus Yellow steppe lemming

Georychus luteus Eversmann, 1840. N.W. of Aral Sea, Kazakhstan. Incl. *migratorius, przewalskii* (Sinkiang).

RANGE (Map 48). Formerly widespread in Kazakhstan to S. Mongolia and N. Sinkiang; now extinct in Kazakhstan. The extreme reduction in range has been described and mapped by Kalabukhov (1970).

Genus ***PROMETHEOMYS***

Prometheomys Satunin, 1901. Type-species *P. schaposchnikowi* Satunin.

A monospecific genus of doubtful affinities. Vorontsov (1966) placed it in the tribe Fibrini (along with *Clethrionomys* and *Dolomys*; Gromov (1972) gave it a very isolated position within the Microtinae.

Prometheomys schaposchnikowi Long-clawed mole-vole

Prometheomys schaposchnikowi Satunin, 1901. Central Caucasus.

RANGE (Map 56). Caucasus and extreme N.E. Asia Minor (Spitzenberger & Steiner, 1964).

Genus ***ELLOBIUS***

Mole-voles

Ellobius Fischer, 1814. Type-species *Mus talpinus* Pallas.

A distinctive genus of microtines strongly adapted for a subterranean life and confined to the steppes of central Asia. The genus was made the sole member of a tribe Ellobiini by Gromov (1972) who queried its inclusion in the Microtinae. Three species were listed by E & M-S but most recent authors agree in recognizing only two.

Temporal ridges fuse to form saggital crest in adult; interparietal usually absent ***E. fuscocapillus***

Temporal ridges remain apart; interparietal present . . . . . . ***E. talpinus***

Ellobius talpinus Northern mole-vole

Mus talpinus Pallas, 1770. R. Volga, Russia. Incl. *albicatus* (Sinkiang), *ciscaucasica, coenosus* (Tien Shan), *fusciceps* (Samarkand), *kashtchenkoi* (W. Siberia), *larvatus* (Mongolia), *murinus, ognevi, orientalis, rufescens, tancrei* (Altai), *transcaspiae, ursulus*.

*Ellobius talpinus tanaiticus** Zubko, 1940. Atamanskii, Salsk District, N. of Caucasus.

?*Ellobius alaicus** Vorontsov, Liapounova *et al.*, 1969. Between Sary-Tashem and Bardabo, Alayski Valley, S. Kirghizia (3300 m).

RANGE (Map 56). Steppes from Ukraine and Crimea through Russian Turkestan to Mongolia and Sinkiang; north to Sverdlovsk, south to N. Afghanistan.

REMARKS. *E. alaicus* was diagnosed on the basis of its karyotype alone. It is tentatively included here as a marginal chromosome race of *E. talpinus*.

Ellobius fuscocapillus Southern mole-vole

Georychus fuscocapillus Blyth, 1842. Quetta, Baluchistan.

E. f. fuscocapillus. Eastern part of range. Incl. *farsistani, intermedius*.

E. f. lutescens Thomas, 1897. Eastern part of range, type-locality in E. Asia Minor. Incl. *legendrei, woosnami*.

RANGE (Map 56). Baluchistan and Afghanistan through Iran and S. Turkmenistan to Kurdistan.

REMARKS. E & M-S treated *lutescens* and *fuscocapillus* as specifically distinct but they have been considered conspecific by most recent authors, e.g. Ognev (1950), Kuznetsov (1965) and Harrison (1972).

Subfamily **GERBILLINAE**

Gerbils and jirds

The gerbillines constitute a rather clearly defined group of small rodents characteristic especially of the desert and dry steppe zone from central Asia to the Sahara but with representatives in dry habitats throughout Africa and in India. The group is sometimes given family rank but is more often treated as a subfamily of the Cricetidae. The generic classification is rather unstable and the distinctions are slight. Nine genera are recognized here; an additional seven occur in Africa south of the Sahara. The few Indian species are all included here.

There has been no comprehensive revision of the group subsequent to Ellerman (1941). The structure of the auditory region, which is used extensively in classification, has been reviewed by Lay (1972).

1. Upper incisors ungrooved; (head and body up to *c.* 180 mm; tail shorter than head and body; ears very short, under 18 mm; bullae moderately large, cheek-teeth hypsodont) ***PSAMMOMYS*** (p. 128)
- Upper incisors each with one groove . . . . (2)
- Upper incisors each with two grooves, close together; (very large: head and body up to 200 mm, hind feet *c.* 40 mm; tail shorter than head and body, with a terminal black tuft; bullae moderately large; cheek-teeth evergrowing). . . ***RHOMBOMYS*** (p. 129)
2. Tail with a distichous tuft extending for the distal half to two-thirds, dark with a white tip; (tail much longer than head and body; medium size: head and body up to *c.* 120 mm; bullae moderately large; cheek-teeth hypsodont) . . . . ***SEKEETAMYS*** (p. 124)
- Tail much less tufted or untufted, tip not white . . . . (3)
3. Tail short, under 50 % of head and body, thick and club shaped; (bullae greatly enlarged) ***PACHYUROMYS*** (p. 124)
- Tail longer, at least 80 % of head and body . . . . (4)
4. Rostrum of skull very short, nasals shorter than frontals; (tail *c.* 90 % of head and body; soles of hind feet entirely haired) . . . . ***BRACHIONES*** (p. 128)
- Rostrum longer, nasals longer than frontals . . . . (5)
5. Cheek-teeth moderately to very hypsodont, M^1 and M^2 bilaterally symmetrical; (head and body usually over 100 mm, up to 200 mm; tail usually about equal to head and body; hind feet less than 25 % of head and body) . . . . (6)
- Cheek-teeth not or only slightly hypsodont, M^1 and M^2 distinctly asymmetrical; (head and body 70–120 mm) . . . . (7)
6. Cheek-teeth less hypsodont, upper ones usually forming disconnected transverse laminae at the occlusal surface; dorsal pelage rather uniform in colour; zygomatic plates very large and projecting forwards on either side of rostrum; bullae rather small ***TATERA*** (p. 124)
- Cheek-teeth more hypsodont, upper ones with laminae lozenge-shaped and interconnected; dorsal pelage usually with dark tips giving speckled appearance; zygomatic plates normal; bullae often enlarged . . . .***MERIONES*** (p. 125)
7. Tail shorter than head and body, untufted; hind feet usually under 25 % of head and body; cheek-teeth slightly hypsodont, uppers only slightly asymmetrical ***DIPODILLUS*** (p. 123)
- Tail longer than head and body, tufted; hind feet usually over 25 % of head and body; cheek-teeth not hypsodont, uppers more distinctly symmetrical . ***GERBILLUS*** (p. 118)

Genus ***GERBILLUS***

Gerbillus Desmarest, 1804. Type-species *G. aegyptius* Desmarest = *Dipus gerbillus* Olivier. Incl. *Endecapleura* = *Hendecapleura*.

A large genus of small mouse-sized gerbils in the Saharan and Arabian regions and in southwestern Africa. Although about half of the species included here have been consistently allocated to *Gerbillus*, the remainder, characterized especially by having the soles of the feet naked instead of hairy as in *Gerbillus* s.s., have been included in *Dipodillus* which has variously been treated as an independent genus or as a subgenus of *Gerbillus*. However Petter (1959) argued that the type-species of *Dipodillus* (*D. simoni*) should alone be separated generically from *Gerbillus*. The next available name for the naked-soled species, *Hendecapleura*, can then be used for a subgenus within *Gerbillus*.

The number and diagnoses of species are still very provisional. Differences between sympatric species can be very slight whilst there is considerable subspecific variation. Much of the subspecific variation involves dorsal colour, size and tail-length. The correlation between colour of pelage and of soil has been clearly demonstrated in *G. nanus* by Harrison & Seton-Browne (1969). Since all these characters are liable to respond to local conditions and to recur in similar form in different parts of the range it is unlikely that many really discrete subspecies can be recognized. In view of this, and the very tentative limits of the species themselves, subspecies are not generally listed below as such.

The genus is also represented on the southern edge of the Sahara, in East Africa and in southern Africa. These forms, including discrete species as well as races of more northern species, are not listed here. All Asiatic forms are included although some extend into the Indian region. The arrangement of species here is based mainly on the works of Petter (1971b), Ranck (1968) and Harrison (1972). The following key is very imperfect but indicates the characters that have been principally used to distinguish species.

1. Soles of hind feet naked (subgenus *Hendecapleura*) . . . . (2)
– Soles of hind feet hairy (subgenus *Gerbillus*) . . . . (8)
2. Tail evenly haired, without terminal tuft; (large: hind feet over 27 mm, greatest length of skull over 30 mm) . . . . **G. poecilops**
– Tail with hairs increasing in length towards tip . . . . (3)
3. Very small: hind feet under 20 mm, greatest length of skull under 23 mm, maxillary tooth-row under 3 mm . . . . **G. henleyi**
– Larger than above . . . . (4)
4. Bullae large, their greatest length (including mastoid portion) over 32 % of greatest length of skull, mastoid parts conspicuously inflated and extending behind occipital condyles (5)
– Bullae small, under 32 % of length of skull, mastoid parts only slightly inflated, not extending behind condyles . . . . (7)
5. Large: greatest length of skull over 30 mm, hind feet generally over 26 mm; (tail very long with prominent black tuft; S.W. Arabia) . . . . **G. famulus**
– Smaller: skull under 30 mm, hind feet under 26 mm . . . . . . (6)
6. Auditory meatus with anterodorsal rim round; a curtain of bone within the meatus blocking its upper half and concealing the ossicles . . . . **G. dasyurus**
– Auditory meatus with the anterodorsal rim inflated and reflexed; no curtain within the meatus . . . . **G. nanus**
7. Tail scarcely tufted, terminal hairs only slightly longer than others; (hind feet 21–25 mm, ears 13–15 mm, greatest length of skull 26–29 mm) . . . . **G. mesopotamiae**
– Tail moderately tufted; greatest length of skull over 27 mm . . . **G. campestris**
8. Large: head and body generally over 100 mm, hind feet over 32 mm, greatest length of skull 31–34 mm, maxillary tooth-row 4·2–5·0 mm; (bullae relatively small, mastoid part not projecting beyond occiput) . . . . **G. pyramidum** **G. perpallidus**
– Smaller than above . . . . (9)
9. Tail very long, over 150 % of head and body (Pakistan) . . . . . **G. gleadowi**
– Tail under 150 % of head and body . . . . (10)
10. Bullae large, mastoid parts projecting behind occipital condyles; tail longer, usually over 130 % of head and body . . . . **G. cheesmani**

- Bullae smaller, not projecting behind condyles; tail shorter, usually under 130% of head and body . . . . (11)

11. Mid-dorsal pelage clear yellowish brown; sides of feet and digits with prominent fringes of hair; tail distinctly tufted . . . . ***G. gerbillus***
- Mid-dorsal pelage washed with grey; sides of feet and digits lacking long fringes; tail almost completely lacking terminal tuft . . . . (12)

12. Larger: head and body usually over 100 mm, greatest length of skull usually over 29 mm, upper molar row over 4·0 mm; (Libya) . . . . ***G. aureus***
- Smaller: head and body usually under 100 mm, greatest length of skull under 29 mm, upper molar row under 4·0 mm . . . . ***G. andersoni***

Subgenus *HENDECAPLEURA*

Gerbillus campestris — Large North African gerbil

Gerbillus campestris Levaillant, 1857. Algeria. Incl. *cinnamomeus* (Morocco), *deserti*, *dodsoni* (N.W. Libya), *gerbii*, *hilda*†, *minutus*, *patrizii* (Cufra Oasis, Libya), *riparius*, *rozsikae*.
*Gerbillus campestris brunnescens** Ranck, 1968:133·5 km S.E. of Derna, Cyrenaica Province, Libya.
*Gerbillus campestris haymani** Setzer, 1958c:208. Siwa Oasis, Western Desert Governorate, Egypt.
*Gerbillus (Hendecapleura) jamesi** Harrison, 1967. Between Bou Ficha and Enfidaville, Tunisia.
*Gerbillus campestris wassifi** Setzer, 1958c:209. Near Salum (*c*. 200 ft), Western Desert Governorate, Egypt.
Also other synonyms from Sudan and Somalia.

RANGE (Map 57). Northern Sahara from Morocco to Egypt and south to Sudan and Somalia.

REMARKS. There is fairly general agreement about the content and delimitation of this species. Petter (1971b) included also *amoenus* (Egypt) but it is here placed in *G. nanus*. The Moroccan form *hilda*, unallocated by E & M-S, is included following Petter.

Gerbillus poecilops — Large Aden gerbil

Gerbillus (Dipodillus) poecilops Yerbury & Thomas, 1895. Lahej, Aden, Arabia.

RANGE (Map 57). S.W. and W. Arabia.

Gerbillus famulus — Black-tufted gerbil

Gerbillus (Hendecapleura) famulus Yerbury & Thomas 1895. Lahej, Aden.

RANGE (Map 57). S.W. Arabia.

REMARKS. This form is given specific rank following Harrison (1972) and E & M-S. Petter (1971b) treated it as a race of *G. nanus*.

Gerbillus nanus — Baluchistan gerbil

Gerbillus nanus Blanford, 1875. Baluchistan. Incl. *amoenus* (Giza, Egypt), *arabium* (N.W. Arabia), ?*garamantis* (Algeria), *grobbeni* (Cyrenaica, Libya), *indus* (Pakistan), *mackilligini* (S.E. Egypt), *mimulus* (S.W. Arabia), *quadrimaculatus* Bodenheimer, *vivax* (Fezzan, Libya).
*Gerbillus nanus setonbrownei** Harrison, 1968c. Suwera, Batinah Coast, Oman.

RANGE (Map 57). From Pakistan through most of Arabia and in N. Africa from Egypt to Algeria.

REMARKS. The extent to which the forms in north and northwestern Africa should be considered conspecific with *G. nanus* is debatable. E & M-S included only *garamantis* from Algeria, making a very discontinuous distribution. Petter (1971b) likewise included *garamantis* but also the intervening Libyan forms *vivax* and *grobbeni*. Ranck (1968) associated *vivax* with the Egyptian *amoenus* and tentatively with *garamantis* and did not believe that *G. nanus* extended west of the Nile. Harrison (1972) thought, in spite of Ranck's views, that some of these African forms should be allocated to *G. nanus*. Since it does not seem possible to diagnose Ranck's *G. amoenus* against *G. nanus* they are here tentatively united.

Correlation of colour with that of the soil was demonstrated by Harrison & Seton-Browne (1969) and illustrated by colour photographs.

Gerbillus dasyurus Wagner's gerbil

Meriones dasyurus Wagner, 1842. W. coast of Arabia. Incl. *dasyuroides, lixa.*
*Gerbillus dasyurus gallagheri** Harrison, 1971. Masifi, Oman.
*Gerbillus (Dipodillus) dasyurus leosollicitus** von Lehmann, 1966c. Deir-el-Hajar, 25 km S.E. of Damascus, Syria.
*Gerbillus (Dipodillus) dasyurus palmyrae** von Lehmann, 1966c. Palmyra, Syria.

RANGE (Map 57). Sinai to Syria and through N. and E. Arabia, with possibly isolated segments in Oman and S.W. Arabia.

REMARKS. The content of this species has been the subject of much confusion. E & M-S included *simoni* (Algeria) which subsequent authors (followed here) have agreed represents a separate genus, *Dipodillus*. They also included other African forms (*amoenus* and *vivax*), the Indian *indus* and *mimulus* (Aden) all of which should be referred to *G. nanus* rather than *G. dasyurus* on the basis of the characters of the auditory meatus described by Harrison (1972).

_Gerbillus mesopotamiae_* Mesopotamian gerbil

Gerbillus (Dipodillus) dasyurus mesopotamiae Harrison, 1956b. Near Amiraya, W. bank of R. Euphrates, S.W. of Faluja, Iraq.

RANGE (Map 57). Valley of lower Tigris and Euphrates.

REMARKS. Harrison (1972) subsequently gave this form specific rank and maintained that it is in places sympatric with *G. dasyurus* (and *G. nanus*).

Gerbillus henleyi Pygmy gerbil

Dipodillus henleyi de Winton, 1903. Wadi Natron, Egypt. Incl. *jordani* (Algeria), *mariae* (N.E. Egypt).
*Gerbillus henleyi makrami** Setzer, 1958c. 3 miles N. of Bir Kansisrob, Sudan Government Administrative Area, Egypt.

RANGE (Map 57). W. Arabia, S. Israel and Jordan through N. Egypt and the coastal strip of Libya to Algeria; S.E. Egypt.

REMARKS. There is fairly general agreement about the content of this species.

Subgenus *GERBILLUS*

Gerbillus gerbillus

Dipus gerbillus Olivier, 1800. Giza Prov., Egypt. Incl. *aegyptius, foleyi* (W. Algeria), *?hirtipes* (Algeria), *latastei* (Tunisia), *?longicaudatus.*
*Gerbillus gerbillus asyutensis** Setzer, 1960. Wadi el Asyuti, 13 miles S.E. of Asyut, Eastern Desert, Egypt.
*Gerbillus gerbillus aeruginosus** Ranck, 1968: 103. El Giof, Cufra Oasis, Cyrenaica, Libya.
*Gerbillus gerbillus discolor** Ranck, 1968: 106. Ghat, Fezzan Province, Libya.
*Gerbillus gerbillus psammophilous** Ranck, 1968: 112. Gialo Oasis, Cyrenaica Province, Libya.
Also other synonyms from further south in Africa.

RANGE (Map 58). Sahara from Algeria and N. Nigeria to Sudan, Egypt and S. Israel. Mainly in desert.

REMARKS. Petter (1971b) included in this species the Asiatic forms here placed in *G. cheesmani*. I tentatively follow Harrison (1972) who kept them separate while recognizing their close relationship. Several forms included in *G. gerbillus* by E & M-S have subsequently been shown to be specifically distinct and these are treated separately below (*G. andersoni*). A southern African group (*paeba*) included in *G. gerbillus* by E & M-S has been allocated to the genus or subgenus *Gerbillurus* by more recent authors (e.g. Herold & Niethammer, 1963).

Gerbillus andersoni†

Gerbillus andersoni de Winton, 1902. Mandara, Egypt. Inc. *allenbyi* (Israel), *bonhotei, eatoni* (Libya).
*Gerbillus eatoni inflatus** Ranck, 1968: 97. 10 km S.W. of Fort Capuzzo, Cyrenaica Prov., Libya.
*Gerbillus eatoni versicolor** Ranck, 1968: 98. 2 km N. of Coefia, Cyrenaica Prov., Libya.

*Gerbillus andersoni blanci** Cockrum, Vaughan & Vaughan, 1976. 2 km N.E. of Bordj Cedria, Tunisia (*c.* 30°42′N, 10°25′E). (*Mammalia* 40:470).

RANGE (Map 58). Mainly coastal sand-dunes from Israel through Egypt to Libya and Tunisia.

REMARKS. Included in *G. gerbillus* by E & M-S, the above forms have been excluded by all more recent authors. The forms *allenbyi*, *andersoni* and *eatoni* have been given separate specific rank, e.g. by Ranck (1968) and Harrison (1972) but were united by Cockrum *et al.* (*Mammalia* 40:467–73, 1976). The karyotype (2n = 40, all biarmed) is very distinct from that of *G. gerbillus*.

Cockrum *et al.*, considered that other described forms from the Middle East and eastern Africa might also be conspecific but they did not name these. This species is sympatric with *G. gerbillus* in Sinai and elsewhere and, according to Ranck (1968), with *G. aureus* in Libya although its relationship to *G. aureus* is still uncertain.

Gerbillus gleadowi — Indian hairy-footed gerbil

Gerbillus gleadowi Murray, 1886. Upper Sind, Pakistan.

RANGE (Map 58). Arid parts of Punjab and Sind.

Gerbillus pyramidum — Greater Egyptian gerbil

Gerbillus pyramidum Geoffroy, 1825. Giza Prov., Egypt. Incl. *burtoni*, *floweri* (Sinai), ?*hesperinus* (Morocco), ?*pygargus* (Upper Egypt), *riggenbachi* (Rio de Oro), *tarabuli* (Libya).
*Gerbillus pyramidum elbaensis** Setzer, 1958c. 2 miles N. of Kansisrob, Sudan Govt. Area, S.E. Egypt.
*Gerbillus pyramidum hamadensis** Ranck, 1968. 5 km E. of Derg, Tripolitania Province, Libya.
Also other synonyms from south of the Sahara.

RANGE (Map 58). Much of the Sahara from Morocco and Niger east to Somalia and Egypt; Sinai and S. Israel. Mainly in subdesert habitats.

REMARKS. E & M-S also included *hirtipes* (Algeria) in this species but it is here tentatively allocated to *G. gerbillus*. There is considerable variation in karyotype, both within and between populations (Zahavi & Wahrman, 1957). The Moroccan form *hesperinus* was considered specifically distinct by Lay (1975).

Gerbillus perpallidus*

Gerbillus perpallidus Setzer, 1958c. Bir Victoria, Western Desert Governorate, Egypt.

RANGE (Map 58). N. Egypt, west of the Nile.

REMARKS. According to Setzer (1958c) this species is sympatric with *G. pyramidum* at several localities. The two species are not separated in the key since the diagnosis given by Setzer does not hold in its entirety in other parts of the range of *G. pyramidum*.

Gerbillus aureus*

Gerbillus pyramidum aureus Setzer, 1956a:179. 12 km W. of Zliten, Tripolitania Province, Libya.
*Gerbillus pyramidum favillus** Setzer, 1956a:180. 2 km E. of Sirte, Tripolitania Province, Libya.
*Gerbillus aureus nalutensis** Ranck, 1968a:90. 40 km E.N.E. of Nalut, Tripolitania Province, Libya.

RANGE (Map 58). Coastal plain of N.W. Libya.

REMARKS. This form is given specific rank following Ranck (1968) who considered that in the form of *favillus* it is sympatric with *G. pyramidum* and *G. eatoni*. Ranck also suggested that *hirtipes* (Algeria) might be the prior name for this species but the lectotype of *hirtipes* does not agree well with Ranck's description of *aureus* (in particular it has a heavily tufted tail) and it is here tentatively included in *G. gerbillus*.

Gerbillus cheesmani Cheesman's gerbil

Gerbillus cheesmani Thomas, 1919. Near Basra, Lower Euphrates, Iraq. Incl. *arduus* (central Arabia).
*Gerbillus cheesmani maritimus** Sanborn & Hoogstraal, 1953. 3 miles S.E. of Hodeida, Yemen.
*Gerbillus cheesmani aquilus** Schlitter & Setzer, 1973. 60 km W. of Kerman, Kerman Prov., Iran.
*Gerbillus cheesmani subsolanus** Schlitter & Setzer, 1973. 56 km E. of Nok Kundi, Kalat Division, Pakistan.

RANGE (Map 59). Most of the Arabian peninsula (except northwest), the Euphrates and Tigris valleys; also in E. Iran and adjacent parts of Afghanistan and Pakistan if *aquilus* and *subsolanus* are included.

REMARKS. Since the description of *aquilus* and *subsolanus* it has been maintained by Lay & Nadler (1975) that these together represent a species distinct from *G. cheesmani.*

Gerbillus hoogstraali*

Gerbillus hoogstraali Lay, 1975:90. 7 km S. of Taroudannt, Morocco.

RANGE (not mapped). Known only from the type-locality.

REMARKS. See under *G. occiduus* below.

Gerbillus occiduus*

Gerbillus occiduus Lay, 1975:94. Aoreora, 80 km W.S.W. of Goulimine, Morocco.

RANGE (not mapped). Known only from the type-locality.

REMARKS. Lay (1975) considered that *G. occiduus*, *G. hoogstraali* and *G. hesperinus* formed a series of closely related allopatric species, differing clearly in karyotype, cranium and pelage. They are not included in the key to species of *Gerbillus.*

Genus *DIPODILLUS*

Dipodillus Lataste, 1881. Type-species *Gerbillus simoni* Lataste.

As used here a genus of three (possibly two) species differing from *Gerbillus* in more hypsodont teeth and shorter untufted tail. *Dipodillus* was for long treated as a subgenus of *Gerbillus*, including all those species with the soles of the feet naked. However Petter (1959) has argued that *simoni*, the type-species of *Dipididlus*, should be separated generically from other *Gerbillus* and has been followed by Harrison (1967) and by Schlitter & Setzer (1972). The validity of this distinction seems very dubious but I follow it pending a more comprehensive revision. Of the three species listed below *D. maghrebi* is quite distinctive but *D. simoni* and *D. kaiseri* might be judged conspecific as done by Petter (1971b).

The genus was reviewed by Cockrum *et al.* (1976) who described a further new insular species from Tunisia, *D. zakariai.*

1. Large: head and body 106–120 mm, hind feet 26–27 mm . . . . . ***D. maghrebi***
– Small: head and body 80–85 mm, hind feet 19–22 mm . . . . (2)
2. Auditory bullae small, greatest length about 6·5 mm . . . . . ***D. simoni***
– Auditory bullae large, greatest length 7·0–7·5 mm . . . . . . ***D. kaiseri***

Dipodillus simoni† Lesser short-tailed gerbil

Gerbillus simoni Lataste, 1881. Oued Magra, N. of Hodna, Algeria.

RANGE (Map 59). N.E. Algeria and Tunisia (Harrison, 1967).

Dipodillus kaiseri* Egyptian short-tailed gerbil

Gerbillus kaiseri Setzer, 1958c. Mersa Matruh, Western Desert Governorate, Egypt.

RANGE (Map 59). Coastal plains of Libya and Egypt.

REMARKS. Very close to *D. simoni* and possibly conspecific. I have not seen specimens.

Dipodillus maghrebi* Greater short-tailed gerbil

Dipodillus maghrebi Schlitter & Setzer, 1972. 15 km W.S.W. of Taounate, Fes Province, Morocco (34°29′N, 4°48′W).

RANGE (Map 59). Known only from the type-locality in N. Morocco.

REMARKS. This seems a clearly distinct species but I have not seen specimens.

Genus *TATERA*

Tatera Lataste, 1882. Type-species *Dipus indicus* Hardwicke.

A predominantly Ethiopian genus with many species in Africa south of the Sahara and one in S.W. Asia. The African species pose many taxonomic problems but the one Asiatic species with which we are concerned here seems clearly distinct. Setzer (1956c) implied, in a distribution map, that *T. robusta*, a species widely distributed south of the Sahara, might occur in S.E. Egypt but this does not seem to have been confirmed by precise records.

Tatera indica Indian gerbil

Dipus indicus Hardwicke, 1807. N. India. Incl. *bailwardi*, *monticola*, *persica*, *pitmani*, *scansa*, *taeniurus* (Syria).
Also many other synonyms from the Indian region.

RANGE (Map 59). All the drier parts of western India and westwards through Pakistan, Afghanistan, Iran and Iraq to Syria and Asia Minor. Also in Sri Lanka.

REMARKS. The Palaearatic form is often considered subspecifically distinct (*T. i. taeniura*) but this is based only on average measurements and is unlikely to be precisely definable.

Genus *PACHYUROMYS*

Pachyuromys Lataste, 1880. Type-species *P. duprasi* Lataste.

A monospecific genus, characterized by the most extreme enlargement of the auditory bullae in any member of the Gerbillinae.

Pachyuromys duprasi Fat-tailed gerbil

Pachyuromys duprasi Lataste, 1880. Laghouat, Algeria. Incl. *faroulti* (W. Algeria), *natronensis* (N. Egypt).

RANGE (Map 59). Northern part of Sahara, from W. Morocco to Egypt.

Genus *SEKEETAMYS*†

Sekeetamys Ellerman, 1947. Type-species *Gerbillus calurus* Thomas.

A monospecific genus. The generic allocation of this species has been very uncertain. Originally described in *Gerbillus* it was placed in *Meriones* by Ellerman (1941). Wassif (1954) argued that it was not *Meriones* and allocated it to *Gerbillus* (*Dipodillus*). Ellerman (in E & M-S) placed *Sekeetamys* as a subgenus of *Meriones* but most subsequent authors have given it full generic rank following Petter (1956).

Sekeetamys calurus Bushy-tailed jird

Gerbillus calurus Thomas, 1892. Sinai.
S. c. calurus. Sinai.
*S. c. makrami** Setzer, 1961. S.E. Egypt.
Sekeetamys makrami Setzer, 1961. Wadi Gumbiet, S.E. Desert Governorate, Egypt.

RANGE (Map 60). E. Egypt, Sinai, S. Israel and central Arabia (for the last see Nader, 1974).

REMARKS. The form *makrami* was reduced to subspecific rank by Petter (1971b) and Harrison (1972).

Genus *MERIONES*

Meriones Illiger, 1811. Type-species *Mus tamariscinus* Pallas. =*Idomeneus*. Incl. *Cheliones*, *Pallasiomys*, *Parameriones*.

The dominant genus of gerbils in the Palaearctic Region, especially in Asia. Differences between the species are slight and the classification is far from definitive. The genus itself is moderately stable except for the inclusion or exclusion of *Sekeetamys* (here excluded). Most of the species are represented in the Arabian region and have been described in detail by Harrison (1972) whose classification is followed here as far as possible. The comprehensive review of the genus by Chaworth-Musters & Ellerman (1947) requires considerable modification as a result of subsequent work but no later revision of the entire genus is available. The African forms are particularly confusing and cannot yet be diagnosed satisfactorily. The following key is based upon that of Chaworth-Musters & Ellerman. It does not include *M. zarudnyi* which has not been seen but appears to resemble *M. tristrami*.

1. Soles of hind feet entirely naked . . . . (2)
– Soles of hind feet at least partly haired . . . . (3)
2. Ventral pelage entirely white; tail usually longer than head and body, heavily tufted **M. persicus**
– Ventral pelage tinged with buff; tail about equal to length of head and body, less heavily tufted . . . . **M. rex**
3. Ears small, under 10% of length of head and body (less than 13 mm) . **M. hurrianae**
– Ears larger, over 10% of head and body, usually over 13 mm . . . . . (4)
4. Bullae small, length in horizontal plane (excluding mastoid part) less than or equal to diastema . . . . (5)
– Bullae larger, length considerably exceeding that of diastema . . . . . (7)
5. Sole of hind foot fully haired, with a darker central streak . . . . . (6)
– Sole of hind foot with a distinct naked patch near the heel, rest uniformly coloured **M. tristrami**
6. Tail sharply bicoloured, dark above and pale below; sole of hind foot with central streak dark brown . . . . **M. tamariscinus**
– Tail not sharply bicoloured, distal part entirely dark; centre of sole reddish brown **M. vinogradovi**
7. Sole of hind foot fully haired; greatest length of skull usually under 36 mm . (8)
– Sole of hind foot naked near the heel; greatest length of skull usually over 36 mm (9)
8. Ventral pelage wholly white; claws pale; bullae larger, mastoid chambers projecting behind occipital condyles . . . . **M. meridianus**
– Ventral pelage grey with white tips; claws dark; bullae smaller, mastoid chambers usually level with condyles . . . . **M. unguiculatus**
9. Auditory meatus with a curtain of bone in the upper part, obscuring the main bodies of the ossicles; claws usually pale . . . . (10)
– Auditory meatus without such a curtain so that the ossicles are clearly visible; claws usually dark . . . . (12)
10. Bullae smaller, mastoid parts less swollen and not projecting much behind occipital condyles, suprameatal triangle (i.e. surface of anterior mastoid chamber) widely open behind; (tail rather lightly tufted) . . . . **M. shawi**
– Bullae very large, mastoid parts greatly swollen and extending well behind condyles, suprameatal triangle closed behind or almost so . . . . (11)
11. Tail heavily tufted with a black dorsal line on at least the distal half; larger: hind feet 36–41 mm, greatest length of skull 41–49 mm . . . . **M. sacramenti**
– Tail less tufted, no black line on distal half; smaller: hind feet 25–35 mm, greatest length of skull 32–42 mm . . . . **M. crassus**
12. Tail prominently tufted; process in front of auditory meatus greatly inflated, in contact with posterior process of zygomatic arch . . . . **M. caudatus**
– Tail less tufted; process in front of meatus less inflated . . . . . **M. libycus**

Meriones persicus Persian jird

Gerbillus persicus Blanford, 1875. Qohrud, 72 miles N. of Isfahan, Iran. Incl. *ambrosius, baptistae* (Baluchistan), *gurganensis, rossicus, suschkini* (Turkmenistan).

RANGE (Map 60). Iran and adjacent parts of Transcaucasian USSR, Turkey, Iraq, Turkmenistan, Afghanistan and Pakistan.

Meriones rex King jird

Meriones rex Yerbury & Thomas, 1895. Lahej, near Aden, S. Arabia.
M. r. rex. Coastal plains.
M. r. buryi Thomas, 1902. Mountains N. of Aden. Incl. *philbyi.*

RANGE (Map 60). S.W. Arabia from Mecca to Aden.

Meriones hurrianae Indian desert gerbil

Gerbillus hurrianae Jerdon, 1867. Punjab, India. Incl. *collinus.*

RANGE (Map 60). Semidesert from Punjab and Kathiawar to S. Afghanistan and S.E. Iran.

Meriones vinogradovi Vinogradov's jird

Meriones vinogradovi Heptner, 1931. Persian Azerbaijan.

RANGE (Map 61). Armenia and adjacent parts of Asia Minor and Iran; Syria. A record from Palestine (Dobroruka, 1959) was probably erroneous (Harrison, 1972).

Meriones tamariscinus Tamarisk gerbil

Mus tamariscinus Pallas, 1773. Mouth of R. Ural, Kazakhstan. Incl. *ciscaucasicus* (N. Caucasus), *jaxartensis* (E. of Aral Sea), *kokandicus* (Fergana Valley), *satschouensis* (Kansu).

RANGE (Map 61). Russian Turkestan from S.W. of the lower Volga to the Altai Mts and through northern Sinkiang (Dzungaria) to the extreme western extremity of Kansu.

Meriones tristrami† Tristram's jird

Meriones tristrami Thomas, 1892. Dead Sea region, Palestine. Incl. *blackleri* (W. Asia Minor), *bodenheimeri* (Syria), *bogdanovi* (Azerbaijan), *intraponticus, kariateni, lycaon.*

RANGE (Map 61). Asia Minor to the Caspian Sea, Iraq and N.W. Iran; south through Syria and Israel to N. Sinai.

REMARKS. E & M-S treated *tristrami* as the easternmost race of the African *M. shawi* and used *M. blackleri* as the name for the remaining Asiatic forms. More recent study, especially of karyotypes, has clarified the distinction between these species and shown that *tristrami* belongs to the Asiatic species for which it is the earliest name (see diagnoses by Harrison, 1972).

Meriones unguiculatus Mongolian gerbil, Clawed jird

Gerbillus unguiculatus Milne-Edwards 1867. N. Shansi, China. Incl. *chihfengensis, koslovi* (W. Mongolia), *kurauchii* (Manchuria).
*Pallasiomys unguiculatus selenginus** Heptner, 1949. Kyakhta Mts, S. of L. Baikal, USSR.

RANGE (Map 61). Most of Mongolia and adjacent parts of USSR to the north, and of China from Sinkiang through Inner Mongolia to Manchuria.

REMARKS. This is the species of gerbil most frequently used as an experimental animal and as a pet. For a general account see Gulotta (1971).

Meriones meridianus Midday gerbil

Mus meridianus Pallas, 1773. Uralsk Region, Kazakhstan. Incl. *auceps* (Shansi), *brevicaudatus, buechner* (Sinkiang), *cryptorhinus, fulvus, heptneri, karelinae* (N.W. Kazakhstan), *lepturus, littoralis, massagetes* (Aral Sea), *nogaiorum* (N. Caucasus), *penicilliger* (Kara-Kum Desert), *psammophilus* (Inner Mongolia), *roborowskii, shitkovi, urianchaicus* (Mongolia), *uschtaganicus.*
*Meriones meridianus dahli** Shidlovskyi, 1962: 115. Nakhichevansk ASSR, Armenia. (Original reference not seen – further described by Gambaryan & Papanyan, 1964.)

RANGE (Map 62). From N. of the Caucasus throughout Russian and Chinese Turkestan and Mongolia to Hopei; south to N.E. Iran, Afghanistan, Chinghai and Shansi. Also an apparently isolated population south of the Caucasus in Armenia.

Meriones shawi Shaw's jird

Gerbillus shawi Duvernoy, 1842. Oman, Algeria. =*savii.* Incl. *albipes, auziensis, crassibulla, grandis* (Morocco), *isis* (Egypt), *laticeps, longiceps, richardii, sellysii, trouessarti.*

RANGE (Map 62). N.W. Africa from Morocco through N. Algeria to Tunisia and Egypt (north of the range of *M. libycus*).

REMARKS. E & M-S included *tristrami* (Dead Sea) in this species but it has been demonstrated that it is specifically distinct (Harrison, 1972). The extent to which this species is distinct from *M. libycus* and if so whether it occurs east of Tunisia is debatable. Petter (1971b) followed the arrangement of E & M-S by including the Egyptian form *isis* in *M. shawi* but Setzer (1961) followed by Ranck (1968) believed that *M. shawi* could not be recognized in Egypt and Libya and doubted if it was distinct from *M. libycus* even in N.W. Africa.

Meriones libycus Libyan jird

Meriones libycus Lichtenstein, 1823. Near Alexandria, Egypt. Incl. *aquilo* (Sinkiang), *arimalius*† (Arabia), *caucasius* (Azerbaijan), *collium* (S.E. Kazakhstan), *edithae, erythrourus* (Afghanistan), *evelynae, eversmanni* (E. of Caspian Sea), *gaetulus, guyonii, iranensis*† (Iran), *marginae, mariae* (Rio de Oro), *maxeratis* (Kopet Dag), *melanurus, oxianus* (Bokhara), *renaultii, schousboeii, sogdianus* (Fergana), *syrius* (Syria), *turfanensis* (Sinkiang).
*Meriones shawi azizi** Setzer, 1956b. 5 km S.E. of Derna, Cyrenaica, Libya.
*Meriones libycus auratus** Ranck, 1968. Gheminez, Cyrenaica Province, Libya.
*Meriones erythrourus farsi** Schlitter & Setzer, 1973. 3 km N. of Bariz, 50 km N. of Lar, Fars Prov., Iran.

RANGE (Map 62). North Africa from Rio de Oro to Egypt and through N. Arabia, Iraq, Iran, Afghanistan and southern Russian Turkestan to Sinkiang.

REMARKS. The Libyan form *caudatus*, included in this species by E & M-S, was considered specifically distinct by Ranck (1968) whose conclusions are followed here. The Iranian form *iranensis* was not allocated to species by E & M-S but was considered a synonym of *libycus* by Lay (1967).

***Meriones caudatus*†**

Meriones libycus caudatus Thomas, 1919. Tripolitania, Libya. Incl. *confalonierii, tripolius.*
*Meriones caudatus amplus** Ranck, 1968:165. El Gatrum, Fezzan Prov., Libya.
*Meriones caudatus luridus** Ranck, 1968:173. Bahr el Tubat, 21 km E.S.E. of Giarabub, Cyrenaica Prov., Libya.

RANGE (not mapped). N. and W. Libya; probably also adjacent parts of Tunisia and Egypt.

REMARKS. E & M-S included *caudatus* and *confalonierii* in *M. libycus* and *tripolius* in *M. crassus*. Ranck (1968) first treated *M. caudatus* as a separate species which he diagnosed and considered to be sympatric with *M. libycus* in some coastal areas of Libya. He also considered that the range of *M. caudatus* 'probably included other portions of North Africa and adjacent Southwest Asia'. In the Arabian region Harrison (1972) was unable to detect any evidence of a species separable from *M. libycus*. If Ranck was correct in recognizing this species as distinct from *M. libycus* then it is very probable that its existence has been overlooked elsewhere. However it is also possible that this form is conspecific with *M. libycus* and that the Libyan forms assigned by Ranck to *M. libycus* should really be allocated to *M. shawi*.

Meriones crassus

Meriones crassus Sundevall, 1842. Sinai. Incl. *charon* (Iran), *ismahelis, longifrons* (W. Arabia), *pallidus* (Sudan), *pelerinus, swinhoei* (Afghanistan).

*Meriones crassus asyutensis** Setzer, 1961:82. Wadi Asyuti, 13 miles S.E. of Asyut, Eastern Desert Governorate, Egypt.
*Meriones crassus perpallidus** Setzer, 1961:68. Cairo-Alexandria road, 4 km from Cairo, Western Desert Governorate, Egypt.

RANGE (Map 62). N. Africa from Algeria and Niger to Egypt and Sudan, and through Arabia and Iran to Afghanistan.

REMARKS. The forms *sacramenti* from Israel and *zarudnyi* from the Afghan/USSR frontier, included here by E & M-S, are considered distinct species (see below). For a general account of the species see Koffler (1972).

Meriones sacramenti Buxton's jird

Meriones sacramenti Thomas, 1922. Ten miles S. of Beersheba, Israel. Incl. *legeri*.

RANGE (Map 61). Apparently confined to a small area in S. Israel. The record from N. Syria by Misonne (1957) was an error according to Harrison (1972).

REMARKS. Included in *M. crassus* by E & M-S, this form has been considered specifically distinct on both morphological and cytological grounds (see Harrison, 1972).

Meriones zarudnyi†

Meriones zarudnyi Heptner, 1937. Afghan/Turkmenistan frontier.

RANGE (Map 61). N. Afghanistan, N.E. Iran and S. Turkmenistan.

REMARKS. Considered as a race of *M. crassus* by E & M-S. Kuznetsov (1965) tentatively treated it as specifically distinct whilst Lay (1967), on the basis of additional material from Afghanistan, was emphatic that it was distinct from both *M. crassus* and *M. tristrami*.

Genus *BRACHIONES*

Brachiones Thomas, 1925. Type-species *Gerbillus przewalskii* Büchner.

A monospecific genus, related to *Meriones* but distinguished by greatly reduced rostrum and abnormally wide frontal region of the skull.

Brachiones przewalskii Przewalski's gerbil

Gerbillus przewalskii Büchner, 1889. Lob Nor, Sinkiang. Incl. *arenicolor* (Yarkand), *callichrous* (N. Kansu).

RANGE (Map 63). Deserts of Sinkiang from Yarkand to N. Kansu (not in Mongolia).

REMARKS. E & M-S gave the range as 'Chinese Turkestan, Mongolia' and the type-locality of *callichrous* as in Mongolia. This locality is however in Kansu or Ningsia province of China and the species is not included in the fauna of the present state of Mongolia by Bannikov (1954).

Genus *PSAMMOMYS*

Psammomys Cretzschmar, 1828. Type-species *P. obesus* Cretzschmar.

A monospecific genus closely related to *Meriones* and characterized by ungrooved upper incisors and rather heavy build. The Libyan form *vexillaris* has been considered specifically distinct from *P. obesus* by Setzer (1957b) and Ranck (1968) but the characters employed seem very variable and they are here united.

Psammomys obesus Fat sand rat

Psammomys obesus Cretzschmar, 1828. Alexandria, Egypt. Incl. *algiricus*, *dianae* (S. Arabia), *edusa*, *nicolli* (Nile Delta), *roudairei* (Algeria), *terraesanctae* (Dead Sea), *tripolitanus*, ?*vexillaris* (N.W. Libya).

RANGE (Map 63). North Africa from Algeria and Tunisia to the coastal area of Egypt and into Palestine and parts of Arabia; also on the coast of Sudan.

REMARKS. This species is very variable, especially in colour, but the available samples are small and difficult to assess. Examination of the type of *vexillaris* does not allow the cranial differences between it and *obesus* given by Setzer (1957b) to be substantiated.

Genus *RHOMBOMYS*

Rhombomys Wagner, 1841. Type-species *R. pallidus* Wagner = *Meriones opimus* Lichtenstein.

A monospecific genus unique amongst the Gerbillinae in having rootless cheek-teeth.

Rhombomys opimus Great gerbil

Meriones opimus Lichtenstein, 1823. Near Bokhara, Uzbekistan. Incl. *alaschanicus*, *dalversinicus*, *fumicolor* (Fergana), *giganteus*, *nigrescens* (Mongolia), *pallidus*, *pevzovi* (Sinkiang), *sargadensis* (Iran), *sodalis*.

RANGE (Map 63). From the Caspian Sea through Russian Turkestan, southern Mongolia and Sinkiang; south to parts of Iran, Afghanistan and W. Pakistan.

Family SPALACIDAE
Blind mole-rats

A distinctive family whose relationship to other muroid families is obscured by the very high degree of specialization for subterranean life. This is carried to a greater degree than in other groups of mole-rats such as the Rhizomyidae, the Myospalacinae (Cricetidae) and the Bathyergidae. Petter (1961d) argued in favour of their close relationship with, and indeed inclusion in, the Cricetidae. This has not been generally adopted but there is fairly general agreement that they should be included in the Muroidea, although Schaub (1958) excluded them.

There is fairly general agreement amongst recent authors that only one genus should be recognized (leaving aside the possible inclusion of the Ethiopian and Oriental genera normally placed in the Rhizomyidae, which has been advocated, with good reason, by Schaub and others). The recent classification of the Spalacidae (*sensu strictu*) by Topačevski (1969) recognized two genera, *Spalax* and *Microspalax*, but all his taxa were ranked more highly than is generally accepted.

Genus *SPALAX*

Spalax Güldenstaedt, 1770. Type-species *Spalax microphthalmus* Güldenstaedt. = *Macrospalax*. Incl. *Glis* Erxleben (not of Brisson), *Myospalax* Hermann (not of Laxmann), *Talpoides* = *Aspalax* = *Anotis*, *Microspalax* = *Nannospalax* = *Ujhelyiana*, *Mesospalax*.

A very distinctive genus confined to the eastern Mediterranean region and southern Russia. The classification within the genus is very unstable. In a recent comprehensive revision Topačevskii (1969) has used a rather split classification reflecting in some respects the excessively split classification of Mehely (1913) and recognizing eight recent species in two genera. I prefer to reduce all Topačevski's taxa in rank and recognize three congeneric species. All recent Russian authors recognize the eastern *S. giganteus* as specifically distinct from *S. microphthalmus* and therefore this course is followed here although these were considered conspecific by E & M-S. Amongst the smaller forms, E & M-S treated the southern *ehrenbergi* as specifically distinct from *S. leucodon*, but Harrison (1972) was unable to find any clear-cut distinguishing character in the area of contact and I therefore follow him in treating them as conspecific.

All these species show considerable local variation, and cytological work with *S. leucodon* has demonstrated apparently discrete, allopatric forms with distinctive karyotypes, e.g. in Israel (Wahrman *et al.*, 1969) and in Yugoslavia (Soldatović *et al.*, 1967). In each of these cases it is arguable that the different chromosome forms should be considered as sibling species but unless and until they can be recognized morphologically or are proven to be sympatric it seems better to retain a taxonomy based on morphologically discrete species and to indicate karyotypically distinct populations in terms of locality and chromosome number.

1. Small foramen present above each occipital condyle; width of rostrum (in front of zygomatic arches) under 11 mm . . . . **S. leucodon**
– No foramina above condyles; width of rostrum over 11 mm . . . . . (2)
2. Smaller: hind feet under 30 mm, condylobasal length and maximum width of skull under 59 mm; dorsal pelage dark brown. . . . . **S. microphthalmus**
– Larger: hind feet over 30 mm, condylobasal length and maximum width of skull over 58 mm; dorsal pelage light greyish yellow . . . . **S. giganteus**

Spalax microphthalmus

Spalax microphthalmus Güldenstaedt, 1770. Nobochopersk Steppes, S. Russia.
S. m. microphthalmus. Volga west to Dnieper. Incl. *pallasii, typhlus.*
S. m. zemni (Erxleben, 1777). Ukraine. Incl. *podolicus, polonicus.*
S. m. arenarius Reshetnik, 1938. S. Ukraine.
S. m. graecus Nehring, 1898. Greece, Rumania. Incl. *antiquus, istricus, mezosegiensis.*

RANGE (Map 64). Ukraine and S. Russia southwest to Bulgaria and Greece and east to the Volga and the plains north of the central Caucasus.

REMARKS. The races separately listed above were given specific rank by Topačevski (1969), who, however, used the name *polonicus* instead of *zemni.*

Spalax giganteus

Spalax giganteus Nehring, 1898. Petrovsk, Caspian Sea, USSR. Incl. *uralensis.*

RANGE (Map 64). The plains to the N.W. of the Caspian Sea (abutting the range of *S. microphthalmus*) and in Kazakhstan east of the R. Ural.

REMARKS. This form was included in *S. microphthalmus* by E & M-S, but has been given specific rank by all recent Russian authors including Kuznetsov (1965). It has not been seen.

Spalax leucodon

Spalax typhlus leucodon Nordmann, 1840. Near Odessa, Ukraine.
S. l. leucodon. Hungary to W. side of Black Sea. Incl. *dolbrogeae, hellenicus, hercegovinensis, hungaricus, monticola, serbicus, syrmiensis, thermaicus, transsylvanicus.*
*Spalax leucodon martinoi** Petrov, 1971a. Cesta Suma, Deliblatska Pescara, near Deliblato, South Banat, Yugoslavia.
*Spalax leucodon peleponnesiacus** Ondrias, 1966:37. Agios Vasilios, Corinthia, Peloponnesus, Greece.
*Spalax leucodon thessalicus** Ondrias, 1966:39. Amphiklia, Phthiotis, Greece.
S. l. insularis Thomas, 1917. Island of Limnos, Greece.
S. l. xanthodon Nordmann, 1840. Caucasus and Asia Minor. Incl. *anatolicus, armeniacus, captorum, cilicicus, corybantium, labaumei, ?nehringi, turcicus.*
S. l. ehrenbergi Nehring, 1898. Syria, Israel (type-locality Jaffa). Incl. *berytensis, intermedius, kirgisorum.*
S. l. aegyptiacus Nehring, 1898. Egypt and Libya.

RANGE (Map 64). The Danube basin to Greece and S. Ukraine; Caucasus through Asia Minor, N. Iraq and Syria to Israel; coastal Egypt and Libya west to Benghazi.

REMARKS. E & M-S suggested that *leucodon* might be conspecific with *S. microphthalmus* while retaining *S. ehrenbergi* as a separate species. All Russian authors treat *S. microphthalmus* and *S. leucodon* separately (their ranges abut in Ukraine) but *ehrenbergi* and *aegyptiacus* are here included in *S. leucodon* following Harrison (1972). The Armenian form *nehringi* was considered specifically distinct by Orlov (1969).

Family MURIDAE

Typical mice and rats

This is the dominant family of rodents in the Oriental Region and, to a lesser extent, in the less arid parts of the Ethiopian Region. Only two genera, *Micromys* and *Apodemus*, are predominantly Palaearctic; one, *Nesokia*, occupies a transitional range between the

Palaearctic and Oriental regions; four Ethiopian genera extend into the southwestern parts of the Palaearctic; and two mainly Oriental genera, *Mus* and *Rattus*, are widespread in the Palaearctic by association with man.

Most recent authors recognize this family in the restricted sense used here, i.e. excluding the voles, gerbils, hamsters etc. that were included in the Muridae by Ellerman (1941) (see p. 88). For a detailed discussion of classification and evolution within this group see Misonne (1969). The generic classification followed here is that of Misonne but differs from that of Ellerman only in separating *Praomys* from *Rattus*. Most classifications give very great weight to small differences in the pattern of cusps on the molar teeth but since these may be very difficult to appreciate in a worn tooth-row I have not used them in the key below.

1. Dorsal pelage, except on head and neck, consisting entirely of spines; bony palate extending far behind teeth, margin about half way between teeth and bullae *ACOMYS* (p. 142)
– Dorsal pelage not spiny or with some rather weak spines mixed with normal hair; palate not extending far behind teeth . . . . (2)
2. Dorsal pelage with multiple pale longitudinal stripes (N.W. Africa) *LEMNISCOMYS* (p. 138)
– Dorsal pelage unstriped or with a single dark median stripe . . . . . . (3)
3. Larger: head and body usually in range 140–250 mm, hind feet over 25 mm (usually over 30 mm), upper molar row over 6 mm . . . . (4)
– Smaller: head and body under 140 mm, hind feet under 30 mm, upper molar row under 6 mm (6)
4. Cheek-teeth with transverse laminae and almost no trace of separate cusps *NESOKIA* (p. 143)
– Cheek-teeth with clearly defined cusps . . . . (5)
5. Banding pattern of hairs producing an overall speckled effect in the dorsal pelage; upper molars, especially M^1, short and wide, length of M^1 only slightly exceeding its width *ARVICANTHIS* (p. 138)
– Dorsal pelage rather uniform in colour; M^1 almost twice as long as wide *RATTUS* (p. 138)
6. Very small: hind feet under 16 mm, maxillary tooth-row under 3·2 mm; M^1 with 5 roots *MICROMYS* (p. 131)
– Larger: hind feet over 16 mm, maxillary tooth-row over 3·2 mm; M^1 with 3 or 4 roots (7)
7. M^3 very small, not more than half length and width of M^2; upper incisors usually appear notched when viewed from the side; (M^1 with 3 roots, i.e. only one lingual root) *MUS* (p. 141)
– M^3 larger; upper incisors not notched; M^1 with 3 or 4 roots . . . . (8)
8. Anterior palatal foramina very long, extending back between first molars; outer margins of pterygoid plates strongly convex; (M^1 with a single root on the lingual side) *PRAOMYS* (p. 140)
– Anterior palatal foramina stopping short of or level with anterior margin of M^1; outer margins of pterygoid plates straight or faintly convex; (M^1 with one or two roots on lingual side) . . . . *APODEMUS* (p. 132)

Genus *MICROMYS*

Micromys Dehne, 1841. Type-species *M. agilis* Dehne = *Mus soricinus* Hermann.

A monospecific genus, generally recognized as such but closely related to the Oriental genera *Vandeleuria* and *Chiropodomys* (Misonne, 1969).

Micromys minutus Harvest mouse

Mus minutus Pallas, 1771. R. Volga, Russia.

M. m. minutus. W. Palaearctic range; Britain. Incl. *agilis, arundinaceus, avenarius, campestris, fenniae* (Finland), ?*flavus, meridionalis, messorius, minatus, minimus, oryzivorus* (Italy), *parvulus, pendulinus, pratensis* (Hungary), *pumilus, sareptae, soricinus* (E. France), *subobscurus, triticeus* (England).

*M. m. danubialis** Simonescu, 1971. Danube Delta.

Micromys danubialis Simonescu, 1971. Danube Delta.

M. m. ussuricus (Barrett-Hamilton, 1899). E. Siberia (type-locality in Ussuri Region). Incl. *batarovi* (Lake Baikal), *kytmanovi*.
M. m. erythrotis (Blyth, 1855). Assam (type-locality) and S. China.
M. m. hondonis Kuroda, 1933. Honshu, Japan.
M. m. japonicus Thomas, 1906. Shikoku (type-locality) and Kyushu, Japan.
M. m. aokii Kuroda, 1922. Tsushima Island, Japan.
*M. m. hertigi** Johnson & Jones, 1955. Quelpart Is., Korea.
Micromys minutus hertigi Johnson & Jones, 1955a. 2 miles S.E. of Mosulp'o, Cheju Do (Quelpart Island), Korea.
M. m. takasagoensis Tokuda, 1941. Taiwan.

RANGE (Map 64). Woodland and moister steppe zones from N.W. Spain (Garzon, 1973) through most of Europe (but absent from Scandinavia and much of the Mediterranean region) and across Siberia to the Ussuri region and Korea, north to about 65°N in European Russia and in Yakutia, south to the northern edge of the Caucasus and N. Mongolia. An apparently isolated segment in S. China (S. Shensi and Nanking southwards) and west through Yunnan to Assam and S.E. Tibet. Island populations on Britain; Honshu, Shikoku, Kyushu and Tsushima (Japan); Quelpart Is. (Korea); and Taiwan.

REMARKS. The recently described *danubialis* from the Danube Delta seems rather clearly differentiated from neighbouring European forms by its large size and long tail but since in these respects it is equalled by some Asiatic forms (e.g. *hertigi*) I tentatively reduce it to subspecific rank.

Studies of seasonal and individual variation in colour, e.g. by Szunyoghy (1958) and Böhme (1969), suggest that colour cannot be used to distinguish subspecies in Europe. Simonescu (1971) reviewed variation throughout the Palaearctic part of the range. He upheld the validity of *soricinus*, *pratensis* and *ussuricus* as subspecies but only on the basis of average differences. The Japanese races listed above were considered distinct by Imaizumi (1960) but I have not been able to review them.

M. m. mehelyi Bolkay, 1925 was shown by Mirić (1966) to be based on a young *Mus musculus*.

For a general review of the species see Piechocki (1958).

Genus *APODEMUS*

Apodemus Kaup, 1829. Type-species *Mus agrarius* Pallas. Incl. *Sylvaemus* = *Nemomys*, *Alsomys*, *Petromys* Martino.
*Karstomys** Martino, 1939. New name for *Petromys* Martino, 1934, preoccupied by *Petromys* Smith, 1834.

A fairly distinctive genus of about 12 species, confined to the Palaearctic and northern part of the Oriental Regions. It is most nearly related to certain African genera, e.g. *Thallomys* and *Thamnomys*. It is the dominant group of murid rodents in the Palaearctic Region. Classification within the genus is rather unstable. E & M-S reduced it to five species, two of them, *A. sylvaticus* and *A. flavicollis*, with ranges almost coincident with that of the genus. Subsequent work has shown this to be an oversimplification.

It is generally agreed that *A. agrarius* is a distinctive species and some would retain it as the sole member of *Apodemus*, placing the remaining species in *Sylvaemus*. This seems excessive splitting and it seems better to treat the division as of no more than subgeneric rank.

The eastern Mediterranean *A. mystacinus* is also a clearly recognized species.

In contrast to Ellerman's allocation of the remaining forms to two widespread Palaearctic species, Zimmermann (1962) argued that there was a clear distinction between those of eastern Asia and the western forms, giving these groups subgeneric rank as *Sylvaemus* in the west and *Alsomys* in the east. The characters he used do indeed seem valid for the recognition of species but there seem to be no characters sufficiently invariable amongst the eastern group to justify uniting them as a distinct subgenus.

However I fully agree with Zimmerman that none of the eastern Asiatic forms should be allocated to *A. sylvaticus* or *A. flavicollis*.

The following treatment includes the entire genus although some species are entirely or partially in the Oriental Region and the treatment of these should be considered particularly provisional.

1. M^3 with 2 internal lobes; M^2 lacking antero-external cusp (i.e. with only one, lingual, cusp in first lamina); narrow black mid-dorsal stripe (except in W. China) . ***A. agrarius***
– M^3 with 3 internal lobes; M^2 usually with an antero-external cusp (vestigial in *A. speciosus* in Japan); no mid-dorsal stripe . . . . (2)
2. Dorsal pelage very grey, lacking clear yellow or reddish tones . . . . (3)
– Dorsal pelage yellowish brown . . . . (5)
3. M^1 with two clearly separated lingual roots (E. Mediterranean) . . . . (4)
– M^1 with one lingual root (Himalayas) . . . . ***A. gurka***
4. Larger: condylobasal length 26–30 mm, hind feet 24–28 mm, upper molar tooth-row 4·7–5·4 mm; anterior margin of masseteric plate oblique . . . . ***A. mystacinus***
– Smaller: condylobasal length 23–25 mm, hind feet 22–24 mm, upper molar row 3·8–4·5 mm; anterior margin of masseteric plate vertical (Island of Krk, Yugoslavia) ***A. krkensis***
5. Supraorbital ridges absent or weakly developed, never forming a protruding edge: M^1 with two clearly separated lingual roots (visible without removing the tooth) . . (6)
– Supraorbital ridges well developed, often forming a protruding edge: M^1 with only one lingual root (occasionally double or treble in S. China (*latronum*) and Taiwan (*semotus*) but not widely separated) . . . . (9)
6. Very small: hind feet under 20 mm, maxillary tooth-row 3·2–3·8 mm; 3 or 4 pairs of mammae . . . . (7)
– Larger: hind feet over 20 mm, tooth-row usually over 3·8 mm; 3 pairs of mammae (8)
7. Tail usually longer than head and body; 4 pairs of mammae; mesopterygoid fossa wide, more than half width between last molars; masseteric plate scarcely projecting in front of anterior border of zygomatic arch (Japan) . . . . ***A. argenteus***
– Tail usually shorter than head and body; 3 pairs of mammae; mesopterygoid fossa less than half width between molars; masseteric plate projecting in front of zygomatic arch as is usual in the genus (E. Europe) . . . .***A. microps***
8. Yellow mark on chest usually large, sometimes forming a complete collar; ventral pelage usually paler grey and dorsal pelage redder than in *A. sylvaticus* where they occur together; larger: hind feet usually over 23 mm, maxillary tooth-row usually 3·8–4·2 mm; supraorbital ridges moderately developed . . . . ***A. flavicollis***
– Yellow mark on chest small or absent (occasionally enlarged longitudinally but not transversely); ventral pelage darker grey and dorsal pelage less reddish than in *A. flavicollis*; smaller: hind feet usually under 23 mm, tooth-row usually 3·6–4·0 mm; supraorbital ridges scarcely developed. . . . . ***A. sylvaticus***
9. Ears dark, contrasting with lighter dorsal pelage, or entire dorsal pelage very dark brown, almost but not quite obscuring yellowish tones (S. China and Taiwan) . . (11)
– Ears light, not contrasting with dorsal pelage which always shows strong yellow or reddish tones (N.E. Asia and Japan) . . . . (10)
10. Yellow pectoral spot usually present; M^2 with antero-external cusp rudimentary, absent after slight wear; larger: hind feet usually 24–28 mm, maxillary tooth-row usually 4·3–5·8 mm (Japan) . . . . ***A. speciosus***
– No yellow pectoral spot; M^2 with antero-external cusp moderately developed: smaller: hind feet usually 24 mm or less, tooth-row usually 3·8–4·3 mm . . ***A. peninsulae***
11. Larger: hind feet 24–27 mm, ears 19–21 mm, maxillary tooth-row usually over 4·0 mm; lingual root of M^1 broad and sometimes divided into two or three parts ***A. latronum***
– Smaller: hind feet 20–23 mm, ears less than 19 mm, tooth-row usually under 4·0 mm; lingual root of M^1 narrow and single . . . . ***A. draco***
(*A. semotus* on Taiwan is somewhat intermediate between these last two species.)

Apodemus mystacinus — Broad-toothed mouse

Mus mystacinus Danford & Alston, 1877. Zebil, Bulgar Dagh, Asia Minor.
A. m. mystacinus. Asia and perhaps Crete. Incl. *euxinus*, *pohlei*† (Syria), ?*rhodius* (Rhodes), *smyrnensis*.
A. m. epimelas (Nehring, 1902). European range.

RANGE (Map 65). S.E. Europe (Croatia, Yugoslavia through Albania and S. Bulgaria to Greece); Palestine, Asia Minor and adjacent parts of Iraq and Georgia; the islands of Rhodes, Crete (Zimmermann, 1953a) and several inshore Aegean islands.

REMARKS. Clearly distinct from all other forms of *Apodemus*. E & M-S tentatively listed *pohlei* under *A. flavicollis*. It is placed here following Harrison (1972).

Apodemus flavicollis Yellow-necked mouse

Mus flavicollis Melchior, 1834. Sielland, Denmark. Incl. *brauneri* (Yugoslavia), *brevicauda, cellarius, fennicus* (Finland), *parvus*, *ponticus* (Caucasus), *princeps*, *samariensis* (E. European Russia), *saturatus* (N. Asia Minor), *wintoni* (England).

*Apodemus tauricus argyropuloi** Heptner, 1948a. New name for *Apodemus flavicollis parvus* Argyropulo, 1945, preoccupied by *Mus sylvaticus parvus* Bechstein, 1796.

*Apodemus flavicollis alpinus** Heinrich, 1951. Allgäu, Osterachtal, S. Germany.

*Apodemus flavicollis alpicola** Heinrich, 1952. New name for *A. f. alpinus* Heinrich, 1951, preoccupied by *Mus sylvaticus alpinus* Burg, 1921.

*Apodemus tauricus geminae** Lehmann, 1961. Monte Gargano, Apulia, Italy.

*Apodemus tauricus dietzi** Kahmann, 1964. Lake Stymphale, Corinthia, Peloponese, Greece.

?*Mus sylvaticus* var. *tauricus** Pallas, 1811. 'Chersonesus tauricus', i.e. mountains of Crimea, Ukraine.

RANGE (Map 65). Europe from N.W. Spain, E. France, Denmark and S. Scandinavia through European Russia to the Urals, south to Italy, the Balkans, Asia Minor and Palestine; England and Wales. Mainly montane on the southern edge of the range.

REMARKS. E & M-S also included in this species many eastern Asiatic forms and many insular forms from Scotland. The eastern Asiatic forms are excluded following Zimmermann (1962) – see above under genus *Apodemus*. This course was also taken by Kuznetsov (1965). Recent students of British *Apodemus*, e.g. Cranbrook (1957) and Berry *et al.* (1967), agree that the forms on the Channel Isles, Hebrides etc. are referable to *A. sylvaticus*.

There have been many claims (e.g. by Amtmann, 1965) that in parts of its range, especially in southern Europe, this species hybridizes to some extent with *A. sylvaticus*. However I am inclined to accept the conclusion of Niethammer (1969b) that this has not been proven. The characters that separate the species are however slight, especially on the southern edge of the range, and show considerable geographical variation within each species and especially in *A. flavicollis*. However it is unlikely that any very discrete races can be defined.

Since 1948 several authors have used the name *A. tauricus* for this species, following Heptner (1948). However this has not been generally adopted – most Russian authors continue to use *A. flavicollis* and this name also preponderates in the very extensive ecological literature on the species. It therefore seems desirable to retain the name *flavicollis*, and the inadequacy of the original description of *tauricus*, which amounts only to 'multo major et elegantissimi velleris' (relative to *A. sylvaticus*), would seem to justify its rejection as not certainly determinable.

Apodemus sylvaticus Wood mouse

Mus sylvaticus Linnaeus, 1758. Upsala, Sweden.

Synonyms from main continental range: *albus*, *alpinus* Burg, *arianus* (N. Iran), *baessleri* (Crimea), *balchaschensis* (L. Balkhash), *callipides* (N.W. Spain), *candidus*, *chorassanicus* (Kopet Dag), *ciscaucasicus* (N. Caucasus), *erythronotus* Blanford, *flaviventris* (Serbia), *flavobrunneus*, *fulvipectus* (N. Caucasus), *griseus*, *isabellinus*, *leucocephalus*, ?*maximus*, *microtis* (S.E. Kazakhstan), *mosquensis*, *niger*, *nigritalus*† (Altai Mts), *pallipes* (Pamirs), *parvus*, *pecchioli*, *pentax* (Punjab), *planicola*, *rusiges*† (Kashmir), *spadix*, *stankovici* (Yugoslavia), *tauricus* Barrett-Hamilton, *tokmak* (E. Kirghizia), *uralensis* (S. Urals), *varius*, *wardi*† (Ladak), *witherbyi*.

*Apodemus sylvaticus milleri** de Beaux, 1926. Borzoli, Genoa, Italy.

*Apodemus sylvaticus hessei** Miric, 1960. Crni Kamen, Sar Planina Mts, near Kacanik, Yugoslavia. (Name from *Sylvaemus sylvaticus sylvaticus* morpha *hessei* Martino, 1933.)

*Apodemus sylvaticus dichruroides** Miric, 1960. No locality. (Name from *Sylvaemus sylvaticus sylvaticus* morpha *dichruroides* Martino, 1933.)

*Apodemus sylvaticus clanceyi** Harrison, 1948. Bagnacavallo, near Lugo, Emilia, N. Italy.

*Apodemus sylvaticus iconicus** Heptner, 1948a. New name for *Mus sylvaticus tauricus* Barrett-Hamilton, 1900, preoccupied by *Mus sylvaticus* var. *tauricus* Pallas, 1811.
*Apodemus sylvaticus kilikiae** Kretzoi, 1964. New name for *Mus sylvaticus tauricus* Barrett-Hamilton, 1900.
Synonyms from islands and other isolates: *algirus* (Algeria), *bergensis* (Norway), *butei* (Is. of Bute, Scotland), *celticus* (Ireland), ?*chamaeropsis*, *creticus* (Crete), *cumbrae* (Gt Cumbrae Is., Scotland), *fiolagan* (Is. of Arran, Scotland), *fridariensis*† (Fair Is., Scotland), *ghia* (Is. of Gigha, Scotland), *grandiculus*, *granti*† (Is. of Yell, Scotland), *hamiltoni*† (Is. of Rhum, Scotland), *hayi* (Morocco), *hebridensis* (Is. of Lewis, Scotland), *hirtensis*† (Is. of Hirta, St Kilda, Scotland), *intermedius* (England), *islandicus* (Iceland), *larus* (Is. of Jura, Scotland), *maclean* (Is. of Mull, Scotland), *nesiticus* (Is. of Mingulay, Scotland), *thuleo*† (Is. of Foula, Scotland), *tirae* (Is. of Tiree, Scotland), *tural* (Is. of Islay, Scotland).
*Apodemus sylvaticus ifranensis** Saint Girons & Bree, 1962. Between Ifrane and Boulemane, Middle Atlas, Morocco.
*Apodemus sylvaticus rufescens** Saint Girons & Bree, 1962. Temmerkennit, Algeria.
*Apodemus sylvaticus ilvanus** Kahmann & Niethammer, 1971. Island of Elba, Italy.
*Apodemus sylvaticus hermani** Felten & Storch, 1970. Pantelleria Island, Italy.

RANGE (Map 66). Widespread in the W. Palaearctic, north to S. Scandinavia and about 60°N in W. Siberia; east to the Altai Mts, Pamirs and to central Nepal in the Himalayas; south to Afghanistan, Iran and Palestine; also in N.W. Africa and on many islands including Iceland, Britain, Ireland and most Mediterranean islands.

REMARKS. E & M-S included in this species many eastern Asiatic forms which are here excluded following Zimmermann (1962) and most subsequent authors. All the described forms from small islands around Britain are here included in *A. sylvaticus* although E & M-S included some in *A. flavicollis* (see remarks under the latter).

No attempt has been made to delimit subspecies. There is considerable variation in size and colour but it is unlikely that a detailed study would reveal many forms that were sufficiently discrete and definable to be worth recognizing as subspecies, even amongst the insular forms. In the continental range the white-bellied form from the arid parts of the Arabian region (*arianus*) is one of the most distinctive. Of the insular forms, that on St Kilda, Scotland is the most distinctive and fully justifies subspecific rank – *A. s. hirtensis* (Barrett-Hamilton). The remaining Scottish insular forms do not justify subspecific rank (Berry *et al.*, 1967).

The question of reputed hybridization with *A. flavicollis* has been referred to under that species.

*Apodemus krkensis**

Apodemus krkensis Miric, 1968. Baska, Krk Island, Quarner, Yugoslavia.

RANGE (Map 65). Known only from the island of Krk, Yugoslavia.

REMARKS. When it was first described this species was compared only with *A. mystacinus* which it resembles approximately in its rather pale, greyish dorsal pelage. However examination of a specimen (by courtesy of Dr D. Carter of Texas Tech University) suggests to me that it is very close to *A. sylvaticus* and has no particular affinity with *A. mystacinus*. It apparently coexists with normally coloured *A. sylvaticus* on Krk.

*Apodemus microps**

Apodemus microps Kratochvil & Rosicky, 1952b. Saca, Kosic, Czechoslovakia.

RANGE (Map 66). Eastern Europe, mainly in Czechoslovakia and Rumania but also in adjacent parts of Poland, Austria, Hungary, Yugoslavia and Bulgaria.

REMARKS. It has been suggested, e.g. by Kratochvil (1962), that this species might be conspecific with *A. microtis* Miller, 1912, from S.E. Kazakhstan, here tentatively placed in *A. sylvaticus*. For a detailed comparison between this species and *A. sylvaticus* see Kratochvil & Zejda (1962).

***Apodemus argenteus*†** Small Japanese field mouse

Mus argenteus Temminck, 1845. Japan.
A. a. argenteus. Honshu, Shikoku and Kyushu. Incl. *geisha*.
A. a. hokkaidi (Thomas, 1906). Hokkaido.
A. a. celatus (Thomas, 1906). Dogo Island.
A. a. sagax Thomas, 1908. Tsushima Island.
A. a. tanei Kuroda, 1924. Tanegashima Island.
A. a. yakui (Thomas, 1906). Yakushima Island.

RANGE (Map 65). Japan, including all four main islands and the smaller ones detailed above.

REMARKS. Of the eastern Asiatic *Apodemus* this species most closely resembles the western *A. sylvaticus*. Zimmermann (1962) included it in his eastern subgenus *Alsomys* for which the only constant character he gave was the presence of only three roots in M^1. However all the specimens of *A. argenteus* I have examined have the lingual root of M^1 clearly double, as in the European species.

The subspecies listed above were recognized by Imaizumi (1960). They seem very doubtfully distinct although *celatus* is rather short-tailed.

Apodemus speciosus Large Japanese field mouse

Mus speciosus Temminck, 1845. Japan.
A. s. speciosus. Honshu, Shikoku, Kyushu and adjacent small islands. Incl. *dorsalis* (Yakushima Is.), *insperatus* (Oshima Is.), *navigator* (Dogo Is.), *sadoensis* (Sado Is.), *tusimaensis* (Tsushima Is.).
A. s. ainu (Thomas, 1906). Hokkaido.
*A. s. miyakensis** Imaizumi, 1969. Miyake Is., Seven Islands of Izu, Honshu.
Apodemus miyakensis Imaizumi, 1969. Miyake Is., Seven Islands of Izu, Honshu, Japan.

RANGE (Map 65). The four main islands of Japan and many small islands (but south only to Yakushima).

REMARKS. Imaizumi (1969) gave *ainu* and *miyakensis* specific rank and the other small-island forms subspecific rank in *A. speciosus*. They are uniformly reduced in rank here since Imaizumi distinguished the subspecies only by mean values of the characters concerned. See also Tsuchiya (1974).

This species has been considered to include also the continental forms here allocated to *A. peninsulae*. Apart from differences in size, pelage and teeth, the specific separation of these forms is also indicated by their sympatry on Hokkaido (see under *A. peninsulae*).

***Apodemus peninsulae*†** Korean field mouse

Micromys speciosus peninsulae Thomas, 1906. Mingyong, 110 miles S.E. of Seoul, Korea.
A. p. peninsulae. Korea, Manchuria and mainland of E. Siberia. Incl. *major* Radde, *majusculus* (Transbaikalia), *praetor* (Manchuria), *rufulus*, *tscherga* (Altai).
A. p. giliacus (Thomas, 1907). Islands of Sakhalin (type-locality) and Hokkaido.
*A. p. sowerbyi** Jones, 1956. Central China.
Apodemus peninsulae sowerbyi Jones, 1956. 30 miles W. of Kuei-hua-cheng, N. Shansi, China (7000 ft).

RANGE (Map 66). S.E. Siberia from Altai to Ussuri and south through Korea and Manchuria to Kansu and Shensi; Sakhalin and Hokkaido. Perhaps also south to Szechuan and Yunnan.

REMARKS. The above forms were divided between *A. sylvaticus* and *A. flavicollis* by E & M-S and were considered as races of the Japanese *A. speciosus* by Kuznetzov (1965). The sympatry on Hokkaido of *A. speciosus* and a smaller species (as well as the very small *A. argenteus*) was shown by Kobayashi & Hayata (1971). They considered the smaller form identical to *giliacus* from Sakhalin to which they gave specific rank. However *giliacus* seems scarcely separable from the continental *A. peninsulae*. As used here *A. peninsulae* is characterized by the complete and constant absence of a yellow pectoral spot, single lingual root of M^1 and smaller size than *A. speciosus*.

I have not attempted to explore in any detail the relationship of this species with the southern Chinese forms *draco*, *orestes*, *latronum* etc. Allen (1940) maintained that *A. peninsulae* could be recognized as far south as Yunnan, independently of these southern species.

***Apodemus draco*†**

Mus sylvaticus draco Barrett-Hamilton, 1900. Kuatun, Fukien, S.E. China.
A. d. draco. S.E. China. Incl. *argenteus* Swinhoe, *bodius* Swinhoe.
A. d. orestes Thomas, 1911. Szechuan (type-locality) to Yunnan and Assam. Incl. *ilex* (Yunnan).

RANGE (Map 66). China from S. Kansu and Shensi south to Burma and Assam. Perhaps north as far as Peking.

REMARKS. This form could conceivably be conspecific with the northern *peninsulae* and with *semotus* on Taiwan. Allen (1940) recorded it north to Peking, distinguishing it from *A. peninsulae* on the basis of its darker ears.

***Apodemus latronum*†**

Apodemus speciosus latronum Thomas, 1911. Tatsienlu, Szechuan, China.

RANGE (Map 65). Szechuan, Yunnan and Upper Burma.

REMARKS. This form is unlikely to be conspecific with the smaller, northern *A. peninsulae*. It is sympatric with *A. draco orestes* in Szechuan.

***Apodemus semotus*†**

Apodemus semotus Thomas, 1908. Taiwan.

RANGE. Taiwan.

REMARKS. This form is close to *A. draco* and could be conspecific with it.

***Apodemus gurkha*†**

Apodemus gurkha Thomas, 1924. Nepal.

RANGE (Map 65). Nepal.

REMARKS. Martens & Niethammer (1972) have shown that this species co-exists with *A. sylvaticus* in Nepal. It is clearly related to the eastern *A. draco* or *A. latronum* rather than to *A. flavicollis*, but it is rather distinctive and may prove to warrant specific rank.

Apodemus agrarius Striped field mouse

Mus agrarius Pallas, 1771. Simbirsk, R. Volga, Russia.
A. a. agrarius. Europe and W. Asia. Incl. *albostriatus*, *caucasicus* (Caucasus), *maculatus*, *nikolskii*, *ognevi* (W. Siberia), *rubens*, *septentrionalis*, *tianschanicus* (Tien Shan), *volgensis*.
*Apodemus agrarius karelicus** Ehrström, 1914. Near Viborg, Karelia, S.E. Finland.
*Apodemus agrarius kahmanni** Malec & Storch, 1963. Banja Bansko, Macedonia, Yugoslavia (265 m).
*Apodemus agrarius henrici** Lehmann, 1970. Münnerstadt, Kissingen district, W. Germany.
A. a. ningpoensis (Swinhoe, 1870). E. Asiatic part of range except S.W. China (type-locality in Chekiang, S.E. China). Incl. *coreae* (Korea), *gloveri*, *harti*, *insulaemus* (Taiwan), *mantchuricus* (Manchuria), *pallidior* (Shantung).
*Apodemus agrarius pallescens** Johnson & Jones, 1955b:169. 8 miles S.W. of Kunsan, Korea.
*Apodemus agrarius chejuensis** Johnson & Jones, 1955b:171. 10 miles N.E. of Mosulp'o, Cheju Do (i.e. Quelpart Island), Korea.
A. a. chevrieri (Milne-Edwards, 1868). S.W. China (type-locality in Szechuan). Incl. *fergussoni* (S. Kansu).

RANGE (Map 67). From W. Germany and N. Italy through E. Europe and W. Siberia to L. Baikal, south to Thrace, the Caucasus and the Tien Shan; in E. Asia from the Amur south through Korea and most of China as far as Yunnan and Fukien; Quelpart Is. (Korea) and Taiwan.

REMARKS. Most of the named forms have been separated on the basis of slight differences in colour or mean size. The eastern and western forms are not very clearly differentiated. The southern Chinese form *chevrieri* completely lacks the black dorsal stripe that otherwise distinguishes the species. However in Hunan the stripe is very faint, intermediate between the situation in Szechuan to the west (no stripe) and Fukien to the east (stripe prominent). The insular forms (Quelpart and Taiwan) are rather large but are not very distinctive.

Genus *ARVICANTHIS*

Arvicanthis Lesson, 1842. Type-species *Lemmus niloticus* Geoffroy. Incl. *Isomys*.

An African genus represented in the Palaearctic only along the lower Nile and in S.W. Arabia by two forms that are here tentatively considered conspecific. In its strictest sense it contains about three or four very closely similar species whose separation still causes some confusion. The African species usually included in *Pelomys*, *Mylomys* and *Lemniscomys* are very similar and their generic separation from *Arvicanthis* (the earliest name) is debatable.

Arvicanthis niloticus Nile grass rat

Arvicola niloticus Desmarest, 1822. Egypt.
A. n. niloticus. Egypt. Incl. *variegatus, discolor, major, minor*.
A. n. naso Pocock, 1934. S.W. Arabia.
Probably also many named forms from south of the Sahara.

RANGE (Map 67). Nile Valley and S.W. Arabia. Probably also farther south in Africa but range cannot be defined until the taxonomy of the genus has been clarified.

Genus *LEMNISCOMYS*

Lemniscomys Trouessart, 1881. Type-species *Mus barbarus* L.

A purely African genus of probably four species, one of which occurs north of the Sahara. Closely related to, and only doubtfully distinct from, the earlier named genus *Arvicanthis*.

Lemniscomys barbarus Barbary striped mouse

Mus barbarus Linnaeus, 1767. 'Barbaria' = Morocco.
L. b. barbarus. N.W. Africa. Incl. *ifniensis*.
Also several other races from south of the Sahara.

RANGE (Map 67). Morocco, Algeria and Tunisia; the dry steppe and open savanna zones south of the Sahara from Senegal to Sudan and south to Tanzania.

Genus *RATTUS*

Rattus Fischer, 1803. Type-species *Mus decumanus* Pallas = *Mus norvegicus* Berkenhaut. Incl. *Epimys*.

A large and confusing genus with many species in the Oriental Region. E & M-S included many African groups as subgenera but there seem to be good grounds for excluding these (see under *Praomys* below). Four species are represented in the Palaearctic Region, but all have probably spread from the Oriental Region by association with man. Three of these were included in a detailed revision by Schwarz & Schwarz (1967) which is valuable but leaves many important questions unanswered. These authors included *norvegicus* and *rattoides* in *R. rattus* but they are here kept separate as explained under *R. rattus*.

1. Distal half of tail well haired, hairs obscuring scales towards the tip; small: hind feet usually under 30 mm; (ventral pelage pure white; dorsal pelage often including spines; tail longer than head and body; E. Asia) . . . . ***R. niviventer***
– Tail sparsely haired throughout, scales not obscured; larger: hind feet usually over 30 mm (2)
2. Tail shorter than head and body; ears short (fall short of eye when laid forwards); temporal ridges on brain-case usually parallel . . . . ***R. norvegicus***
– Tail as long as or longer than head and body; ears longer (reach at least as far as eye when laid forwards); temporal ridges more curved . . . . (3)
3. Tail sharply bicoloured, dark above and lighter below; palate shorter, extending little if any distance behind posterior margins of M^3 . . . . ***R. rattoides***
– Tail not bicoloured; palate extending considerably behind posterior margins of M^3 ***R. rattus***

Rattus rattus Ship rat, Black rat

Mus rattus Linnaeus. 1758. Sweden.

R. r. rattus. Mediterranean region to India, sporadically in ports elsewhere. =*domesticus.* Incl. *albus, alexandrino-rattus, alexandrinus* (Egypt), *ater, brookei, caeruleus, chionagaster* Cabrera, *flaviventris* (Arabia), *fuliginosus, fulvaster, fuscus, intermedius, jurassicus, latipes, leucogaster, nemoralis, nericola, picteti, ruthenus, sueirensis* (Morocco), *sylvestris, tectorum* (N. Italy), *varius.*

R. r. tanezumi (Temminck, 1845). China and Japan. =*nezumi.* Incl. *flavipectus* (S. China).

*R. r. diardi** (Jentink, 1879). E. Indies, but also on the island of Hachijo, Japan according to Imaizumi (1967). *Mus diardi* Jentink, 1879. W. Java.

Also many other synonyms of the above races, and many other races, from outside the Palaearctic Region.

RANGE (Map 68). Widespread throughout the tropics and throughout the southern part of the Palaearctic Region including the Mediterranean region of Europe, S. Russia, S.W. Asia, N. Africa, S. China and Japan. Sporadically further north, mainly in ports, e.g. in Britain and Scandinavia.

REMARKS. The taxonomy of this group of rats was reviewed in some detail by Schwartz & Schwartz (1967) who included in *R. rattus* many forms that are normally considered specifically distinct, including *norvegicus* and *rattoides.* This action was based on the belief that these diverse forms have arisen from different geographical races of a single species in the Indonesian region. However it ignores the fact that these forms now co-exist and behave as good species.

Within *R. rattus* in the more restricted sense Schwarz & Schwarz (1967), along with many other authors, recognized the three principal colour forms found in the Western Palaearctic as subspecies (*rattus* – dark grey above and below; *alexandrinus* – brown above, grey below; *frugivorous* – brown above, white below). However these co-exist in many areas – in the Mediterranean and S.W. Asia *alexandrinus* tends to be the form found in and around buildings whilst *frugivorus* is the dominant form in fields. Amongst urban populations of *rattus* type, individuals of the other colour forms occur. These have been shown in the laboratory to be interfertile colour morphs (Caslick, 1956) and therefore it is better to treat them as consubspecific. There is however both large-scale and local variation in morph frequency.

Many of the Oriental forms generally considered conspecific with *R. rattus* need further investigation and some may prove to be specifically distinct.

Rattus rattoides

Mus rattoides Hodgson, 1845. Nepal.

R. r. rattoides. Nepal.

*Rattus rattus khumbuensis** Biswas & Khajuria, 1955. Nepal.

R. r. vicerex (Bonhote, 1903). W. Himalayas. Incl. *shigarius*†.

R. r. turkestanicus (Satunin, 1903). Turkestan, Iran and Afghanistan.

RANGE (Map 68). From S. Russian Turkestan, N.E. Iran and Afghanistan through Pakistan and N. India at least to Sikkim. Perhaps also in S. China and Burma.

REMARKS. This species was called *R. rattoides* by E & M-S. Schlitter & Thonglongya (1971) demonstrated that this name is preoccupied by *Mus rattoides* Pictet & Pictet, 1844, believed to be a synonym of *R. rattus*, and used *R. turkestanicus* for this species. This is an unnecessary change – *Mus rattoides* Pictet & Pictet should be considered a *nomen oblitum*.

The above arrangement of subspecies follows Niethammer & Martens (1975). Schwarz & Schwarz (1967) included all these forms in their '*brunneus*' group of *R. rattus* (based on *R. brunneus* (Hodgson, 1845) from Nepal). Many other forms in S. China, Burma and Thailand have been associated with this group but their relationship remains doubtful.

Rattus norvegicus Common rat, Brown rat

Mus norvegicus Berkenhaut, 1769. Britain.

R. n. norvegicus. W. Palaearctic, sporadically elsewhere. Incl. *caspius*, *decumanus*, *discolor*, *hibernicus* (Ireland), *hybridus*, *surmolottus*.

R. n. caraco (Pallas, 1779). E. Asia, type-locality Transbaikalia. Incl. *griseopectus* (Szechuan), *humiliatus* (N. China), *insolatus*† (Shensi), *otomoi* (Honshu, Japan), *plumbeus*, *primarius*, *socer* (Kansu), *sowerbyi*.

RANGE (Map 68). The whole of Europe, including most islands; Asia Minor; eastwards across southern Siberia to the Pacific; most of China and Japan. Also in most of temperate North America and locally, mainly in ports, in many other parts of the world.

REMARKS. Within the Palaearctic Region this species and *R. rattus* co-exist and behave as clearly distinct species. Schwarz & Schwarz (1967) considered them conspecific, mainly because of the intermediate condition of the form *nitidus* (W. Himalayas etc.). Clearly in a group like this, whose present distribution is the result of multiple introduction by man, many unorthodox relationships must occur and it is conceivable for example that two forms may behave as distinct species in one area whilst elsewhere other populations of apparently the same forms will interbreed.

Rattus niviventer White-bellied rat

Mus (Rattus) niviventer Hodgson, 1836. Katmandu, Nepal. Incl. *chihliensis* (near Peking), *sacer* (Shantung) and many other synonyms from S. China and elsewhere in the Oriental Region.

RANGE (Map 68). Throughout most of China from the region of Peking southwards, the eastern Himalayas and south into the Malayan peninsula.

REMARKS. This species is clearly distinct from the *R. rattus* group which includes the three previous species. Some forms from the Malaysian area that have been included in this species may prove to be specifically distinct.

Genus *PRAOMYS**

Soft-furred rats

Praomys Thomas, 1915:477. Type-species *Epimys tullbergi* Thomas = *P. morio* Trouessart. Incl. *Mastomys*.
*Myomys** Thomas, 1915:477. Type-species *Epimys colonus* A. Smith.
*Myomyscus** Shortridge, 1942. Type-species *Mus verroxi* A. Smith.

A mainly African genus, included in *Rattus* by E & M-S, but with several characteristics, e.g. soft pelage, that seem to justify generic separation (see Davis, 1965; Misonne, 1969). One species occurs north of the Sahara and another in S.W. Arabia. The genus is sometimes further split, placing these species in *Mastomys* and *Myomys* respectively, but this seems unnecessary.

Ventral pelage pure white; 5 pairs of mammae, in distinct anterior and posterior groups; tail considerably longer than head and body (Arabia) . . . . . . ***P. fumatus***
Ventral pelage grey; at least 6 pairs of mammae, evenly spaced; tail about equal to head and body (N.W. Africa) . . . . ***P. erythroleucus***

Praomys erythroleucus Western multimammate rat

Mus erythroleucus Temminck, 1853. Guinea.
P. e. peregrinus (de Winton, 1898). Morocco. Incl. *calopus*.
Also many races in Africa south of the Sahara.

RANGE (Map 67). Savanna and cultivated areas of Africa south of the Sahara and in S.W. Morocco.

REMARKS. This and closely related species, often separated in the genus *Mastomys*, are the dominant parasitic mice in many parts of Africa. They are sometimes united as *P. natalensis* (Smith, 1834) but recent work on chromosomes suggests that *P. erythroleucus* (2n = 38) is specifically distinct from *P. natalensis* (2n = 36) (Matthey, 1966) and that on this basis the Moroccan form is referable to *P. erythroleucus* (Tranier, 1974).

Praomys fumatus* Rock rat

Mus fumatus Peters, 1878. Ukamba, Kenya.
*P. f. yemeni** (Sanborn & Hoogstraal, 1953). Arabia.
Myomys fumatus yemeni Sanborn & Hoogstraal, 1953. Kariet Wadi Dhahr, 6 miles N.W. of San'a, Yemen.
Also several races in E. Africa.

RANGE (Map 67). Mountains of Yemen, S.W. Arabia; dry steppes of E. Africa from Somalia to Tanzania.

REMARKS. Harrison (1972) supported the view that the Arabian form is subspecifically distinct from those in Africa.

Genus *MUS*

Mus Linnaeus, 1758. Type species *Mus musculus* L. = *Musculus*.

A genus represented by several species in the Oriental Region and in Africa south of the Sahara, although the latter (along with some Oriental species) are frequently separated in the genus *Leggada*. In addition the genus includes *M. musculus*, the house mouse, which is now a ubiquitous parasite of man and may originally have had a southern Palaearctic distribution. It is the only species that reaches the Palaearctic unless one considers *Mus poschiavinus* Fatio, 1869 a distinct species. The latter is a dark form known only in the Puschlav Valley in southern Switzerland. The difference in chromosome complement between it and normal *M. musculus* has recently led to claims that it should be considered specifically distinct although it can be interbred with *M. musculus* in captivity (see for example Radbruch, 1973).

The relationship of *M. musculus* to the purely Oriental species has been clarified by Marshall (1972, 1977).

Mus musculus House mouse

Mus musculus Linnaeus, 1758. Sweden. Incl. *abbotti* (Asia Minor), *acervator*, *acervifex*, *airolensis*, *albicans*, *albinus*, *albidiventris*, *albus*, *albula*, *amurensis*, *ater*, *azoricus* (Azores), *bactrianus* (Afghanistan), *bicolor*, *bieni*, *borealis*, *candidus* Laurent, *canicularius*, *caniculator*, *caoceii* (Sardinia), *caudatus*, *cinereo-maculatus*, *decolor*, *domesticus* (Ireland), *faeroensis* (Faeroes), *far*, *flavescens*, *flavus*, *formosovi*, *funereus*, *gansuensis*, *gentilis* (S. Egypt), *gentilulus* (Aden), *germanicus*, *hapsaliensis*, *helveticus*, *helvolus*, *heroldii*, *hispanicus*, *hortulanus* (N. Caucasus), *jamesoni*, *kaleh-peninsularis*, *kurilensis*, *kuro*, *longicauda*, *lusitanicus*, *lynesi*, *maculatus*, *major* Severtzov, *manchu* (Manchuria), *melanogaster*, *mogrebinus*, *mollissinus*, *molossinus* (Japan), *mongolium*, *muralis* (St Kilda, Scotland), *mykinessiensis*, *niger*, *niveus*, *nogaiorum*, *nordmanni*, *nudoplicatus*, *orientalis* Cretzschmar, *orii*, *oxyrrhinus*, *pachyceros*, *polonicus*, ?*poschiavinus* (Switzerland), *praetextus* (Syria), *raddei*, *reboudia* (Algeria), *rifensis*, *rotans*, *rubicundus*, *rufiventris*, *sareptanicus*, *sergii*, *severtzovi*, *spicilegus* (Hungary), *spretus* (Algeria), *striatus* Billberg, *subcaeruleus* Fritsche, *subterraneus*, *takayamei*, *tataricus*, *tomensis* (Altai), *variabilis*, *varius*, *vinogradovi*, *wagneri* (Volga-Ural region), *yamashinai* (S. Korea), *yesonis*.
*Mus musculus hanuma** Miric, 1960. Sarajevo, Bosnia. (Name from *M. m. spicelegus* morpha *hanuma* Ognev, 1948.)
*Mus musculus helgolandicus** Zimmermann, 1953b. Island of Heligoland.

RANGE (Map 69). Nearly worldwide by association with man. In the Palaearctic absent only from N.E. Siberia.

REMARKS. Subspecific variation is very considerable but even the broadest outlines have not been adequately described in spite of a number of attempts. The main difficulty is the apparent co-existence of dissimilar forms. In these cases it has not been adequately demonstrated whether the two forms represent genetically distinct morphs within a freely interbreeding population (as in *Rattus rattus*); or distinct geographical races that have been brought together by introduction and behave, at least locally, as distinct species with a complete or partial reproductive barrier.

The revision by Schwarz & Schwarz (1943) defined the main subspecies and provided synonymies, but is difficult to apply because of the very subtle nature of the differences between the major groups they recognized and because of the lack of information on the interrelationships of sympatric forms. The following major geographical groups can be recognized:

1. 'Wild forms', living out of doors throughout the southern Palaearctic from N.W. Africa and Iberia through the steppes of S. Russia and Turkestan to Mongolia and Manchuria. Characterized by light dorsal colour, white ventral pelage (with or without grey bases to the hairs) and short tail (considerably shorter than head and body). Including the forms *wagneri* Eversmann, 1848 (Volga-Ural region), *spicilegus* Petenyi, 1882 (Hungary), *spretus* Lataste, 1883 (Algeria) and *manchu* Thomas, 1909 (Manchuria).

2. Western 'commensal forms', more closely associated with man, all with longer tails and darker pelage, but varying from southern forms with pale bellies (e.g. *praetextus* Brants, 1827 from Syria) to darker northern forms. There are grounds for believing that the latter have colonized N.W. Europe from both the east (*musculus*) and the west (*domesticus* Rutty, 1772) and that there is a zone of only partial interbreeding, e.g. in Denmark where it is reflected in the biochemical polymorphism as well as in morphology (Selander *et al.*, 1969).

3. Eastern 'commensal forms' in Japan and Korea (*molossinus* Temminck, 1845), probably derived from eastern wild forms but with darker pelage and longer tail.

The various strains of laboratory mice are derived from this species, but Marshall (1972) has pointed out that they have been bred from both European (*musculus*) and Japanese (*molossinus*) forms.

Genus *ACOMYS*

Spiny mice

Acomys Geoffroy, 1838. Type-species *Mus cahirinus* Desmarest. Incl. *Acosminthus*, *Acanthomys*.

A very distinctive genus of spiny mice in Africa and S.W. Asia. Classification within the genus is very problematical. In the Palaearctic Region there is one clearly distinct species, *A. russatus*, and another group of closely related forms that may represent a single species (*A. cahirinus*) but has frequently been split. Variation within this latter group involves both chromosomes and colour, some forms closely associated with man being melanic. Since no satisfactory specific diagnoses can be given if the group is split it is here provisionally treated as a single species.

Soles of feet black; tail distinctly shorter than head and body; spiny pelage of back extending forwards to back of head . . . . ***A. russatus***

Soles of feet pale; tail at least as long as head and body; spiny pelage less extensive, not reaching back of head . . . . ***A. cahirinus***

Acomys russatus — Golden spiny mouse

Mus russatus Wagner, 1840. Sinai.

A. r. russatus. Sinai and Egypt. Incl. *aegyptiacus*, *affinis*.

*A. r. lewisi** Atallah, 1967. N. Jordan.

Acomys lewisi Atallah, 1967. 3 km N.W. of Azraq-Shishan, Syrian Desert, Jordan.

*A. r. harrisoni** Atallah, 1970. S. Jordan.

Acomys russatus harrisoni Atallah, 1970. ½ km S. of Qumran Caves, near Ain Faschka, Jordan.

RANGE (Map 69). N.E. Egypt to Jordan and E. Arabia.

REMARKS. The very dark form *lewisi* was included in *A. russatus* by Harrison (1972).

Acomys cahirinus

Mus cahirinus Desmarest, 1819. Cairo, Egypt. Incl. *chudeaui* (Mauritania), *cineraceus* (Sudan), *dimidiatus* (Sinai), *flavidus* (Sind), *hispidus*, *homericus* (Aden), *hunteri* (N. Sudan), *megalotis*, *minous* (Crete), *nesiotes* (Cyprus), *sabryi*, *seurati* (Hoggar), *viator* (Libya).

*Acomys dimidiatus megalodus** Setzer, 1959. Gebel el Galala el Bahriya, Wadi Sayal (60 km S. of Suez), Eastern Desert, Egypt.

RANGE (Map 69). Semi-desert habitats from Mauritania across the southern edge of the Sahara, with apparently isolated populations in some of the montane oases of the Sahara; throughout much of Egypt and Arabia, north through Palestine to southern Asia Minor and east through S. Iran to Sind; the islands of Crete and Cyprus; also in Sudan and much of E. Africa.

REMARKS. Two species were recognized in this group by Petter (1954) who used the names *A. cahirinus* and *A. cineraceus* and maintained that they were sympatric near Khartoum, Sudan; and by Setzer (1959, 1971) who used the names *A. cahirinus* and *A. dimidiatus* and considered them sympatric in parts of Egypt. However the diagnoses given by these authors cannot be applied throughout the range of the group which is therefore treated here provisionally as a single species, without denying the probability that the situation is more complex.

The complex variation in chromosome complement has been reviewed by Matthey & Baccar (1967).

Genus *NESOKIA*

Nesokia Gray, 1842. Type-species *Arvicola indica* Gray. =*Nesocia*. Incl. *Spalacomys*.

A distinctive, monospecific genus, characterized especially by very hypsodont teeth with transverse laminae almost completely obscuring the cusp pattern.

Nesokia indica Short-tailed bandicoot rat

Arvicola indica Gray, 1830. India.

N. i. indica. India to Iraq and Turkestan. Incl. *bailwardi* (N. Iran), *beaba*, *boettgeri*, *buxtoni* (Iraq), *dukelskiana*, *griffithi*, *hardwickei*, *huttoni* (Afghanistan), *insularis*, *legendrei*, *myosurus* (Iraq), *satunini* (Turkmenistan).

*Nesokia indica chitralensis** Schlitter & Setzer, 1973. 4 miles N. of Chitral, 4800 ft, Chitral State, Pakistan.

N. i. bacheri Nehring, 1897. Palestine (type locality) and Egypt. Incl. *suilla* (Suez).

N. i. scullyi Wood-Mason, 1876. Sinkiang. Incl. *brachyura*.

RANGE (Map 69). Dry steppes and cultivated areas from N.W. India (Kumaon and Rajputana) through Afghanistan and Iran to N. Arabia, Iraq, Syria and southern Russian Turkestan; Palestine and lower Egypt, west to the Bahariya Oasis (Wassif, 1959); S. Sinkiang (Yarkand and Lob Nor).

REMARKS. Many races have been described on the basis of differences in colour and size. Within *N. i. indica* as here recognized, the western forms, in Iraq, are especially pale and yellow but intergrade with darker animals in Iran (Lay, 1967).

Family **GLIRIDAE**

Dormice

This family includes six genera (mostly monospecific) in the Palaearctic Region and one (*Graphiurus*) in the Ethiopian Region. The aberrant genus *Selevinia* was included as a subfamily by E & M-S but excluded as the sole member of a separate family by Simpson (1945) and by most Russian authors, and this latter course is followed here. The two Oriental genera *Platacanthomys* and *Typhlomys* have usually been included in this family, in a subfamily Platacanthomyinae, but this has also been accorded family rank and more recently, as a result of palaeontological studies, has been considered a subfamily of Cricetidae (Mein & Freudenthal, 1971).

The generic classification has for long been stable although the differences are in some cases rather slight.

The name of the family is controversial, Muscardinidae being frequently used and Myoxidae less often. If the generic name *Glis* can be retained (which is highly desirable for stability although it is at present technically invalid) then Gliridae is clearly the correct name for the family. Muscardinidae was used by Ellerman (1940) and by E & M-S, at first because Gliridae was thought to be preoccupied (refuted by Simpson,

1945) and later because of the instability of *Glis*. Gliridae was used by Simpson and by many subsequent authors.

1. Tail uniformly bushy throughout its length . . . . (2)
– Tail short-haired, at least in proximal half . . . . (5)
2. Pelage orange-brown; crowns of cheek-teeth flat; M^1 much longer than M^2 ***MUSCARDINUS*** (p. 144)
– Pelage not orange-brown; teeth not as above . . . . (3)
3. Dark dorsal stripe; bullae very small (Japan) . . . . . . ***GLIRULUS*** (p. 146)
– No dorsal stripe; bullae inflated . . . . (4)
4. Larger: head and body over 130 mm, hind feet over 24 mm; upper molars with 5 or 6 low cusps on labial side; angular process of mandible perforated . . . ***GLIS*** (p. 144)
– Smaller: head and body under 130 mm, hind feet under 24 mm; upper molars with 2 main cusps on labial side; angular processes not perforated . . . ***DRYOMYS*** (p. 146)
5. Tail short-haired; no black pattern on face . . . . . ***MYOMIMUS*** (p. 147)
– Tail long-haired in distal half; bold black pattern on face . . . ***ELIOMYS*** (p. 145)

Genus *GLIS*

Glis Brisson, 1762. Type-species *Sciurus glis* L. = *Myoxus*. Incl. *Elius* (in part). (The availability of Brisson (1762) has been queried. Its rejection would make it necessary to use the name *Myoxus* Zimmermann, 1780 for this genus but this seems neither necessary nor desirable.)

A monospecific and moderately distinct genus.

Glis glis — Fat dormouse

Sciurus glis Linnaeus, 1766. Germany.
G. g. glis. Europe except for Mediterranean region. Incl. *avellanus, esculentus, ?giglis, vulgaris.*
G. g. pyrenaicus Cabrera, 1908. Pyrenees and N. Spain.
G. g. italicus Barrett-Hamilton, 1898. Italy, Slovenia and Sicily (type-locality Sienna). Incl. *abruttii, insularis* (Sicily), *intermedius* Altobello.
*Glis glis vagneri** Martino & Martino, 1941. Vrhpolje, Kamnik, Kamniske Alpe, Slovenia, Yugoslavia.
G. g. postus Montagu, 1923. Yugoslavia.
G. g. minutus Martino, 1930. Serbia and Bulgaria.
*Glis glis intermedius** Martino & Martino, 1941. Presaca, Donji Milanovav, N.E. Serbia. (Preoccupied by *G. italicus intermedius* Altobello, 1920.)
*Glis glis martinoi** Miric, 1960. New name for *Glis glis intermedius* Martino & Martino, 1941.
*G. g. pindicus** Ondrias, 1966. Greece.
Glis glis pindicus, Ondrias, 1966. Moni Stomiou, near Konitsa, Epirus, Greece (1600 m).
G. g. orientalis (Nehring, 1903). Asia Minor and Caucasus. Incl. *spoliatus, tschetshenicus.*
G. g. persicus (Erxleben, 1777). S. shores of Caspian Sea. Incl. *caspicus, caspius, petruccii.*
G. g. melonii Thomas, 1907. Sardinia (type-locality) and Corsica.
*G. g. argenteus** Zimmermann, 1953. Crete.
Glis glis argenteus Zimmermann, 1953a. White Mts, near Samaria, Crete (1000 m).

RANGE (Map 70). Woodland parts of Europe from the Mediterranean to the Baltic (except for most of Iberia, N. France, the Low countries and Denmark) and east to the R. Volga; also in the Caucasus, N. Asia Minor and N. Iran. Present on Crete, Corfu, Sicily, Corsica, Sardinia but not the Balearics.

REMARKS. For a monograph of the species see Veitinghoff-Riesch (1960). The subspecies in S.E. Europe were diagnosed and mapped by Ondrias (1966) and his arrangement is followed above although there is likely to be considerable integradation and the subspecific boundaries are probably arbitrary.

Genus *MUSCARDINUS*

Muscardinus Kaup, 1829. Type-species *Mus avellanarius* L.

A very distinctive, monospecific genus most closely related to *Glis* but characterized by rather finely ridged teeth and very prehensile feet.

Muscardinus avellanarius — Hazel dormouse

Mus avellanarius Linnaeus, 1758. Sweden.
M. a. avellanarius. Most of Europe. Incl. *anglicus* (England), *corilinum, muscardinus.*

*Muscardinus avellanarius kroecki** Niethammer & Bohmann, 1950. Vitosa, 10 km S. of Sofia, Bulgaria (1500 m).
M. a. pulcher Barrett-Hamilton, 1898. Italy and Sicily. Incl. *niveus, speciosus.*
M. a. zeus Chaworth-Musters, 1932. Greece (including Corfu).
M. a. trapezius Miller, 1908. Asia Minor.

RANGE (Map 70). Europe from the Mediterranean to the Baltic (except for Iberia and Denmark) and east to 50°E in Russia; isolates in England and Wales, S. Sweden, Sicily, Corfu and northern Asia Minor.

REMARKS. The arrangement of subspecies follows that of Witte (1962) and Roesler & Witte (1969).

Genus *ELIOMYS*

Eliomys Wagner, 1840. Type-species *E. melanurus* Wagner. Incl. *Bifa.*

A very distinctive genus confined to the western Palaearctic. Treated here as monospecific although two species were recognized by E & M-S and by Harrison (1972).

Eliomys quercinus Garden dormouse

Mus quercinus Linnaeus, 1766. Germany.
'*quercinus*' group of subspecies:
E. q. quercinus. Mainland European range except for southern Iberia, Italy and Balkans. Incl. *gotthardus, hamiltoni, hortualis, jurassicus, nitela, raticus, ?superans* (Russia).
*E. q. valverdei** Palacios *et al.*, 1974. N.W. Spain.
Eliomys quercinus valverdei Palacios *et al.*, 1974. Penas Apañadas, Picos dos Tres Obispos, Donis, Sierra de los Ancares, Prov. de Lugo, Spain (1700 m).
E. q. gymnesicus Thomas, 1903. Minorca and Majorca. (See Kahmann & Tiefenbacher (1969) for a detailed account.)
E. q. munbyanus (Pomel, 1856). Morocco and Algeria. Incl. *lerotina.*
'*lusitanicus*' group of subspecies:
E. q. lusitanicus Reuvens, 1890. S. Iberia. Incl. *amori.*
E. q. dichrurus† (Rafinesque, 1814). Sicily (type-locality) and S. Italy. Incl. *cincticauda* (S. Italy), *pallidus.*
E. q. sardus Barrett-Hamilton, 1901. Sardinia (type-locality) and Corsica.
E. q. ophiusae Thomas, 1925. Formentera, Balearic Is.
*E. q. liparensis** Kahmann, 1960, Lipari Island, N. of Sicily.
Eliomys quercinus liparensis Kahmann, 1960. Lipari Is.
*E. q. dalmaticus** Dulic & Felten, 1962. Dalmatian coast.
Eliomys quercinus dalmaticus Dulic & Felten, 1962. Gipfel Kosa (940 m), Mosor Mountains (14 km E. of Split), Dalmatia, Yugoslavia.
'*melanurus*' group of subspecies:
E. q. melanurus† Wagner, 1840. Asia Minor to Sinai.
E. q. cyrenaicus Festa, 1922. Cyrenaica, Libya.
E. q. tunetae Thomas, 1903. Tunisia and adjacent parts of Libya and Algeria; and in Rio de Oro. Incl. *occidentalis* (Rio de Oro).
*E. q. denticulatus** Ranck, 1968. S.W. Libya.
Eliomys quercinus denticulatus Ranck, 1968. El Gatrun, Fezzan Prov., Libya.

RANGE (Map 71). Europe from the Mediterranean to just short of Denmark and the Baltic coast, and east through the wooded parts of European Russia to the southern Urals and north to Finland; N. Africa from Rio de Oro to N.W. Libya; Cyrenaica, N. Egypt (Hoogstraal *et al.*, 1955) and N. Arabia to S. Asia Minor; Sicily and most of the W. Mediterranean islands.

REMARKS. The Asiatic *melanurus*, given specific rank by E & M-S and by Harrison (1972), was treated as a subspecies of *E. quercinus* by Niethammer (1959) and by Ranck (1968) because of the intermediate character of some of the north African forms.

Variation in the species is extensive and complex. The three groups of subspecies listed above were recognized by Niethammer (1959) mainly on the basis of the pattern of pigmentation of the tail. The map given by Niethammer shows zones of integradation between these groups. Any zoogeographical interpretation of this situation should take into account the possibility that man may have played some part in the process.

Petter (1961c) treated *E. lusitanicus* as a species overlapping with *E. quercinus* in central Iberia. The allocation of *Musculus dichrurus* Rafinesque, 1814 to *E. quercinus* rather than to *Apodemus sylvaticus* follows Bree (1974).

Genus *DRYOMYS*

Dryomys Thomas, 1906. Type-species *Mus nitedula* Pallas. = *Dyromys*.

A genus of two clearly distinct species closely related to *Eliomys*.

Prominent black mask on face; larger: hind feet 19–22 mm, upper tooth-row 3·5–4·0 mm . . . ***D. nitedula***

No black on face; smaller: hind feet about 18 mm, upper tooth-row about 3·4 mm; (dorsal pelage very pale, thick and soft) . . . . ***D. laniger***

Dryomys nitedula Forest dormouse

Mus nitedula Pallas, 1779. Lower Volga, Russia.

D. n. nitedula. European Russia. Incl. *carpathicus* (Upper Silesia), *daghestanicus* (Dagestan, N.E. Caucasus), *dryas*, *obelenskii* (Voronezh, R. Don), *tanaiticus*.

D. n. robustus (Miller, 1910). Bulgaria.

*D. n. ravijojla** Paspalev *et al.*, 1952. Macedonia.
Dryomys nitedula ravijojla Paspalev *et al.*, 1952. Senecki Suvati, Bistra Mts, Macedonia.

D. n. wingei (Nehring, 1902). Greece.

*D. n. diamesus** Lehmann, 1959. Montenegro.
Dryomys nitedula diamesus Lehmann, 1959. Ivanova Korita, Montenegro, Yugoslavia.

D. n. intermedius (Nehring, 1902). Eastern Alps.

*D. n. aspromontis** Lehmann, 1963. S. Italy.
Dryomys nitedula aspromontis Lehmann, 1963. Gambarie d'Aspromonte, Calabria, Italy.

D. n. phrygius (Thomas, 1907). Asia Minor to Palestine.

D. n. ognevi (Heptner & Fomosov, 1928). Caucasus, type-locality in S. Dagestan. Incl. *caucasicus*.

D. n. pictus (Blanford, 1875). N. Iran and Armenia. Incl. *kurdistanicus* (W. Azerbaijan), *tichomirowi* (Tbilisi).

D. n. angelus (Thomas, 1906). Russian Turkestan. Incl. *bilkjewiczi* (Kopet Dag Mts), *pallidus* (Uzbekistan), *saxatilis* (Pamirs).

D. n. milleri (Thomas, 1912). Bogdo-Ola Mts, Sinkiang.

RANGE (Map 71). Widespread in deciduous woodland from the Balkans and Carpathians through European Russia as far north as Moscow and south to the Caucasus, Asia Minor and N. Iran. Peripheral distribution probably very fragmented, and mainly in montane forest: in the eastern Alps, in Calabria, Palestine, central Iran, Afghanistan, much of Russian Turkestan to the extreme eastern end of the Tien Shan in Sinkiang.

REMARKS. Those races occurring in the USSR have been listed above following the review of variation by Rossolimo (1971a). There has been no recent revision, with diagnoses, of the European forms, but it is unlikely that all those listed will prove valid. They are discussed by Roesler & Witte (1969).

Dryomys laniger*

Dryomys laniger Felten & Storch, 1968. Çiğlikara, Bey Mts (2000 m), 20 km S.S.E. of Elmali, Antalya Prov., Asiatic Turkey.

RANGE (Map 71). Known only from the type-locality in S.W. Asia Minor.

REMARKS. This seems to be a very distinctive species although described from only a single individual. A coloured photograph has been published by Felten *et al.* (1971a). The species has not been seen.

Genus *GLIRULUS*

Glirulus Thomas, 1906. Type-species *Graphiurus elegans* Temminck = *Myoxus japonicus* Schinz.

A distinctive monospecific genus, probably most closely related to *Dryomys*.

Glirulus japonicus Japanese dormouse

Myoxus javanicus Schinz, 1845 (*lapsus* for *japonicus*). Japan. Incl. *elegans* Temminck, *lasiotis*.

RANGE (Map 71). Honshu, Shikoku and Kyushu, Japan.

Genus *MYOMIMUS*

Myomimus Ognev, 1924. Type-species *M. personatus* Ognev.

A distinctive genus, here considered monospecific (but see under Remarks). A synonym based on fossil material from Palestine is *Philistomys* Bate, 1937.

Myomimus personatus Mouse-tailed dormouse

Myomimus personatus Ognev, 1924. The Turkmenistan-Iran border near the Caspian Sea.

RANGE (Map 70). Extreme S.W. Turkmenistan: W. Asia Minor (Mursaloglu, 1973); S.E. Bulgaria, Thrace. (Known also from subfossil material in S. Asia Minor and Palestine.)

REMARKS. The measurements of tooth-rows of Bulgarian animals given by Peshev *et al.* (1964) (4·5–4·8 mm for both upper and lower rows) would seem to be incompatible with the values of 4·0 and 3·8 given by Ognev (1947) for the upper and lower rows respectively of the type of *personatus*, and might suggest a specific difference. However, two of the Bulgarian specimens concerned are now in the British Museum and although I agree with Peshev *et al.* with respect to the other cranial measurements, my measurements of the teeth differ from theirs and agree instead with those of the type. Both have values of 3·9 mm for the upper tooth-row and 4·0 mm for the lower. (These specimens are nos. 795 and 802 of Peshev; BM 66. 5592 and 3 respectively.) There is therefore no reason to suspect that more than one species is involved. More recently however specimens from Bulgaria and from N.W. Iran have been described as distinct species: *M. bulgaricus* Rossolimo, 1976a and *M. setzeri* Rossolimo, 1976b.

Family SELEVINIIDAE

This family contains a single species which was placed in a subfamily of Gliridae by Ellerman (1940 and in E & M-S) but has been given family rank by almost all other authors. It is characterized especially by the absence of premolars, simple rudimentary molars, greatly enlarged bullae and deeply grooved upper incisors.

Genus *SELEVINIA*

Selevinia Belosludov & Bashanov, 1939. Type-species *S. betpakdalensis* Belosludov & Bashanov. (Dated 1938 by E & M-S, but date of publication given as 5 February 1939 by Bashanov & Belosludov, 1941.)

Selevinia betpakdalensis Desert dormouse

Selevinia betpakdalensis Belosludov & Bashanov, 1939. Betpakdala Desert, Kazakhstan.
*Selevinia paradoxa** Argyropulo & Vinogradov, 1939 (17 March). Betpakdala Desert.

RANGE (Map 70). Deserts to the west and north of Lake Balkhash, Kazakhstan.

REMARKS. For an English version of the original description see Bashanov & Belosludov (1941). This species has not been seen.

Family ZAPODIDAE

Jumping mice

A small family represented by two genera of mouse-like species in the Palaearctic and two in the Nearctic Region. The group is closely related only to the Dipodidae with which it is sometimes united, e.g. by E & M-S. Ellerman (1940) argued this course in relation to characters of the hind feet and skull, but the group that he particularly considered intermediate, the pygmy jerboas (*Cardiocranius* and *Salpingotus*), only resemble the zapodids in lacking actual fusion of the metatarsals – in other respects, including the overall elongation of the hind legs and feet, they are typical of the bipedal dipodids. The overall adaptations to bipedal and quadripedal (albeit sometimes saltatorial) locomotion seem to divide the groups into two rather clear-cut parts, justifying family rank by analogy with other groups of rodents. This division was used by Simpson (1945) and by Kuznetsov (1965).

The Zapodidae further divides into two groups, one containing the wholly Palaearctic genus *Sicista*, the other the Nearctic *Zapus* and *Napaeozapus* along with the single species *Eozapus setchuanus* in China. These last three genera are very similar and doubtfully justify separate recognition.

Upper incisors grooved; ventral pelage white; hind feet over 25 mm **EOZAPUS** (p. 149)
Upper incisors ungrooved; ventral pelage grey or buff; hind feet under 20 mm **SICISTA** (p. 148)

Subfamily SICISTINAE
Genus *SICISTA*
Birch mice

Sicista Gray, 1827. Type-species *Mus subtilis* Pallas. Incl. *Sminthus*.

A clearly defined genus usually treated as the sole recent member of the subfamily Sicistinae. It is confined to the Palaearctic Region except for an extension into the Himalayas and S.W. China (included below). Classification within the genus is rather unstable. The two striped species seem well defined, but the unstriped forms are inadequately known. Five species in all are listed below, but some Russian authors would recognize three more by splitting *S. concolor* as used here.

1. Black mid-dorsal stripe present . . . . (2)
– No mid-dorsal stripe . . . . (3)
2. Tail about 150 % of head and body; hind feet over 15·8 mm; back of uniform colour apart from central dark stripe . . . . ***S. betulina***
– Tail about 120–130 % of head and body; hind feet usually under 15·8 mm; dark dorsal stripe flanked by obscure stripes rather lighter than rest of pelage . . . . ***S. subtilis***
3. Tail over 150 % of head and body; penis with numerous small spines only ***S. concolor***
– Tail under 150 % of head and body; penis with one or two pairs of very large, backwardly directed spines in addition to numerous small spines . . . . (4)
4. Dorsal and ventral pelage more reddish; two pairs of very large spines on penis ***S. napaea***
– Dorsal and ventral pelage lacking reddish colour; one pair of large spines on penis ***S. pseudonapaea***

Sicista subtilis Southern birch mouse

Mus subtilis Pallas, 1773. Upper Tobol, W. Siberia. Incl. *lineatus*, *loriger*, *nordmanni* (Ukraine), *pallida*, *severtzovi* (Voronej Prov., S. European Russia), *sibirica* (Altai), *trizonus* (Hungary), *vaga* (R. Ural).

Range (Map 72). Steppes from E. Austria (Bauer, 1954). Hungary and Rumania through S. Russia, N. Kazakhstan and southern part of W. Siberia to the Altai, L. Balkash and L. Baikal.

Remarks. Ognev (1948) and Kuznetsov (1965) diagnosed five subspecies on the basis of slight differences in colour but admitted extensive intergradation.

Scista betulina Northern birch mouse

Mus betulina Pallas, 1779. R. Ischim, W. Siberia. Incl. *montana* (Hungary), *norvegica* (Norway), *strandi* (Caucasus), *tatricus*.
*Sicista betulina taigica** Stroganov & Potapkina, 1950. R. Ket (trib. of Ob), W. Siberia (*fide* Gromov *et al.*, 1963).

Range (Map 72). Boreal and montane forest zone from Norway and Denmark to the Ussuri region of S.E. Siberia, but range very fragmented. North to the Arctic Circle at the White Sea, but not found east of the Yenesei except around Lake Baikal and again in the Ussuri region. South to Austria, the Carpathians, Caucasus and Sayan Mts.

Sicista napaea Altai birch mouse

Sicista napaea Hollister, 1912. Tapucha, Altai Mts.

RANGE (Map 72). Northwestern Altai Mts and foothills to the north.

REMARKS. Not seen.

Sicista pseudonapaea*

Sicista pseudonapaea Strautman, 1949. Southern Altai (N.V. – *fide* Kuznetsov, 1965).

RANGE (Map 72). Taiga of southern Altai.

REMARKS. Kuznetsov (1965) treated this as a distinct species but Vinogradov & Gromov (1952) and Gromov *et al.* (1963) considered it a subspecies of *S. betulina*. I have not seen it.

Sicista concolor Chinese birch mouse

Sminthus concolor Büchner, 1892. Guiduisha, N. side of Sining Mts, Kansu, China.
S. c. concolor. China. Incl. *weigoldi* (Szechuan).
S. c. leathemi (Thomas, 1893). Kashmir. Incl. ?*flavus*.
S. c. tianschanica (Salensky, 1903). Tien Shan.
S. c. caudata† Thomas, 1907. Sakhalin.
?*S. c. caucasica*† Vinogradov, 1925. Caucasus.

RANGE (Map 72). Mountains of W. China (Kansu and Szechuan), Kashmir, Tien Shan, the island of Sakhalin and the N. slopes of the Caucasus.

REMARKS. Kuznetsov (1965) included *tianschanica* and *caudata* in *S. concolor*, and although he treated *caucasica* as specifically distinct there is an editorial footnote to the effect that it might be conspecific. One of the characters given by Kuznetsov to separate *caucasica* from the rest – whether the tail is or is not sharply bicoloured – does not hold on British Museum material, and the other differences, in the penis, seem slight although it has not been possible to check them.

E & M-S gave *concolor* and *caucasica* specific rank; Ognev (1948) and Vinogradov & Gromov (1952) gave them and *tianschanica* separate specific rank.

Subfamily ZAPODINAE

Genus *EOZAPUS*

Eozapus Preble, 1899. Type-species *Zapus setchuanus* Pousargues.

A monospecific genus, closely related to *Zapus* of North America. *Zapus* is the prior name if they should be considered congeneric, as they were for example by Simpson (1945).

Eozapus setchuanus Szechuan jumping mouse

Zapus setchuanus Pousargues, 1896. Tatsienlu, W. Szechuan.
E. s. setchuanus. Szechuan.
E. s. vicinus (Thomas, 1912). Kansu.

RANGE (Map 72). Montane forest of W. China from W. Szechuan to S. Kansu and S.E. Chinghai (i.e. scarcely entering the Palaearctic Region as delimited here).

Family DIPODIDAE

Jerboas

A moderately large family represented throughout the arid zone of the southern Palaearctic Region from the Gobi Desert to the Sahara but with the greatest diversity in Russian Turkestan. Sometimes considered to include the genera included above in Zapodidae, but here limited to the true jerboas as explained under the heading of Zapodidae.

Ten genera are included. The classification of these is fairly stable, the only deviation from the generic arrangement used here being that the species listed here as *Jaculus lichtensteini* is sometimes separated generically as *Eremodipus*.

All species of the family occur in the Palaearctic Region as defined, with the exception of *Salpingotus michaelis* in Pakistani Baluchistan (and possibly of *S. thomasi*). All are included below in their entirety.

The great majority of species are dealt with in detail by Ognev (1948a).

1. Very small: head and body under 60 mm, hind feet under 28 mm; three central metatarsals not fused; (auditory bullae grossly inflated) . . . . (2)
– Larger: head and body over 70 mm; hind feet over 28 mm; three central metatarsals fused to form a cannon bone . . . . (3)
2. Hind feet with 5 toes; zygomatic arch simple . . . ***CARDIOCRANIUS*** (p. 157)
– Hind feet with 3 toes; zygomatic arch with a prominent ventral process ***SALPINGOTUS*** (p. 157)
3. Muzzle rather long and pig-like; ears very long, about half length of head and body; anterior and posterior parts of jugal bones meeting at an obtuse angle ***EUCHOREUTES*** (p. 158)
– Muzzle short; ears shorter, not more than a third length of head and body; jugals bent through a right angle . . . . (4)
4. Hind feet with 5 or (in one species) 4 toes; upper incisors pro-odont; bullae and mastoid bones relatively little inflated . . . . (5)
– Hind feet with 3 toes; upper incisors not pro-odont; bullae and mastoids grossly inflated (7)
5. Tail very long, over 125 % of head and body, slender and with a well developed 'flag' at the tip (6)
– Tail shorter, about equal to head and body, thick throughout, with or without a poorly developed flag . . . . ***PYGERETMUS*** (p. 156)
6. Ears long, reaching beyond snout when laid forwards (perhaps not in *A. bobrinskii*); molars complex, upper ones with 3 external folds; P^4 usually present (but small) ***ALLACTAGA*** (p. 153)
– Ears short, not reaching snout; molars simpler, upper ones with 2 external folds; P^4 absent ***ALACTAGULUS*** (p. 156)
7. Tip of tail dark, terminal 'flag' poorly developed; (upper incisors white and grooved) ***STYLODIPUS*** (p. 153)
– Tip of tail white, terminal flag well developed . . . . (8)
8. Upper incisors grooved; palate extending behind last molars; root of lower incisor forming a prominent process on the mandible . . . . (9)
– Upper incisors ungrooved; palate not extending behind teeth; root of lower incisor not forming a process on the mandible . . . . . . ***PARADIPUS*** (p. 151)
9. Incisors yellow; P^4 present (but small) . . . . ***DIPUS*** (p. 150)
– Incisors white; P^4 absent . . . . ***JACULUS*** (p. 151)

Subfamily **DIPODINAE**

Genus ***DIPUS***

Dipus Zimmermann, 1780. Type-species *Mus sagitta* Pallas. = *Dipodipus*.

A monospecific genus, distinguished from *Jaculus* especially by the much less inflated mastoid components of the bullae.

Dipus sagitta Northern three-toed jerboa

Mus sagitta Pallas, 1773. Yamuishevskaya, R. Irtish, Siberia.
D. s. sagitta. Upper Irtish and Cis-Altai Steppes.
D. s. zaissanensis Selewin, 1934. S.E. Kazakhstan.
D. s. lagopus Lichtenstein, 1823. S. Russian Turkestan.
D. s. innae (Ognev, 1930). Lower Volga and Ural. Incl. *kalmikiensis*.
D. s. nogai Satunin, 1907. N.E. Causasus.
D. s. deasyi Barrett-Hamilton, 1900. Sinkiang.

D. s. sowerbyi Thomas, 1908. N. China and E. Mongolia (type-locality in N. Shensi). Incl. *halli* (Hopei).
*D. s. ubsanensis** Bannikov, 1947. N.W. Mongolia.
Dipus sagitta ubsanensis Bannikov, 1947a. Borig-Del Sands, near Baga-Nor, Ubsa-Nor, N.W. Mongolia.

RANGE (Map 73). Desert, steppe and dry woodland from the R. Don and the N.W. coast of the Caspian Sea, through most of Russian Turkestan, Sinkiang and Mongolia to S. Manchuria and Shensi.

Genus *PARADIPUS*

Paradipus Vinogradov, 1930. Type-species *Scirtopoda ctenodactyla* Vinogradov.

A monospecific genus most closely resembling *Dipus* but generally regarded as distinct.

Paradipus ctenodactylus Comb-toed jerboa

Scirtopoda ctenodactyla Vinogradov, 1929. Repetek, Turkmenistan.

RANGE (Map 73). S.W. Turkestan from the Caspian Sea northeast to the Syr Darya.

Genus *JACULUS*

Jaculus Erxleben, 1777. Type-species *J. orientalis* Erxleben. Incl. *Scirtopoda* =*Haltomys*, *Eremodipus*. *Jaculus* Erxleben, 1777 was placed on the Official List of Generic Names by Opinion 730 of the International Commission on Zoological Nomenclature.

A genus of five species extending from Turkestan to the western Sahara, all included here. These seem clearly defined and the only uncertainty is whether *J. jaculus* as used here should be subdivided.

1. Fenestrae on dorsal surface of skull, above bullae, very small and elongate; lateral process of each parietal bone directed outwards and downwards; alveolar process of incisor on mandible small; (hind feet over 25 mm, ear under 20 mm) . . . ***J. lichtensteini***
– Fenestrae very large and triangular; lateral processes of parietals forming a vertical crest; alveolar process on mandible prominent . . . . (2)
2. Penis with a pair of long curved spines attached to the baculum; hind feet usually over 66 mm (3)
– Penis without such spines; hind feet 55–66 mm . . . . ***J. jaculus***
3. Bullae relatively small, space between them at occiput (above foramen magnum) about 48–52% of interorbital width; maxillary tooth-row 5·5–6·5 mm (N. Africa) ***J. orientalis***
– Bullae of medium size, separated by space about 45% of interorbital width; maxillary tooth-row about 5·5 mm (Turkestan) . . . . ***J. turcmenicus***
– Bullae large, space between them 35–45% of interorbital width; maxillary tooth-row usually under 5 mm (Iran and Baluchistan) . . . . ***J. blanfordi***

Jaculus lichtensteini Lichtenstein's jerboa

Scirtopoda lichtensteini Vinogradov, 1927. Merv, Turkmenistan.

RANGE (Map 74). S.W. Turkestan from the Caspian Sea to the Aral Sea; also south of Lake Balkhash.

REMARKS. Not seen.

Jaculus jaculus Lesser Egyptian jerboa

Mus jaculus Linnaeus, 1758. Giza, Egypt.
J. j. jaculus. Egypt to Morocco. Incl. *aegyptius*, *centralis* (central Algerian Sahara), *darricarrerei*, *deserti* (N.E. Algeria), ?*hirtipes*, *macromystax*, ?*macrotarsus*, *sefrius* (N.W. Algeria).
*Jaculus jaculus elbaensis** Setzer, 1955:183. Wadi Darawena, Jebel Elba, Sudan Government Administrative Area, Egypt.
*Jaculus jaculus favillus** Setzer, 1955:184. Bir Bosslanga, Salum, W. Desert Governorate, Egypt.
*Jaculus jaculus whitchurchi** Ranck, 1968:242. 10 km S. of Agedabia, Cyrenaica Province, Libya.
*Jaculus jaculus cufrensis** Ranck, 1968:238. El Giof, Cufra Oasis, Cyrenaica Province, Libya.
*Jaculus jaculus collinsi** Ranck, 1968:236. Tazerbo Oasis, Cyrenaica Province, Libya.
*Jaculus deserti vastus** Ranck, 1968:230. Wadi er Rueis, Gebel el Harug el Asued, 340 km N.W.N. of Tazerbo Oasis, Cyrenaica Province, Libya.
*Jaculus deserti rarus** Ranck, 1968:228. Ain Zueia, Gebel Uweinat, Cyrenaica Province, Libya.

*Jaculus deserti fuscipes** Ranck, 1968:226. 7 km S. of El Gheddahia, Tripolitania Province, Libya.
*Jaculus jaculus arenaceous** Ranck, 1968:234. Edri, Fezzan Province, Libya.
*Jaculus jaculus tripolitanicus** Ranck, 1968:240. 25 km N. of Gharian, Tripolitania Province, Libya.

J. j. schlueteri (Nehring, 1901). Sinai to near Jaffa.

J. j. vocator Thomas, 1921. Arabia, Syria and Iraq west of R. Euphrates. Incl. *florentiae*, *oralis* (Kuwait), *syrius*.

J. j. loftusi (Blanford, 1875). Iraq east of R. Euphrates and S.W. Iran.

Also the following forms from south of the Sahara.

J. j. favonicus Thomas, 1913. S.W. Mauritania.

J. j. airensis Thomas & Hinton, 1921. Air, Niger.

J. j. butleri Thomas, 1922. Sudan (around Khartoum).

J. j. gordoni Thomas, 1903. Sudan (Kordofan).

J. j. vulturnus Thomas, 1913. N. Somalia.

RANGE (Map 74). Desert and semidesert throughout the Saharan region from Mauritania and Morocco to N. Somalia and Egypt, and throughout the Arabian region north to central Syria and east to S.W. Iran. In the Sahara particularly the range is likely to be very fragmented.

REMARKS. This species was treated as such by E & M-S and by Harrison (1972). However, on the basis especially of Libyan material Ranck (1968) maintained that there were two species involved which were sympatric in parts of Libya. In *J. jaculus* he specifically included, besides the nominate form from Egypt, only his five new Libyan forms (*whitchurchi*, *cufrensis*, *collinsi*, *arenaceous*, and *tripolitanicus)*, but he gave the distribution as virtually the entire area given here. He diagnosed it as having pale pelage, white fringes on the hind toes, a single foramen in the angular process of the mandible, larger skull and less inflated bullae. In *J. deserti* he included three new races in Libya (*fuscipes*, *rarus* and *vastus)*, *favillus* (coastal desert of Egypt) and the Arabian forms *schlueteri*, *vocator* and *loftus*; and gave the range from Iraq to Algeria but excluding the southern part of the Sahara. He diagnosed it as having darker pelage, brown fringes on the toes, two similar-sized foramina in the mandible, smaller skull and larger bullae.

Harrison (1972) maintained that no such distinction could be made in the extensive collections available from the Arabian region and I am bound to agree with this conclusion. The dorsal colour, the colour of the foot fringes and the perforation of the mandible are certainly highly variable but appear quite uncorrelated in material from Arabia and Iraq. Specimens examined from Egypt have the fringes consistently white and the perforation of the mandible mostly single but such a correlation is less apparent in Algeria.

Ranck's argument for treating these forms as specifically distinct therefore depends very largely on his claim that they are sympatric in Libya. But although he claimed that 'in several localities in Libya jerboas of the *deserti* type and those typical of *J. j. jaculus* occur sympatrically and show no evidence of interbreeding', his maps show only one such locality. This is near Marble Arch, on the coast, from where only one specimen of each form was available. The two-species hypothesis must be considered unproven and, judging by variability outside Libya, improbable although by no means impossible.

Subspecific variation is very extensive and confusing. Ranck (1968) admitted intergradation between several of the Libyan races he recognized, and likewise Setzer (1958b) between *jaculus* and each of *favillus*, *schlueteri* and *elbaensis* in Egypt. Harrison (1972) postulated intergradation between *schlueteri* and *vocator* in the Negev Desert, but showed that the River Euphrates acts as a barrier between *vocator* and *loftusi* which are distinctly different on either side of the river. However, there are probably many isolated populations within the species some of which may be distinctive. There may well be some discrete discontinuities within the group included above under *J. j. jaculus*, but I have not attempted to analyse these.

Jaculus blanfordi Blanford's jerboa

Dipus blanfordi Murray, 1884. Bushire, Iran.

RANGE (Map 74). E. and S. Iran and Baluchistan (Mirza, 1965).

REMARKS. E & M-S implied that this species was most closely related to *J. jaculus*, but it has been shown by Didier & Petter (1960) that the very complex structure of the penis allies it very clearly with *J. orientalis* rather than with *J. jaculus*.

Jaculus orientalis Greater Egyptian jerboa

Jaculus orientalis Erxleben, 1777. Egypt. Incl. *bipes, gerboa, locusta, mauritanicus* (Oran, Algeria).

RANGE (Map 74). North Africa from Morocco to Sinai and into S. Israel, but avoiding the more extreme arid habitats.

Jaculus turcmenicus*

Jaculus turcmenicus Vinogradov & Bondar, 1949. Tchagil sands, S. coast of Kara-Bogaz Gol, Nebitagsk dist., Turkmenia, USSR.

RANGE (Map 74). From the S.E. coast of the Caspian Sea through Turkmenia to the Kyzyl-Kum Desert.

REMARKS. For a further account of this species see Stalmakova (1957). It has not been seen.

Genus *STYLODIPUS*

Stylodipus G. Allen, 1925. Type-species *S. andrewsi* Allen. Incl. *Halticus* Brandt, *Scirtopoda* (in part).

A genus with one, or possibly two, species, most closely related to *Jaculus*. This genus is called *Scirtopoda* by most Russian authors. E & M-S argued that *Scirtopoda* was incorrect because of the valid, if unwise, designation of *Dipus mauritanicus* Duvernoy as its type-species, and this is indeed the case according to the International Code. I am inclined to follow the Code strictly here, especially since Kuznetsov (1965) has also adopted *Stylodipus*.

The Mongolian form *andrewsi*, characterized especially by the normal retention of rudimentary P^4, is given specific rank by some. I have not seen it but tentatively give it subspecific rank as did E & M-S.

Stylodipus telum Thick-tailed three-toed jerboa

Dipus telum Lichtenstein, 1823. Aral Sea region.
S. t. falzfeini (Brauner, 1913). Ukraine.
S. t. turovi (Heptner, 1934). Don and Volga steppes.
S. t. amankaragai (Selewin, 1934). N.W. Kazakhstan. Incl. ?*birulai*.
S. t. telum. Lower Ural to Aral Sea. Incl. *halticus* Brandt, *proximus*.
S. t. karelini (Selewin, 1934). E. Kazakhstan.
S. t. andrewsi Allen, 1925. Mongolia.

RANGE (Map 73). From lower Dnieper in Ukraine across N. Kazakhstan to Mongolia, south to Turkmenia. Kowalski (1968) provided additional records for S. Mongolia.

Genus *ALLACTAGA*

Allactaga Cuvier, 1836. Type-species *Mus jaculus* Pallas = *Dipus sibiricus major* Kerr. = *Alactaga, Scirtetes*. Incl. *Scarturus, Scirtomys, Allactodipus*.

A genus of about eight species, wholly Asiatic except for one species in N.E. Africa. The last has been separated generically because of the presence of only one lateral digit on each hind foot instead of two as in the other species. The genus is closely related only to *Alactagulus* and *Pygeretmus* but these are fairly generally treated as distinct.

The species seem fairly clearly defined although some are very little known. It has recently been demonstrated that *williamsi* should be treated as conspecific with *A.*

euphratica (Atallah & Harrison, 1968); otherwise classification at the species level is that used by E & M-S and by all Russian authors.

1. Four digits on hind feet, i.e. only one small lateral one in addition to the three large ones (Africa) . . . . *A. tetradactyla*
– Five digits on hind feet, i.e. two small and three large (Asia) . . . . . (2)
2. Auditory bullae large: length (from anterior surface to suture with mastoid) about equal to width of post-orbital constriction . . . . (7)
– Auditory bullae small, length less than three-quarters of post-orbital width . . (3)
3. Upper premolar as large as last molar; (hind feet 60–80 mm; maxillary tooth-row 6·5–8·0 mm) . . . . *A. sibirica*
– Upper premolar no more than half width of last molar . . . . (4)
4. Hind feet over 75 mm; condylobasal length over 36 mm . . . . (5)
– Hind feet under 75 mm; condylobasal length under 36 mm . . . . . . (6)
5. Black part of tail flag divided below by a white median stripe; hind feet under 83 mm *A. severtzovi*
– Black part of tail flag without a white stripe; hind feet over 83 mm . . *A. major*
6. Hind feet over 60 mm; maxillary tooth-row over 5·7 mm. . . . . *A. euphratica*
– Hind feet under 60 mm; maxillary tooth-row under 5·7 mm . . . . . *A. elater*
7. Flag of tail black or dark brown and white; brushes of hair on digits of hind feet poorly developed (length of hair not more than twice depth of digits) . . . . . (8)
– Flag of tail dark and light greyish brown; brushes of hair on digits very well developed, length of hair more than twice depth of digit, up to 12 mm long (as in *Dipus*); (maxillary tooth-row about 6 mm, ear about 23 mm) . . . . *A. bobrinskii*
8. Ears short (*c.* 30 mm); black area of tail flag preceded by white area; maxillary tooth-row *c.* 6·4 mm . . . . *A. bullata*
– Ears long (*c.* 40 mm); no subterminal white area on tail; maxillary tooth-row *c.* 5·4 mm *A. hotsoni*

Allactaga sibirica — Mongolian five-toed jerboa

Yerbua sibirica Forster, 1778. Transbaikalia.
A. s. sibirica. Transbaikalia and N.E. Mongolia. Incl. *alactaga, brachyurus, halticus, medius, mongolica, saliens.*
A. s. annulata (Milne-Edwards, 1867). Kansu and Shansi. Incl. *longior.*
A. s. saltator (Eversmann, 1848). Altai. Incl. *grisescens.*
*A. s. semideserta** Bannikov, 1947. Mongolia except N.E.
Allactaga sibirica semideserta Bannikov, 1947. Lower Bukhu-Muren River, near Lake Ochit-Nor, W. Mongolia.
A. s. suschkini Satunin, 1900. S. Kazakhstan. Incl. *rückbeili.*
A. s. altarum Ognev, 1946. Tien Shan.
*A. s. dementievi** Toktosunov, 1955. Kirghizia.
Allactaga saltator dementievi Toktosunov, 1958. N. of Lake Issyk-Kul, Kirghizia. (*Fide* Kuznetsov (1965) – original reference not seen.) Perhaps a *nomen nudum.*

RANGE (Map 75). Dry steppes from R. Ural and Caspian Sea to the Tien Shan and Altai, and through Mongolia to the Upper Amur, Manchuria, Hopei, Kansu and Shansi.

REMARKS. The races given above are those recognized by Kuznetsov (1965) in the USSR plus *annulata* and *semideserta.* The differences seem slight, involving mainly colour and size, and it seems unlikely that many of these really represent discrete subspecies.

According to Jones & Johnson (1965) the specimen listed by Allen (1938–40) from Korea (and mentioned by E & M-S) was in fact from Jehol, Manchuria and there is no valid record for Korea.

Allactaga elater — Small five-toed jerboa

Dipus elater Lichtenstein, 1825. E. Kazakhstan.
A. e. elater. Kazakhstan. Incl. *dzungariae* (Zungaria), *vinogradovi.*
A. e. indica Gray, 1842. Iran to Baluchistan (type-locality in Afghanistan). =*bactriana.* Incl. *strandi* (Turkmenistan), *turkmeni* (N. Iran).

A. e. caucasica Nehring, 1900. E. Asia Minor to S.W. Caspian Sea. Incl. *aralychensis* (Mt Ararat, Armenia).
A. e. kizljaricus Satunin, 1907. N.W. of Caspian Sea.

RANGE (Map 76). From lower Volga and E. Asia Minor through the desert and semidesert zones of Turkestan, Iran and Afghanistan to Sinkiang and Baluchistan.

REMARKS. The above arrangement of subspecies follows Kuznetsov (1965). The differences are slight ones of colour and size and it is possible that no very discrete subspecies exist.

Allactaga euphratica Euphrates jerboa

Allactaga euphratica Thomas, 1881. Iraq. Incl. *laticeps* (W. Turkey), *schmidti* (Caucasus), *williamsi*† (E. Turkey).
*Allactaga williamsi caprimulga** Ellerman, 1948. Shiber Pass, Kabul-Bamian Road, Afghanistan.

RANGE (Map 76). Steppe and semidesert from Jordan and N. Saudi Arabia through Iraq to Asia Minor and the Caucasus, and east through N. Iran and Afghanistan.

REMARKS. Atallah & Harrison (1968) demonstrated convincingly that *williamsi* and *euphratica* are conspecific. They recognized all the named forms listed above except *laticeps*, as subspecies, but admitted that *schmidti* was apparently identical with *williamsi* and that *williamsi* and *euphratica* formed a 'perfect cline'. They diagnosed *caprimulga* solely on the basis of greater average ear length in spite of the fact that this measurement overlaps extensively with that of the western forms. There therefore seem no good reasons for recognizing discrete subspecies.

Allactaga hotsoni Hotson's jerboa

Allactaga hotsoni Thomas, 1920. Kant, Persian Baluchistan.

RANGE (Map 75). Known only from the type-locality in Persian Baluchistan.

REMARKS. Ellerman (1961) maintained that, apart from its exceptionally large bullae, this species 'is not very remote from *A. elater*'. Lay (1967) maintained that the type 'bears striking resemblance to specimens of *A. williamsi* in the British Museum (NH) from Shibar Pass, Afghanistan . . . suggesting that detailed study may show *A. hotsoni* to be synonymous with *A. williamsi*'. However, the very large bullae, short tooth-row and short, wide incisive foramina place it well outside the range of variation otherwise known in *A. euphratica* (including *williamsi*), and it therefore seems probable that it does indeed represent a distinct species.

Allactaga bullata

Allactaga bullata G. Allen, 1925. Tsagan-Nor, Mongolia.

RANGE (Map 75). Deserts of southern and western Mongolia and adjacent parts of Inner Mongolia, Kansu and probably Sinkiang.

Allactaga bobrinskii

Allactodipus bobrinskii Kolesnikov, 1937. Kizil-kum Desert, Uzbekistan.

RANGE (Map 75). Kizil-kum and Kara-kum Deserts, in Uzbekistan and Turkmenistan.

REMARKS. Some Russian authors have suggested that there might be a close relationship between this form and *A. hotsoni*, both having large bullae. However, the type of *hotsoni* has the hind digits normally haired, without the brushes of long hair described for *A. bobrinskii*. Although this feature suggests an affinity with *Dipus*, Ognev (1948) argued that in all other respects this species is a typical *Allactaga*. On the other hand Shenbrot (1974) supported its separation from *Allactaga* as the sole member of the genus *Allactodipus*. I have not seen it.

Allactaga severtzovi Severtzov's jerboa

Allactaga severtzovi Vinogradov, 1925. Kopal District, extreme S.E. Kazakhstan.

RANGE (Map 75). Most of Turkestan from the Caspian Sea to Lake Balkash anc Tashkent.

Allactaga major Great jerboa

Dipus sibiricus major Kerr, 1792. Kazakhstan.
A. m. macrotis Brandt, 1844. European Russia except N. Caucasian region. Incl. *jaculus* Pallas (not of Linnaeus).
A. m. fuscus Ognev, 1924. N. Caucasian steppes.
A. m. major. Central Kazakhstan.
*Alactaga jaculus intermedius** Ognev, 1948:129. Dolinskoe, Karaganda, central Kazakhstan.
A. m. vexillarius (Eversmann, 1840). W. and S. Kazakhstan. Incl. *chachlovi*, *flavescens*, *hochlovi*.
A. m. spiculum (Lichtenstein, 1825). N. Kazakhstan and adjacent parts of Siberia. Incl. *brachyotis*, *decumanus* (Urals), *nigricans*.

RANGE (Map 77). Steppes and deserts from Moscow and Kiev to north of the Caucasus; eastwards through Kazakhstan and W. Siberia (north to about 57°) as far as the R. Ob; southeast as far as the foothills of the Tien Shan and almost to Tashkent.

REMARKS. The arrangement of subspecies follows Kuznetsov (1965), but variation is likely to be continuous over most of the range.

Allactaga tetradactyla Four-toed jerboa

Dipus tetradactylus Lichtenstein, 1823. Near Alexandria, Egypt. Incl. *brucii*.

RANGE (Map 77). Coastal gravel plains of Egypt and Libya from near Alexandria to the Gulf of Sidra (to at least 18°E) (Ranck, 1968).

Genus *ALACTAGULUS*

Alactagulus Nehring, 1897. Type-species *Dipus acontion* Pallas =*Dipus sibiricus pumilio* Kerr.

A monospecific genus very closely related to *Pygeretmus* and only marginally distinct from it although treated as distinct in all recent classifications.

Alactagulus pumilio

Dipus sibiricus pumilio Kerr, 1792. Between Caspian Sea and R. Irtish, Kazakhstan.
A. p. pumilio. Main part of range including Kazakhstan. Incl. *acontion*, *minor*, *minutus*, *pygmaea*.
A. p. dinniki Satunin, 1920. N. Caucasus.
*Alactagulus pygmaeus tanaiticus** Ognev, 1948:266. Village of Remontnoe, Salsk District, S. European Russia.
*A. p. aralensis** Ognev, 1948. Between Aral Sea and Lake Balkhash.
Alactagulus pygmaeus aralensis Ognev, 1948:268. Sary-Su River, Telekul landmark area, 200 km east of Perovsk, Kazakhstan.
A. p. turcomanus Heptner & Samarodov, 1939. Kara-Kum Desert, Turkmenistan.
A. p. potanini Vinogrodov, 1926. S. Mongolia and Inner Mongolia.

RANGE (Map 73). From the R. Don in S. European Russia through most of Kazakhstan to the R. Irtish and south to the extreme north of Iran (east of the Caspian Sea); also in S. Mongolia and the adjacent parts of Inner Mongolia. Iranian records were listed by Lay (1967).

REMARKS. Russian authors call this *A. pygmaeus*, based on *Mus jaculus* var. *pygmaea* Pallas 1779:284, or '*A. acontion* Pallas, 1778'. The former is clearly invalid, being preoccupied by *Mus citellus* var. *pygmaeus* Pallas, 1779:122 (Article 59a of International Code). It does not seem that *acontion* can be dated prior to 1811.

The subspecies listed above are those recognized by Kuznetsov (1965), with the addition of *A. p. potanini*.

Genus *PYGERETMUS*

Pygeretmus Gloger, 1841. Type-species *Dipus platyurus* Lichtenstein. =*Platycercomys*, *Pygerethmus*.

A genus of two species in central Asia, closely related only to *Alactagulus* and *Allactaga* but generally recognized as distinct. For a detailed account of the genus (in Russian) see Vorontsov *et al.* (1969).

Tail over 90 mm, terminal hairs long and white, subterminal band black; hind feet over 40 mm **_P. shitkovi_**

Tail under 90 mm, terminal hairs short and black; hind feet under 40 mm **_P. platyurus_**

Pygeretmus platyurus Lesser fat-tailed jerboa

Dipus platurus Lichtenstein, 1823 (corrected to *D. platyurus* by Lichtenstein, 1828). E. of Aral Sea.
*Pygerethmus vinogradovi** Vorontsov, 1958. S. side of Lake Zaisan, S.E. Kazakhstan.

RANGE (Map 77). W. Kazakhstan from Ural River to just east of the Aral Sea; S.E. Kazakhstan, on S. side of Lake Zaisan.

Pygeretmus shitkovi Greater fat-tailed jerboa

Alactagulus shitkovi Kuznetsov, 1930. N.W. of Lake Ala-Kul, S.E. Kazakhstan. =*zhitkovi*.

RANGE (Map 77). E. Kazakhstan, in region of Lake Balkhash.

REMARKS. The spelling *zhitkovi* is more frequently used by Russian authors but this is not a valid emendation (Article 32a (ii) of International Code).

Subfamily CARDIOCRANIINAE

Genus *CARDIOCRANIUS*

Cardiocranius Satunin, 1903. Type-species *C. paradoxus* Satunin.

A distinctive, monospecific genus, closely related only to *Salpingotus*.

Cardiocranius paradoxus Five-toed pygmy jerboa

Cardiocranius paradoxus Satunin, 1903. Nan Shan, N.W. Kansu, China.

RANGE (Map 76). Nan Shan (Kansu), Gobi Desert in southern Mongolia, Upper Yenesei and adjacent Tesin-Naryn region (Tuva ASSR) and north of Lake Balkhash (Kazakhstan) (last reported by Smirnov, 1971). Kowalski (1968) gave many additional locality records for Mongolia.

REMARKS. Very little known. For a recent account see Kryzhanovsky & Shcherbak (1969). It has not been seen.

Genus *SALPINGOTUS*

Salpingotus Vinogradov, 1922. Type-species *S. kozlovi* Vinogradov.

A distinctive genus of five little-known species, closely related only to *Cardiocranius*. For a recent review, with distribution map, see Vorontsov & Smirnov (1969).

1. Tail about $2\frac{1}{2}$ times length of head and body (*c.* 125 mm), with scattered long hairs throughout its length and with hairs of the sparse, terminal tuft over 5 mm in length; greatest length of skull over 26 mm; maxillary tooth-row *c.* 3·8 mm. . . . . . . **_S. kozlovi_**
– Tail about twice length of head and body (up to 105 mm), short-haired except at tip; greatest length of skull under 25 mm; maxillary tooth-row under 3·5 mm . . . . . (2)
2. Hind feet over 22·5 mm; wide anterior part of zygomatic arch not sharply delimited from narrow posterior part . . . . **_S. thomasi_**
– Hind feet under 22·5 mm; anterior and posterior parts of zygomatic arch sharply delimited (3)
3. Proximal half of tail swollen; hind feet over 19·5 mm; greatest length of skull over 20 mm; maxillary tooth-row over 2·8 mm . . . . (4)
– Proximal half of tail not swollen; hind feet under 19·5 mm; greatest length of skull under 20 mm; maxillary tooth-row under 2·8 mm . . . . **_S. michaelis_**
4. Tail with a dense, black terminal tuft; in dorsal view outline of mastoid part of bulla meets that of tympanic part at a distinct angle; greatest width of skull over 17 mm; angular region of mandible without a perforation . . . . **_S. heptneri_**

– Tail with only a sparse terminal tuft; outlines of mastoid and tympanic parts of bulla collinear; greatest width of skull under 16·5 mm; mandible with a perforation in the angular region . . . . ***S. crassicauda***

Salpingotus kozlovi Kozlov's pygmy jerboa

Salpingotus kozlovi Vinogradov, 1922. Khara-Khoto, Gobi Desert, Mongolia.

RANGE (Map 76). Gobi Desert of southern Mongolia. E & M-S quoted a record from the Irtish River on the USSR-Chinese boundary from Elizaryeva (1949) but this seems to be treated as *S. crassicauda* by subsequent Russian authors.

REMARKS. Not seen.

Salpingotus crassicauda Thick-tailed pygmy jerboa

Salpingotus crassicauda Vinogradov, 1924. Shara-in-Sumu, Gobi-Altai, W. Mongolia.

RANGE (Map 76). Deserts of S. Mongolia; in W. Mongolia (south of the Altai) and the adjacent part of the USSR in the upper Irtish valley; also to the south of Lake Balkhash and north of the Aral Sea (last recorded by Lobachev, 1971).

Salpingotus thomasi Thomas's pygmy jerboa

Salpingotus thomasi Vinogradov, 1928. ?Afghanistan.

RANGE (not mapped). Known only from the type which is of doubtful origin. According to Ognev (1948) Vinogradov thought it more likely to have come from S. Tibet.

Salpingotus heptneri* Heptner's pygmy jerboa

Salpingotus heptneri Vorontsov & Smirnov, 1969. 80 km N.E. of Takhta-Kupir, Karakalpaksk ASSR, N.W. Kizil-Kum Desert, USSR.

RANGE (Map 76). Known only from the type-locality, just south of the Aral Sea, in Uzbekistan.

REMARKS. This can be considered a western vicariant of *S. crassicauda* but it seems moderately well differentiated. It has not been seen.

Salpingotus michaelis* Baluchistan pygmy jerboa

Salpingotus michaelis Fitzgibbon, 1966. Desert plateau of Nushki, N.W. Baluchistan, Pakistan (*c.* 29°N, 66°E, 3500 ft).

RANGE (Map 76). N.W. Baluchistan.

Subfamily **EUCHOREUTINAE**

Genus ***EUCHOREUTES***

Euchoreutes Sclater, 1891. Type-species *E. naso* Sclater.

A monospecific and very distinctive genus, placed by Ognev (1948) and by Ellerman (1941) in a separate subfamily.

Euchoreutes naso Long-eared jerboa

Euchoreutes naso Sclater, 1891. Yarkand, Sinkiang.
E. n. naso. Sinkiang.
E. n. alaschanicus Howell, 1928. Alashan Desert, Inner Mongolia.

RANGE (Map 75). W. Sinkiang; Chinghai (Chang & Wang, 1963), and the Alashan Desert (Inner Mongolia and extreme southern part of Mongolia).

Family **HYSTRICIDAE**

Old-world porcupines

A clearly defined family represented by about four genera throughout the Ethiopian and Oriental regions. Two species of *Hystrix* extend into the Palaearctic Region. For a review of the family see Mohr (1965).

The relationship of this family to the South American 'hystricomorph' rodents has been long debated. For opposing views see Wood (1950) and Lavocat (1951), who postulated parallel evolution, and Landry (1957) who argued for a closer phyletic relationship.

Genus *HYSTRIX*

Crested porcupines

Hystrix Linnaeus, 1758. Type-species *H. cristatus* L. =*Histrix*. Incl. *Acanthion*, *Oedocephalus*.

A genus of about eight species in Africa and S. Asia, two of which reach the Palaearctic Region. These are crested porcupines, subgenus *Hystrix*. *Acanthion* and *Thecurus*, including the crestless and short-crested species in China and S.E. Asia, are variously given subgeneric or generic rank. For a review of the subgenus *Hystrix* see Corbet & Jones (1965). The two species included here are very distinct.

Crest predominantly white; short spines on rump dark; nasal region more inflated: length of nasals 58–68 % of occipitonasal length; premaxillae narrow . . . ***H. cristata***

Crest predominantly brown; short spines on rump forming a conspicuous white patch when quills are raised; nasal region less inflated: nasals 45–50 % of occipitonasal length: premaxillae wide . . . . ***H. indica***

Hystrix indica Indian crested porcupine

Hystrix cristata var. *indica* Kerr, 1792. India. Incl. *aharonii* (Palestine), *hirsutirostris* (S.W. of Caspian Sea), *leucurus*, *mersinae* (Asia Minor), *mesopotamica*, *naryensis* (Kirgizia), *satunini* (Turkmenia), *schmidtzi*. Also other synonyms from India and Sri Lanka.

Range (Map 78). Asia Minor, Palestine and Arabia to S. Turkestan and most of western India; Sri Lanka.

Remarks. E & M-S queried the inclusion of *hirsutirostris* because of a skull illustrated by Kuznetsov (1944) which is clearly *H. cristata*. However, there is no reason to believe that *H. cristata* occurs in Asia as correctly shown by Heptner (1952) who considered *hirsutirostris* a race of *H. indica*. Kuznetsov (1965) continued to allocate it to *H. cristata*.

Hystrix cristata North African crested porcupine

Hystrix cristata Linnaeus, 1758. Italy. =*europaea*. Incl. *alba*, *cuvieri* (N. Africa), ?*daubentoni*, *occidanea*. Also other synonyms from south of the Sahara including *galeata* from East Africa.

Range (Map 78). N. Africa (Morocco to Libya and probably Egypt); Sicily, Italy, Albania and N. Greece; South of the Sahara in the steppe and savanna zones from Senegal to Ethiopia and N. Tanzania. Perhaps introduced to Europe.

Family **CAPROMYIDAE**

Genus *MYOCASTOR*

Myocastor coypus Coypu

Mus coypus Molina, 1782. Chile.

Range (not mapped). Southern S. America. Established by escape from fur farms and by deliberate introduction in several parts of Europe (e.g. E. England, France, Germany), in the USSR (Caucasus, S. Turkestan), E. Asia Minor and Japan.

Family **CTENODACTYLIDAE**

Gundis

A distinctive family with no near relatives, confined to northern Africa. Four genera are generally recognized, all monospecific. Two occur north of the Sahara; the other two, *Felovia* and *Pectinator*, occupy small ranges in Senegal and in Somalia and Ethiopia respectively.

Upper molars kidney-shaped, without incision on inner sides; tail (with hairs) shorter than hind feet when these are laid backwards . . . . *CTENODACTYLUS*

Upper molars 8-shaped, with incisions on outer and inner sides; tail about equal to hind feet when laid backwards . . . . *MASSOUTIERA*

Genus *CTENODACTYLUS*

Ctenodactylus Gray, 1830. Type-species *C. massonii* Gray.

A monospecific genus. Two species, *C. gundi* and *C. vali*, were recognized by Petter (1961b) and Misonne (1971), but I prefer to follow Ranck (1968) who treated them as conspecific, believing that Petter had misinterpreted locality data and wrongly supposed the two forms to be sympatric.

Ctenodactylus gundi Gundi

Mus gundi Rothman, 1776. Gharian, 80 km S. of Tripoli, Libya.

C. g. gundi. Atlas of Morocco to Tunisia and extreme N.W. Libya. Incl. *arabicus*, *massonii*, *typicus*.

C. g. vali Thomas, 1902. N.W. Libya: 'transitional desert between the Gebel es Soda and the Gulf of Sirte' (Ranck, 1968); Saharan Morocco and Algeria. Incl. *joleaudi* (N. Algerian Sahara).

RANGE (Map 78). Atlas Mts from E. Morocco to N.W. Libya and immediately adjacent subdesert fringe to the south.

Genus *MASSOUTIERA*

Massoutiera Lataste, 1885. Type-species *Ctenodactylus mzabi* Lataste.

A monospecific genus, closely related to *Ctenodactylus*.

Massoutiera mzabi Lataste's gundi

Ctenodactylus mzabi Lataste, 1881. Ghardaia, N. Algerian Sahara. Incl. *harterti* (W. Algerian Sahara), *rothschildi* (Air).

RANGE (Map 78). Western Sahara from Ghardaia (Algeria) south through the Tassili des Ajjer and the Hoggar Mountains to Air (Niger); Tibesti Mountains (Chad) (specimens in British Museum).

REMARKS. E & M-S included Morocco in the range, probably following Cabrera (1932), but the latter did not list *Ctenodactylus*, and later authors (e.g. Petter & Saint Girons, 1965) seem to be agreed that only *Ctenodactylus* occurs in Morocco.

Order CARNIVORA

Carnivores

If the Pinnipedia are excluded, this order contains seven families of which six are represented by indigenous species in the Palaearctic Region whilst the seventh, the Procyonidae, is included because of an established introduction. The order is of world-wide distribution except for Australasia.

The Carnivora of the USSR were well reviewed by Heptner & Naumov (1967) and by Novikov (1956/62); of Siberia by Stroganov (1962); and of India by Pocock (1939, 1941), the last being of interest for the higher classification as well as dealing with many species of the southern Palaearctic.

1. Hind feet with 4 digits, predominantly digitigrade . . . . (2)
– Hind feet with 5 digits, predominantly plantigrade . . . . (4)
2. Pollex close to other toes; claws retractile (only partly so in *Acinonyx*); muzzle short; 3 or 4 upper cheek-teeth (cats) . . . . *FELIDAE* (p. 179)
– Pollex remote from other toes or absent; claws not retractile; muzzle elongate; at least 5 upper cheek-teeth . . . . (3)
3. Pollex present (except in *Lycaon*); 6 upper cheek-teeth (2 upper molars on each side) (dogs) *CANIDAE* (p. 161)
– Pollex absent; 5 upper cheek-teeth (1 upper molar) (hyaenas) . *HYAENIDAE* (p. 179)
4. Very large (head and body over 1 metre); tail invisible (bears) . . *URSIDAE* (p. 165)
– Small to medium (head and body under 1 metre); tail conspicuous . . . . (5)

5. Bullae divided into two compartments, the division showing externally as a groove between the smaller, anterior ectotympanic around the aperture and the larger, posterior endotympanic (mongoose and genets) . . . . . . *VIVERRIDAE* (p. 177)
– Bullae undivided, without an external groove, usually rather flat with a protruding meatus (6)
6. Carnassial teeth (P^4 and M_1) well differentiated (weasels, otters, badgers) *MUSTELIDAE* (p. 167)
– Carnassial teeth not differentiated (raccoons) . . . *PROCYONIDAE* (p. 166)

Family **CANIDAE**

Dogs and foxes

Contains ten genera of which six are represented in the Palaearctic Region, one of them very marginally. The range of the family is almost that of the Carnivora. There has been no monographic revision since that of Mivart (1890), but the classification has been reviewed by Clutton-Brock *et al.* (1976).

1. Pollex absent; pelage boldly patterned black and yellow (Africa) *LYCAON* (p. 165)
– Pollex present; pelage not boldly patterned . . . . (2)
2. Cheeks black, contrasting conspicuously with surrounding pelage, and bearing ridges of long hair . . . . *NYCTEREUTES* (p. 164)
– Cheeks not contrasting in colour and lacking ridges of long hair . . . . . (3)
3. Size large (head and body of adult over 800 mm, shoulder height over 450 mm) . (4)
– Size smaller (head and body under 800 mm, shoulder height under 450 mm) . . (5)
4. Pelage entirely yellow-brown, except for white ventral parts; 6 lower cheek-teeth on each side *CUON* (p. 164)
– Dorsal pelage greyish brown; 7 lower cheek-teeth on each side . . *CANIS* (p. 161)
5. Ears rounded, scarcely projecting above pelage; pelage grey or white (Arctic regions) *ALOPEX* (p. 162)
– Ears pointed, projecting above pelage; pelage at least partly yellowish or reddish brown *VULPES* (p. 162)

Genus *CANIS*

Dogs

Canis Linnaeus, 1758. Type-species *Canis familiaris* L. Incl. *Alopedon*, *Dieba*, *Lupulus* Gervais, *Lupus*, *Oxygous*, *Sacalius*, *Thos*, *Vulpicanis*.

In addition to the two species included here there are a further three in Africa and two in N. America.

Larger: head and body up to 1·4 m, hind feet over 180 mm, condylobasal length up to 260 mm; neck greyish brown; cingulum of M^1 indistinct. . . . . ***C. lupus***
Smaller: head and body under 900 mm, hind feet under 170 mm, condylobasal length under 170 mm; side of neck reddish brown; cingulum of M^1 prominent . . . ***C. aureus***

Canis lupus Wolf

Canis lupus Linnaeus, 1758. Sweden. Incl. *albus* (Arctic Siberia), *altaicus* (Altai), *arabs* (S. Arabia), *argunensis*, *campestris* (Kirghizia), *canus*, *chanco* ('Chinese Tartary'), *communis*, *coreanus* (Korea), *cubanensis* (Caucasus), *dietanus*, *desertorum* (Turkestan), *dybowskii* (Kamchatka), *ekloni*, *filchneri* (Kansu), *flavus*, *fulvus*, *hattai* (Hokkaido), *hodophilax* (Honshu, Japan), *hodopylax*, *italicus*, *japonicus*, *kamtschaticus*, *karanorensis* (Gobi), *kurjak*, *laniger* Hodgson (not of Smith), *lycaon*, *major*, *minor*, *niger*, *orientalis*, *pallipes* (India), *rex*, *signatus* (Spain), *tschiliensis* (Hopei), *turuchanensis*.
*Canis lupus bactrianus** Laptev, 1929. Surkhan-Dar'inskaya Oblast, Uzbekistan, USSR (*fide* Heptner & Naumov, 1967).
*Canis lupus dorogostaiskii** Skalon, 1936. Borzya, S. Transbaikalia (*fide* Novikov, 1956).
Also many synonyms based on N. American forms, of which the earliest is *occidentalis* Richardson, 1829.

RANGE (Map 79). Formerly the entire Palaearctic Region except for N. Africa, the more extreme areas of desert and areas of unbroken dense taiga; and extending south throughout China and to Dharwar in India. Now extinct in Europe west of Russia except in Iberia, Italy, some parts of Scandinavia and the Balkans. Extinct in Japan. Present in N. America including most of the Arctic islands and Greenland, south to

20°N in the Mexican highlands but absent from the south-eastern United States where it is replaced by *C. rufus*.

REMARKS. Geographical variation is considerable, but in view of the continuity of distribution I doubt whether any discrete races can be recognized. Northern animals tend to be large, grey and long-haired; southern ones smaller, brown and short-haired. Imaizumi (1970c) gave the extinct Japanese wolf specific rank (*C. hodophilax*) on the basis of cranial differences.

It is now fairly generally agreed that the domestic dog was derived from the wolf, especially from the smaller, southern forms.

Canis aureus — Northern jackal

Canis aureus Linnaeus, 1758. Lar, Persia. Incl. *algirensis* (Algeria), *barbarus* Smith (not of Shaw), *caucasica*, *cruesemanni* (Thailand), ?*dalmatinus*, *doederleini*, *ecsedensis*, *graecus*, ?*greyi*, *hadramauticus* (Arabia), *hungaricus* Ehik (not of Margo), *indicus*, *kola* (Gujerat), *lanka* (Sri Lanka), *lupaster* (Egypt), *minor* Mojsisovico (not of Wagner nor of Ogérien), *moreotica* (Greece), *moroccanus*, *naria* (S. India), *nubianus* (Upper Egypt), *sacer*, *soudanicus* (Sudan), *studeri*, *syriacus*, *tripolitanus* (Libya), *typicus*, ?*variegatus* Cretzschmar (not of Gmelin), *vulgaris*.
Also the following from further south in Africa: *anthus* Cuvier, 1820 (Senegal), incl. *seneghalensis*; *riparius* Hemprich & Ehrenberg, 1832 (coast of Ethiopia), incl. *hagenbecki*, *mengesi* and *somalicus*, all from Somalia; *thooides* (Sudan); and *bea* Heller, 1914 (Kenya).

RANGE (Map 79). Primarily in the steppe zone from Morocco through N. Africa and S.W. Asia to India and as far as Thailand; North to Asia Minor, the Caucasus and Aral Sea; south to S. Kenya in Africa and west from there through the steppe zone to Senegal; isolates in S.E. Europe (north to Hungary) and on Sri Lanka.

REMARKS. Most subspecies have been based on average differences and there is wide intergradation. The earliest names for the isolates in Europe and Sri Lanka are *moreotica* and *lanka* respectively, but existing data do not support the recognition of these as subspecies.

Genus *ALOPEX*

Alopex Kaup, 1829. Type-species *Canis lagopus* L. =*Leucocyon*.

A monospecific genus marginally distinct from *Vulpes* with which it is sometimes united, e.g. by Bobrinskii (1965).

Alopex lagopus — Arctic fox

Canis lagopus Linnaeus, 1758. Lapland.
A. l. lagopus. Range except Komandorskiye and Medny Islands. =*arctica*, *typicus*. Incl. *argenteus*, *caerulea*, *fuliginosus* (Iceland), *spitzbergenensis* (Spitzbergen).
A. l. beringensis Merriam, 1902. Bering Is., Komandorskiye Is., Bering Sea. A very large race.
Vulpes lagopus var. *beringianus** Suvorov, 1912. Bering Is.
*A. l. semenovi** Ognev, 1931. Mednyi Is., Komandorskiye Is.
A [*lopex*] *beringensis semenovi* Ognev, 1931:265. Mednyi Is., Komandorskiye Is., Bering Sea.
Also the following names based on North American specimens: *groenlandicus*, *hallensis*, *innuitus*, *kenaiensis*, *pribilofensis*, *ungava*.

RANGE (Map 79). Circumpolar, occupying the entire tundra zone including most of the arctic islands. In Map 79 the broken lines indicate the southward extension of range in winter.

Genus *VULPES*

Foxes

Vulpes Frisch, 1775. Type-species *Canis vulpes* L. Incl. *Cynalopex*, *Fennecus*†, *Megalotis*. (The work of Frisch, 1775 was rejected by opinion 258 of the International Commission (1954) but a proposal to conserve *Vulpes* Frisch, 1775 has been made by Clutton-Brock & Corbet (1975) in order to avoid replacement of *Vulpes* by *Fennecus* if these are united.)

A genus of 11 or 12 species of which six are represented in the Palaearctic Region. Two others come close to the southern border of the region, namely *V. pallida* on the southern edge of the Sahara and *V. bengalensis* in India. The remaining species occur in

S. Africa and N. America. Reasons for including *Fennecus* (and the American *Urocyon*) in *Vulpes* were given by Clutton-Brock *et al.* (1976) who demonstrated that *V. bengalensis* is the most typical member of the genus with *V. vulpes* a rather aberrant member.

1. Back of the ears black or dark brown, contrasting with the colour of the head and back; legs marked with black; (condylobasal length over 120 mm) . . . . . ***V. vulpes***
– Back of ears not contrasting; legs without black . . . . (2)
2. Tail less than half length of head and body; ear less than or equal to half length of hind foot (3)
– Tail more than half length of head and body; ear more than half length of hind foot (4)
3. Muzzle long and slender, length of upper tooth-row twice width across upper carnassials; condylobasal length about 145 mm . . . . ***V. ferrilata***
– Muzzle shorter, length of tooth-row less than twice width across carnassials; condylobasal length under 110 mm . . . . ***V. corsac***
4. Pelage predominantly grey; (tip of tail dark) . . . . ***V. cana***
– Pelage sandy brown . . . . (5)
5. Ears very large, considerably larger than skull; tympanic bullae greatly enlarged, one quarter length of skull; very small: hind feet under 100 mm . . . . . . ***V. zerda***
– Ears smaller, usually a little shorter than skull; bullae smaller, about one fifth length of skull; larger: hind feet up to 115 mm . . . . ***V. rueppelli***

Vulpes vulpes Red fox

Canis vulpes Linnaeus, 1758. Sweden. Incl. *acaab*, *aegyptiacus* (Egypt), *alba* Borkhausen, *algeriensis*, *alopex*, *alpherakyi* (Turkestan), *alticola*, *anadyrensis*, *anatolica* (W. Asia Minor), *anubis*, *arabica* (Muscat), *atlantica*, *aurantioluteus* (Szechuan), *barbarus* (N.W. Africa), *beringiana* (N.E. Siberia), *caucasica*, *cinera*, *communis*, *crucigera* (Germany), *daurica*, *diluta*, *dolichocrania*, *eckloni*, *europaeus*, *ferganensis*, *flavescens* (N. Iran), *griffithi* (Afghanistan), *himalaicus*, *hoole* (S. China), *huli*, *hypomelas*, *ichnusae* (Sardinia), *indutus* (Cyprus), *jacobi*, *jakutensis* (Yakutsk), *japonica* (Japan), *kamtschadensis*, *kamtschatica*, *karagan* (Khirgizia), *kiyomasai*†, *krimeamontana* (Crimea), *kurdistanica*, *leucopus*, *lineatus*, *lineiventer*, *lutea*, *melanogaster*, *melanotus*, ?*meridionalis*, *montana* (Himalayas), *nepalensis*, *nigra*, *nigro-argenteus*, *nigrocaudatus*, *niloticus*, *ochroxantha*, *palestrina* (Palestine), *pamirensis*, *peculiosa* (Korea), *persicus*, *pusillus* (Punjab), *schrencki* (Sakhalin), *silaceus* (Spain), *splendens*, *splendidissima* (Kurile Is), *stepensis*, *tobolica* (W. Siberia), *tschiliensis* (N.E. China), ?*ussuriensis*, *variegatus*, *vulgaris*, *vulpecula*, *waddelli* (Tibet).
*Vulpes vulpes tarimensis** Matschie, 1907. Tarim, Sinkiang, China.
*Vulpes vulpes septentrionalis** Brass, 1911. Norway.
*Vulpes vulpes crymensis** Brauner, 1914. Crimea.
*Vulpes vulpes toschii** Lehmann, 1969a. Calvo Dist., Monte Gargano, Apulia, Italy.
Also many N. American synonyms of which the earliest is *fulvus* Desmarest, 1820 from Virginia.

RANGE (Map 80). Throughout the entire Palaearctic Region except for Iceland, the Arctic Islands, some parts of the Siberian tundra and areas of extreme desert (but present in subdesert steppe). Also in S. China to about 20°N; through most of N. America south to about 30°N.

REMARKS. This is one of the most widespread and ubiquitous of Palaearctic mammals. Geographical variation is considerable but is confused by extensive individual variation. The continuity of range is such that it is doubtful whether any discrete, definable subspecies can be recognized. Isolates are on Britain, Ireland, the Mediterranean islands, N.W. Africa and the Japanese islands. The N. American and Palaearctic forms are clearly conspecific and even doubtfully distinct subspecifically (see Churcher, 1959, for a discussion).

Vulpes corsac Corsac fox

Canis corsac Linnaeus, 1768. Steppes between Ural and Irtisch rivers, USSR. Incl. *eckloni* Przewalski, *kalmykorum* (Astrakhan), *nigra* Kastschenko, *turkmenica* (Turkestan).
*Vulpes corsac scorodumovi** Dorogostaiski, 1935. S. Transbaikalia.

RANGE (Map 80). The dry steppe and subdesert zone of central Asia from the lower Volga to Inner Mongolia, including most of Turkestan, Mongolia, Sinkiang and Tibet.

REMARKS. Most Russian authors, e.g Heptner & Naumov (1967), recognize three races but admit intergradation.

Vulpes rueppelli Sand fox

Canis rüppelii Schinz, 1825. Sudan. Incl. *caesia* (Asben, W. Sahara), *cyrenaica* (Libya), *famelicus*, *sabaea* (Arabia), *somaliae** (Somalia), *zarudneyi* (Persian Baluchistan).
*Vulpes rüppelli cufrana** de Beaux, 1939. Cufra Oasis, S.E. Libya.

RANGE (Map 80). The desert zone from the western Sahara (Air, Hoggar, Morocco) to Somalia and through Egypt and Arabia to Iran and E. Afghanistan.

Vulpes cana Afghan fox

Vulpes canus Blanford, 1877. Gwadar, Baluchistan. Incl. *nigricans* (Bokhara Mts, Uzbekistan).

RANGE (Map 80). Mountain steppe of Baluchistan, Iran, Afghanistan and adjacent part of USSR.

REMARKS. A little-known species. Lay (1967) listed the few known localities.

Vulpes ferrilata Tibetan sand fox

Vulpes ferrilatus Hodgson, 1842. Near Lhasa, Tibet.

RANGE (Map 79). The Tibetan plateau, at about 4000 to 5000 m; the Nan Shan Plateau in Chinghai, China.

Vulpes zerda Fennec

Canis zerda Zimmermann, 1780. Sahara. Incl. *arabicus*, *aurita*, *brucei*, *cerda*, *cerdo*, *denhamii*, *fennecus*, *saarensis*.

RANGE (Map 81). Desert zone from S. Morocco, Hoggar and Air west to Sudan and Egypt; a single record from Kuwait.

REMARKS. Frequently treated as the sole member of a distinct genus *Fennecus* but this is scarcely justifiable when the whole range of *Vulpes* species is considered (Clutton-Brock *et al.*, 1976).

Genus *NYCTEREUTES*

Nyctereutes Temminck, 1839. Type-species *N. viverrinus* Temminck.

A rather distinctive, monospecific genus. The resemblance to the raccoon (*Procyon lotor*) goes beyond the pattern of the pelage, as was discussed in detail by Frechkop (1959).

Nyctereutes procyonoides Raccoon-dog

Canis procyonoides Gray, 1834. Near Canton, S. China.
N. p. procyonoides. Mainland. Incl. *amurensis*, *koreensis* (Korea), *orestes* (Yunnan), *sinensis*, *stegmanni*, *ussuriensis* (Ussuri R.).
*Nyctereutes procionoides kalininensis** Sorokin, 1958. Kalinin District (N.W. of Moscow), USSR.
N. p. viverrinus Temminck, 1844. Honshu, Shikoku, Kyushu.
N. p. albus Beard, 1904. Hokkaido.

RANGE (Map 81). Woodland zones of E. Asia from the R. Amur to Yunnan and N. Vietnam; west to Shansi and E. Szechuan; all the main islands of Japan. Introduced into European Russia and now widespread in many habitats north to 63°N and south to the Caucasus, and has recently spread as far as Sweden, W. Germany, Czechoslovakia and Hungary. See Röben (1975) for an account of its spread throughout W. Germany.

Genus *CUON*

Cuon Hodgson, 1838. Type-species *Canis primaevus* Hodgson. =*Cyon*. Incl. *Chrysaeus*, *Anurocyon*.

A monospecific genus distinguished from *Canis* by the shorter tooth-row and absence of M_3. Its status as a distinctive genus was supported by a numerical analysis using a large number of characters by Clutton-Brock *et al.* (1976).

Cuon alpinus Dhole, Red dog

Canis alpinus Pallas, 1811. Amur district of E. Siberia. Incl. *hesperius* (Semiryechensk Region, E. Turkestan), *jason* (Altai). Also many synonyms from the Oriental Region.

RANGE (Map 81). From the mountains of central Asia (Pamirs, Tien Shan, Altai and Sayan) and the Amur south, in forested areas, through India, China and Malaya to Sumatra and Borneo.

Genus *LYCAON*

Lycaon Brookes, 1827. Type-species *L. tricolor* Brookes = *Hyaena picta* Temminck.

A distinctive monospecific genus.

Lycaon pictus African hunting dog

Hyaena picta Temminck, 1820. Mozambique. Incl. *ebermaieri* (L. Chad), *sharicus* (Lower Shari R.). Also many other synonyms mainly from eastern and southern Africa.

RANGE (Map 81). Throughout the savanna and subdesert zones of Africa south of the Sahara, extending northwards into the Sahara as far as Tanezrouft, S. Algeria.

Family **URSIDAE**

Bears

A clearly defined family except that the inclusion of the giant panda, *Ailuropoda*, is controversial. Excluding *Ailuropoda* there are seven rather clearly defined species, but the generic classification is unstable. Six genera are commonly recognized of which three are represented, each by a single species, in the Palaearctic Region.

1. Pelage entirely white; ears inconspicuous; profile of forehead entirely convex; molars relatively small (combined length of M^1 and M^2 less than width of palate at M^1) ***Thalarctos maritimus***
– Pelage not entirely white; ears conspicuous; profile of forehead with slight concavity; molars large (length of M^1 and M^2 greater than width of palate) . . . . . . (2)
2. Pelage normally brown (but black or grey in China and Tibet); ears short and rounded; rostrum longer (anterior margin of orbit to anterior point of premaxilla greater than width across post-orbital processes) . . . . ***Ursus arctos***
– Pelage black with a white band across the chest; ears long and tufted; rostrum shorter (anterior margin of orbit to point of premaxilla less than width across post-orbital processes) . . . . ***Selenarctos thibetanus***

Genus *SELENARCTOS*

Selenarctos Heude, 1901. Type-species *Ursus thibetanus* Cuvier. = *Arcticonus*. Incl. *Euarctos* Allen (in part).

A monospecific genus marginally distinct from *Ursus*.

Selenarctos thibetanus Asiatic black bear

Ursus thibetanus Cuvier, 1823. Sylhet, Assam. = *torquatus*. Incl. *clarki*, *formosanus* (Taiwan), *gedrosianus* (Baluchistan), *japonicus* (Japan), *laniger* (Kashmir), *leuconyx* Heude, *macneilli*, *melli* (Hainan), *mupinensis* (Szechuan), *rexi*, *ussuricus* (Ussuri Region, E. Siberia), *wulsini* (Hopei).

RANGE (Map 82). Forested regions of E. Asia from the Ussuri Region of S.E. Siberia south through China to Cambodia and Thailand, and east through the Himalayan foothills to Kashmir, E. Afghanistan (Povolny, 1966) and Baluchistan. The Japanese islands of Honshu, Shikoku and Kyushu; Taiwan.

REMARKS. There is variation in size and in the density and length of the pelage, and some of the names listed above may represent discrete races, but no clear diagnoses are available.

Genus *URSUS*

Ursus Linnaeus, 1758. Type-species *Ursus arctos* L. Incl. *Melanarctos*, *Mylarctos*, *Myrmarctos*, *Ursarctos*.

This genus is variously considered monospecific or to contain one or even all of the other species of bears. The other species most commonly included in *Ursus* is the American black bear, *U. americanus*.

Ursus arctos Brown bear

Ursus arctos Linnaeus, 1758. Sweden.

U. a. arctos. Northern Palaearctic range. =*cadaverinus*, *fuscus*, *normalis* Gray, *ursus*. Incl. *albus*, *alpinus*, *annulatus*, *argenteus*, *aureus*, *badius*, *baikalensis* (L. Baikal), *beringiana* (Shantar Is., E. Siberia), *bosniensis*, *brunneus*, *cavifrons*, *collaris* (Siberia), ?*euryrhinus*, *eversmanni*, *falciger*, *ferox* Temminck, *formicarius*, *fuscus* Gmelin, *grandis*, *griseus*, *jeniseensis* (Upper Yenesei), *kolymensis* (N.E. Siberia), *lasiotus* (N. China), *longirostris*, *major*, *mandchuricus* (Vladivostock), *marsaricus*, *melanarctos*, *minor*, *myrmephagus*, *niger*, *norvegicus*, *piscator* (Kamchatka), *polonicus*, *pyrenaicus*, *rossicus*, *rufus*, *scandinavicus*, *sibiricus*, *stenorostris*, *yesoensis* (Hokkaido, Japan).

U. a. syriacus Hemprich & Ehrenberg, 1828. Syria, Iraq, Caucasus and Iran (type-locality in Lebanon where now extinct). Incl. *caucasicus*, *dinniki*, *lasistanicus*, ?*meridionalis* (Caucasus), *persicus* (N. Iran), *schmitzi* (Palestine), *smirnovi*.

U. a. isabellinus Horsfield, 1826. Himalayas to Tien Shan (type-locality in Nepal). Incl. *leuconyx* (Tien Shan), *pamirensis* (Pamirs).

U. a. pruinosus Blyth, 1854. S. and E. Tibet and adjacent parts of Chinghai. Incl. *lagomyiarius*.

U. a. crowtheri Schinz, 1844. Atlas Mts, N.W. Africa (extinct).

?*U. a. shanorum* Thomas, 1906. Reputedly from Upper Burma, but locality probably mistaken (see Pocock, 1941).

Also many synonyms based on N. American specimens, of which the earliest is *horribilis* Ord, 1815.

RANGE (Map 82). Formerly the entire forested parts of the Palaearctic Region, except the southern Japanese islands, but including the Himalayas and S.E. Tibet; and western N. America south in the Rockies to New Mexico. Now extinct in the British Isles, much of W. Europe and in N. Africa; and elsewhere the range is probably much more fragmented than formerly.

REMARKS. This species is now generally considered to include the brown and grizzly bears of N. America. It is doubtful whether even the few subspecies separately listed above have any validity, but a problem is posed by the fact that the former, probably continuous, range is now fragmented by extinction. The most clear-cut aspects of variation are the tendencies to large size in N.E. Siberia; to black colour in N. China; to pale colour in S.W. Asia; and to grey in Tibet. Central Asian animals also tend to have a white collar. There seems no value in recognizing further subspecies until they can be given exclusive diagnoses.

A recent monograph, with emphasis on the Pyrenees, is that of Couturier (1954).

Genus ***THALARCTOS***

Thalarctos Gray, 1825. Type-species *T. polaris* Gray = *Ursus maritimus* Phipps. =*Thalassarctos*, *Thalassiarchus*.

A monospecific genus marginally distinct from *Ursus*.

Thalarctos maritimus Polar bear

Ursus maritimus Phipps, 1774. Spitzbergen. Incl. *jenaensis*, *marinus*, *polaris* (Siberia), *spitzbergensis*. Also several names based upon N. American specimens.

RANGE (Map 82). Circumpolar, confined to the arctic coasts and sea-ice. Breed mainly on the arctic islands of Siberia. Occasionally animals have drifted on ice to Iceland and to Japan.

REMARKS. Eastern Siberian animals are larger than western ones and are treated as subspecifically distinct by Heptner & Naumov (1967). Manning (1971) on the other hand considered the variation clinal and did not recognize subspecies.

Family **PROCYONIDAE**

Raccoons etc.

The Procyonidae includes six genera of medium-sized carnivores in the Americas and also, according to some authorities, the two species of pandas in the Himalayas and

S.W. China. One American species, *Procyon lotor*, has been introduced and is well established in parts of Europe and the USSR.

Genus *PROCYON*

Procyon Storr, 1780. Type-species *Ursus lotor* L. =*Campsiurus*, *Lotor*.

Contains two species, in N. and S. America, plus some insular forms that are doubtfully distinct.

Procyon lotor Raccoon

Ursus lotor Linnaeus, 1758. Pennsylvania. Including many N. American synonyms.

RANGE (Map 82). The whole of N. and Central America except for the tundra and most of the boreal forest zone. Recent escapes from captivity have resulted in well established feral populations in W. Germany (Hessen, Eifel); adjacent parts of France and the Netherlands; White Russia; Caucasian region; and in the Fergana region of Turkestan.

Family **MUSTELIDAE**

Weasels etc.

A large family of small and medium-sized carnivores found in all regions except Australasia and Madagascar, but especially well represented in the Holarctic Region. The limits of the family are fairly clearly defined, but classification within the family is very unstable and there has been no comprehensive revision since that of Pocock (1921). Ten genera are represented in the Palaearctic Region.

1. Highly modified for aquatic life: tail very thick proximally, thinly haired; feet webbed; vibrissae very prominent (otters) . . . . (2)
– Not highly modified for aquatic life: tail slender but usually thickly haired; feet at most only partly webbed; vibrissae less prominent . . . . (3)
2. Tail less than half length of head and body; hind feet paddle-like with digits almost equal in length; 2 pairs of lower incisors (E. Asia, marine). . . . . ***ENHYDRA*** (p. 176)
– Tail more than half length of head and body; digits of hind feet unequal; 3 pairs of lower incisors . . . . ***LUTRA*** (p. 176)
3. Ventral pelage, including abdomen, black or dark brown, boldly contrasting with dorsal pelage which is white, grey, mottled or striped . . . . (4)
– Ventral pelage not entirely black, if dark then not contrasting sharply with dorsal pelage (8)
4. Body stout; head and body over 400 mm; tail under a quarter length of head and body; dorsal pelage grey or white . . . . (5)
– Body slender; head and body under 400 mm; tail more than half length of head and body; dorsal pelage mottled or striped . . . . (7)
5. Ears with pinnae; a bold black stripe from eye to ear, bordered below by white . (6)
– Ears without pinnae; no detached black stripe on head . . ***MELLIVORA*** (p. 175)
6. Throat black; palate terminating about half way between last molars and tips of pterygoid processes . . . . ***MELES*** (p. 175)
– Throat white; palate terminating near tips of pterygoid processes, close to bullae (China) ***ARCTONYX*** (p. 175)
7. Dorsal pelage striped black or dark brown and white . ***POECILICTIS*** (p. 174)
– Dorsal pelage mottled yellow and brown . . . . . . ***VORMELA*** (p. 171)
8. Body large (head and body up to 850 mm) combined with short tail (under a quarter of head and body); a yellow stripe from flank to rump; sagittal crest strongly developed, overhanging occiput . . . . ***GULO*** (p. 174)
– Head and body under 600 mm (or larger with tail about two-thirds length of head and body); no yellow stripe from flank to rump; sagittal crest not overhanging occiput . . (9)
9. Throat pale (white to orange), contrasting with darker dorsal and ventral pelage; hind feet over 70 mm; post-canine teeth 5/6 . . . . ***MARTES*** (p. 172)
– Throat dark, or pale and concolorous with rest of ventral pelage; hind feet under 70 mm; post-canine teeth 4/5 . . . . ***MUSTELA*** (p. 168)

Subfamily MUSTELINAE

Genus *MUSTELA*

Mustela Linnaeus, 1758. Type-species *M. erminea* L. =*Arctogale*. Incl. *Eumustela, Foetorius, Gale, Hydromustela, Ictis* Kaup, *Kolonokus, Lutreola, Mustelina, Putorius.*

In the widest sense in which this genus is now employed it contains about 12 species in the Holarctic and Oriental Regions of which seven occur in the Palaearctic (plus one that has been introduced). However, the genus is frequently split, the names *Putorius, Lutreola* and *Kolonokus* being used for separate genera.

1. Light ventral pelage (white or yellow) sharply demarcated from dark dorsal pelage (except in white winter pelage) . . . . (2)
– Ventral and dorsal pelage without a sharp boundary . . . . (3)
2. Distal half of tail black or very dark brown; tail, including hair, about half length of head and body or longer; size larger: head and body 170–380 mm, condylobasal length 40–52 mm in males, 36–45 mm in females . . . . **M. erminea**
– Tail concolorous or with a few dark hairs at the tip; tail less than half length of head and body; size smaller: head and body not over 300 mm, condylobasal length not over 45 mm in males, 36 mm in females (except in Egypt and some other parts of the Mediterranean region where the size is comparable to that of *M. erminea*) . . . . . . **M. nivalis**
3. Pattern of light and dark markings on face; legs and feet darker than body pelage; body pelage yellowish brown, more or less obscured by dark guard hairs . . . (4)
– Pelage concolorous except for white marks on lips and sometimes on throat; colour uniform dark brown or light reddish brown . . . . (5)
4. Tail entirely black; body pelage predominantly dark brown (Europe) . **M. putorius**
– Only distal half of tail black; body pelage predominantly straw-coloured **M. eversmanni**
5. Pelage predominantly light reddish brown . . . . (6)
– Pelage uniform dark chocolate brown (except in some varieties of feral mink) . . (7)
6. White lips and chin contrasting sharply with surrounding darker pelage; larger: head and body 250–400 mm, condylobasal length 55–68 mm (males), 50–56 mm (females) **M. sibirica**
– White lips and chin shading into adjacent pelage; smaller: head and body 190–260 mm, condylobasal length 46–53 (males), 40–47 mm (females) . . . . **M. altaica**
7. Upper lip white; usually no white on chest; lower margin of auditory meatus, in ventral view, does not obscure roof of meatus . . . . **M. lutreola**
– Upper lip dark; often white marks on chest; lower margin of meatus expanded, obscuring roof of meatus in ventral view . . . . **M. vison**

Mustela erminea — Stoat

Mustela erminea Linnaeus, 1758. Sweden.

M. e. erminea. Mainland range and Britain. Incl. *aestiva* (Germany), *algiricus* (reputedly Algeria but see below under 'Remarks'), *alpestris, baturini* (Gt Shantar Is., E. Siberia), *birulae* Martino, *digna, ferghanae* (Ferghana, E. Turkestan), *giganteus, hyberna, kamtschatica, karaginensis* (Karaginski Is, Kamchatka), *lymani* (Altai), *maculata, major, martinoi, minima, mongolica* (Altai), *naumovi, ognevi, orientalis* (Kolyma, E. Siberia), *ricinae* (Islay, Scotland), *schnitnikovi, stabilis* (England), *tobolica, transbaikalia* (L. Baikal), *whiteheadi* (N. India).

*Putorius kaneii** Baird, 1857. Arikamchechen Is., Bering Sea (type-locality fixed by Hall, 1944).

*Mustela erminea teberdina** Kornejev, 1941. Teberdina region, Caucasus, USSR.

*Mustela erminea balkarica** Baziev, 1962. S.W. part of Kabardino-Balkan ASSR, central Caucasus, USSR.

M. e. hibernica (Thomas & Barrett-Hamilton, 1895). Ireland.

M. e. nippon Cabrera, 1913. Honshu, Japan.

RANGE (Map 83). The tundra, boreal forest and deciduous forest zones of the Palaearctic and Nearctic regions. In Europe including Britain and Ireland but excluding most of the Mediterranean zone; in Asia extending south in the steppe and montane forests to the Caucasus (perhaps an isolated segment), to the W. Himalayas, N. Mongolia and Manchuria. Sakhalin, Hokkaido and N. Honshu. In N. America south to California in the west and to Maryland in the east.

REMARKS. Some of the southern Asiatic populations may be isolated and distinct, but no satisfactory racial diagnoses are available. This species is almost certainly not represented in N. Africa. The form *algiricus* Thomas, 1895 was based upon a specimen in the British Museum that was part of a collection purchased in 1856 which reputedly came from Algeria. But this collection contained other European species, such as *Microtus arvalis*, which, like *M. erminea*, have never subsequently been found in N. Africa, and it is therefore very likely that the entire collection had been erroneously labelled Algeria.

Mustela nivalis — Weasel

Mustela nivalis Linnaeus, 1766. Sweden.

M. n. nivalis. Mainland range in Eurasia; British and some Mediterranean islands. =*minor*, *typicus*. Incl. ?*albipes* Mina Palumbo, *alpinus*, *caraftensis* (Sakhalin), *caucasicus* (Caucasus), ?*corsicanus* (Corsica), *dinniki*, *dombrowskii*, ?*fulva* Mina Palumbo, *gale*, ?*galinthias* (Crete), *hungarica*† Vasarhelyi (not of Ehik), *ibericus* (S. Spain), *italicus*, *kamtschatica*, *major* Fatio, *meridionalis* (S. Italy), *minutus*, *monticola*, *mosanensis* (Korea), *nikolskii* (Crimea), *pallidus* (Ferghana, E. Turkestan), *punctata* (E. Transbaikalia), *pusillus*, *pygmaeus* (N.E. Siberia), *russelliana* (Szechuan, China), *siculus* (Sicily), *stoliczkana* (Sinkiang), *trettaui*, *varsarhelyi*† (status *fide* Frechkop, 1963), *vulgaris*.

*Mustela nivalis kerulenica** Bannikov, 1952. Near Undurkhan, R. Kerulen, Mongolia.

*Mustela nivalis heptneri** Morozova-Turova, 1953. Islim Tchesme, R. Egri-Geke, tributary of Kushki, Badkhiz, S. Turkmenia, USSR.

M. n. boccamela Bechstein, 1800. Sardinia.

M. n. numidica Pucheran, 1855. N.W. Africa, type-locality Tangier; ?Azores. Incl. *africana* Gray (not of Desmarest), *atlas*.

M. n. subpalmata Hemprich & Ehrenberg, 1833. Egypt.

M. n. namiyei Kuroda, 1921. Honshu, Japan; perhaps also Hokkaido. Incl. ?*yesoidsuma* (Hokkaido).

M. n. rixosa (Bangs, 1896). N. America. Incl. *allegheniensis*, *campestris*, *eskimo*.

RANGE (Map 84). The entire Palaearctic Region with the exception of the Arabian peninsula, Ireland, Iceland and the Arctic islands; present on the Azores and apparently on the island of Sao Tomé in W. Africa (specimen in British Museum). Also in much of N. America if *rixosa* is conspecific. The distribution on the Mediterranean and Atlantic Islands is probably influenced by introductions by man.

REMARKS. The above treatment is very tentative. There are two areas of considerable uncertainty in this species which have not yet been adequately resolved. One is the possibility that in Europe a small species coexists with the more abundant species, the small species having been referred to as *M. minuta* or *M. rixosa*. The existence of such a situation has never been clearly shown and seems unlikely, although there is undoubtedly very considerable variation in size and in characters of the pelage. The other area of uncertainty is the Mediterranean region where the very large forms in N. Africa and on some of the islands have been considered to be specifically distinct. Frechkop (1963) for example treated *M. boccamela* as a species confined to Sardinia, and *M. numidica* as a North African species with the Egyptian *subpalmata* as a race. In a more recent detailed study Mazak (1970) concluded that all European forms were conspecific.

Variation in size is very considerable, the smallest forms being found in the north, in mountains and especially in E. Asia and N. America; large forms in the south-west, reaching the maximum size in Egypt.

Mustela altaica — Mountain weasel

Mustela altaica Pallas, 1811. Altai Mts.

M. a. altaica. Altai. Incl. *alpinus*.

M. a. raddei (Ognev, 1928). Sayan Mts, N.E. Mongolia, S.E. Siberia and probably Manchuria and Korea. Type-locality in S.E. Transbaikalia.

M. a. sacana Thomas, 1914. Tien Shan (type-locality Przhevalsk) and Pamir. Incl. *birulai* (Pamir).

M. a. temon Hodgson, 1857. Himalayas, Tibet and W. China; type-locality in Sikkim. Incl. *astutus* (Szechuan), *longstaffi*.

RANGE (Map 84). Forested mountain regions of E. Asia from Altai south to Himalayas and Tibet and through E. China to Korea and S.E. Siberia.

Mustela sibirica Siberian weasel, Kolinsky

Mustela sibirica Pallas, 1773. W. Altai.
M. s. sibirica. W. and central Siberia. Incl. *miles* (Transbaikalia), *australis* (W. Siberia).
M. s. manchurica Brass, 1911. Amur region, Manchuria and Korea. Incl. *charbinensis*, *coreana* (Korea), *katsura*, *peninsulae*. A large form, but probably confluent with *M. s. sibirica.*
M. s. subhemachalana Hodgson, 1837. China, except Manchuria, to the Himalayas; Taiwan. Incl. *canigula* (Tibet), *davidianus* (Kiangsi), *fontanieri* (Peking), *hamptoni*, *hodgsoni*, *horsfieldi*, *humeralis*, *major*, *melli*, *moupinensis* (Szechuan), *noctis*, *stegmanni* (Shantung), *tafeli*, *taivana* (Taiwan).
M. s. itatsi Temminck, 1844. Japan, introduced on Sakhalin and Iriomote Is., Ryukyu Islands. =*natsi*. Incl. *asaii* (Oshima Is., Japan).
M. s. sho (Kuroda, 1924). Yakushima Is., Japan.
M. s. quelpartis (Thomas, 1908). Quelpart Is., Korea.
M. s. lutreolina Robinson & Thomas, 1917. Java.

RANGE (Map 83). Taiga, montane forest and cultivated regions from eastern European Russia to E. Siberia and south through Korea and China to the Himalayas and Upper Burma; all the main islands of Japan; Taiwan and Java. Introduced to Kirgizia, to Sakhalin and, in 1965, to Iriomote Is., Ryukyu Islands (Obaba, 1967).

REMARKS. In spite of its very isolated position the Javan form seems very similar indeed to other *M. sibirica* and is quite distinct from the intervening *M. nudipes* in Malaya, Sumatra and Borneo.

Mustela lutreola European mink

Viverra lutreola Linnaeus, 1761. Finland.
M. l. lutreola. N. and E. Europe, and W. Siberia. =*europaea*, *fulva*, *minor*. Incl. ?*alba*, *albica*, ?*alpinus*, *borealis* Novikov, *budina*, *cylipena*, *glogeri*, *hungarica* Ehik, 1932, *novikovi*, *transsylvanica*, *varina*, *wyborgensis*.
*Lutreola lutreola ehiki** Kretzoi, 1942. New name for *M. l. hungarica* Ehik, 1932 (not Ehik, 1928).
M. l. biedermanni Matschie, 1912. W. France. Incl. *aremorica*. Doubtfully distinct from *M. l. lutreola.*
*M. l. turovi** Kuznetzov, 1939. Caucasus.
Mustela (*Lutreola*) *lutreola turovi* Kuznetzov, in Novikov, 1939:47. Caucasus.
Incl. *binominata*, *borealis* Novikov, *caucasicus* Novikov, *novikovi.*

RANGE (Map 84). From Finland, Poland and Hungary east to about 75°E in W. Siberia and south to the Caucasus (the Caucasian population possibly isolated); also in W. France although formerly the population was probably continuous from there eastwards, and other isolated populations may persist in Germany.

REMARKS. The report in the *Zoological Record* for 1950 of a new subspecies, *borealis*, from Kazakstan described by Sludski was an error, the abbreviation 'Nov', having been mistaken for 'nova' when 'Novikov' was intended.

Although this species is superficially very similar to the American *M. vison*, the distinctness of these species is supported by the very considerable difference in karyotype, *M. lutreola* more closely resembling *M. sibirica* in this respect (Volobuev *et al.*, 1974).

Mustela vison* American mink

Mustela vison Schreber, 1777: pl. 127b. Eastern Canada.
*Mustela vison tatarica** Popov, 1949. Tatariya, USSR.
Mustela (*Putorius*) *vison domestica** Haltenorth, in Doderlein, 1955. For domestic mink.
*Lutreola vison altaica** Ternovski, 1958. Central course of Sary-Kokshi, Altai.

RANGE. Indigenous throughout most of N. America. Deliberately introduced in many forested parts of USSR, and present in many parts of Europe as a result of accidental escapes of domestic animals, e.g. in Britain, Iceland, Scandinavia and Germany.

Mustela putorius European polecat

Mustela putorius Linnaeus, 1758. Sweden. =*foetens*, *foetidus*, *infectus*, *iltis*, *vulgaris*. Incl. *albus*, *anglia* (Wales), *aureola* (Spain), *caledoniae* (Scotland), ?*flavicans*, *furo-putorius*, *manium*, *orientalis* Brauner, *rothschildi*, *stantschinskii*, *verus*, ?*vison* de Selys Longchamps.
Putorius putorius L. f. *piriformis** Kostron, 1948: 52. Rajhrad, Moravia, Czechoslovakia. (For status see Kratochvil, 1952a.)
*Putorius putorius ognevi** Kratochvil, 1952a. Central European Russia. Preoccupied by *M. erminea ognevi* Jurgenson, 1932.
*Putorius putorius orientalis** Polushina, 1955. East of a line from Pskov through Minsk, Zhitomir to Vinnitsa, European Russia. Preoccupied by *M. erminea orientalis* Ognev, 1928 and *Putorius putorius orientalis* Brauner, 1929.
Mustela (*Putorius*) *putorius mosquensis** Heptner, 1965. Savvino, 20 km E. of Moscow.

RANGE (Map 83). The forest zones of Europe, except most of Scandinavia, to the Urals. Present in Britain (now confined to Wales and the adjacent parts of England), but has never occurred in Ireland. It is sympatric with *M. eversmanni* in much of S. European Russia and the adjacent countries of E. Europe.

REMARKS. E & M-S included in this species *M. eversmanni* and all the forms here allocated to that species. But all Russian authors, through whose region the boundary runs, recognize two species.

The domestic ferret is of uncertain origin. Cranially it more closely resembles *M. eversmanni* but the karyotype is identical to that of *M. putorius* and differs from that of *M. eversmanni* (Volobuev *et al.*, 1974).

***Mustela eversmanni*†** Steppe polecat

Mustela eversmanni Lesson, 1827. Turkestan. Incl. *admirata* (Hopei, N.E. China), *amurensis* (Amur Basin), *aurea* (European Russia), *hungarica* (Hungary), *larvatus* (S.E. Tibet), *lineiventer* (Little Altai), *michnoi* (Transbaikalia), *talassicus*, *tiarata* (Kansu, China), *tibetanus*.
*Mustela eversmanni robusta** Ehik, 1927. Hungary.
*Putorius eversmanni satunini** Migulin, 1928. Nagaiski Steppes, Dagestan ASSR, USSR.
*Putorius eversmanni occidentalis** Brauner, 1929. Kherson Bay, Ukraine.
*Putorius eversmanni nobilis** Stroganov, 1958: 150. Kolchetava, Kazakhstan.
*Putorius eversmanni pallidus** Stroganov, 1958: 150. Kargat, between Novosibirsk and Lake Chany, W. Siberia.
*Putorius eversmanni tuvinicus** Stroganov, 1958: 152. Chaa Khol', Tuvinskaya Oblast, USSR.
*Putorius eversmanni dauricus** Stroganov, 1958: 154. Smolensk, near Chita, Transbaikalia, USSR.
*Putorius eversmanni heptopotamicus** Stroganov, 1960. Ili, S. Prebalkhash, Kazakhstan, USSR.

RANGE (Map 83). The entire steppe and subdesert zone of USSR, Mongolia, Tibet and E. China, with probably isolated segments in E. Europe (Hungary, Austria, Czechoslovakia, Poland, E. Germany).

REMARKS. In view of the continuity of range it seems unlikely that many of the above names represent discrete subspecies. *M. nigripes* (Audubon & Bachman, 1851), the black-footed ferret from the prairie zone of N. America, is very similar and was considered conspecific with *M. eversmanni* by Kostron (1948).

This species was considered conspecific with *M. putorius* by E & M-S, but has been considered specifically distinct by all subsequent authors working in the boundary zone. The relationship between the two species was reviewed by Heptner (1964).

Genus *VORMELA*

Vormela Blasius, 1884. Type-species *Mustela sarmatica* Pallas = *Mustela peregusna* Guldenstaedt.

A monospecific genus characterized especially by the boldly patterned pelage.

Vormela peregusna Marbled polecat

Mustela peregusna Guldenstaedt, 1770. River Don, USSR.
V. p. peregusna. Western part of range. Incl. *alpherakii* (Transcaspia), *euxina* (Rumania), *koshewnikovi*, *ornata* (see below under 'Remarks'), *sarmatica*, *syriaca* (Syria), *tedshenika*.
  *Vormela peregusna pallidior** Stroganov, 1948: 129. Semireche, Kopalski district, Kirgizia.
  *Vormela peregusna obscura** Stroganov, 1948: 131. Dolina Vakhsha, Tadzhikistan.
V. p. negans Miller, 1910. Mongolia and W. China; type-locality Ordos Desert, Inner Mongolia, China.

RANGE (Map 84). Steppe and subdesert zones from S.E. Europe (Bulgaria and Rumania) to W. China, south to Palestine and Baluchistan.

REMARKS. Pocock (1936) described *V. p. ornata* with type-locality near Lake Baikal. This species has not subsequently been found closer to L. Baikal than central Mongolia, and it seems very probable that the locality of the type was mistaken. It was acquired by Dr C. Hose during a journey on the Trans-Siberian railway, but no further details of its origin are available. The other specimens mentioned by Pocock are labelled simply 'Siberia'. Pocock in 1937 wrote on the label of the type: 'Two in Bombay coll. from Kurdistan apparently same as this'. The name *ornata* can therefore be considered a synonym of *peregusna* with the type-locality unknown.

Genus *MARTES*

Martens

Martes Pinel, 1792. Type-species *M. domestica* Pinel = *Mustela foina* Erxleben. Incl. *Charronia, Lamprogale, Zibellina.*

Contains probably seven species of which four are Palaearctic, two are Nearctic and one is predominantly Oriental but extends north to E. Siberia. This last species, *M. flavigula*, is often separated in the genus *Charronia*, but the differences are slight. Four forms here given specific rank, *martes*, *zibellina*, *melampus* and *americana*, constitute a closely related allopatric series and could perhaps be considered conspecific as suggested by Hagmeier (1961). For a detailed account of the evolution of the genus see Anderson (1970).

1. Large: head and body up to 800 mm, condylobasal length usually over 90 mm; length of tail about two-thirds that of head and body; yellow on neck bordered above by a clear dark line . . . ***M. flavigula***
– Smaller: head and body under 600 mm, condylobasal length usually under 90 mm; tail about half length of head and body or shorter; yellow on neck, if present, not abutted by a dark band . . . (2)
2. Muzzle and face wholly or partly black (Japan) . . . ***M. melampus***
– Muzzle and face not black . . . (3)
3. Tail relatively short, about 40% of head and body, terminal hair about level with extended hind feet; light patch on throat, if present, not sharply demarcated from surrounding pelage . . . ***M. zibellina***
– Tail longer, about 50% of head and body, terminal hair reaching far beyond extended hind feet; throat white or yellow, sharply demarcated from surrounding pelage . . . (4)
4. Throat pure white; P^3 biconvex when viewed with the cusp central; M^1 small, greatest diameter less than length of labial margin of P^4 . . . ***M. foina***
– Throat yellow or creamy; P^3 with a labial concavity and prominent lingual convexity forming a slight postero-lingual concavity; M^1 larger, greatest diameter greater than labial margin of P^4 . . . ***M. martes***

Martes martes — Pine marten

Mustela martes Linnaeus, 1758. Upsala, Sweden.
M. m. martes. Range except Caucasian region. =*sylvestris, sylvatica, vulgaris*. Incl. *abietum, latinorum* (Sardinia), *notialis* (S. Italy), *ruthena*.
*Martes martes uralensis** Kuznetzov, 1941:126. Near Miassa, S. Urals, USSR.
*Martes martes sabaneevi** Yurgenson, 1947. Valley of R. Petchora, N. Urals, USSR.
M. m. lorenzi Ognev, 1926. Caucasus.

RANGE (Map 85). Forest throughout most of Europe except southern Iberia and Greece, east to about 80°E in W. Siberia; the islands of Ireland, Britain (limited to Wales, Lake District and N.W. Scotland), Sardinia, Corsica (Verbeek, 1974) and Sicily; the Caucasian region, Elburz Mts and N.E. Asia Minor.

Martes foina Beech marten

Mustela foina Erxleben, 1777. Germany.
M. f. foina. Europe and S.W. Asia. Incl. *alba, bosniaca, domestica, mediterranea* (Spain), *mehringi* (Caucasus), *rosanowi* (Crimea), *syriaca* (Syria).
M. f. bunites Bate, 1906. Crete. Very doubtfully distinct.
M. f. milleri Festa, 1914. Rhodes.
*M. f. ognevi** Laptev, 1946. Kopet Dag, Turkestan.
Martes foina ognevi Laptev, 1946. Central Kopet Dag, USSR.
M. f. intermedia Severtzov, 1873. Mountains of Central Asia, type-locality in Kirgizia. Incl. *altaica* (Altai), *leucolachnaea* (Sinkiang).
M. f. kozlovi Ognev, 1931. E. Tibet.

RANGE (Map 85). The deciduous woodland and Mediterranean zones of S. and central Europe east through the Caucasus and Asia Minor to the Altai and the Himalayas; the islands of Crete, Rhodes and Corfu; possibly also in W. China.

REMARKS. Although this species is very similar to *M. martes*, the two are clearly distinct, although sympatric, in Europe.

Martes melampus Japanese marten

Mustela melampus Wagner, 1841. Japan.
M. m. melampus. Honshu, Shikoku and Kyushu, Japan. Incl. *japonica, melanopus, bedfordi*.
M. m. tsuensis Thomas, 1897. Tsushima Islands, Japan.
M. m. coreensis Kuroda & Mori, 1923. Mainland of S. Korea.

RANGE (Map 85). Japan: Honshu, Kyushu, Shikoku, Tsushima, introduced on Sado Is.; S. Korea. Perhaps also in parts of China.

REMARKS. See remarks under *M. zibellina* below about the relationship of these two species in Korea.

Martes zibellina Sable

Mustela zibellina Linnaeus, 1758. Tomsk district, Siberia. Incl. *alba, amurensis, asiatica* (Kamchatka), *baicalensis, brachyura* (Hokkaido, Japan), *coreensis* (Korea), *ferruginea, fusco-flavescens, hamgyenensis, kamtschadalica, kamtchatica, maculata, ochracea, princeps* (attributed by Bannikov, 1954 to Birula, 1916, not 1922), *rupestris, sajanensis, sahalinensis* (Sakhalin Is.), *sylvestris, yeniseensis*.
*Martes zibellina tungussensis** Kuznetsov, 1941:116. Lower Tunguska, Siberia.
*Martes zibellina averini** Bazhanov, 1943. Katon-Karagai region, E. Kazakhstan oblast, S. Altai, USSR.
*Martes zibellina altaica** Yurgenson, 1947. Oyrotskaya region, Altai, USSR. (Reference *fide* Heptner & Naumov, 1967.) Preoccupied by *M. foina altaica* Satunin, 1914.
*Martes zibellina tomensis** Timofeev & Nadeev, 1955:37. Kuznetsk Alatau, R. Tutuyas, tributary of Tom′, W. Siberia.
*Martes zibellina angarensis** Timofeev & Nadeev, 1955:41. Valley of Angara R. at Boguchaevskom, Krasnoyarsk Basin, Upper Yenesei, USSR.
*Martes zibellina ilimpiensis** Timofeev & Nadeev, 1955:44. Source of Kochechumo, tributary of R. Kotui, Central Siberian Plateau.
*Martes zibellina vitimensis** Timofeev & Nadeev, 1955:47. Valley of R. Mam′, right tributary of R. Vitim, Yakutia, E. Siberia.
*Martes zibellina obscura** Timofeev & Nadeev, 1955:47. Upper R. Chikoi, right tributary of R. Selengi, Zabaykal′e, central Siberia.

RANGE (Map 85). Originally the entire taiga zone from Scandinavia to E. Siberia, south to the Altai and N. Korea; the islands of Sakhalin and Hokkaido. Extinct in Scandinavia and Finland. Range in USSR complicated due to local extermination and subsequent reintroductions (mapped in detail by Heptner & Naumov, 1967).

REMARKS. This species was held by Heptner & Naumov (1967) to include also the Japanese and Korean martens here listed as *M. melampus*. The South Korean form of *M. melampus* (if the type of *coreensis* Kuroda & Mori is typical) is very different from the few Siberian specimens of *M. zibellina* in the British Museum and there are no geographical grounds for supposing it to have been long isolated. On the other hand the possible introduction of *M. melampus* to S. Korea must be considered.

M. zibellina overlaps slightly with *M. martes* in the Ural region and some hybridization occurs.

No attempt is made here to distinguish races. Very few of the names listed are likely to represent valid races in view of the former continuity of range. There is a monograph by Pavlinin (1966).

Martes flavigula Yellow-throated marten

Mustela flavigula Boddaert, 1785. Nepal.

M. f. flavigula. Himalayas, and mainland of E. Asia. =*typica.* Incl. *aterrima* (E. Siberia), *borealis, koreana* (Korea), *leucotis, melina, quadricolor.*

Also many other names from S. China and S.E. Asia of which the following represent isolates: *gwatkinsi* (S. India), *chrysospila* (Taiwan), *henrici* (Sumatra), *saba* (Borneo), *robinsoni* (Java).

RANGE (Map 85). Forest and open woodland from the Amur valley south throughout China and S.E. Asia, and west along the Himalayan foothills and valleys as far as Kashmir and N.W. Pakistan. Isolates in S. India (if *gwatkinsi* is conspecific) and on the islands of Taiwan, Sumatra, Java and Borneo.

Martes – *Incertae sedis*

Mustela ?toufoeus Hodgson. Based on two trade skins reputedly from Llasa, Tibet. These might represent *M. foina* as suggested by E & M-S, but the pale greyish head is suggestive of *M. melampus* with which Pocock (1941) associated it.

Genus ***POECILICTIS***

Poecilictis Thomas & Hinton, 1920. Type-species *Mustela libyca* Hemprich & Ehrenberg.

A monospecific genus which should probably be reunited with *Ictonyx* from which it differs chiefly by having the bullae hypertrophied.

Poecilictis libyca Saharan striped weasel

Mustela libyca Hemprich & Ehrenberg, 1833. Libya.

P. l. libyca. Libya.

P. l. vaillanti (Loche, 1856). Tunis to Morocco, type-locality Algeria.

P. l. multivittata (Wagner, 1841). Central Sudan.

P. l. oralis Thomas & Hinton, 1920. Coastal Sudan.

P. l. rothschildi Thomas & Hinton, 1920. S.W. Sahara, type-locality in N. Nigeria.

*P. l. alexandrae** Setzer, 1958 Lower Egypt.

Poecilictis libyca alexandrae Setzer, 1958a. El Qatta, Imbaba, Giza Province, Egypt.

RANGE (Map 86). Circum-Saharan in semidesert, including the whole of North Africa, Sudan, Tchad, Nigeria, Niger and Mauritania.

Genus ***GULO***

Gulo Storr, 1780. Type-species *Mustela gulo* L. Dated by E & M-S from Frisch, 1775 which has subsequently been rejected by the International Commission (Opinion 258, 1954).

A monospecific genus, if the Palaearctic and Nearctic forms are considered conspecific which seems reasonable.

Gulo gulo Wolverine, Glutton

Mustela gulo Linnaeus, 1758. Lapland. =*arcticus, arctos, borealis, vulgaris.* Incl. *albus* (Kamchatka), *biedermanni* (Altai), *kamtschaticus, luscus* Trouessart, *sibirica, wachei.*

Also several names based on N. American specimens of which the earliest is *luscus* Linnaeus, 1758 (over which *gulo* has page priority).

RANGE (Map 82). The entire Palaearctic taiga zone and the southern part of the tundra from Norway to N.E. Siberia, south to about 58°N in Europe and W. Siberia, to N. Mongolia and the Ussuri Region in central and E. Asia; the island of Sakhalin; also in the entire boreal forest and tundra of N. America with an isolated population in California (now extinct in the Rocky Mts of USA). Formerly south to 50°N in European Russia and west to Germany.

REMARKS. Kurten & Rausch (1959) made a detailed cranial comparison of Palaearctic and Nearctic *Gulo* and concluded that they are conspecific. Although they concluded that they were subspecifically distinct, they did not describe characters that would support their separation even at that level.

Subfamily **MELLIVORINAE**

Genus *MELLIVORA*

Mellivora Storr, 1780. Type-species *Viverra ratel* Sparrman = *Viverra capensis* Schreber. = *Melitoryx*, *Ratellus*. Incl. *Lipotus*, *Ursitaxus*.

A distinctive, monospecific genus, usually placed as the sole member of subfamily Mellivorinae.

Mellivora capensis Ratel, Honey badger

Viverra capensis Schreber, 1776. Cape of Good Hope.

M. c. indica (Kerr, 1792). India to Turkestan. Incl. *mellivorus*, *ratel*, *ratelus*.

M. c. inaurata (Hodgson, 1836). Southern foothills of Himalayas.

M. c. wilsoni Cheesman, 1920. S.W. Iran (type-locality), Iraq, N. Saudi Arabia and possibly Palestine.

M. c. pumilio Pocock, 1946. S. Arabia (probably intergrading with *M. c. wilsoni* according to Harrison (1968a).

M. c. leuconota Sclater, 1867. S. Morocco and in equatorial West Africa.

RANGE (Map 86). All the drier regions from S. Nepal, S. and E. India and Pakistan through Iran and Turkestan to Arabia and Palestine; formerly in lower Egypt; S. Morocco and throughout all the dry savanna and steppe zones of Africa south of the Sahara.

REMARKS. The allocation of Moroccan specimens to *M. c. leuconota* is dubious.

Subfamily **MELINAE**

Badgers

Genus *MELES*

Meles Brisson, 1762. Type-species *Ursus meles* L. = *Taxus*, *Melesium*. Incl. *Meledes*.

A fairly distinctive, monospecific genus, related to the Oriental genera *Arctonyx* and *Melogale* and to the North American *Taxidea*.

Meles meles Badger

Ursus meles Linnaeus, 1758. Sweden. = *communis*, *europaeus*, *typicus*, *vulgaris*. Incl. *alba*, *altaicus* (Altai), *amurensis*, *anakuma* (Japan), *arcalus* (Crete), *arenarius* (Caucasus), *blanfordi* (Sinkiang), *britannicus* (Britain), *canescens* (Iran), *caninus*, *caucasicus*, *chinensis*, *danicus*, *hanensis*, *heptneri*, *leucurus* (Tibet), *leptorhynchus* (N. China), *maculata*, *marianensis* (Spain), *mediterraneus*, *melanogenys*, *minor*, *ponticus*, *raddei* (Transbaikalia), *rhodius* (Rhodes), *severtzovi* (Turkestan), *sibiricus* (Tomsk), *schrenkii*, *siningensis*, *tsingtauensis*, *talassicus*, *tauricus*, *taxus*, *tianschanensis*.

*Meles meles aberrans** Stroganov, 1962. Bogembai, Akmolinskaya dist., Kazakhstan, USSR.

RANGE (Map 86). All the woodland and steppe zones of the Palaearctic except in N.E. Siberia, N. Africa and the Arabian region. The northern limit is about 65° in Scandinavia but near the Arctic Circle in European Russia. The southern boundary runs through Palestine, Iran, Tibet and southern China. There are not likely to be any major discontinuities in the continental range except perhaps in central Asia. Insular isolates are on Ireland, Britain, the Balearic islands, Crete, Rhodes, Quelpart (Korea) and all the large islands of Japan.

REMARKS. Geographical variation is considerable but has not been comprehensively described and no attempt has been made above to distinguish between valid and invalid subspecies. Eastern Asiatic forms tend to have the dorsal pelage brown instead of pure grey. The insular populations are not strongly differentiated.

Genus *ARCTONYX*

Arctonyx Cuvier, 1825. Type-species *A. collaris* Cuvier. Incl. *Trichomanis*.

A fairly distinctive, monospecific genus.

Arctonyx collaris Hog-badger

Arctonyx collaris Cuvier, 1825. Bhutan Duars, E. Himalayas.
A. c. leucolaemus (Milne-Edwards, 1807). N. China (type-locality Peking). Incl. *milne-edwardsi* (S. Kansu). Also other races in the Oriental part of the range of which the principal are *albogularis* (S. China), *dictator* (Thailand), *consul* (Burma) and *hoevenii* (Sumatra).

RANGE (Map 86). Forest zones from Peking south throughout southern China and Indochina to peninsular Thailand and perhaps Perak; west to Sikkim; the island of Sumatra.

Subfamily **LUTRINAE**
Otters
Genus ***LUTRA***

Lutra Brisson, 1762. Type-species *Mustela lutra* L. (Can be dated from Brünnich, 1771 if Brisson, 1762 is considered invalid). *Lutris*, *Lutrix*. Incl. *Lutrogale*, *Lutronectes*.

The generic classification within the subfamily Lutrinae is very unstable and badly needs revision. As most frequently used in recent years, i.e. including *Lutrogale* and *Hydrictis*, the genus *Lutra* includes about eight species with a world-wide distribution. There is only one Palaearctic species, but an Oriental species extends to S. Iraq.

Tail only slightly flattened, without lateral keels; upper margin of rhinarium sinuous; molar teeth more trenchant . . . . ***L. lutra***
Tail more strongly flattened, with lateral keels; upper margin of rhinarium straight; molars wider, with lower cusps . . . . ***L. perspicillata***

Lutra lutra Otter

Mustela lutra Linnaeus, 1758. Upsala, Sweden. =*piscatoria*, *vulgaris*. Incl. *amurensis*, *angustifrons* (Algeria), *baicalensis* (L. Baikal), ?*fluviatilis*, *japonica*, *kamtschatica*, *oxiana*, *meridionalis* (N. Iran), *marinus* Billberg, *nudipes*, *roensis* (Ireland), *seistanica* (Seistan, Iran), *splendida*, *stejnegeri*, *whitleyi* (Japan).
Also other synonyms in the Oriental Region of which the most important are *bareng* (Sumatra), *chinensis* (S. China), *monticola* (Nepal) and *nair* (S. India).

RANGE (Map 87). The entire Palaearctic Region except for the Siberian tundra, N. Africa east of Algeria, Arabia and S. Iran, but including the British and Japanese islands. Also through the Himalayas and S. China to Malaya; isolates in S. India, Sri Lanka, Taiwan and Sumatra.

REMARKS. Variation is slight and there is probably little discontinuity of distribution or variation. The subspecies most often recognized in the Palaearctic Region are *L. l. seistanica* and *L. l. meridionalis*.

Lutra perspicillata Indian smooth-coated otter

Lutra perspicillata Geoffroy, 1826. Sumatra.
*L. p. maxwelli** (Hayman, 1957). Iraq.
Lutrogale perspicillata maxwelli Hayman, 1957. Abusakhair, 5 miles W. of Persian frontier, 35 miles S.E. of Amara, Iraq.
L. p. sindica (Pocock, 1940). Indus Valley, Pakistan.
L. p. perspicillata. India to Sumatra and doubtfully Borneo. Incl. several synonyms.

RANGE (Map 87). The marshes of the lower Tigris, Iraq; otherwise entirely Oriental, from the Indus through most of India to Indochina and Malaya; on Sumatra and, doubtfully, Borneo.

REMARKS. *L. p. maxwelli* appears to be a clear-cut, isolated subspecies characterized by very dark dorsal pelage and grey throat.

Subfamily **ENHYDRINAE**
Genus ***ENHYDRA***

Enhydra Fleming, 1822. Type-species *Mustela lutris* L. =*Latax*, *Enhydris*. Incl. *Pusa* Oken.

A monospecific genus placed by E & M-S, following Simpson (1945), in the subfamily

Lutrinae. This has presumably been done on cladistic grounds, but *Enhydra lutris* differs from the other otters in so many respects and to such a large degree that its allocation to a separate subfamily, as was done by Pocock (1921), seems fully justified.

Enhydra lutris Sea otter

Mustela lutris Linnaeus, 1758. Kamchatka. Incl. *gracilis*, *kamtschatica*, *marina*, *nereis* (California), *orientalis*, *stelleri*.

RANGE (Map 87). Completely marine, in shallow inshore waters on rocky coasts. Now restricted to the Kurile and Komandorskiye Islands, formerly also on the east coast of Kamchatka; in N. America formerly from the Aleutian Islands to Lower California but now confined to the western Aleutians and a short stretch of the coast of central California.

REMARKS. A monograph of the species in the USSR is available in Russian and English translation (Anon., 1947).

Family **VIVERRIDAE**

Civets etc.

A large family of small and medium-sized carnivores represented throughout the tropics of Africa and Asia with a few species reaching the southern fringe of the Palaearctic Region. The family includes the genets, mongooses and civets and is the ecological equivalent of the Procyonidae in the Americas and of the Mustelidae although there is considerable overlap with the latter. Simpson (1945) listed 36 genera of which four are represented in the Palaearctic Region, two of these very marginally.

1. Ears pointed, erect, conch simple; tail terete and only very slightly tapering; anus not enclosed in a glandular sac; no bony auditory meatus . . . . (2)
– Ears rounded, short, conch with complex valves; tail tapering; anus in a glandular sac; bony auditory meatus well developed . . . . (3)
2. Body spotted; tail ringed (Europe, Africa and Arabia) . . . . ***GENETTA*** (p. 177)
– Body not spotted; tail not ringed (E. Asia) . . . . . . ***PAGUMA*** (p. 178)
3. Ears projecting above head; frontal region of skull as high as braincase; (terminal half of tail white; dorsal pelage grey) . . . . ***ICHNEUMIA*** (p. 179)
– Ears not projecting above head; frontal region of skull lower than braincase ***HERPESTES*** (p. 178)

Genus ***GENETTA***

Genetta Cuvier, 1816. Type-species *Viverra genetta* L. =*Odmaelurus*.

A genus of about ten species confined to Africa south of the Sahara except that one species extends into the Mediterranean region and Arabia.

Genetta genetta Genet

Viverra genetta Linnaeus, 1758. Spain.
G. g. genetta. European mainland. =*vulgaris*. Incl. ?*communis*, *gallica*, *hispanica*, *melas*, *peninsulae*, *rhodanica*.
*Genetta genetta pyrenaica** Bourdelle & Dezilière, 1951. Pyrenees, France.
G. g. balearica Thomas, 1902. Majorca, Balearic Is.
G. g. afra Cuvier, 1825. N. Africa. Incl. *barbara*, *bonaparti*.
G. g. terraesanctae Neumann, 1902. Palestine.
G. g. granti Thomas, 1902. S.W. Arabia.
Also several other named forms from south of the Sahara of which *G. g. senegalensis* (Fischer) reaches N.E. Sudan and Senegal.

RANGE (Map 87). France and Iberia; Majorca; Palestine; S.W. Arabia; N. Africa from Morocco to Cyrenaica; the savanna zones of Africa south of the Sahara.

REMARKS. On zoogeographical grounds it seems highly probable that the European population is the result of human introduction.

Genus *PAGUMA*

Paguma Gray, 1831. Type-species *Gulo larvatus* Hamilton-Smith. Incl. *Ambliodon*.

A monospecific genus of arboreal civets related to *Paradoxurus*.

Paguma larvata — Masked palm-civet

Gulo larvatus Hamilton-Smith, 1827. Type-locality unknown. There are many synonyms, all from the Oriental Region.

RANGE (Map 88). Near Peking (probably an isolated population) (Leroy, 1948); the entire mainland of the Oriental Region except for peninsular India; the islands of Hainan, Sumatra, Borneo, Andamans; introduced to central Honshu, Japan (Udagawa, 1954).

Genus *HERPESTES*

Herpestes Illiger, 1811. Type-species *Viverra ichneumon* Gmelin. Incl. *Calogale*, *Mangusta*.

A genus of about 14 species throughout Africa and the Oriental Region. One of the four African species and two of the Oriental species extend into the southern part of the Palaearctic Region.

1. Total length under 600 mm; condylobasal length under 70 mm . . . ***H. auropunctatus***
– Total length over 600 mm; condylobasal length over 70 mm . . . . . . (2)
2. Total length under 800 mm; condylobasal length under 80 mm; tip of tail white or almost so ***H. edwardsi***
– Total length up to 1·1 m; condylobasal length 90–100 mm; tip of tail darker than rest ***H. ichneumon***

Herpestes ichneumon — Egyptian mongoose

Viverra ichneumon Linnaeus, 1758. Egypt.
H. i. ichneumon. Asia Minor to Egypt and perhaps E. Libya. Incl. *aegyptiae*, *major*, *pharaon*.
H. i. numidicus (Cuvier, 1834). Morocco to Tunisia. Incl. *sangronizi*.
H. i. widdringtoni Gray, 1842. Iberia. Incl. *dorsalis*, *ferruginea* Seabra, *grisea* Seabra.
Also many other named forms from south of the Sahara.

RANGE (Map 88). S. Spain and Portugal; Morocco to Tunisia; Egypt (and perhaps E. Libya) through Palestine to S. Asia Minor; most of the savanna zone of Africa south of the Sahara. Introduced on the island of Mljet, Yugoslavia, in central Italy and in Madagascar.

REMARKS. On zoogeographical grounds it seems highly probable that the European population is the result of human introduction, perhaps in antiquity.

Herpestes auropunctatus — Small Indian mongoose

Mangusta auropunctata Hodgson, 1836. Nepal.
H. a. pallipes (Blyth, 1845). Iraq to W. India, type-locality Kandahar, Afghanistan. Incl. *helvus*, *persicus*.
Also other named forms in the Oriental Region.

RANGE (Map 88). From Iraq and Iran through N. India to Malaya; the island of Hainan. Introduced on Mafia Island (E. Africa), the Hawaiian Islands, most of the Caribbean Islands and on the coast of S. America from Guyana to French Guiana.

Herpestes edwardsi — Indian grey mongoose

Ichneumon edwardsii Geoffroy, 1818. Madras.
H. e. ferrugineus Blanford, 1874. E. Arabia to Indus Valley, type-locality in Sind. Incl. *andersoni*, *montanus*, *pallens*.
Also many other named forms in the Oriental Region.

RANGE (Map 88). From N.E. Arabia through Iran to Baluchistan and the whole of India; Bahrain; Sri Lanka; introduced in Malaya, Mauritius, and some of the Ryukyu Islands.

Genus *ICHNEUMIA*

Ichneumia Geoffroy, 1837. Type-species *Herpestes albicaudus* Cuvier. =*Lasiopus* Geoffroy (not of Dejean).

A monospecific genus distinguished by more digitigrade feet and less trenchant teeth than those of *Herpestes*.

Ichneumia albicauda White-tailed mongoose

Herpestes albicaudus Cuvier, 1829. Senegal. Incl. many synonyms from Africa south of the Sahara, few, if any, of which are likely to represent good subspecies.

RANGE (Map 88). The northeastern coastal region of Oman, Arabia; throughout most of the savanna zones of Africa south of the Sahara, north to Senegal and N. Sudan.

Family **HYAENIDAE**

Hyaenas

Contains the three species of hyaena, usually placed in two genera, and by some extended to include the monospecific genus *Proteles*, although this is frequently placed in a separate family. An Ethiopian group with one species extending north of the Sahara and into S.W. Asia. *Proteles cristatus* extends north in Sudan to the Egyptian border, including the Gebel Elba which was once Egyptian territory.

Genus *HYAENA*

Hyaena Brisson, 1762 (or Brünnich, 1771 if Brisson, 1762 is considered invalid). Type-species, in either case, *Canis hyaena* L. =*Euhyaena*.

Contains two clearly defined species, one listed below, the other, *H. brunnea*, confined to S. Africa.

Hyaena hyaena Striped hyaena

Canis hyaena Linnaeus, 1758. Benna Mountains, Laristan, S. Iran.

H. h. hyaena. Asiatic part of the range and perhaps Egypt. =*antiquorum*, *fasciata*, *orientalis*, *striata*, *virgata*. Incl. *bilkiewiczi*, *bokcharensis* (Turkestan), *indica* (India), *satunini*, *syriaca* (Syria), ?*vulgaris* (?Egypt), *zarudnyi*.

H. h. sultana Pocock, 1934. S. Arabia. Doubtfully distinct.

H. h. barbara Blainville, 1844. N.W. Africa, type-locality Oran, Algeria.

Also several synonyms from south of the Sahara, of which *dubbah* Meyer, 1791 (Sudan) is the earliest.

RANGE (Map 89). From W. Bengal and S. India through Iran and S. Turkestan to Arabia, Asia Minor and lower Egypt; Cyrenaica to Morocco; the savanna zone south of the Sahara from Senegal to Sudan and south to central Tanzania.

Family **FELIDAE**

Cats

A clearly defined family of almost world-wide distribution, containing about 36 species. Classification within the family is very unstable; almost every species has at some time been placed in a separate genus, whilst the opposite procedure of placing all the small cats in a single genus *Felis*, as was done by Simpson (1945), followed by E & M-S, is also unsatisfactory. Pending a revision of the family that takes into account all the known species and all available characters, I propose to follow the generic classification used by E & M-S. Amongst the Palaearctic species here included in the genus *Felis*, the lynx is most often given generic rank (genus *Lynx*). Fourteen species occur (or did occur until recently) in the Palaearctic Region.

1. Claws fully retractile into cutaneous sheaths; space between upper canines and first large premolars (ignoring rudimentary premolars) greater than diameter of canines (2)

– Claws only partially retractile, without cutaneous sheaths; space between upper canines and first large premolars less than diameter of canines (legs very long; pelage spotted) *ACINONYX* (p. 185)

2. Head and body over 1 m (if near 1 m tail almost as long as head and body); suspenders of hyoid ligamentous and elastic . . . . ***PANTHERA*** (p. 184)
– Head and body under 1 m (if near 1 m tail much shorter than head and body); suspenders of hyoid ossified . . . . ***FELIS*** (p. 180)

Genus *FELIS*

Felis Linnaeus, 1758. Type-species *Felis catus* L. =*Catus*. Incl. *Caracal*, *Catolynx*, *Cervaria*, *Chaus*, *Eremaelurus*, *Eucervaria*, *Galeopardus*, *Leptailurus*, *Lynceus*, *Lynchus*, *Lynx*, *Oncoides*, *Otocolobus*, *Pardina*, *Poliailurus*, *Prionailurus*, *Profelis*, *Serval*, *Servalina*, *Trichaelurus*, *Urolynchus*.

In its widest sense (as used here) this is a worldwide genus of about 28 species of which nine occur in the Palaearctic Region. Of the synonyms listed above the following are most frequently used for distinct genera: *Lynx*, *Otocolobus*, *Caracal*, *Prionailurus*, *Leptailurus* and *Profelis*. To assist in correlating synonyms, these are used here as subgeneric headings, but it must be emphasized that no stability of nomenclature will be achieved in this group until a comprehensive revision has been undertaken giving revised diagnoses of the genera and/or subgenera based on a broad spectrum of characters and on all the known species.

For a classification of the entire family see Pocock (1917). A more detailed and recent revision of *Felis* s.s. and *Otocolobus* is that of Pocock (1951), and of *Felis* s.s. that of Haltenorth (1953, 1957).

1. Tail short, under 40 % of head and body; hind foot in adults over 140 mm; greatest length of skull over 110 mm. . . . . (2)
– Tail long, over 40 % of head and body; hind foot under 140 mm; greatest length of skull under 110 mm. . . . . (5)
2. Ears prominently tufted (tufts over 20 mm long) . . . . (3)
– Ears without, or with very short tufts (under 20 mm) . . . . . . (4)
3. Pelage completely unspotted; tail uniformly coloured; cheeks without ruffs of long hair; smaller (greatest length of skull under 130 mm) . . . . ***F. caracal***
– Pelage usually with some spots, especially on legs; tip of tail black; cheeks with ruffs of long hair; larger (skull over 130 mm) . . . . ***F. lynx***
4. Pelage spotted, tail with dark rings; legs long (shoulder height *c.* 560 mm); ears rounded ***F. serval***
– Pelage unspotted; tail without dark rings; legs shorter (shoulder height *c.* 410 mm); ears pointed . . . . ***F. chaus***
5. Ears very short and rounded, set widely apart on sides of head; pelage of forelegs unmarked; pelage very long and soft; (body unmarked except for a few transverse dark lines on rump) (central Asia) . . . . ***F. manul***
– Ears more conspicuous, set more closely together; forelegs always with at least one transverse dark line; pelage not exceptionally long . . . . (6)
6. A white spot on the back of each ear; stripes extending across chest; tail (like body) spotted (E. Asia) . . . . ***F. bengalensis***
– No white spots on ears; no stripes on chest; tail ringed (sometimes very faintly) (7)
7. Soles of feet thickly haired, concealing pads; body almost unmarked except for two dark bands across forelegs and several fainter bands on hindlegs; auditory bullae very large, extending forwards beyond posterior margin of glenoid cavity . . ***F. margarita***
– Soles less thickly haired, pads visible; body usually distinctly spotted or striped; bullae not enlarged, not reaching glenoid cavity . . . . ***F. silvestris***
F. bieti

Subgenus *FELIS*

Felis silvestris Wild cat

Felis (*catus*) *silvestris* Schreber, 1777. Germany. (This name was validated by Opinion 465 of the International Commission on Zoological Nomenclature, 1957.)

F. s. silvestris. European mainland and Scotland. Incl. *euxina* (Rumania), *ferox*, *ferus*, *grampia* (Scotland), *morea* (Greece), *molisana* (Italy), *tartessia* (S.W. Spain).

F. s. jordansi Schwarz, 1930. Majorca, perhaps also the other Balearic Islands.

F. s. reyi Lavauden, 1929. Corsica.

*F. s. cretensis** Haltenorth, 1953. Crete.
Felis silvestris cretensis Haltenorth, 1953. Kanea, Crete.
F. s. caucasica Satunin, 1905. Caucasus and Asia Minor. Incl. *trapezia*.
F. s. tristrami Pocock, 1944. Palestine. Incl. *syriaca* Tristram, *maniculata* Yerbury & Thomas (Aden).
F. s. iraki Cheesman, 1921. Kuwait and Iraq.
*F. s. gordoni** Harrison, 1968. Oman, Arabia.
Felis silvestris gordoni Harrison, 1968a. Wadi Suwera, 6 miles west of Sohar, Batinah coast of Oman.
F. s. caudata Gray, 1874. Turkestan, type-locality Ferghana. Incl. *griseoflava*, *issikulensis*, *kozlovi* (E. Tien Shan), *longipilis* Zukovsky, *maimanah*† (Afghanistan), ?*matschiei* (S.E. of Caspian Sea), *murgabensis*, *schnitnikovi*.
F. s. ornata Gray, 1832. Iraq (doubtful) and Iran to India. Incl. ?*nesterovi* (S. Iraq), *servalina*, *torquata*.
F. s. chutuchta† Birula, 1917. S. Mongolia.
F. s. vellerosa† Pocock, 1943. Borders of Ordos and N.E. Shensi, China.
F. s. libyca Forster, 1780. N. Africa (type-locality in Tunis), ?Sardinia, Sicily. =*lybiensis*. Incl. ?*bubastis* (sacred cat of ancient Egypt), *cyrenarum* (Libya), *mauritana* (Morocco), *mediterranea*, ?*sarda* (Sardinia), *torquata*.
Also other named forms south of the Sahara of which *ocreata* Gmelin, 1791 (Ethiopia) and *cafra* Desmarest, 1822 (S. Africa) are the earliest.

RANGE (Map 89). Deciduous woodland, savanna and steppe zones from W. Europe to W. China and central India, and throughout these zones of Africa. In Europe it does not reach the northern limit of the deciduous woodland and is now largely confined through persecution to isolated areas of montane, and often coniferous, forest. There is probably also a considerable amount of fragmentation of the range in the Arabian region and in central Asia since it avoids desert. It is present in Scotland, and on all the main Mediterranean islands. The distribution and variation in the Mediterranean region has no doubt been considerably influenced by man and by interbreeding with domestic cats.

REMARKS. The recognition of *silvestris* and *libyca* as conspecific follows Haltenorth (1953) and seems very reasonable. As in the case of *Lepus* spp. the transition between these two forms takes place in the Middle East where human interference has been greatest, in this case complicated by domestication, and it may well be that the 'natural' relationship of the two forms in this region can no longer be determined. The listing of subspecies is based, with some modification, upon that of Haltenorth (1953, 1957) who gave diagnoses. The Sardinian form, *sarda*, was considered by Pocock (1951) to occur also in Morocco and Algeria, but in view of the fact that Pocock recognized transition to the paler *libyca* in N. Africa these two are here tentatively united. Haltenorth (1953) discussed in detail the reasons for transferring the Mongolian and Chinese forms *chutuchta* and *vellerosa* from *F. bieti* to *F. silvestris*.

The domestic cat is believed to have developed from Mediterranean forms of this species with later interbreeding with other forms. For this reason some authors call the wild species *F. catus* Linnaeus, 1758, i.e. the earliest name given to the domestic cat. For the arguments against this see the remarks on the nomenclature of domestic animals on p. 8.

Felis bieti Chinese desert cat

Felis bieti Milne-Edwards, 1892. Vicinity of Tongola and Tatsienlu, Szechuan, China. Incl. *pallida* (Kansu), *subpallida*.

RANGE (Map 89). Dry steppe from central Szechuan to S. Mongolia.

REMARKS. Pocock (1951), followed by E & M-S, included the forms *chutuchta* (Mongolia) and *vellerosa* (Shensi) in this species. Haltenorth (1953) considered them more likely to represent *F. silvestris*. In fact *F. bieti* is very poorly known and its distinctness from *F. silvestris* seems very doubtful

Felis chaus Jungle cat

Felis chaus Güldenstaedt, 1776. Terek R., Caucasus. =*typica*. Incl. *affinis* (N. India), *catolonyx*, *chrysomelanotis*, *erythrotus*, *fulvidina*, *furax* (Palestine), *jacquemontii*, *kelaarti* (Sri Lanka), *kutas* (Bengal), *nilotica* (Egypt), *prateri* (Sind), *rüppelii* Brandt.
Felis (*Felis*) *chaus oxiana** Heptner, 1969. Tigrovoya Balka Reserve, lower Vakhsh R. (tributary of Amu Darya), Russian Turkestan.

RANGE (Map 90). Thickets and dry woodland from Egypt (W. to Mersa Matruh) through Iraq to the Volga delta, Aral Sea, Chinese Turkestan, and eastwards through most of India to Thailand and Vietnam; Sri Lanka.

REMARKS. E & M-S listed eight races that were diagnosed by Pocock (1951), but these were recognized entirely on the bases of average differences and it is very doubtful if any are worth retaining.

The name *shawiana* Blanford from Yarkand, Chinese Turkestan is retained in the synonymy of *F. chaus* following Pocock (1939) who showed that the type was composite and selected the skull as the type. It would seem that this must be accepted unless it is shown that *F. chaus* does not occur in this area. Haltenorth (1953) accepted the skin, which is of *F. silvestris*, as the type and listed *shawiana* as a race of *F. silvestris*.

Felis margarita Sand cat

Felis margarita Loche, 1858. Near Negonça, Algeria.
F. m. margarita. African part of range. =*marginata*, *margueritei*. Incl. *meinertzhageni*.
*Felis margarita airensis** Pocock, 1938. In-Abbangarit, west of Air, Niger.
F. m. thinobia (Ognev, 1926). Turkestan.
*F. m. scheffeli** Hemmer, 1974. Pakistan.
Felis margarita scheffeli Hemmer, 1974b. Nushki Desert, Pakistan.
*F. m. harrisoni** Hemmer *et al.*, 1976. Oman.
Felis margarita harrisoni Hemmer *et al.*, 1976. Northern edge of Umm as Samin, Oman, 21°55′N, 55°50′E.

RANGE (Map 89). A desert species known from a small number of specimens. S. Morocco, Senegal, Algeria, Niger; Sinai; S. and E. Arabia, N. Iran, Turkestan and Baluchistan. E & M-S included Sinai in the range. Harrison (1968a) considered that an error, but a specimen was reported by Hemmer (1974a).

REMARKS. In spite of the fact that the conspecificity of *thinobia* and *margarita* has been disputed (e.g. by Haltenorth, 1953) it is difficult from the literature to find differences that would justify even a subspecific separation. No examples of *thinobia* have been examined. For a detailed review of the species see Hemmer *et al.* (1976).

Subgenus *OTOCOLOBUS*

Felis manul Pallas's cat

Felis manul Pallas. Jida R., S. of L. Baikal, Siberia. Incl. *nigripecta* (Tibet), *mongolicus*, *satuni*, *ferruginea* (Kopet Dag, Turkestan).

RANGE (Map 90). Steppe and semidesert, especially montane, from the Caspian Sea through S. Turkestan, Iran and Afghanistan to Transbaikalia, Mongolia, W. China, Tibet and Kashmir.

Subgenus *LYNX*

Felis lynx Lynx

Felis lynx Linnaeus, 1758. Sweden.
F. l. lynx. Boreal forest and Carpathians. Incl. *alba*, *baicalensis*, *borealis*, *cervaria*, *dinniki* (Caucasus), *guttata*, *kattlo*, *lupulinus*, *lynculus*, *melinus* (Volga), *orientalis* Satunin, *virgata*, *vulgaris*, *vulpinus*, *wrangeli* (E. Siberia).
*Lynx lynx kozlovi** Fetisov, 1950b. Barun-Burinkhan, Salenginskiy region, Buryatskaya ASSR, USSR.
*Lynx lynx neglectus** Stroganov, 1962. Glazkovka, Supunskiy reserve, Primorsk Territory, Lake Baikal, USSR (pre-occupied by *Felis neglecta* Gray, 1838).
Felis (*Lynx*) *lynx stroganovi** Heptner, 1969. New name for *Lynx lynx neglectus* Stroganov, 1962.
F. l. pardina Temminck, 1824. Iberia. Incl. *pardella* Miller.
F. l. sardiniae Mola, 1908. Sardinia.

F. l. isabellina Blyth, 1847. Mountains of central Asia; type-locality Tibet. Incl. *tibetanus*, *wardi* (Altai), *kamensis*.
Also other forms in North America where the earliest name is *canadensis* Kerr, 1792.

RANGE (Map 90). The entire taiga forest from Scandinavia to E. Siberia; montane forest in Europe (formerly widespread but now confined to Iberia, Balkans and Carpathians), Caucasus, Asia Minor, Kopet Dag, and in all the main ranges of central Asia from the Altai to Kashmir and east to Manchuria, Kansu, Tsaidam and S.E. Tibet; the island of Sakhalin and perhaps Sardinia; the taiga zone of North America. To the south it is replaced by the related but clearly distinct *F. caracal* in S.W. Asia and by *F. rufus*, the bobcat, in North America. A detailed account of the present and former distribution in Europe was given by Kratochvil (1968).

REMARKS. Kurten & Rausch (1959) made a detailed cranial comparison of Scandinavian and North American forms and concluded that they were probably subspecifically distinct, but they did not consider the geographically intermediate Siberian animals. The Caucasian population is probably quite isolated, but previous subspecific descriptions have been based only on average differences in size and seem invalid. *F. l. isabellina* in central Asia may be equally invalid. The degree of spotting of the pelage is very variable in all parts of the range, but all the southern forms tend to be more heavily spotted.

Subgenus *CARACAL*

Felis caracal — Caracal

Felis caracal Schreber, 1776. S. Africa.
F. c. algira Wagner, 1841. N.W. Africa. Incl. *berberorum*, *corylinus*, *medjerdae*, *spatzi*.
F. c. schmitzi Matschie, 1912. Egypt and Asiatic range. Incl. *aharonii*, *bengalensis* Fischer, *michaelis* (Turkmenia).
Also other named forms from south of the Sahara.

RANGE (Map 90). Steppe and savanna zones from Turkestan and N.W. India to Egypt; Algeria and Morocco; most of the savanna zone south of the Sahara, south to Cape Province.

REMARKS. It is quite probable that no subspecies can be upheld in this species. E & M-S used *schmitzi* for the Asiatic race. Harrison (1968a), although using *schmitzi*, pointed out that Arabian animals differed from those from N.E. Africa only in average size and average colour in which case I would consider them consubspecific and use the name *F. c. nubicus* Fischer, 1829 (Sudan) were it not that a study of African forms might well show that *nubicus* is not discretely separable from the South African *F. c. caracal*. This species was allocated by Leyhausen (1973) to the genus *Profelis*.

Subgenus *LEPTAILURUS*

Felis serval — Serval

Felis serval Schreber, 1776. Cape of Good Hope, S. Africa.
F. s. constantina Forster, 1780. Algeria. Incl. *algiricus*.
Also many other named forms south of the Sahara.

RANGE (Map 90). Algeria and High Atlas region of Morocco; practically the entire savanna zone south of the Sahara.

Subgenus *PRIONAILURUS*

Felis bengalensis — Leopard cat

Felis bengalensis Kerr, 1792. Bengal, India.
F. b. euptilura Elliot, 1871. E. Siberia, Manchuria, Korea, ?Hopei. =*raddei*, *undata* Radde (not of Desmarest). Incl. *decolorata* (Hopei), *manchurica* (Manchuria), ?*microtis*.
Also many other named forms in the Oriental Region.

RANGE (Map 89) From the lower Amur in E. Siberia through Korea and Manchuria and almost throughout the Oriental Region, west to Baluchistan and southeast to Java and Borneo.

REMARKS. Heptner (1971) considered that *euptilura* should be considered specifically distinct from *F. bengalensis*.

Genus *PANTHERA*

Panthera Oken, 1816. Type-species *Felis pardus* L. = *Pardus*. Incl. *Leo, Tigris, Uncia*. An application (Corbet *et al*., 1974) is before the International Commission on Zoological Nomenclature to place *Panthera* Oken on the Official List of available names, since Oken's *Lehrbuch der Naturgeschichte* has been rejected by the Commission (Opinion 417, 1956). If *Panthera* Oken is not validated, the correct name for this genus will be *Leo* Brehm, 1829.

The genus *Panthera* contains five recent species of 'great cats' with a very disjoint distribution – four species in the Old World (all included below) and one, *P. onca*, the jaguar, in South America. However, the relict nature of this pattern is proven by the presence of the extinct Pleistocene species *P. atrox* in N. America and *P. spelaea* in northern Eurasia. It represents a fairly well defined taxon although *Neofelis* is sometimes included, but its rank has varied in recent classifications from subgenus (of *Felis*) to subfamily.

1. Pelage pattern of spots in clusters or rings . . . . (2)
– Pelage unpatterned or striped . . . . (3)
2. Ground colour yellow or buff (or occasionally black); pelage short; most clusters consisting of discrete spots . . . . **P. pardus**
– Ground colour grey; pelage long with a dense underfur; clusters of spots forming continuous rings . . . . **P. uncia**
3. Striped; without a mane . . . . **P. tigris**
– Unstriped; males with a mane . . . . **P. leo**

Panthera pardus Leopard

Felis pardus Linnaeus, 1758. Egypt. Incl. *barbarus* Blainville, *bedfordi, chinensis, ciscaucasica* (Caucasus), *fontanierii, grayi, hanensis, japonensis* (N. China), *jarvisi* (Sinai), *leopardus* Sclater (Iran), *orientalis* (Korea), *nimr* (S.W. Arabia), *palearia, panthera* (Algeria), *saxicolor, tulliana* (Asia Minor), *villosa* (Amur), *vulgaris*.
*Panthera pardus dathei** Zukowsky, 1959. Central Iran.
*Panthera pardus transcaucasica** Zukowsky, 1964. Armenia.
Also many other named forms from the Oriental and Ethiopian Regions of which the earliest are *P. p. fusca* (Meyer, 1794) from Bengal and *P. p. leopardus* (Schreber, 1777) from Senegal.

RANGE (Map 91). North Africa, S.W. Asia north to the Caucasus and Kopet Dag, throughout the entire Oriental Region and north through China to Korea and the lower Amur region; throughout Africa south of the Sahara. The leopard occurs in all habitats from forest to semidesert but is now scarce, especially in populous areas.

REMARKS. In view of the great mobility and wide habitat tolerance of this species there are unlikely to be many discrete races. Of the names listed above *P. p. nimr* from S. Arabia seems most likely to represent a recognizable subspecies characterized by small size and very pale pelage.

Panthera tigris Tiger

Felis tigris Linnaeus, 1758. Bengal.
P. t. virgata Illiger, 1815. Caucasus to Lake Balkash; type-locality N. Iran. Incl. *septentrionalis, trabata* (Ili Valley, L. Balkash).
P. t. altaica (Temminck, 1844).
*Felis tigris altaicus** Temminck, 1844. Korea. Range: N. Korea, Manchuria and S.E. Siberia. Incl. *amurensis, coreensis, longipilis, mandshurica, mikado, mikadoi, mongolica*.
P. t. lecoqi (Schwarz, 1916). Lob Nor region, Sinkiang, W. China.
Also other named forms from the Oriental Region of which the mainland ones at least probably constitute a single subspecies, *P. t. tigris*.

RANGE (Map 91). Forest and dense riverine thicket throughout the Oriental Region except Borneo and the Philippines; probably isolated populations in the southern

Caucasus, E. Asia Minor, Elburz Mts, N. Afghanistan, Tien Shan (?E. to Lob Nor) and the Amur–Ussuri region. It formerly occurred throughout the Amur basin, as far west as Lake Baikal, and with occasional strays as far north as Yakutia, but it is now limited to a small area on the lower Amur, Ussuri and extreme N. Korea.

REMARKS. A detailed account of the Siberian subspecies, including maps of former and present distribution, was given by Mazak (1967). Leyhausen (1973) has included the tiger in the genus *Neofelis*, i.e. along with the clouded leopard *N. nebulosa* of the Oriental Region. If there is taxonomic justification for this then the generic name *Tigris*, probably dated from Gray, 1843, would have priority over *Neofelis* Gray, 1867.

Panthera leo Lion

Felis leo Linnaeus, 1758. Algeria.
P. l. leo. Formerly N.W. Africa. =*nobilis*. Incl. *africanus, barbaricus, barbarus, nigra*.
P. l. persica Meyer, 1826. Gir Forest, formerly to Iran and Asia Minor. Incl. *asiaticus, bengalensis* Bennett, *goojratensis, indicus*.
Also many synonyms from south of the Sahara which probably refer to *F. l. leo*. If the lions from south of the Sahara represent a distinct race the earliest name is *P. l. senegalensis* (Meyer, 1826).

RANGE (Map 91). Extinct in the Palaearctic Region. Formerly in all non-desert parts of North Africa and S.W. Asia, persisting in Morocco until the 1920's and last seen in Iran in 1942 (Heaney, 1944). Apparently present in Greece until about A.D. 100. Now present in most of the savanna zone of Africa south of the Sahara, and in the Gir Forest, Kathiawar, N.W. India.

Panthera uncia Ounce, Snow leopard

Felis uncia Schreber, 1775. Altai Mts. =*irbis*. Incl. *uncioides* (Nepal).

RANGE (Map 91). Mountains of central Asia from the Altai through the Tien Shan and Pamir ranges to Kashmir, Nepal and parts of Tibet and Chinese Turkestan.

REMARKS. For a detailed review see Hemmer (1972).

Genus *ACINONYX*

Acinonyx Brookes, 1828. Type-species *A. venator* Brooks=*Felis venatica* Smith. Incl. *Cynailurus, Cynaelurus*.

A clearly defined monospecific genus.

Acinonyx jubatus Cheetah

Felis jubata Schreber, 1775. S. Africa. Incl. ?*guttata* (?Egypt), *raddei* (Transcaspia), *venatica* (India), *venator*.

Also many synonyms from south of the Sahara.

RANGE (Map 91). Steppe and savanna zones from Baluchistan through Iran and Turkestan to N.E. Arabia. Formerly in most dry areas of India and in Egypt and Morocco where it may survive; widespread south of the Sahara.

REMARKS. *A. j. venatica* is usually listed as a subspecies extending from Asia to N. Africa, but it is very doubtful if it is distinct from the southern African form.

Order PINNIPEDIA

Seals etc.

The pinnipedes are a very clearly defined and stable group but their relationship with the Carnivora is controversial. They are frequently considered a suborder of Carnivora, the remaining carnivores then comprising the suborder Fissipedia. The three families are all represented in Palaearctic waters. Recent comprehensive works on the group are by Scheffer (1958) and King (1964). Scheffer gave a synoptic key to all genera but used a highly split classification, his definition of a genus being close to that normally employed for species.

1. Hind limbs turned forwards and used for walking; ears with external pinnae (reduced to a small flap in walrus); fore limbs long (over a quarter length of head and body) and almost naked; post-canine teeth with one prominant cusp, any additional cusps rudimentary (2)
– Hind limbs used only for swimming, not capable of being turned forwards; ears without external pinnae; fore limbs short (less than a quarter length of head and body) and densely haired; post-canine teeth with at least 3 distinct cusps . . . . ***PHOCIDAE***
2. With prominent protruding tusks; body scantily haired; adult lacking incisors and with flat post-canine teeth . . . . ***ODOBENIDAE***
– Without protruding tusks; body densely haired; adult with incisors and conical post-canine teeth . . . . ***OTARIIDAE***

Family **OTARIIDAE**

Eared seals, sea-lions

A clearly defined family of about 12 species placed in six or seven genera. Some of the genera are poorly defined with consequent instability of classification. Three genera, all monospecific, occur in the North Pacific, including the Asiatic coasts, the remainder being confined to the southern hemisphere. A fourth genus, *Neophoca*, was erroneously included by E & M-S on the basis of its supposed presence in Japan; this error was exposed by Silvertsen (1954).

1. Pelage soft with dense underfur; usually 6 post-canine teeth in each upper jaw ***CALLORHINUS***
– Pelage harsh, without underfur; usually 5 post-canine teeth in each upper jaw . . (2)
2. Conspicuous gap between 4th and 5th post-canine teeth, about the width of two teeth; larger (males up to 3·5 m, females up to 2·3 m) . . . . ***EUMETOPIAS***
– Post-canine teeth evenly spaced; smaller (males up to 2·4 m, females up to 1·8 m) ***ZALOPHUS***

Genus *CALLORHINUS*

Callorhinus Gray, 1859. Type-species *Phoca ursina* L. =*Callotaria*. Incl. *Arctocephalus* Gill.

A monospecific genus, closely related only to *Arctocephalus*, which comprises about seven species of fur seals in the southern hemisphere.

Callorhinus ursinus Northern fur seal

Phoca ursina Linnaeus, 1758. Bering Island, N.E. Siberia. Incl. *curilensis, krachenninikowii, ?nigra.*

RANGE (Map 92). Breeds in three discrete colonies: on Robben Island, Sakhalin; on both of the Komandorski Islands; and on both of the main Pribilof Islands, Bering Sea. They migrate southwards in coastal waters in winter and spring, the Komandorski and Sakhalin animals moving into Japanese waters south to about 35°N, and the Pribilof animals reaching about 33°N in California, although some mixing of these populations takes place.

Genus *ZALOPHUS*

Zalophus Gill, 1866. Type-species *Otaria gillespii* MacBain = *O. californiana* Lesson.

A monospecific genus closely related to *Otaria*, the southern sea-lion.

Zalophus californianus Californian sea-lion

Otaria californiana Lesson, 1828. California. Incl. *gillespii, japonica* (Japan), *wollebaeki** (Galapagos). *Eumetopias elongatus** Gray, 1873:766. Japan.

RANGE (Map 92). Formerly on the southern coasts of Honshu, Japan and Take Shima (sea of Japan) where it is now extinct; surviving in California and Mexico; Galapagos. A closely inshore species, rarely going beyond 15 km.

REMARKS. The Japanese population was never well known, but does not appear to have differed from the Californian animals. The Galapagos form is possibly subspecifically distinct.

Genus *EUMETOPIAS*

Eumetopias Gill, 1866. Type-species *Arctocephalus monteriensis* Gray = *Phoca jubata* Schreber.

A monospecific genus doubtfully distinct from *Otaria* Peron, 1816.

Eumetopias jubatus Steller's sea-lion

Phoca jubata Schreber, 1776. Kamchatka. Incl. *leonina* Pallas, *stellerii.*

RANGE (Map 92). From Hokkaido and the Sea of Okhotsk to the Komandorski, Aleutian, Pribilof and St Lawrence Islands, and from the south coast of Alaska to California. An abundant species, breeding throughout the range but with the greatest concentration on the Aleutian Islands.

Family **ODOBENIDAE**

A very distinctive family containing a single genus and species, although it has recently been treated as a subfamily within the Otariidae (Mitchell, 1968).

Genus *ODOBENUS*

Odobenus Brisson, 1762. Type-species *Odobenus* Brisson = *Phoca rosmarus* L. Incl. *Rosmarus*, *Trichechus* Linnaeus, 1766, not of Linnaeus, 1758. *Odobenus* Brisson, 1762 was validated by Opinion 467 of the International Commission (1957) and *Rosmarus* Brunnich, 1771 placed on the Official Index of Rejected Names.

A monospecific genus which is the sole representative of the family Odobenidae.

Odobenus rosmarus Walrus

Phoca rosmarus Linnaeus, 1758. N. Atlantic.
O. r. rosmarus. N. Atlantic. Incl. *arcticus.*
O. r. divergens Illiger, 1815. N. Pacific and E. Siberia, type-locality in Alaska. Incl. *cookii*, *obesus*, *orientalis.*
*Odobenus rosmarus laptevi** Chapski, 1940. Laptev Sea, N.E. Siberia.

RANGE (Map 92). Arctic coasts and sea-ice from Spitzbergen to the north of the Yenesei and from the Laptev Sea east to Alaska and the northern parts of the Bering Sea; also in N.E. Canada and W. Greenland. Occasional vagrants south to Iceland, Scotland, Norway, Hokkaido and Kodiak Is.

Family **PHOCIDAE**

Seals

A large, clearly defined family represented on all temperate and polar coasts. Both Scheffer (1958) and King (1964) recognized 18 species in 13 genera, but some of these genera can scarcely be justified and the classification at the generic level is very unstable. Six genera are here recognized from Palaearctic waters. *Pusa*, *Histriophoca* and *Pagophilus* were all included in *Phoca* by E & M-S and more recently by Burns & Fay (1970). All authors are agreed that *Pusa* and *Phoca* are very closely related; they are here treated as congeneric, but *Histriophoca* and *Pagophilus* are more distinctive and are here retained as genera.

1. Three upper incisors on each side; digits of hind flippers subequal . . . . (2)
– Two upper incisors on each side; 1st and 5th digits of hind flippers clearly longer than others (6)
2. Mystacial vibrissae smooth, thick and straight; 3rd digit of fore flipper longest; 2 pairs of mammae . . . . ***ERIGNATHUS*** (p. 190)
– Mystacial vibrissae nodular, slender and curled; 3rd digit of fore flipper shorter than the adjacent ones; 1 pair of mammae . . . . (3)
3. Snout long, distance from nose to eye about twice that from eye to ear; profile of forehead straight or convex. . . . . ***HALICHOERUS*** (p. 190)

– Snout shorter, distance from nose to eye considerably less than twice that from eye to ear; profile of forehead concave . . . . (4)
4. Adult pelage usually spotted (rarely in Baikal seal); bone of nasal septum stops far short of posterior margin of palate which is deeply incised. . . . . ***PHOCA*** (p. 188)
– Adult pelage usually unspotted but with dark bands; nasal septum reaches margin of palate which is almost straight and lacks a median incision . . . . (5)
5. Fore flipper and adjacent part of body dark; condylobasal length in adults over 200 mm; upper tooth-rows curved in both vertical and horizontal planes ***HISTRIOPHOCA*** (p. 189)
– Fore flipper and adjacent part of body pale; condylobasal length under 200 mm; upper tooth-rows straight. . . . . ***PAGOPHILUS*** (p. 189)
6. One lower incisor on each side; males with an inflatable sac on top of the snout; anterior nares of skull vertical; premaxillae not reaching nasals (Arctic) . ***CYSTOPHORA*** (p. 191)
– Two lower incisors on each side; males without a nasal sac; anterior nares horizontal; premaxillae making contact with nasals (Mediterranean) . ***MONACHUS*** (p. 190)

Genus *PHOCA*

Phoca Linnaeus, 1758. Type-species *P. vitulina* L. =*Callocephalus*. Incl. *Halicyon*, *Haliphilus*, *Pagomys*, *Pusa*.

As used here *Phoca* includes five species which all occur in the Palaearctic Region. Three of these are commonly separated as genus *Pusa*, whilst the monospecific genera *Histriophoca* and *Pagophilus* have often been included in *Phoca*. However they are considerably more different from *P. vitulina* than are the species of *Pusa* and are here excluded.

1. Nasal bones wide, combined anterior width 25–30 % of length; interorbital width over 7 mm; infraorbital foramen small, diameter less than three-quarters that of alveolus of canine (2)
– Nasal bones narrow, combined anterior width about 20 % of length; interorbital width under 7 mm; infraorbital foramen about equal in diameter to canine or larger . . (3)
2. Pelage pale with a grey saddle and dark spots; young born on ice in white pelage (N. Pacific) ***P. largha***
– Pelage (in N. Pacific) usually very dark with pale spots and rings; white coat shed before or at birth . . . . ***P. vitulina***
3. Zygomatic arches narrow, invisible when skull is viewed from behind; usually spotted (4)
– Zygomatic arches projecting, visible when skull is viewed from behind; usually unspotted (Lake Baikal) . . . . ***P. sibirica***
4. Infraorbital foramen equal in diameter to canine; adults with ring-shaped spots (Arctic) ***P. hispida***
– Infraorbital foramen larger than canine; adults usually with large solid spots (Caspian Sea) ***P. caspica***

Phoca vitulina Common seal

Phoca vitulina Linnaeus, 1758. Gulf of Bothnia.
P. v. vitulina. Atlantic coasts. =*variegata*. Incl. *canina*, *linnaei*, *littorea*, *scopulicola*, *thienemanni* (Iceland).
P. v. stejnegeri Allen, 1902. Pacific coast.
*Phoca ochotensis kurilensis** Inukai, 1942. S. Kurile Islands.
*Phoca insularis** Belkin, 1964. Cape Dokuchaeva, Iturup Island, Kuriles.

RANGE (Map 93). Iceland; European coasts from N. Portugal to Arctic Norway including Baltic but not Gulfs of Bothnia and Finland; N. Pacific coasts from Shantung and S. Japan to Bering Straits; Alaska (including N. coast) and Aleutian Islands south to California; W. Atlantic from New York to Ellesmere Island and Hudson Bay (including the freshwater Seal Lakes east of Hudson Bay).

REMARKS. Scheffer (1958) retained five subspecies while admitting that the only morphological distinction possible was between Atlantic and Pacific populations (and that rather tenuous). McLaren (1966), however, has argued that *largha* is specifically distinct (W. Pacific) whilst retaining the eastern Pacific form *richardi* in *P. vitulina*.

***Phoca largha*†** Larga seal

Phoca largha Pallas, 1811. E. Kamchatka. Incl. *chorisii, macrodens, nummularis* (part), *pallasi, petersi*.
*Phoca tigrina** Lesson, 1827. Kamchatka.

RANGE (Map 93). N.W. Pacific coasts from N. China to the Bering Straits, breeding on ice. Map shows breeding areas.

REMARKS. The taxonomy of northern Pacific *Phoca* has been reviewed by Shaughnessy & Fay (*J. Zool. Lond.* 182:385–419, 1977) who show that this form probably deserves specific rank whilst retaining *stejnegeri* and the eastern Pacific *richardi* in *P. vitulina*. *P. largha* breeds on ice, in dispersed pairs, about two months before *P. vitulina* and the pups are born in a white pelage which is retained for about a month.

Phoca hispida Ringed seal

Phoca hispida Schreber, 1775. Greenland and Labrador.
P. h. hispida. Arctic Ocean. Incl. *anellata, birulai, foetida, pomororum* (Novaya Zemblya), *pygmaea, rochmistrovi*.
P. h. botnica Gmelin, 1788. Baltic and adjacent lakes. Incl. *ladogensis* (L. Ladoga), *octonata, saimensis* (L. Saima, Finland), *undulata*.
P. h. ochotensis Pallas, 1811. Sea of Okhotsk, E. Siberia. Incl. *gichigensis, nummularis*† (Japan).
P. h. krascheninikovi Naumov & Smirnov, 1935. Bering Sea.

RANGE (Map 93). Whole of Arctic Ocean, as far north as there is open water, breeding on ice; the Bering and Okhotsk Seas south to Hokkaido; Baltic Sea including the Gulfs of Bothnia and Finland; Lake Saima and adjacent lakes (Finland) and Lake Ladoga (Russia); vagrants south to Iceland and British Isles. There is unlikely to be contact between the Lake Saima population and the Lake Ladoga one, but there may be some interchange between the latter and the Baltic.

REMARKS. *Phoca nummularis* Temminck, 1834, listed by E & M-S as a synonym of *P. vitulina*, was considered a synonym of *P. hispida ochotensis* by King (1961).

Phoca caspica Caspian seal

Phoca vitulina var. *caspica* Gmelin, 1788. Caspian Sea.

RANGE (Map 93). Caspian Sea, migrating between the ice-covered northern end in winter when they breed to the deeper southern end in summer.

Phoca sibirica Baikal seal

Phoca vitulina var. *sibirica* Gmelin, 1788. Lake Baikal. Incl. *baicalensis, oronensis*.

RANGE (Map 93). Lake Baikal.

Genus *HISTRIOPHOCA*

Histriophoca Gill, 1873. Type-species *Phoca fasciata* Zimmermann.

A monospecific genus frequently included in *Phoca*. It is very close to *Pagophilus*.

Histriophoca fasciata Ribbon seal

Phoca fasciata Zimmermann, 1783. Kurile Islands. Incl. *equestris*.

RANGE (Map 94). Okhotsk and Bering Sea from Sakhalin to Cape Navarin; occasional in Alaska and the Kurile Islands. This is a rare species throughout its range.

Genus *PAGOPHILUS*

Pagophilus Gray, 1844. Type-species *Phoca groenlandica* Erxleben. =*Pagophoca*.

A monospecific genus closely related to *Histriophoca* and often included with it in *Phoca*.

Pagophilus groenlandicus Harp seal

Phoca groenlandica Erxleben, 1777. Greenland and Newfoundland. Incl. ?*albicauda, dorsata,* ?*leuscopla, oceanica* (White Sea), *semilunaris*.

RANGE (Map 94). N. Atlantic and adjacent parts of Arctic Ocean from around Baffin Island to Severnaya Zemblya and south to Newfoundland. They migrate south in spring when they breed on drifting pack ice. Breeding is concentrated in three areas: off Newfoundland and the Gulf of St Lawrence; N. of Jan Mayen; and in the White Sea. Stragglers occur as far as the Mackenzie River, Virginia, Scotland and the Baltic.

Genus *HALICHOERUS*

Halichoerus Nilsson, 1820. Type-species *H. griseus* Nilsson = *Phoca grypus* Fabricius.

A monospecific genus, fairly well differentiated from *Phoca* which it most closely resembles.

Halichoerus grypus — Grey seal

Phoca grypus Fabricius, 1791. Greenland. Incl. *atlantica, baltica, griseus, halichoerus* (Norway), *macrorhynchus* (Baltic), *pachyrhynchus*.

RANGE (Map 92). In three main segments: the eastern part of the Baltic and the gulfs of Bothnia and Finland; around the British Isles (mainly west and north) and extending to Iceland and along the Norwegian coasts to the White Sea (probably not breeding east of Murmansk); the Gulf of St Lawrence and around Newfoundland. Scheffer (1958) wrote 'It is now rare in Greenland', but King (1964) maintained that there were no reliable records for Greenland.

Genus *ERIGNATHUS*

Erignathus Gill, 1866. Type-species *Phoca barbata* Erxleben.

A rather clearly defined, monospecific genus.

Erignathus barbatus — Bearded seal

Phoca barbata Erxleben, 1777. S. Greenland. Incl. *albigena, lepechenii, leporina, nautica* (Okhotsk Sea, E. Siberia), *parsonii*.

RANGE (Map 94). Throughout the Arctic Ocean at the edge of the ice, south in the W. Atlantic to the Gulf of St Lawrence and in the Pacific to the Sea of Okhotsk. An abundant species, breeding on ice and feeding in shallow waters. Occasional vagrants as far as France.

REMARKS. Although *E. b. nautica* is frequently listed there seems to be no evidence that it represents a discretely definable subspecies.

Genus *MONACHUS*
Monk seals

Monachus Fleming, 1822. Type-species *Phoca monachus* Hermann. = *Pelagios* = *Pelagocyon* = *Rigoon*. Incl. *Heliophoca*.

A very distinctive genus more closely related to the southern genera of Phocidae than to the northern ones. Three species are usually recognized, in the Mediterranean, Caribbean and Hawaii, but they are very similar, and Scheffer (1958) remarked that 'the three forms would be regarded as subspecies if it were not for the evidence of their complete isolation'. Isolation provides no grounds for specific separation, but since the single form dealt with here is the earliest named, its nomenclature is not affected by the dubious status of the other two forms.

Monachus monachus — Mediterranean monk seal

Phoca monachus Hermann, 1779. Mediterranean. Incl. *albiventer, atlantica* (Madeira), *bicolor, hermanii, leucogaster, mediterraneus*.

RANGE (Map 94). The Mediterranean and the southern parts of the Black Sea; the Madeira and Canary Islands and the Atlantic coast of Morocco and Rio de Oro south to Cape Blanco. It is now scarce in the Mediterranean and Black Seas, but still occurs on the Turkish and Bulgarian coasts. The largest remaining breeding colonies are on the coasts of Morocco and Rio de Oro.

Genus *CYSTOPHORA*

Cystophora Nilsson, 1820. Type-species *C. borealis* Nilsson = *Phoca cristata* Erxleben. = *Cystophoca*. Incl. *Stemmatopus*.

A distinctive, monospecific genus more closely related to the southern seals than to the other northern genera.

Cystophora cristata Hooded seal

Phoca cristata Erxleben, 1777. S. Greenland and Newfoundland. = *borealis*. Incl. *isidorei*†, *mitrata*.

RANGE (Map 93). Edge of drifting ice from Spitzbergen to Newfoundland and Baffin Island. The main breeding areas are off S.W. Greenland and near Jan Mayen. Vagrants have been recorded as far south as Florida and the Bay of Biscay.

REMARKS. E & M-S listed *Phoca isidorei* Lesson with a query under *Monachus monachus*. According to Miss J. E. King (in litt.) this name was based on a young specimen of *C. cristata*, the skull being figured by Gervais (1848–52: pl. 42).

Order **PROBOSCIDEA**
Elephants

Family **ELEPHANTIDAE**

Contains two monospecific genera, neither of which now occurs in the Palaearctic Region, although both survived in the region into early historical time.

Ears much smaller than lateral area of head; profile of back convex; molars (in adults) with over 12 transverse laminae . . . . ***Elephas maximus***
Ears larger than lateral area of head; profile of back concave; molars with up to 10 transverse laminae . . . . ***Loxodonta africana***

Genus *LOXODONTA**

Loxodonta Cuvier, 1827. Type-species *Elephas capensis* Cuvier = *E. africanus* Blumenbach. = *Loxodon*.

Probably monospecific, but the small form, *cyclotis*, has been given specific rank.

Loxodonta africana African elephant

Elephas africanus Blumenbach, 1797. S. Africa.
L. a. cyclotis (Matschie, 1900). West Africa (type-locality in Cameroon) and perhaps formerly in N.W. Africa.
Also other forms, probably all consubspecific with the nominate race, in E. and S. Africa.

RANGE. The savanna and forest zones of Africa. Extinct north of the Sahara but believed to have survived in N.W. Africa until at least the first century A.D.

REMARKS. These were the elephants that were domesticated by the Carthaginians and Romans and used, for example, by Hannibal in his European campaigns (Zeuner, 1963).

Genus *ELEPHAS*

Elephas Linnaeus, 1758. Type-species *E. maximus* L.

Monospecific.

Elephas maximus Indian elephant

Elephas maximus Linnaeus, 1758. Sri Lanka. Incl. many synonyms from the Oriental Region.

RANGE. At present from India to Malaya and on the islands of Sri Lanka and Sumatra. This species appears to have been present in Syria until about 800 B.C., but even during the period when they are known from archaeological and fossil evidence to have been in Syria, about 1500–800 B.C., it is likely that this population was quite isolated from the main part of the range in India. Zeuner (1963) gave an account of the archaeological data. An alternative view (Brentjes, 1969) is that these Syrian animals were mammoths, i.e. the totally extinct *Mammuthus primigenius*.

Order **HYRACOIDEA**

Hyraxes, dassies

A very distinctive order containing a single family, confined to Africa and the Arabian region.

Family **PROCAVIIDAE**

Contains two genera, *Procavia* and *Dendrohyrax*, although the latter is frequently split, *Heterohyrax* being considered distinct by some authors. Both genera are represented throughout Africa south of the Sahara, but only *Procavia* occurs north of the Sahara. Records of *Dendrohyrax* in the Saharan mountains, mainly under the name *Heterohyrax*, seem erroneous.

Genus ***PROCAVIA***

Procavia Storr, 1780. Type-species *Cavia capensis* Pallas. =*Hyrax*. Incl. *Euhyrax*.

This genus is represented almost throughout the range of the order. It seems fairly well differentiated from *Dendrohyrax* (including *Heterohyrax*) by the more hypsodont teeth, but the two genera have frequently been confused. No satisfactory classification within the genus *Procavia* exists. The taxonomy is complicated by great variability with age, the permanent dentition being late in developing, and apparently also by colour dimorphism.

E & M-S considered the genus likely to be monospecific (*P. capensis*), whilst others (Hahn, 1934; Bothma, 1971) have recognized up to five species. I would recognize at least five species of which only one occurs north of the Sahara.

Procavia syriaca — Syrian hyrax

Hyrax syriaca Hemprich & Ehrenberg, 1828. Mt Sinai. Incl. *antineae*, *bounhioli* (Hogger Mts), *burtoni* (Egypt), *ehrenbergi*, *jayakari* (S. Arabia), *schmidtzi*, *sinaiticus*.
Probably also other synonyms from further south, including *ruficeps*.

RANGE (Map 95). Rocky hillsides from Lebanon and Arabia through Sinai to Egypt; probably south through Sudan to N. Tanzania and west throughout the savanna zone south of the Sahara; isolates in the Saharan massifs of Hoggar, Tassili (including S.W. Libya), Air and probably Tibesti.

REMARKS. I am inclined to consider conspecific all those forms of *Procavia* occurring from N. Tanzania (*matschiei*) northwards having a predominantly yellow streak covering the dorsal gland, i.e. excluding only *P. habessinica* of the Ethiopian Plateau.

A uniformly dark brown morph appears to be frequent in the Hoggar population.

The name *syriaca* was attributed by E & M-S to Schreber, 1784, but that account is so confused that it seems sensible to follow Roche (1972) and date it from Hemprich & Ehrenberg, 1828.

Order **SIRENIA**

Sea-cows

A clearly defined order with two families, one of which is marginally represented on Palaearctic shores. The other, Trichechidae, includes the manatees of the Atlantic coasts of Africa and the Americas. A general account of the order was given by Mohr (1957).

Family **DUGONGIDAE**

Contains only one living species, but a second, generically distinct, is only recently extinct.

Smaller, total length up to 2·9 m; cheek-teeth present . . . . . . ***Dugong dugon***
Larger, up to 8 m; teeth entirely absent . . . . ***Hydrodamalis gigas***

Genus ***DUGONG***

Dugong Lacepède, 1799. Type-species *D. indicus* Lacepède. =*D. dugon* Müller. =*Dugungus*. Incl. *Dugongidus* =*Halicore* =*Platystomus*.

A monospecific and clearly defined genus.

Dugong dugon Dugong

Trichecus dugon Müller, 1776. Cape of Good Hope to Philippines. =*dugung, dugong*. Incl. *cetacea, hemprichi* (Red Sea), *indicus, lottum, tabernaculi*.
Also other names from tne Oriental and Australian regions.

RANGE (Map 95). The shores of the Red Sea; also along the coast of East Africa south to Natal; in S. India and Sri Lanka; the Indonesian islands, Philippines and N. Australia; in the Riukiu Islands and perhaps occasionally in the southern islands of Japan. The range has decreased considerably during the course of this century. It is a marine species of shallow, coastal waters.

REMARKS. A detailed study of this species in the Red Sea was published by Gohar (1957).

Genus ***HYDRODAMALIS****

Hydrodamalis Retzius, 1794. Type-species *H. stelleri* Retzius. =*Nepus*. Incl. =*Rhytina* =*Rytina* =*Stellerus*.

Contains a single species which is now extinct. The genus is clearly distinct from *Dugong*. It is frequently called *Rhytina* Illiger, 1811, but *Hydrodamalis* has clear priority and was used by Allen (1942), by Simpson (1945) and by Heptner & Naumov (1967).

Hydrodamalis gigas* Steller's sea-cow

Manati gigas Zimmermann, 1780:426.
Manati balaenurus Boddaert, 1785:173.
Trichechus manatus, var. *borealis* Gmelin, 1788:61.
Hydrodamalis stelleri Retzius, 1794.
Rytina cetacea Illiger, 1815.

RANGE (Map 95). Extinct. Formerly around the Komandorski Islands (Bering Is. and Copper Is.) in the Bering Sea; latest certain record in 1768.

Order **PERISSODACTYLA**

Odd-toed ungulates

Contains three families of which only one, the Equidae, is represented in the Palaearctic Region. A second, Rhinocerotidae, was represented in North Africa until the Neolithic period.

Family **EQUIDAE**

Horses

A clearly defined family of about seven species in Africa and Eurasia. These are now most often referred to a single genus, *Equus*, but the genus has frequently been split, the asses being placed in the genus *Asinus* and the zebras in *Hippotigris*. This was done, for example, by Groves & Mazak (1967), but the characters involved seem more appropriate to specific or subgeneric, rather than generic, rank. Consequently only one genus is recognized here, as was done by E & M-S, and this classification has the advantage of concordance with that of Ansell (1971a) for the African species.

Genus *EQUUS*

Equus Linnaeus, 1758. Type-species *E. caballus* L., the domestic horse. Added to the Official List of Generic Names by Opinion 271 of the International Commission, 1954. Incl. *Asinus, Hemionus, Microhippus, Onager*.
*Asinohippus** Trumler, 1961. As subgenus of *Hemionus*, type-species *Equus khur* Lesson.

In addition to the species listed below, this genus includes the three species of zebra in Africa. The present status, with respect to conservation, of all the species listed here is described in the Red Data Book (IUCN, 1966).

1. 'Chestnuts' (epidermal callosities) on inner surfaces of fore and hind legs; whole of tail long-haired; auditory meatus of skull short, invisible from above . . . . . ***E. ferus***
– 'Chestnuts' confined to fore legs; tail tufted, long hair confined to distal half; auditory meatus longer, visible from above . . . . (2)
2. Ears very long, over 200 mm in adults (Africa) . . . . . . ***E. africanus***
– Ears shorter, usually under 150 mm . . . . (3)
3. Dorsal pelage dark reddish brown in summer, dark brown in winter; white ventral pelage with a prominent dorsal extension behind the shoulder. . . . . . . ***E. kiang***
– Dorsal pelage pale yellowish brown; no white patch behind the shoulder ***E. hemionus***

Equus ferus — Wild horse, Tarpan

*Equus ferus** Boddaert, 1785. Russia.
E. f. ferus. Formerly the steppes of European Russia, now extinct.
*Equus gmelini** Antonius, 1912. Sagrodovskaya Steppe, right bank of Dnepr, near Kharson, Ukraine.
*E. f. silvestris** von Brincken, 1828. Bialowieza, Poland.
*Equus gmelini silvatica** Vetulani, 1927. Bialowieza, Poland.
E. f. przewalskii Poliakov, 1881. Mongolia and Sinkiang. Incl. *hagenbecki*.

RANGE (Map 95). Now restricted to a small area on the border of S.W. Mongolia and Sinkiang; original range probably covered the whole steppe zone from Mongolia through N. Turkestan and S. European Russia to E. Poland and Hungary, but avoiding the drier steppes and deserts of S. Turkestan occupied by *E. hemionus*.

REMARKS. This is the ancestor of the domestic horse. There is some doubt as to whether the 'wild' horses that remained in Poland until about 1812 and in Ukraine until about 1850 were really unaffected by domestic stock, but it seems likely that they were basically genuine remnants of the indigenous species. There seem no grounds for separating these subspecifically. It is likely that *E. f. ferus*, the tarpan of southern Russia, was the principal ancestor of domestic breeds. For remarks on the nomenclature of domesticated species see p. 8, and for a discussion of the relationship of wild and domesticated horses see Groves (1974) on which the above synonymy is based.

Equus hemionus — Kulan, Onager

Equus hemionus Pallas, 1775. Transbaikalia.
E. h. hemionus. L. Balkash and Mongolia to Manchuria. =*typicus*.
Incl. *castaneus* (Mongolia), *finschi* (E. Kazakhstan).
E. h. luteus Matschie, 1911. Gobi Desert to Kansu. Incl. *bedfordi*.
E. h. onager Boddaert, 1785. Iran and adjacent parts of Afghanistan and Pakistan. Incl. *hamar*.
*Asinus dzigguetai** Wood, 1879. Iran.
*Microhippus hemionus bahram** Pocock, 1947. Yezd, central Iran.
E. h. khur Lesson, 1827. Rann of Kutch and adjacent parts of India and Pakistan. Incl. *indicus*.
*Microhippus hemionus blanfordi** Pocock, 1947. Sham Plains, Baluchistan.
E. h. hemippus Geoffroy, 1855. Formerly Syria and Iraq, probably extinct. Incl. *syriacus*.
*E. h. kulan** (Groves & Mazak, 1967). S. Turkestan and N.W. Afghanistan.
Asinus hemionus kulan Groves & Mazak, 1967. Badkhyz Reserve, Turkmenia, USSR (*c.* 35°50′N, 61°40′E).

RANGE (Map 95). Desert and dry steppe zones from W. Manchuria and Kansu through Mongolia, Sinkiang and S. Turkestan to Baluchistan and the Rann of Kutch, and through Iran to (formerly) Iraq and Syria.

REMARKS. The above arrangement of subspecies follows Groves & Mazak (1967) who gave diagnoses. It is probable that the range, and the variation, were at one time virtually continuous, but in view of the fragmentation of the range by local extinction it may be that some of the subspecies are now isolated and discrete. For a further account of subspecies, with maps, see Tour (1975).

***Equus kiang*†** Kiang

Equus kiang Moorcroft, 1841. Ladak, Kashmir.
E. k. kiang. Ladak and adjacent parts of Tibet. Incl. *equioides*, *kyang*.
E. k. holdereri Matschie, 1911. E. Tibet and W. Szechuan. Incl. *tafeli*.
E. k. polyodon (Hodgson, 1847). Border of Tibet and Sikkim.
*Hemionus kiang nepalensis** Trumler, 1959. 'Nepal', according to Groves & Mazak (1967) 'more probably the region of Tibet just north of the Sikkhim border'.

RANGE (Map 95). Plateau steppes of Tibet and Chinghai, and adjacent parts of Szechuan, Sikkim and Kashmir.

REMARKS. The above arrangement follows Groves & Mazak (1967). E & M-S treated this form as a subspecies of *E. hemionus*.

Equus africanus* African ass

Asinus africanus Fitzinger, 1857. Nubia (Sudan).
E. a. atlanticus P. Thomas, 1884. Extinct; formerly Algeria and Morocco.
E. a. africanus. Probably extinct; formerly S. Egypt and N. Sudan. Incl. ?*taeniopus* (Red Sea coast), *dianae*.
*Equus asinus hippagrus** Schomber, 1963. Sahara.
E. a. somaliensis Noack, 1884. Somalia and E. Ethiopia. Incl. *aethiopicus*, *somalicus*.

RANGE. Now probably confined to Danakil region of Ethiopia and to Somalia. Formerly in E. Sudan where it has probably become extinct during the present century; and in N.W. Africa where it probably became extinct during Roman times.

REMARKS. This is the ancestor of the domestic ass which is usually called *Equus asinus* L. It is probable that the domestic form was developed from *E. a. africanus* in Upper Egypt in predynastic times. It is doubtful whether wild asses now found in Sudan and in mountains of the eastern Sahara are of indigenous wild stock or, as is more likely, are feral animals of domestic stock. The origin of the domestic ass was discussed by Zeuner (1963). For a detailed account of this species see Groves *et al.* (1966) on which the above synonymy is based.

Order **ARTIODACTYLA**

Even-toed ungulates

The three suborders of this order are all represented in the Palaearctic Region. The most recent comprehensive classification of the order, by Haltenorth (1963), does not differ at the family level, and differs only slightly at lower levels, from that of Simpson (1945) which was followed by E & M-S.

1. Upper incisors present . . . . (2)
– Upper incisors absent (suborder Ruminantia) . . . . (4)
2. Teeth hypsodont; 2 digits on each foot; metapodials fused to form a cannon-bone except at distal end where they diverge: (suborder Tylopoda) . . ***CAMELIDAE*** (p. 197)
– Teeth bunodont; 4 digits on each foot; metapodials free; (suborder Suiformes) (3)
3. Muzzle long, terminating in a flat disk around nostrils; feet small, completely hooved; lateral toes not reaching ground . . . . ***SUIDAE*** (p. 196)
– Muzzle short and broad; feet large, with nails; lateral toes touching ground ***HIPPOPOTAMIDAE*** (p. 196)

4. Without antlers or horns; without facial glands and pedal glands . . . *MOSCHIDAE* (p. 198)
– With antlers or horns at least in male (except in *Hydropotes*); usually with facial and pedal glands . . . . (5)
5. With deciduous antlers in male (rarely also in female) . . . *CERVIDAE* (p. 198)
– With non-deciduous horns, frequently in both sexes . . . . *BOVIDAE* (p. 204)

Suborder **SUIFORMES**

Family **SUIDAE**

Pigs

A clearly defined family of five genera distributed throughout the Ethiopian and Oriental Regions, with one species extending from the Oriental Region into the Palaearctic.

Genus ***SUS***

Sus Linnaeus, 1758. Type-species *S. scrofa* L. Incl. *Centuriosus*, *Scrofa*, *Sinisus*.

A genus of about four species confined to Eurasia, only one of which reaches the Palaearctic Region.

Sus scrofa Wild boar

Sus scrofa Linnaeus, 1758. Germany. =*setosus*, *europaeus*, *scropha*. Incl. *algira*, *aper*, *attila* (Rumania), *baeticus*, *barbarus* (N. Africa), *canescens*, *castilianus* (Spain), *celtica*, *continentalis*, *coreanus*, *falzfeini*, *ferus*, *gigas*, *leucomystax* (Japan), *japonica*, *libycus* (Asia Minor), *majori*, *mandchuricus*, *meridionalis* (Sardinia), *nigripes* (Chinese Tien Shan), *raddeanus* (Mongolia), *reiseri*, *sahariensis* (Algeria), *sardous*, *songaricus*, *ussuricus* (Ussuri R., E. Siberia).
*Sus leucomystax sibiricus** Staffe, 1922. Tunkinsk Mts, Sayan Mts, Siberia.
Also other forms in the Oriental Region of which the earliest name is *cristatus* Wagner, 1839 from India.

RANGE (Map 96). Formerly the entire steppe and broadleaved woodland zones of the Palaearctic and Oriental Regions, including Britain, Ireland and Japan. Now extinct in the British Isles and Scandinavia and considerably reduced in parts of continental Europe. Isolates are still present on Honshu, Sardinia, Corsica and N.W. Africa. Still present throughout most of the continental area, but range probably more fragmented than formerly.

REMARKS. Many races have been named on the basis of difference in size and colour, but these have never been shown to have discrete boundaries. This is the ancestor of the domestic pig in Europe and S.W. Asia; the domesticated pig of China is likely to have arisen from *S. s. cristatus* or one of the other races in S.E. Asia.

Family **HIPPOPOTAMIDAE**

Contains two monospecific genera in Africa, one of which was formerly represented in Palaearctic Africa.

Genus ***HIPPOPOTAMUS***

Hippopotamus Linnaeus, 1758. Type-species *H. amphibius* L.

A monospecific genus clearly separable from the only other recent genus, *Choeropsis*.

Hippopotamus amphibius* Hippopotamus

Hippopotamus amphibius Linnaeus, 1758. Nile.

RANGE. River systems almost throughout Africa south of the Sahara; formerly throughout the Nile, but now extinct in Egypt and the northern Sudan. According to Flower (1932) it became very rare in the Nile Delta during the 18th century, perhaps the last one being killed in 1815, whilst in upper Egypt the last record appears to be from Aswan about 1816. See also Kock (1970a).

Suborder TYLOPODA

Contains a single family.

Family CAMELIDAE

Camels

In addition to the Palaearctic camels, this family includes two genera, *Lama* and *Vicugna*, in South America.

Genus *CAMELUS*

Camelus Linnaeus, 1758. Type-species *C. bactrianus* L.

A clearly defined genus of two species, both Palaearctic.

A single dorsal hump; pelage short . . . . *C. dromedarius*
Two dorsal humps; pelage long and shaggy in winter . . . . *C. ferus*

Camelus ferus Bactrian camel

Camelus bactrianus ferus Przewalski, 1883. Sinkiang, China.

RANGE (Map 96). As a wild animal now confined to the western Gobi Desert, probably persisting in two areas near the Lop Nor and in S. Mongolia. Formerly throughout the dry steppe and semi-desert zone from Russian Turkestan to the Gobi Desert. See Heptner & Naumov (1961) for a detailed account.

REMARKS. The use of the name *C. ferus*, the first name given to wild Bactrian camels, follows the general principles adopted for the naming of domesticated animals and their wild ancestors (p. 8), and it also agrees with the most recent Russian work (Heptner & Naumov, 1961). The earlier name *C. bactrianus* L., 1758 referred to the domesticated form which is used extensively throughout central Asia. It is best referred to as '*Camelus ferus* (domestic)' or '*Camelus* (Bactrian)'.

'_Camelus dromedarius_'* One-humped camel or dromedary

Camelus dromedarius Linnaeus, 1758:65. 'Africa'.

RANGE. Extinct in the wild. May have persisted as a wild animal in Arabia until about the beginning of the Christian period, but since domestication had taken place by 1800 B.C. or earlier and remains of wild and domesticated animals cannot easily be distinguished, the history of the species is very uncertain. It almost certainly occurred throughout the Arabian region.

In N.W. Africa there is no good evidence of its occurrence in the wild. A Pleistocene form from Algeria, *C. thomasi* Pomel, has been considered to be more closely related to the bactrian camel (Gautier, 1966), but it is very doubtful if the two species can be separated on skeletal characters.

REMARKS. This species raises a nomenclatorial difficulty in being the only domesticated species completely unknown in the wild even as fossil material. Bohlken (1961) assumed conspecificity with *C. ferus* and treated *dromedarius* as a synonym of *bactrianus*. The two domesticated forms can be interbred, but the male hybrids have been reported as sterile (Gray, 1954), and therefore there seem to be good reasons for treating them as distinct species. Therefore in the absence of any wild progenitor, the Linnean name *C. dromedarius* seems quite explicit and unambiguous for the domesticated dromedary.

For the history of its domestication see Zeuner (1963).

Suborder RUMINANTIA

Ruminants

Besides the three families represented in the Palaearctic Region this suborder includes the Tragulidae, Giraffidae and Antilocapridae. It is possible that giraffes (*Giraffa camelopardalis)* survived in N.W. Africa north of the Sahara into historical times.

Family **MOSCHIDAE**

Contains only the genus *Moschus* which has more often been placed in a subfamily of the Cervidae. However, in the most recent and comprehensive revision of these groups Flerov (1952) gave the musk deer family rank, arguing that they are closer to the Tragulidae than to the Cervidae, and this arrangement is tentatively accepted here.

Genus ***MOSCHUS***

Moschus Linnaeus, 1758. Type-species *M. moschiferus* L. =*Odontodorcus*.

A very distinctive genus confined to Eastern Asia. The number of species is controversial. E & M-S recognized only one, but Flerov (1952) recognized three, and his view that the forms *sifanicus* and *berezovskii* are specifically distinct in Szechuan is supported by a study by Kao (1963) based on substantial samples. Only one species occurs in the Palaearctic Region, whichever classification is followed.

Moschus moschiferus Musk deer

Moschus moschiferus Linnaeus. 1758. 'Tataria versus Chinam'. This has generally been supposed to be in the region of Mongolia, and Heptner & Naumov (1961) limited it to the Altai. There seem to be no grounds for following Flerov (1952) who took the type-locality as Northern India.
M. m. moschiferus. Mainland. Incl. *altaicus*, *arcticus*, *parvipes* (Korea), *sibiricus*, *turowi* (Amur).
M. m. sachalinensis Flerov, 1929. Sakhalin Island.
Perhaps also other synonyms from S. China and the Himalayas if the forms in these regions are conspecific.

RANGE (Map 96). Forested areas of E. Siberia east of Altai and Yenesei; N. Mongolia, Manchuria and Korea. Forms in S. China and the Himalayas are doubtfully conspecific.

REMARKS. This species was called *M. sibiricus* by Flerov (1952) who treated *M. moschiferus* as a distinct species in the Himalayas. Apart from this change in nomenclature the above arrangement follows Flerov.

Family **CERVIDAE**

Deer

A large family with about 50 species in all the main zoogeographical regions except the Ethiopian and Australasian. The most recent comprehensive classification (Flerov, 1952) listed 11 genera, but there is a little instability at the generic level, especially in the degree to which *Cervus* is subdivided. Seven genera are represented in the Palaearctic and others have been introduced.

1. Posterior end of vomer extending back as far as basioccipital, completely dividing posterior opening of nares, not incised on its posterior edge; main beams of antlers bending sharply forwards in upper half (subfamily Neocervinae) . . . . (2)
– Vomer short, not dividing posterior opening of nares, posterior margin of vomer deeply incised; main beams of antlers not directed forwards . . . . (3)
2. Antlers in both sexes, 1st tines branched; hooves very large and splayed ***RANGIFER*** (p. 203)
– Antlers only in male, 1st tines unbranched; hooves of normal size ***ODOCOILEUS*** (p. 202)
3. Males with tusk-like upper canines; (antlers absent or very short on long pedicels) (4)
– Males without tusk-like canines; antlers present and pedicels short . . . . . (5)
4. Males with short antlers on long pedicels; frontal bones with lateral crests ***MUNTIACUS*** (p. 199)
– Males without antlers; frontal bones without lateral crests ***HYDROPOTES*** (p. 203)
5. Tail very long, hairs reaching heel; all tines of antlers pointing backwards; lateral toes as long as central ones . . . . ***ELAPHURUS*** (p. 201)
– Tail shorter, not reaching heel; antlers with some tines directed forwards; lateral toes shorter than central ones . . . . (6)
6. Muzzle large and inflated, upper lip wide and overhanging mouth; rhinarium almost completely haired; (antlers usually very flattened and palmate) . ***ALCES*** (p. 202)
– Muzzle normal; rhinarium naked . . . . (7)

7. Coronet of antlers (immediately above pedicel) very rugose; tail not visible in living animal **CAPREOLUS** (p. 203)
– Coronet of antlers not rugose; tail clearly visible . . . . . **CERVUS** (p. 199)

Genus *MUNTIACUS*

Muntiacus Rafinesque, 1815. Type-species *Cervus muntjak* Zimmermann. (Validated by Opinion 460, 1957.) =*Cervulus, Muntjaccus, Muntjacus, Stylocerus.* Incl. *Prox, Procops.*

A clearly defined genus, related only to *Elaphodus*, of about five species. These are entirely Oriental, but one species is feral in Europe.

Muntiacus reevesi Reeve's muntjac

Cervus reevesi Ogilby, 1839. Canton, S. China. There are several synonyms from its native range.

RANGE (Map 96). S. China and Taiwan. Introduced to England, where it is spreading widely in the lowlands, and in France.

Genus *CERVUS*

Cervus Linnaeus, 1758. Type-species *C. elaphus* L. =*Elaphus, Strongyloceros, Eucervus* Acloque. Incl. *Axis, Dactylocerus, Dama, Elaphoceros, Harana, Hippelaphus, Machlis, Palmatus, Platyceros, Przewalskium, Pseudaxis, Pseudocervus, Sica, Sika, Sikaillus.*

As used here, a genus of about 15 species in the Palaearctic and Oriental Regions, with one species extending also into North America. The genus is frequently split. E & M-S recognized *Dama* and *Axis* as distinct genera, but Flerov (1952) included these in *Cervus*. Four species are Palaearctic and a further one is included on the basis of introductions.

1. Pelage of adults completely unspotted[1]; shoulder height of adult usually over 1·1 m (2)
– Pelage of adults spotted, at least in summer; shoulder height of adult under 1·1 m . (3)
2. Muzzle entirely brown; bez tine present, i.e. a 2nd, forwardly projecting tine immediately above the 1st, proximal tine of the antler . . . . **C. elaphus**
– End of muzzle white; no bez tine (E. Tibet and adjacent regions) . **C. albirostris**
3. Antlers flattened, especially distally but sometimes only proximally; (tail with a dark dorsal stripe which also tends to divide the brown and white on the buttocks) . **C. dama**
– Antlers not flattened. . . . . (4)
4. Heavily spotted at all seasons, including neck; dorsal surface of tail completely black **C. axis**
– Spotting of pelage much reduced in winter and spots never present on neck; dorsal surface of tail usually predominantly white . . . . **C. nippon**

Cervus dama Fallow deer

Cervus dama Linnaeus, 1758. Sweden (introduced).
C. d. dama. Europe. =*platyceros, plinii.* Incl. *albus, leucaethiops, maura, mauricus, niger, varius, vulgaris.*
C. d. mesopotamica† Brooke, 1875. Luristan, Iran. =*mesopotamiae.*

RANGE (Map 97). Formerly the entire Mediterranean region including probably N.W. Africa and lower Egypt, and east to Iran. Now extinct in Africa and Asia except for a few survivors of *C. d. mesopotamica* in W. Iran. Widely introduced and well established in most of Europe including the British Isles and southern Sweden. Also introduced in New Zealand and in N. and S. America.

REMARKS. The generic name *Dama* of Frisch, 1775, frequently used for this species, was validated by Opinion 581 (1960). E & M-S treated *mesopotamica* as specifically distinct, but most recent authors give it subspecific status, e.g. Harrison (1968a), Heptner & Naumov (1961) and Haltenorth (1959), the last of whom gave a very detailed account of this subspecies.

Cervus axis Chital, Spotted deer

Cervus axis Erxleben, 1777. Ganges, India. Several synonyms from native range.

[1] Except *C. elaphus* in N. Africa.

RANGE. India and Sri Lanka; an introduced colony in Istria, N. Yugoslavia.

REMARKS. See Niethammer (1963) for details of the introduction.

Cervus nippon Sika deer

Cervus nippon Temminck, 1838. Japan.

C. n. nippon. Honshu, Shikoku and Kyushu, Japan and some adjacent small islands. Incl. *aceros, aplodontus, blakistoninus, brachypus, consobrinus, dainius, dejardinus, dolichorhinus, elegans, ellipticus, fuscus, hollandianus, infelix, japonicus, keramae* (Kerama Is., Ryukyu Is.), *latidens, legrandianus, marmandianus, minoensis, minor* Brooke, *mitratus, orthopodicus, orthopus, paschalis, regulus, rex, schizodonticus, schlegeli, sendaiensis, sica, sicarius, sika, typicus, sylvanus, xendaiensis.*

*Cervus centralis** Kishida, 1936:275. Yatsugatake, Nagano Pref., Honshu, Japan.

C. n. yesoensis Heude, 1884. Hokkaido. Incl. ?*matsumotei, rutilus.*

*C. n. yakushimae** Kuroda & Okada, 1950. Islands of Yakushima, Mageshima and Akuneoshima, Kyushu.

Cervus nippon var. *yakushimae* Kuroda & Okada, 1950:60. Yakushima Is., S. Japan.

Cervus nippon var. *mageshimae** Kuroda & Okada, 1950:62. Mageshima Is., Kyushu, Japan.

C. n. hortulorum Swinhoe, 1864. S.E. Siberia, Korea, Manchuria, China. Incl. *andreanus, arietinus, brachyrhinus, cycloceros, cyclorhinus, dugenneanus, dybowskii, euopis, frinianus, gracilis, granulosus, grassianus* (N. Shensi), *grilloanus, hyemalis, ignotus, imperialis, joretianus, kopschi* (Kiangsi, S. China), *lachrymosus, mandarinus* (Hopei), *mantchuricus, microdontus, microspilus, oxycephalus, pouvrelianus,* ?*pseudaxis* Gray, *riverianus, surdescens, yuanus.*

*Cervus nippon swinhoei** Glover, 1956. No locality; based on captive animal described by Swinhoe (1865). Preoccupied by *Cervus swinhoii* Sclater, 1862, a form of *C. unicolor* from Taiwan.

C. n. taiouanus Blyth, 1860. Taiwan. =*taëvanus* =*tai-oranus* =*taivanus*. Incl. *devilleanus, dominicanus, morrisianus, novioninus, schulzianus.*

*C. n. pulchellus** Imaizumi, 1970. Tsushima Is., Japan.

Cervus pulchellus Imaizumi, 1970b. Are, Izuhara, Tsushima Is., Japan.

Cervus nippon incertae sedis: *kematoceros, minutus, modestus.*

RANGE (Map 97). Japan (all main islands), Ussuri region of Siberia south through Korea and Manchuria, and formerly from Hopei south to N. Vietnam; Taiwan. Now probably extinct in China, apart from Manchuria and in a few deer parks elsewhere. Introduced and well established in Ireland, Britain, Denmark and central Europe (mainly *C. n. nippon*), and in European Russia (*C. n. hortulorum*). *C. n. taiouanus* has been introduced to the island of Oshima near Tokyo, Japan.

REMARKS. This is a clearly defined group which seems best treated as a single species as was done by E & M-S. Imaizumi (1970b) treated it as a series of four allopatric species, namely *C. nippon* (including *yakushimae* as a subspecies), *C. hortulorum* (including *yesoensis* as a subspecies), *C. taiouanus* and *C. pulchellus*. These were based mainly on differences of cranial proportions which do not seem to justify specific rank.

The subspecific variation has never been adequately defined. Several races are frequently recognized from mainland China, but these are very doubtfully distinct and have never been known from more than a very few specimens. The situation is further complicated by the fact that many descriptions have been made from park or captive animals of unknown origin and possibly including hybrids.

Lowe & Gardiner (1975) have described hybrids between this species and *C. elaphus* in England and have suggested that all mainland Asiatic forms might represent similar hybrids.

Cervus albirostris Thorold's deer

Cervus albirostris Przewalski, 1883. Nan Shan, W. Kansu, China. Incl. *dyboskii* Sclater, *sellatus, thoroldi.*

RANGE (Map 97). Very marginally Palaearctic. The eastern part of the Tibetan plateau from the Nan Shan Mts (Chinghai) through extreme W. Szechuan to S.E. Tibet.

REMARKS. A clearly defined and distinctive species. It is the type of the generic name *Przewalskium* Flerov which was still given subgeneric rank by Flerov (1952).

Cervus elaphus Red deer, Wapiti

Cervus elaphus Linnaeus, 1758. S. Sweden.

C. e. elaphus. Europe, except S. Spain, Corsica and Sardinia. =*typicus*, *vulgaris*. Incl. *albicus*, *albifrons*, *albus*, *atlanticus* (W. Norway), *bajovaricus*, *balticus*, *bolivari*, *campestris* Botezat, 1903 (preoccupied by *Cervus campestris* Cuvier, 1817, a synonym of *Odocoileus virginianus*), *debilis*, *germanicus*, *hippelaphus* (Belgium), *montanus* (Carpathians), *neglectus*, *rhenanus*, *saxonicus*, *scoticus* (Scotland), *varius*, *visurgensis*.
*Cervus elaphus brauneri** Charlemagne, 1920. Crimean Mts, Ukraine.
Cervus Elafus (sic) *tauricus** Fortunatov, 1925. Ukraine.
*Cervus elaphus carpathicus** Tatarinov, 1956. Ukrainian Carpathians.
C. e. corsicanus Erxleben, 1777. Corsica, Sardinia, S. Spain and N.W. Africa. =*minor*. Incl. *barbarus* (Tunisia), *corsiniacus*, *hispanicus* (S.W. Spain), *mediterraneus*.
C. e. maral Gray, 1850. Asia Minor and Caucasus to N. Iran (type-locality). Incl. *caspius*, *caucasicus*.
C. e. bactrianus Lydekker, 1900. Riverine woodland of Turkestan and N. Afghanistan, type-locality Tashkent. Incl. *hagenbecki*.
C. e. affinis Hodgson, 1841. Himalayas from Kashmir to Yunnan and through W. China to Kansu; type-locality Saul Forest, Nepal. Incl. *cashmeerianus*, *cashmerensis*, *cashmirianus*, *casperianus*, *hanglu*, *kansuensis* (Kansu), *macneilli* (W. Szechuan), *wardi*.
C. e. wallichi Cuvier, 1823. S.W. Tibet, type-locality near Lake Mansarowar. Incl. *nariyanus*, *tibetanus*.
C. e. yarkandensis Blanford, 1892. Tarim Basin, Sinkiang; type-locality Maralbashi Forest.
C. e. alashanicus Bobrinski & Flerov, 1935. Alashan Range, S.E. Mongolia.
C. e. xanthopygus Milne-Edwards, 1867. Amur Basin, Sikhote Alin Range of S.E. Siberia, E. Manchuria (type-locality near Pekin). Incl. *bedfordianus* (Manchuria), *ussuricus* (Ussuri R.).
*C. e. canadensis** Erxleben, 1777. E. Siberia.
[*Cervus elaphus*] *canadensis* Erxleben, 1777:305. Eastern Canada. Incl. *asiaticus* (W. Sayan Mts), *baicalensis* (E. Sayan Mts), *biedermanni*, *eustephanus*, *isubra* (N. Manchuria), *lühdorfi*, *sibirica* Severtzov (not of Schreber), *songarica* (Dzungarian Tien Shan), *wachei* (W. Mongolia).
Also several N. American synonyms some of which may represent valid subspecies.

RANGE (Map 97). Deciduous woodland zone of Europe including Ireland, Britain, S. Scandinavia, Corsica and Sardinia; N.W. Africa where now confined to a small area on the border of Tunisia and Algeria; Asia Minor and the Caucasus to the Elburz Mts; riverine and montane woodlands of Turkestan; most of the montane forests of central Asia from the Himalayas and Tien Shan to the Sayan and Altai; the Amur Basin, Manchuria and parts of W. China south to Yunnan. Also in N. America where it was formerly widespread but is now restricted to the west and to reserves. The present distribution in Eurasia is fragmented due to local extinction. It formerly occurred widely in European Russia including Crimea.

Siberian animals (*C. e. canadensis*) have been introduced to the southern Urals (Kaznevsky, 1956), and Spanish animals to N. Morocco (Lehmann, 1969b).

REMARKS. The above classification is that of Flerov (1952) who gave diagnoses of all subspecies recognized. The treatment of the large Asiatic forms as consubspecific with the North American form is debatable. Although inherently improbable, this is, however, consistent with the earlier classification of Lydekker (1915), who gave *canadensis* specific rank, including these Asiatic forms as subspecies. This was not accepted by Heptner & Naumov (1961) who treated the Asiatic and American wapiti as subspecifically distinct.

If the latter view is accepted, the valid name for the Asiatic subspecies is doubtful. The earliest name, *asiaticus* Severtzov, 1873, has been rejected on the grounds that it was not intended as a scientific name; *sibiricus* Severtzov, 1873 has been supposed to be preoccupied by *Cervus sibiricus* Schreber, 1784, but this was disputed, and the name used, by Heptner & Naumov (1964); the next available name is *songaricus* Severtzov, 1873 from the eastern Tien Shan. In view of the current Russian usage of *sibiricus* Severtzov, and the very dubious validity of '*Cervus Tarandus* Linnaeus *sibiricus*' of Schreber, 1784, *sibiricus* would seem to be the most acceptable name.

Genus *ELAPHURUS*

Elaphurus Milne-Edwards, 1866. Type-species *E. davidianus* Milne-Edwards.

A rather distinctive genus containing only a single species.

Elaphurus davidianus Père David's deer

Elaphurus davidianus Milne-Edwards, 1866. Based on a captive animal in the Imperial Hunting Park, Peking, China. Incl. *menziesianus* (Honan – subfossil), *tarandoides*.

RANGE. Known only in captivity. Original range probably the lowlands of N. China. Animals at present in zoos and parks are derived from a herd established at Woburn, Bedfordshire, England, itself derived from animals received from Peking in the late 19th century.

REMARKS. See Jones (1951) for a detailed account of the history and anatomy of this species.

Genus *ODOCOILEUS**

Odocoileus Rafinesque, 1832. Type-species *Cervus virginianus* Boddaert. Incl. *Dama* Zimmermann, 1780 (not of Frisch, 1775). *Odocoileus* Rafinesque, 1832 was placed on the Official List of Generic Names by Opinion 581 of the International Commission (1960).

A genus of about seven species in North and South America, one of which has been introduced in Europe.

Odocoileus virginianus* White-tailed deer

Dama virginiana Zimmermann, 1780. Virginia, USA. Including many synonyms from its native range.

RANGE (Map 98). Most wooded areas of N. and Central America south of about 52°N, and parts of northern South America. Introduced in 1934 and now well established in S.W. Finland.

REMARKS. The introduced stock came from Minnesota and has been called *O. v. borealis* Miller, although this form is unlikely to be discretely separable from the nominate subspecies. For an account of the introduced population see Niethammer (1963).

Genus *ALCES*

Alces Gray, 1821. Type-species *Cervus alces* L. =*Alcelaphus* Gloger =*Paralces*.

A clearly defined genus of one species if the Nearctic and Palaearctic forms are considered conspecific.

Alces alces Elk, Moose

Cervus alces Linnaeus, 1758. Sweden.

A. a. alces. Europe and W. Siberia. =*alce, antiquorum, jubata, machlis*. Incl. *angusticephalus, coronatus, europaeus, palmatus, tymensis* (Siberia), *typicus, uralensis* (S. Urals).
*Alces machlis meridionalis** Matschie, 1913:156. Samara district, S.W. of Urals.

*A. a. americanus** Clinton, 1822. E. Siberia and N. America.
Cervus americanus Clinton, 1822. Eastern N. America. Incl. *bedfordiae* (E. Siberia), *cameloides* (?Manchuria), *pfizenmayeri, yakutskensis* (N.E. Siberia). Also several synonyms from N. America.

*A. a. caucasicus** Vereshchagin, 1955. Caucasus (extinct).
Alces alces caucasicus Vereshchagin, 1955. Kuban, Terek, N. Caucasus.

RANGE (Map 98). Coniferous forest zone from Scandinavia and E. Poland to E. Siberia; southern boundary through S. Urals, Altai and lower Amur. Also in the corresponding zone of N. America. Formerly further west in Europe and with an isolate in Caucasus (extinct by early 19th century).

REMARKS. The subspecific classification follows that of Flerov (1952). The earliest Asiatic name for the eastern Siberian form is *cameloides* Milne-Edwards, 1867. This was used by Heptner & Naumov (1961) for the Ussuri form along with *A. a. pfizenmayeri* Zukowski, 1910 for the remaining parts of E. Siberia. Flerov gave characters of the two subspecies, but without any information on their variability in the boundary region. For a recent monograph see Heptner & Nasimovitsh (1967).

Genus *RANGIFER*

Rangifer H. Smith, 1827. Type-species *Cervus tarandus* L. (Opinion 91 of the International Commission). =*Tarandus*, *Achlis*.

A very clearly defined genus which is frequently considered monospecific, although the Nearctic and Palaearctic forms are also sometimes given separate specific status.

Rangifer tarandus — Reindeer, Caribou

Cervus tarandus Linnaeus, 1758. Swedish Lappland.

R. t. tarandus. Tundra zone of Palaearctic. =*borealis*, *cilindricornis*, *furcifer*, *lapponum*, *typicus*. Incl. *asiaticus* (Siberia), *lenensis* (mouth of R. Lena), *pearsoni* (Novaya Zemlya), *rangifer*, *sibiricus*.

R. t. platyrhynchos (Vrolik, 1829). Spitzbergen. Incl. *spetsbergensis*, *spitzbergensis*.

R. t. fennicus Lonnberg, 1909. Forest zone of Palaearctic range; type-locality in Torne District of Finnish Lappland. Incl. *angustirostris* (Bargusin Mts, L. Baikal), *buskensis*, *chukchensis*, *phylarchus* (Kamchatka), *setoni* (Sakhalin), *silvicola* (Karelia), *transuralensis*, *valentinae* (Altai Mts), *yakutskensis*.

*Rangifer tarandus dichotomus** Hilzheimer, 1936:157. Seitovski Possad, near Orenberg, USSR.

Also five subspecies in N. America and Greenland, the earliest named being *R. t. groenlandicus* (Linnaeus, 1767) from W. Greenland and the central Canadian tundra, and *R. t. caribou* (Gmelin, 1788) from eastern Canada (woodland caribou).

RANGE (Map 98). The tundra zone of the entire Palaearctic and Nearctic regions and southwards in the more open country of the taiga zone to the Altai Mts and N. Mongolia. Present on most of the Arctic islands and Sakhalin and introduced to Iceland. Some populations migratory, moving south in winter. Range much altered by local extinction and replacement by domesticated reindeer, e.g. in N. Scandinavia. Map shows present distribution (from Banfield, 1961).

REMARKS. The above subspecific classification follows Banfield (1961) who gave detailed descriptions and an account of the taxonomic history of each subspecies. For a monograph on the domesticated reindeer see Herre (1955). The following synonyms refer to late Pleistocene and post-glacial fossil remains of this species in the Palaearctic (from Banfield, 1961): *Cervus guettardi* Desmarest, 1822; *Tarandus priscus schottini* Sternberg, 1828; *Cervus priscus* Cuvier, 1835; *Cervus destrenii*, *C. leufroyi*, *C. rebulii*, *C. tournalii*, all of Serres, 1835; *Cervus leptocerus* Eichwald, 1835; *Cervus bucklandi* Owen, 1846; *Cervus martialis* Gervais, 1852; *Cervus tarandus diluvii* Rutten, 1909; *Rangifer tarandus hibernicus* Scharff, 1918; *Rangifer constantini* Flerov, 1934.

Genus *HYDROPOTES*

Hydropotes Swinhoe, 1870. Type-species *H. inermis* Swinhoe. =*Hydrelaphus*.

A very distinctive, monospecific genus.

Hydropotes inermis — Chinese water-deer

Hydropotes inermis Swinhoe, 1870. Deer Island, Yangtze River, a few miles upstream from Chinkiang, Kiangsu, China.

H. i. inermis. Lower Yangtze Basin. Incl. *affinis*, *kreyenbergi*.

H. i. argyropus Heude, 1884. Korea.

RANGE (Map 96). The lower Yangtze Basin, west to Hupeh; Korea; introduced in England where it is established in Bedfordshire and spreading slowly, and also in France.

Genus *CAPREOLUS*

Capreolus Gray, 1821. Type-species *Cervus capreolus* L. (First proposed by Frisch (1775), but this work has been rejected in Opinion 258 (1954) of the International Commission.) =*Caprea*.

A monospecific and clearly defined genus.

Capreolus capreolus Roe deer

Cervus capreolus Linnaeus, 1758. Sweden.

C. c. capreolus. Europe, east to central European Russia. = *capraea.* Incl. *albicus, albus* (France), ?*armenius* (Asia Minor), *baleni, balticus, canus* (Spain), *cistaunicus,* ?*coxi* (N. Iraq), *decorus, dorcas, europaeus, grandis, italicus, joffrei, niger, plumbeus, rhenanus, thotti* (Scotland), *transsylvanicus, transvosagicus, varius, vulgaris, warthae, whittalli, zedlitzi.*

C. c. pygargus (Pallas, 1771). Siberia and E. European Russia. Incl. *ferghanicus* (Ferghana, Turkestan), *tianschanicus* (Chinese Tien Shan).

?*Capreolus pygargus* var. *caucasica** Dinnik, 1910:73. North Caucasus.

C. c. bedfordi Thomas, 1908. N. China, type-locality in Shansi. Incl. *mantschuricus* Noack, 1889 (not of Swinhoe, 1864), *ochracea* (Korea).

C. c. melanotis Miller, 1911. Kansu (type-locality) to Szechuan and E. Tibet.

RANGE (Map 98). Mainly in the deciduous and Mediterranean woodland zones from W. Europe (including Britain and S. Scandinavia) to S.E. Siberia, and south through W. China to Szechuan and E. Tibet. Also in the Caucasus, Asia Minor, N. Iraq and N. Iran; and from the Altai to the Tien Shan and the mountains of Turkestan.

REMARKS. The subspecific classification above follows Flerov (1952) who gave diagnoses, but there is likely to be considerable continuity of variation between these. The form in N. Iraq may be isolated and worthy of subspecific rank (*C. c. coxi* Cheesman & Hinton, 1923) as was maintained by Harrison (1968). The N. Caucasian form, *caucasica*, was placed in *C. c. pygargus* by Flerov (1952), although on zoogeographical grounds it would seem more likely to be associated with the form in Asia Minor which he placed in *C. c. capreolus.* However, it is probable that any discontinuity of the continental range is due to recent extinction.

Family **BOVIDAE**

One of the dominant families of mammals with about 45 genera and 120 species. It is abundantly represented in the Ethiopian Region, moderately so in the Palaearctic Region, with relatively few species in the Oriental and Nearctic Regions. It has always been clearly defined, especially by the unique character of the horns which are unbranched and non-deciduous on a bony core, and the only very closely related family is the Antilocapridae (the pronghorn of N. America).

Classification at the generic level is less stable whilst the grouping of genera in tribes and subfamilies has been very variable. In the following account subfamily names are used following Simpson (1945) although one of these, the Hippotraginae, is an assemblage of very doubtful validity. Eighteen genera are listed here, but one of these is represented only by an introduced species (*Ovibos*) and one is extinct in the Palaearctic Region (*Alcelaphus*).

1. Horns emerging laterally close to occiput, bases widely separated; size and build of domestic cattle . . . . (2)
– Horns arising from top of skull, bases close together; smaller and more lightly built than domestic cattle . . . . (3)
2. Mane of long hair in shoulder region . . . . ***BISON*** (p. 206)
– Mane on whole length of flanks or wholly short-haired . . . . . ***BOS*** (p. 205)
3. Size of horns showing strong sexual dimorphism, being relatively small or absent in females (4)
– Horns well developed in both sexes . . . . (10)
4. Facial glands and associated pre-orbital fossae present . . . . (5)
– Facial glands and pre-orbital fossae absent . . . . (7)
5. Upper surface of muzzle grossly inflated. . . . . . . ***SAIGA*** (p. 211)
– Muzzle normal . . . . (6)
6. Horns very thick at base, prominently spiral, with fine, closely spaced transverse ridges; pelage long and dense . . . . ***OVIS*** (p. 217)
– Horns slender, no more than slightly spiral, with prominent, widely spaced rings; pelage short . . . . ***GAZELLA*** (p. 208)

7. With a beard or a long mane on throat and/or front of legs . ***CAPRA*** (p. 213)
– Without a beard or mane . . . . (8)
8. Horns thick, strongly spiral; pelage grey. . . . . . . ***PSEUDOIS*** (p. 217)
– Horns slender, straight or weakly spiral; pelage yellowish brown . . . (9)
9. Horns very long, erect and almost straight; I_1 small, considerably narrower than combined width of I_2–C_1 . . . . ***PANTHOLOPS*** (p. 211)
– Horns distinctly curved; I_1 large, equal to combined width of I_2–C_1 ***PROCAPRA*** (p. 210)
10. Horns very long and slender, greatly exceeding length of head; 'basal pillars' (accessory cusps) in centre of inner surface of upper molars and outer surface of lower molars (11)
– Horns short, not longer than head; molars without 'basal pillars' . . . . (12)
11. Horns spiral; legs wholly white . . . . ***ADDAX***[1] (p. 207)
– Horns straight or curved in one plane; legs marked with brown or black ***ORYX*** (p. 206)
12. Larger than sheep or goat . . . . (13)
– About the size of a sheep or goat . . . . (14)
13. Horns spreading laterally and downwards; pelage long and shaggy (Arctic) ***OVIBOS*** (p. 213)
– Horns erect, on common pedicel; pelage short (Africa) . ***ALCELAPHUS*** (p. 207)
14. Horns erect and sharply hooked backwards at tips; skin glands on head behind horns (Europe) . . . . ***RUPICAPRA*** (p. 212)
– Horns sloping backwards, not sharply hooked; no glands on top of head (Asia) . (15)
15. Facial glands and pre-orbital fossae present; (larger: shoulder height over 750 mm) ***CAPRICORNIS*** (p. 212)
– Facial glands and pre-orbital fossae absent; (smaller: shoulder height under 750 mm) (16)
16. Horns slender, terete . . . . ***NEMORHAEDUS*** (p. 212)
– Horns broad at the base, with a keel on the anterior margin ***HEMITRAGUS*** (p. 213)

Subfamily **BOVINAE**
Genus ***TRAGELAPHUS****

Tragelaphus Blainville, 1816. Type-species *Antilope sylvatica* Sparrman. Incl. *Strepsiceros*.

A genus of seven species confined to subsaharan Africa except for the record described below.

Tragelaphus imberbis* Lesser kudu

Strepsiceros imberbis Blyth, 1869. Ethiopia.

RANGE. Somali arid zone of E. Africa, south to Tanzania. A single record from Yemen.

REMARKS. This species was unknown in the Arabian peninsula until Harrison (1972) recorded a single specimen obtained, in the form of horns with fresh skin attached, in S. Yemen in 1967.

Genus ***BOS***

Bos Linnaeus, 1758. Type-species *Bos taurus* L. (domestic ox). = *Taurus*. Incl. *Poephagus*, *Urus*.

A genus of five species in the Palaearctic and Oriental Regions. Besides the two Palaearctic species listed below these are *B. gaurus*, *B. javanicus* and *B. sauveli* in the Oriental Region. The last three are sometimes excluded (in genus *Bibos*, e.g. by Bohlken, 1958), while at the other extreme the bisons are sometimes included in *Bos*, but the classification adopted here seems to have fairly general acceptance.

Short-haired; back straight (neural spines of vertebrae decreasing in size evenly from thoracic to lumbar region); 13 pairs of ribs. . . . . ***B. primigenius***
Long hair on neck and flanks; fore part of back humped (neural spines on thoracic vertebrae very long, giving way abruptly to short ones in lumbar region); 14 pairs of ribs ***B. mutus***

[1] See also *Tragelaphus imberbis* which has spiral horns and vertical white stripes on the body.

Bos primigenius* Urus, Wild ox

Bos primigenius Bojanus, 1827.

RANGE. Extinct in wild, surviving in form of domestic cattle. In historical time confined to Europe (including Britain and S. Scandinavia), N. Africa and the Near East. Survived latest in Poland where reputed to have become extinct in 1627.

REMARKS. This was the ancestor of the domestic cattle, including probably the zebu or '*Bos indicus*'. For an account of its domestication see Zeuner (1963), and for the history of its extinction in the wild see Heptner & Naumov (1961).

The name *primigenius* is used for this species following the principles laid down on p. 8 for the nomenclature of domestic animals. It is predated by *B. taurus* Linnaeus, 1758, the domestic ox of Europe, which is certainly a derivative of this species, and by *B. indicus* Linnaeus, 1758, the domestic ox of India (zebu).

Bos mutus Wild yak

Poephagus mutus Przewalski, 1883. W. Nan Shan, N. Kansu, China.

RANGE (Map 99). As a wild species, in most of Tibet, W. Kansu, Ladak and the Pamirs. Formerly in the Altai, Sayan and Transbaikal Mts where it may have survived until the 18th century.

REMARKS. This is the ancestor of the domesticated yak, which was named *Bos grunniens* Linnaeus, 1766 (=*B. poephagus* Pallas, 1811), but in accordance with the principles explained on p. 8 for the nomenclature of domesticated animals is better referred to in the vernacular or as '*Bos mutus* (domestic)'.

Genus *BISON*

Bison H. Smith, 1827. Type-species *Bos bison* L. (fixed by Opinion 91 of the International Commission). Incl. *Bonasus*.

A genus of two forms, in N. America and Europe, which are here considered conspecific but are frequently given specific rank. Closely related to *Bos* (a limited degree of hybridization is possible) and at one time frequently treated as a subgenus of *Bos*, but most recent authors give it full generic status.

Bison bison Bison, Wisent

Bos bison Linnaeus, 1758. 'Mexico and Florida'.

B. b. bonasus Linnaeus, 1758. Europe, type-locality Bialowieza Forest, Poland. =*europaeus*. Incl. *bison* H. Smith (not of Linnaeus), *nostras*, *urus*.

*B. b. caucasicus** Turkin & Satunin, 1904. Caucasus.

Bos bonasus caucasicus Turkin & Satunin, 1904. Caucasus (extinct). Incl. *caucasia*.

Also other races in N. America.

RANGE (Map 99). In early historical time throughout central Europe, including S. Sweden, east to the Volga and in the Caucasus. By the beginning of this century it was confined to the Bialowieza Forest (Poland/Lithuania) and the Caucasus. Both populations became extinct (Bialowieza in 1921, Caucasus in 1925), but animals from the Bialowieza population survived in captivity and have been reintroduced into the Bialowieza Forest and, with limited freedom, to several areas in European Russia including the Caucasus.

REMARKS. The recognition of the European and American forms as conspecific follows the comprehensive revision by Bohlken (1967).

Subfamily HIPPOTRAGINAE

Genus *ORYX*

Oryx Blainville. 1816. Type-species *Antilope oryx* Pallas =*O. gazella* L. (S. Africa). =*Oryx*. Incl. *Aegoryx*

A distinctive genus of three species, one of which is wholly Ethiopian (*O. gazella*, th gemsbok of southern Africa and the beisa of East Africa). Haltenorth (1963) took th extreme view of treating all three as conspecific – *O. gazella* (L.) – but I prefer to follow Ansell (1971b) who recognized three species.

Horns almost straight; body entirely white; legs dark (Arabia) . . . . ***O. leucoryx***
Horns distinctly curved; body yellowish brown, especially on neck and shoulders; legs pale (Sahara) . . . . ***O. dammah***

Oryx leucoryx Arabian oryx

Antilope leucoryx Pallas, 1777. Arabia. =*asiatica, pallasii.* Incl. *beatrix, latipes.*

RANGE (Map 99). Formerly most of the desert zone of Arabia north to Sinai, N. Jordan, perhaps S. Syria and Iraq west of the Euphrates. Now confined to S.E. Arabia and close to extinction. See Stewart (1963, 1964) for a detailed account, including maps of past and present range. There are captive herds in USA.

Oryx dammah* Scimitar oryx

Antilope dammah Cretzschmar, 1826. 'Haraza', E. Sudan. Incl. *algazel* (not available), *ensicornis**, *leucoryx** Sundevall (not of Pallas), *nubica**, *senegalensis** (the last four from south of the Sahara – see Allen (1939) for details), *tao.*

RANGE (Map 99). Formerly the semidesert zones north and south of the Sahara, but probably now extinct north of the Sahara. Formerly from Egypt to Morocco. Still present in S.W. Sudan, Chad, Niger and probably elsewhere in the western Sahara. See Dolan (1969) and Kock (1970b).

REMARKS. This species was frequently called *O. leucoryx* (which should properly be restricted to the Arabian oryx) and *O. algazel* (which is invalid because of the rejection of Oken's (1816) *Lehrbuch der Naturgeschichte* – Opinion 417, 1956).

E & M-S used *O. tao* H. Smith, 1827 and did not mention *dammah* Cretzschmar except by inference in referring to Allen (1939) for extralimital synonyms (where it is dated 1826). Ansell (1971b) adopted the obvious course of using *O. dammah* Cretzschmar, 1826 on the basis of priority. However, on the copy of Cretzschmar in the British Museum (Natural History) Morrison-Scott has included a list of dates of publication giving 1827 for the name *dammah*. In these circumstances there seems to be a marginal advantage in using *dammah* in order to correlate with the most recent comprehensive work on Africa (Ansell, 1971b) and since the evidence for the later publication date is not clear.

Genus *ADDAX*

Addax Laurillard, 1841. Type-species *Antilope suturosa* Otto =*Cerophorus nasomaculatus* Blainville.

A monospecific and clearly defined genus, related to *Oryx*.

Addax nasomaculatus Addax

Cerophorus (Gazella) nasomaculata Blainville, 1816. Probably from Senegambia, W. Africa. =*mytilopes.* Incl. *gibbosa* (Egypt), *suturosa.*

RANGE (Map 99). At one time throughout most of the Sahara from Senegal and Rio de Oro in the west to Sudan and Egypt. Now much reduced and confined to the southern Sahara from Mauritania and Mali (where relatively abundant) through S. Algeria, Niger and Chad to Darfur in Sudan. This is the most desert-adapted of the antelope. See Kock (1970b).

Genus *ALCELAPHUS**

Alcelaphus Blainville, 1816. Type-species *Antilope bubalis* Pallas = *A. buselaphus* Pallas.
Bubalis Goldfuss, 1820: 367. *Antilope buselaphus* Pallas.
Damalis H. Smith, 1827: 343. ?*Antilope buselaphus* Pallas.
Acronotis H. Smith, 1827: 346. *Antilope buselaphus* Pallas.
Bubalus Ogilby, 1827. *Bubalus mauritanicus* Ogilby = *Antilope bubalis* Pallas.

A genus of two species confined to Africa. At present they are found only south of the Sahara, but one species was until recently represented in North Africa.

Alcelaphus buselaphus * Hartebeest

Antilope buselaphus Pallas, 1766. ?Morocco.
A. b. buselaphus. N. of Sahara.
Antilope bubalis Pallas, 1777. 'Africae . . . borealiori'.
Bubalus mauritanicus Ogilby, 1837. New name for *Antilope bubalis* Pallas.
A [*lcelaphus*] *bubalinus* Flower & Lydekker, 1891.
Bubalis boselaphus Trouessart, 1898. Emendation.
Also several synonyms based on fossil and subfossil material, and many referring to subspecies from south of the Sahara.

RANGE. Most of the short-grass steppe zones of Africa, but now extinct north of the Sahara. Formerly throughout N. Africa from Morocco to Egypt. They survived in S. Tunisia until at least 1902 and in the high plateau of E. Morocco and the adjacent region of Algeria until at least 1933. Probably also survived into the present century in the Adrar region of S. Algeria and Mali.

REMARKS. This species was omitted by E & M-S because it is extinct in the Palaearctic Region. The above synonyms are taken from Allen (1939).

Subfamily ANTILOPINAE

Genus *GAZELLA*

Gazella Blainville, 1816. Type-species *Capra dorcas* L. (fixed by Opinion 108 of International Commission). =*Dorcas*. Incl. *Eudorcas*, *Korin*, *Leptoceros* Wagner, *Matschiea*, *Nanger*, *Trachelocele*, *Tragops* Hodgson, *Tragopsis*.

This genus includes about 13 species ranging in the steppe and desert zones from East Africa to India and central Asia. The Asiatic species here placed in *Procapra* are closely related and were at one time included in *Gazella*, whilst in southern Africa the nearest relative is *Antidorcas marsupialis*, the springbok. Classification within the genus is very problematical, the slight differences between species being obscured by considerable geographical variation within species. Recent work has considerably altered the classification given by E & M-S. The arrangement adopted here is based on the work of Groves (1969) and Gentry (1971), but several problems of the delimitation of species remain.

1. Very large: shoulder height about 900 mm, greatest length of skull over 210 mm; large white area on rump enclosing base of tail and extending forward on flank; preorbital fossae very shallow . . . . ***G. dama***
- Smaller: shoulder height usually under 800 mm, skull under 210 mm (except in *G. rufina*); white on rump more limited, not enclosing base of tail . . . . (2)
2. Large: skull of adult male about 230 mm; dark band on flank separated from white ventral pelage by brown band (N.W. Africa, extinct) . . . . ***G. rufina***
- Smaller: skull under 210 mm; dark flank band (which may be very obscure) abutting directly against white ventral pelage . . . . (3)
3. Pelage rather long and thick; preorbital fossae large . . . . (4)
- Pelage short; preorbital fossae small . . . . (5)
4. Smaller: greatest length of skull not over 190 mm; dark stripe on top of muzzle obscure or lacking, entire muzzle tending to be white; males with a goitre-like swelling on the throat in breeding season (Asia) . . . . ***G. subgutturosa***
- Larger: greatest length of skull about 200 mm; facial stripes well marked; males without a swelling on the throat (N.W. Africa) . . . . ***G. cuvieri***
5. Pelage very pale greyish brown, with flank band very obscure; horns very long, straight and erect (N. Africa) . . . . ***G. leptoceros***
- Pelage darker or, if pale, yellowish rather than greyish brown; horns shorter and more sinuous . . . . (6)
6. Horns of male rather widely separated, about 25 mm apart, and rather short, usually less than 260 mm; nasals making only marginal contact with premaxillae and widely rounded behind, not penetrating far into frontals (Arabian region) . . . . . ***G. gazella***

– Horns of males closer together, usually not more than about 20 mm apart, and longer, frequently over 260 mm; nasals making considerable contact with premaxillae and penetrating more deeply between frontals . . . . ***G. dorcas***

Gazella subgutturosa Goitred gazelle

Antilope subgutturosa Güldenstaedt, 1780. N.W. Persia.
G. s. subgutturosa. Iran, Caucasus and Turkestan. =*typica.* Incl. *persica, seistanica* (Seistan).
*Gazella subgutturosa gracilicornis** Stroganov, 1956c. Bakhshan Valley, Tadjikistan, USSR.
G. s. marica† Thomas, 1897. Arabia. Doubtfully distinct from *G. s. subguttorosa.*
G. s. yarkandensis Blanford, 1875. Sinkiang.
G. s. hillieriana Heude, 1894. Mongolia, W. China, N. Tibet. Incl. *mongolica, reginae* (Tsaidam), *sairensis* (Zungaria).

RANGE (Map 101). Desert and subdesert steppes from Palestine, central Arabia and the eastern Caucasus through Iran, Baluchistan, S. Turkestan and Sinkiang to the Gobi Desert, the Ordos plains and Tsaidam.

REMARKS. The above arrangement follows that of Groves (1969). Heptner & Naumov (1961) also placed *gracilicornis* in the synonymy of *G. s. subgutturosa.* E & M-S treated *marica* as a race of *G. leptoceros*, but there is general agreement amongst more recent authors that it is a form of *G. subgutturosa* and it intergrades with the nominate subspecies in Iraq. Lange (1972) went further and included the N. African *leptoceros* in this species. While this may well reflect the true affinities of *leptoceros* the differences, especially the lack of a throat-swelling in the male of *leptoceros* and the presence of horns in the female, would seem to justify specific separation.

Gazella dorcas Dorcas gazelle

Capra dorcas Linnaeus, 1758. Lower Egypt.
G. d. dorcas. Egypt and N.E. Libya. Incl. *corinna, kevella* Pallas, *sundevalli.*
*G. d. osiris** Blaine, 1913. Sahara (L. Chad, Air, Hoggar, Darfur and N. Sudan).
Gazella littoralis osiris Blaine, 1913. Nakheila, near Junction of Nile and Atbara Rivers, Sudan. Incl. *neglecta* (Algerian Sahara).
G. d. massaesyla Cabrera, 1928, Rif. Morocco to Senegal. Incl. *cabrerai, kevella* Sundevall (not of Pallas), *maculata* (not available).
G. d. saudiya Carruthers & Schwarz, 1935. Arabia (type-locality 150 miles N.E. of Mecca).
G. d. fuscifrons Blanford, 1873. Seistan. Incl. *hayi, kennioni.*
G. d. chrystii Blyth, 1842. Cutch and Kathiawar.
G. d. bennetti (Sykes, 1831). India. Incl. *hazenna.*
Also other races in Sudan, Ethiopia and Somalia.

RANGE (Map 100). Desert and subdesert zones around the Sahara and in the plateaus of Hoggar and Tadmeit; from the Red Sea Hills of Sudan to N. Somalia; W. Arabia and possibly in parts of Palestine, Syria and W. Iraq; E. Iran through Baluchustan to Pakistan and central India.

Harrison (1968a) showed on his map records of *G. dorcas* in S. Turkey, but these are based on very dubious evidence (Kumerloeve, 1969).

REMARKS. The above arrangement of subspecies follows Groves (1969) who gave characters. Gentry (1971) suggested that all the North African forms, along with *neglecta* from the central Sahara, could be considered consubspecific (*G. d. dorcas*).

Gazella gazella Mountain gazelle, Idmi

Antilope gazella Pallas, 1766. Syria.
G. g. gazella. Syria and Palestine. Incl. *merrilli* (N. of Jerusalem).
G. g. arabica (Lichtenstein, 1827). Arabia, type-locality Farsan Is., Red Sea. Incl. *cora* (Persian Gulf), *erlangeri* (near Aden), *hanishi* (Great Hanish Is., Red Sea), ?*rueppelli* (Sinai), *typica.*
G. g. muscatensis Brooke, 1874. Oman.

RANGE (Map 101). Mountains of Oman, S. and W. Arabia, Sinai and Palestine.

REMARKS. The limitation of this species to the forms listed above follows Harrison (1968a) and Groves (1969). E & M-S also included *bennetti* (India) here allocated to *G.*

dorcas, and *cuvieri* (N.W. Africa) which has been considered to be a separate species (Gentry, 1971; Groves, 1969). *G. gazella* is a problematical species and its existence throughout much of Arabia as a species distinct from *G. dorcas* cannot be considered conclusive, although Harrison (1968a) maintained that the two are sympatric in S.W. Arabia. The nomenclature is also problematical in view of the doubtful identity of Pallas' *Antilope gazella.*

***Gazella cuvieri*†** Edmi gazelle

Antilope cuvieri Ogilby, 1841. Mogador, Morocco. Incl. *cineraceus, corinna* Lacepède & Cuvier (not of Pallas), *kavella* Tristram (not of Pallas nor Sundevall), *vera.*

RANGE (Map 101). Atlas Mountains in Morocco, Algeria and Tunis.

REMARKS. E & M-S treated this as a subspecies of *G. gazella*, but both Groves (1969) and Gentry (1971) agreed in giving it specific status and did not consider it closely related to *G. gazella.* Groves (1969) considered many forms from south of the Sahara to be conspecific with *G. cuvieri*, including all the forms commonly placed in *G. rufifrons* and *G. thomsoni*, but Gentry (1971) disagreed and did not consider them to be related even at a superspecies level.

Gazella leptoceros Sand gazelle, Rhim

Antilope leptoceros Cuvier, 1842. 'Sennaar', but more likely lower Egypt.
G. l. leptoceros. W. Egypt. =*cuvieri, typica*. Incl. *harab* (Libya).
G. l. loderi Thomas, 1894. Tunis and E. Algeria. Doubtfully distinct.

RANGE (Map 101). Western desert of Egypt and N.E. Libya; sand deserts of S.W. Tunisia and the adjacent parts of Algeria and N.W. Libya.

REMARKS. E & M-S included the Arabian form *marica* in this species, but Gentry (1971), Groves (1969) and Lange (1972) all emphatically placed *marica* in *G. subgutturosa.* Lange treated *leptoceros* and *loderi* as races of *G. subgutturosa*, mainly on the basis of cranial characters. While there may well be a close relationship here, the lack of a 'goitre' in *leptoceros* and the presence of horns in the female would seem to justify specific rank.

Gazella rufina

Gazella rufina Thomas, 1894. Algeria. Incl. *pallaryi.*

RANGE. Extinct. Known only from specimens bought in markets of Algiers and Oran at the end of last century.

REMARKS. E & M-S included this form tentatively in *G. rufifrons* (from W. Africa south of the Sahara), but Gentry (1971) and Groves (1969) agree in giving it specific status.

Gazella dama Addra gazelle

Antilope dama Pallas, 1766. Lake Chad.
G. d. mhorr (Bennett, 1833). Morocco. =*mohr.*
G. d. lazanoi Agacino, 1934. Rio de Oro.
Also other forms from the southern parts of the Sahara.

RANGE (Map 101). Desert and subdesert zones of the western and southern Sahara from Morocco (where perhaps extinct) and Rio de Oro through Senegal and along the fringe of the savanna zone to Darfur.

Genus *PROCAPRA*

Procapra Hodgson, 1846. Type-species *P. picticaudata* Hodgson. Incl. *Prodorcas.*

A genus of three species closely related to *Gazella* and formerly included in *Gazella.* They do, however, appear to form a natural group at one extreme of the variation found in *Gazella*, and most recent authors agree on giving *Procapra* generic rank. E &

M-S recognized only two species, but Groves (1967), following Stroganov (1949c), recognized three.

1. Small face-glands and carpal tufts present; large inguinal glands present; size larger (shoulder height up to 760 mm, greatest length of adult skull over 225 mm). . ***P. gutturosa***
– Face-glands, carpal tufts and inguinal glands absent; size smaller (shoulder height not over 640 mm, greatest length of adult skull under 225 mm) . . . . (2)
2. Horns curved in two planes, points turned inwards; white rump patch divided by a median line of darker colour . . . . ***P. przewalskii***
– Horns curved in only one plane, points not incurved; white patch continuous over rump ***P. picticaudata***

Procapra picticaudata Tibetan gazelle, Goa

Procapra picticaudata Hodgson, 1846. 'Type-locality said to be Hundes [Tibet], but more likely the district north of Sikkim, where most of Hodgson's specimens were obtained after 1844' (Groves, 1967).

RANGE (Map 100). Most of Tibet and adjacent parts of Ladak and Szechuan, and in the Nan Shan of Chinghai, in plateau grassland and the higher barren zones.

***Procapra przewalskii*†** Przewalski's gazelle

Gazella przewalskii Büchner, 1891. Chagrin-Gol, E. Nan Shan (37°N, 102°E).
P. p. przewalskii. Ordos to Nan Shan. Incl. *gutturosa* Przewalski (not of Pallas), *cuvieri.*
*P. p. diversicornis** Stroganov, 1949. S. Ordos and S.W. Kansu.
Gazella przewalskii diversicornis Stroganov, 1949c. Oasis of Sin-Zhin-Pu, Kansu (*fide* Groves, 1967).

RANGE (Map 100). Subdesert steppes from the Nan Shan and Kukunor to the Ordos Plateau.

REMARKS. According to Stroganov (1949c) (quoted by Groves, 1967) this species is sympatric with *P. picticaudata* in the regions of Nan Shan and Kukunor. Groves (1967) gave the characters of the two subspecies. The type-locality of *G. przewalskii* was erroneously given as S. Ordos Desert by E & M-S.

Procapra gutturosa Mongolian gazelle, Zeren

Antilope gutturosa Pallas, 1777. Transbaikalia. =*orientalis.*
P. g. gutturosa. Eastern part of range.
P. g. altaica Hollister, 1913. Altai.

RANGE (Map 100). Dry steppe and subdesert of most of Mongolia and Inner Mongolia, extending into small areas of the USSR adjacent to N.W. Mongolia and in the southern Nerchinsk Mts.

REMARKS. Groves (1967) upheld the above subspecific separation and gave characters but without considering intergradation or degree of isolation.

Subfamily **CAPRINAE**

Genus ***PANTHOLOPS***

Pantholops Hodgson, 1834. Type-species *Antilope hodgsonii* Abel.

A very distinctive, monospecific genus, most closely related to *Saiga.*

Pantholops hodgsoni Chiru, Tibetan antelope

Antelope hodgsonii Abel, 1826. Arrun Valley, Tibet. Incl. *chiru, kemas.*

RANGE (Map 102). Plateau steppe of Tibet and Chinghai between 3800 and 5500 m, and the immediately adjacent part of Ladak.

Genus ***SAIGA***

Saiga Gray, 1843. Type-species *Capra tatarica* L. =*Siaga.* Incl. *Colus.*

A distinctive, monospecific genus closely related only to *Pantholops.*

Saiga tatarica Saiga

Capra tatarica Linnaeus, 1766. Ural Steppes, Russia.
S. t. tatarica. Russian steppes. Incl. *colus, imberbis, saiga, sayga, scythica.*
S. t. mongolica Bannikov, 1946. W. Mongolia.

RANGE (Map 102). Formerly throughout the steppe zone of southern Russia from Ukraine to Turkestan and as far as Mongolia and Sinkiang. By 1920–1930 it was reduced to a few small isolated areas scattered between the lower Volga and Lake Balkhash, but it has since recovered dramatically under protection and occupies much of its former range.

REMARKS. Monographed by Bannikov (1961).

Genus *NEMORHAEDUS*

Naemorhedus H. Smith, 1827. Type-species *Antilope goral* Hardwicke. =*Caprina* Wagner, *Kemas* Ogilby, *Naemorhaedus, Nemorhedus, Nemorrhedus.* Incl. *Urotragus.*

A genus only marginally distinct from *Capricornis*, confined to eastern Asia. Possibly monospecific, but a form recently described from Upper Burma, *N. cranbrooki* Hayman, 1961, may be specifically distinct (distinguished especially by its bright fox-red pelage), and *N. baileyi* Pocock, 1914 from S.E. Tibet is also of doubtful status (see Hayman, 1961). Neither of these affect the status and name of the widespread species, *N. goral*. See Dolan (1963) for a review.

Nemorhaedus goral Goral

Antilope goral Hardwicke, 1825. Nepal.
N. g. caudatus (Milne-Edwards, 1867). N. China and Amur region of Siberia; type-locality Bureja Mts, Amur region, S.E. Siberia. Incl. *crispa, galeanus* (S. Shensi), *raddeanus, vidianus.*
Also other named subspecies in the Oriental Region, all post-dating *caudatus.*

RANGE (Map 103). Wooded mountain slopes from the Himalayas through W. China to Manchuria, Korea and the Sikhote Mts of S.E. Siberia.

Genus *CAPRICORNIS*

Serows

Capricornis Ogilby, 1837. Type-species *Antilope thar* Hodgson. =*Capricornus.* Incl. *Austritragus, Capricornulus, Lithotragus, Nemotragus.*

A genus of two species in E. Asia, one, *C. sumatraensis*, being wholly in the Oriental Region. The other, *C. crispus* from Japan, has been separated generically in *Capricornulus* but with little justification. The genus is very close to the earlier-named *Nemorhaedus*, and has been reviewed by Dolan (1963).

Capricornis crispus Japanese serow

Antilope crispa Temminck, 1844. Honshu, Japan.
C. c. crispus. Japan. Incl. *pryeri, pryerianus, saxicola.*
C. c. swinhoei Gray, 1862. Taiwan.

RANGE (Map 102). Honshu, Shikoku and Kyushu, Japan; Taiwan.

REMARKS. E & M-S queried the specific allocation of *swinhoei*, but Dolan (1963) treated it as a race of *C. crispus*.

Genus *RUPICAPRA*

Rupicapra Blainville, 1816. Type-species *Capra rupicapra* L. =*Capella, Cemas* Gloger.
A clearly defined monospecific genus related to *Capricornis* and *Nemorhaedus.*

Rupicapra rupicapra Chamois

Capra rupicapra Linnaeus, 1758. Switzerland.
R. r. rupicapra. Alps, Tirol, Slovenia. =*capella, dorcas, europea, tragus.* Incl. ?*faesula.*
R. r. pyrenaica Bonaparte, 1845. Pyrenees.

R. r. cartusiana Couturier, 1938. Massif de la Chartreuse, Isère, France.
R. r. parva Cabrera, 1911. N.W. Spain.
R. r. ornata Neumann, 1899. Abruzzi Apennines, Italy.
R. r. balcanica Bolkay, 1925. Balkans and Greece. Incl. *olympica* (Greece).
R. r. carpatica Couturier, 1938. Carpathians.
R. r. caucasica Lydekker, 1910. Caucasus.
R. r. asiatica Lydekker, 1908. Asia Minor.
*R. r. tatrica** Blahout, 1972. Tatra Mts.
Rupricapra rupricapra tatrica Blahout, 1972. Belanske, High Tatra Mts, Poland.

RANGE (Map 102). Very fragmented, in upper regions of montane forest and adjacent open slopes. N.W. Spain, Pyrenees, Alps, Jura, Abruzzi Apennines, Carpathians, Tatra Mts, Balkan Mts, Caucasus, N. Asia Minor, Taurus Mts.

REMARKS. Monographed in great detail by Couturier (1938). See also Dolan (1963) for a concise taxonomic review and Knaus & Schröder (1975) for a more recent monograph.

Genus *OVIBOS**

Ovibos Blainville, 1816. Type-species *Bos moschatus* Zimmermann, 1780. =*Bosovis* Kowarzik, 1911.

A distinctive, monospecific genus closely related only to the genus *Budorcas* of Asia.

Ovibos moschatus* — Musk ox

Bos moschatus Zimmermann, 1780:86. Manitoba, Canada.
*O. m. wardi** Lydekker, 1900. Greenland and Arctic Islands of Canada.
Ovibos moschatus wardi Lydekker, 1900. Clavering Island, Greenland.

RANGE (Map 102). Tundra of N. America and Greenland. Successfully introduced from Greenland to the Dovrefjell district of Norway (in 1932) and to Spitzbergen (1929). See Lønø (1960a).

REMARKS. This species was present in the Palaearctic tundra during the late Pleistocene and may have survived into the post-glacial period. Monographed by Pedersen (1958).

Genus *HEMITRAGUS*
Tahrs

Hemitragus Hodgson, 1841. Type-species *Capra quadrimammis* vel *jharal* Hodgson = *Capra jemlahica* Smith. Incl. *Kemas* Gray.

A genus of three species, closely related to *Capra*. Only one species is Palaearctic, the others being found in the Himalayas and Peninsular India.

Hemitragus jayakari — Arabian tahr

Hemitragus jayakari Thomas, 1894. Oman.

RANGE (Map 104). Mountains of Oman, S.E. Arabia.

REMARKS. E & M-S, while retaining this form as a species, considered that it might be regarded as a subspecies of *H. jemlahicus* of the Himalayas. Harrison (1968) upheld its specific status.

Genus *CAPRA*
Goats

Capra Linnaeus, 1758. Type-species *Capra hircus* L., the domestic goat. =*Aegoceros*, *Aries* Link, *Hircus*, *Tragus*. Incl. *Ammotragus*, *Eucapra*, *Euibex*, *Orthaegoceros*, *Turocapra*, *Turus*.

A purely Palaearctic genus including the goats and ibex and, following Ansell (1971b), the Barbary sheep, although previous authors, including E & M-S, gave the last generic rank as *Ammotragus*. The genus *Pseudois*, here retained, is also closely related and should perhaps be included.

Classification of the genus *Capra* s.s. is very unstable. E & M-S recognized five species, distinguished primarily by the shape of the horns (without considering the Ethiopian form *walie*). The only recent comprehensive reviews are those of Couturier (1962), who used a highly unorthodox quinquenomial nomenclature, and Haltenorth (1963). Heptner & Naumov (1961) considered the problems involved in dealing with the Russian forms, and other authors have reviewed smaller parts of the genus. Most of the divergence between different classifications involves nomenclature and rank, but a more fundamental problem is the status of the Caucasian forms. Heptner & Naumov (1961) recognized two species in the eastern and western Caucasus respectively with a sympatric zone in the central Caucasus. This is accepted in preference to Couturier's recognition of three allopatric but conspecific races.

In the following key the characters of the horns only apply in full to adult males.

1. Mane of long hair on front of each foreleg; no inter-ramal beard, but a double beard further back on throat; tail long, reaching to hocks; (horns widely divergent, with fine transverse ridges) (N. Africa) . . . . ***C. lervia***
- No mane on forelegs; males with inter-ramal beard: tail short . . . . . (2)
2. Horns terete, without longitudinal ridges and without prominent transverse ridges, widely spreading, with a single open spiral when well developed; beard very short (E. Caucasus) . ***C. cylindricornis***
- Horns with prominent longitudinal and/or transverse ridges; beard longer . . (3)
3. Anterior face of horns rounded or with a keel, without closely spaced knobs . . (4)
- Anterior face of horns (and horn-cores) flattened, with closely spaced transverse ridges or knobs . . . . (5)
4. Longitudinal ridge beginning on medial face of each horn; horns tapering sharply, widely divergent and with a single open spiral (Iberia) . . . . . . ***C. pyrenaica***
- Longitudinal ridge beginning on anterior face of horn; horns gradually tapering, curved only in sagittal plane and with widely spaced knobs on anterior ridge . . ***C. aegagrus***
- Longitudinal ridge beginning on posterior face of horn; horns spiral; beard extending back to form long mane on neck and chest . . . . ***C. falconeri***
5. Horns widely divergent, rather short, with closely spaced, low, transverse ridges on anterior face (W. Caucasus) . . . . ***C. caucasica***
- Horns not widely divergent, very long in old males, with prominent knobs on anterior face ***C. ibex*** (6)
6. Horns with front surface narrow and with a slight longitudinal keel on antero-medial angle (Sudan to Arabia). . . . . ***C. i. nubiana***
- Horns with front surface wide and without trace of a keel . . . . . (7)
7. Bony prominence on forehead (Ethiopia) . . . . ***C. i. walie***
- No bony prominence on forehead . . . . (8)
8. Size larger (up to 1050 mm at shoulder); beard well developed; often a pale saddle patch on back . . . . ***C. i. sibirica*** and other Central Asiatic forms
- Size smaller (up to 860 mm); beard smaller; no pale saddle . . . . . ***C. i. ibex***

Capra aegagrus — Wild goat, Bezoar, Pasang

Capra aegagrus Erxleben, 1777. Daghestan, E. Caucasus.

C. a. aegagrus. Caucasus, Iran, Asia Minor, Oman. Incl. *caucasica* Gray, *cicilica*, *florstedti* (Bulghar Dag, Asia Minor), *gazella* Gmelin (Iran), *persica* (Laristan, Iran).

C. a. cretica Schinz, 1838. Crete, introduced on mainland of Greece. Incl. *cretensis*.

C. a. picta (Erhard, 1858). Aegian islands of Erimomilos (type-locality) and Samothraki.

C. a. blythi Hume, 1875. Sind (type-locality), Baluchistan and E. Iran. Incl. *neglectus*.

*C. a. turcmenica** Tzalkin, 1950. Kopet Dag and Khurasan Mts, Iran.

Capra aegagrus turcmenicus Tzalkin, 1950. 25 km S. of Gjaurs railway station, Tawa Mts, central Kopet Dag, Turkmenistan. Doubtfully distinct from *C. a. blythi* according to Heptner & Naumov (1961).

RANGE (Map 103). Mountains from Asia Minor and Caucasus to Kopet Dag, W. Afghanistan, Baluchistan and Sind. Isolates in Oman (Harrison, 1968a), on some of the Aegean Islands and Crete. Probably occurred formerly in N. Syria, Lebanon and perhaps further south in Palestine.

REMARKS. This is certainly the principal ancestor of the domestic goat. E & M-S accordingly used the name *C. hircus* L. which was based on the domestic goat, but according to the general principles followed for the nomenclature of domesticated animals (see p. 8) this name is passed over in favour of *C. aegagrus* and this name was also used by Heptner & Naumov (1961) and by Harrison (1968a). The taxonomy of this and the related species is confused by the probability that some populations have been derived from, or affected by interbreeding with domesticated goats. This, along with local extinction, e.g. in Palestine, makes it difficult to assess its relationships with other wild goats. It is, for example, possible that it once intergraded with *C. ibex nubiana* in Palestine. It is sympatric with *C. cylindricornis* in the eastern Caucasus but does not come into contact with *C. caucasica* in the western Caucasus.

The following forms, listed by E & M-S, are probably of domestic origin and synonymous with *hircus*: *dorcas* Reichenow, *jourensis*. For other synonyms of the domesticated goat see Couturier (1962:639).

The wild goats of Crete and the Aegean Islands have been reviewed by Schultz-Westrum (1963) who recognized the subspecies listed above.

Feral goats of domestic origin are found in many areas outside the range of any wild species, e.g. in Scotland, Wales, many Mediterranean Islands and on many oceanic islands throughout the world.

Capra ibex — Ibex

Capra ibex Linnaeus, 1758. Switzerland.

C. i. ibex. Alps (present populations derived by reintroduction from Gran Paradiso, Italian Alps, which is the only surviving endemic population). =*alpina*. Incl. *europea*, *graicus*.

C. i. sibirica (Pallas, 1776). Sayan and Altai Mts. =*pallasii*. Incl. *altaica*, *fasciata*, *lydekkeri*, *typica*.
*Capra sibirica lorenzi** Satunin, 1905. 320 verst S. of Nishneudinsk, Sayan Mts (*fide* Heptner & Naumov, 1961).

C. i. alaiana Noack, 1902. Mountains of Turkestan (type-locality Alai Mts). Incl. *almasyi* (Tien Shan), *merzbacheri*, *transalaiana*.
*Capra sibirica formosovi** Tzalkin, 1949. Kyamattu Gorge, 26 km S.W. of Novo-Dimitrievsk, Tien Shan. (Perhaps not separable from the following.)

C. i. hemalayana Hodgson, 1841. W. Himalayas. Incl. *dauvergnii*, *filippii*, *pedri*, *sacin*, *sakeen*, *sakin*, *skyn*, *wardi*.
?*Capra sibirica dementievi* Tzalkin, 1949. Mt Tokhta-Khom, S.W. of Yarkand, Sinkiang, China (*fide* Heptner & Naumov, 1961).

C. i. hagenbecki Noack, 1903. Mongolian Altai. (Considered probably distinct from *C. i. sibirica* by Heptner & Naumov, 1961).

C. i. nubiana F. Cuvier, 1825. Palestine, Arabia, Egypt and Sudan (type-locality in Upper Egypt). =*typica*. Incl. *arabica*, *beden* (S.W. Arabia), *mengesi* (S. Arabia), *sinaitica* (Sinai).

*C. i. walie** Rüppell, 1835. Simien Mts, Ethiopia. =*vali*.
Capra walie Rüppell, 1835. Simen and Gojjam, Ethiopia.

RANGE (Map 103). Mountains of Spain, Pyrenees, Alps and central Asia from E. Sayan Mts (W. of Lake Baikal) to Turkestan, N. Afghanistan and Kashmir; Palestine, Arabia, Sinai, Egypt and Sudan east of Nile; Simien Mts of Ethiopia. Probably extinct in Egypt except perhaps in Sinai.

REMARKS. The above arrangement follows that of E & M-S except that (1) *severtzovi* (W. Caucasus) is provisionally excluded (see below under *C. caucasica*); (2) several races of central Asiatic ibex are listed, following Heptner & Naumov (1961); (3) *walie* (Ethiopia) is tentatively included. The last has frequently been given specific rank, e.g. most recently by Ansell (1971b). It is isolated from the nearest population of *C. i. nubiana* and pending more detailed study the choice of specific or subspecific rank seems rather arbitrary, although it is undoubtedly of at least subspecific rank. Heptner & Naumov gave *C. sibirica* (including all the central Asian forms) specific rank, but this seems to have been based as much on zoogeographical as on morphological grounds.

As here delimited, *C. ibex* makes contact with *C. falconeri* in Kashmir and Afghanistan, and possibly with *C. aegagrus* in Oman. It almost certainly made contact in the past with *C. aegagrus* in Palestine and Lebanon, and records of both types of horns in Upper Palaeolithic cave deposits in Lebanon (see Zeuner, 1963:131) suggest that there was no complete intergradation and support the specific separation of *C. ibex* and *C. aegagrus*.

Capra caucasica West Caucasian tur, Kuban

Capra caucasica Güldenstaedt & Pallas, 1783. E. of Mt Elbruz, central Caucasus. Incl. *dinniki*† (N.W. Caucasus), *raddei*† (S.W. Caucasus), *severtzovi*† (W. of Mt Elbruz).

RANGE (Map 103). W. Caucasus from 39°55′E to headwaters of R. Psygansu at 43°30′E. Eastern end of range overlaps with that of *C. cylindricornis* by about 60 km (see Heptner & Naumov, 1961 for details).

REMARKS. E & M-S, while recognizing two species in the Caucasus, used the name *caucasica* for the eastern species and treated the western one as a race of *C. ibex* (*C. i. severtzovi*). The use of *caucasica* for the western species follows Heptner & Naumov (1961) who gave both forms specific rank. The western one does seem close to *C. ibex*, but less so than any of the forms here included in *C. ibex* and it is therefore provisionally retained as a species.

***Capra cylindricornis*†** East Caucasian tur

Ovis cylindricornis Blyth, 1841. Eastern end of high Caucasus, probably near Mt Kazbek. Incl. *pallasii* Rouillier.

RANGE (Map 103). E. Caucasus from about 48°30′E (E. of Baba Dag) to 43°10′E in the central Caucasus, overlapping that of *C. caucasica* by about 60 km.

REMARKS. This has been generally recognized as a distinct species, but was called *C. caucasica* by E & M-S. According to Heptner & Naumov (1961) individual hybrids with *C. caucasica* are known in the sympatric zone, but there is no population showing intermediate characters.

Capra pyrenaica Spanish ibex

Capra pyrenaica Schinz, 1838. Spanish Pyrenees.
C. p. pyrenaica. Pyrenees. =*typica*.
C. p. hispanica Schimper, 1848. Sierra Nevada, S.E. Spain.
C. p. lusitanica Schlegel, 1872. N. Portugal (extinct).
C. p. victoriae Cabrera, 1911. Sierra de Gredos, central Spain.

RANGE (Map 104). Present, at least until recently, in the central Spanish Pyrenees; Sierra de Gredos; Sierra Morena, Sierra de Ronda, Sierra Nevada, Sierra de Cazorla (Andalusia); Sierra de Cardo (N. Valencia) (Couturier, 1962). In the Pyrenees survival of the indigenous race is very dubious (near Mt Perdido), but animals from the Sierra de Gredos and elsewhere have been released nearby (IUCN, 1966).

REMARKS. This form seems more distinct than any of those here included in *C. ibex*, and it is therefore provisionally retained as a species.

Capra falconeri Markhor

Aegoceros (*Capra*) *falconeri* Wagner, 1839. Astor, Kashmir.
C. f. falconeri. E. Kashmir. =*typica*.
C. f. heptneri Zalkin, 1945. Tadzhikistan, Russian Turkestan. Incl. *ognevi* (S. Uzbekistan).
C. f. cashmiriensis Lydekker, 1898. W. Kashmir to Hindu Kush.
C. f. megaceros Hutton, 1842. N. Pakistan/Afghanistan border in region of Kyber Pass.
C. f. jerdoni Hume, 1875. W. Pakistan, between Peshawar and Quetta.
?*C. f. chialtanensis* Lydekker, 1913. Baluchistan S. of Quetta. Of doubtful status: possibly a form intermediate between *C. falconeri* and *C. aegagrus blythi* which occurs immediately to the south (Roberts, 1977).

RANGE (Map 104). Mountains from E. Kashmir to the Hindu Kush and south in W. Pakistan to region of Quetta; S. Uzbekistan and Tadzhikistan in USSR; according to Couturier (1962) also in N.W. Afghanistan to the west of Herat. The range is now highly fragmented (see map in Schaller & Kahn (1975) for details). It makes contact with the range of *C. ibex* in Kashmir, but the ibex tend to be on higher ground. In the south the range must at one time have abutted on that of *C. aegagrus blythi*, but the nature of the contact is obscure.

REMARKS. The southernmost race, *C. f. chialtanensis*, has been interpreted rather vaguely as the result of hybridization with *C. a. blythi* (Roberts, 1977) or with domesticated goats (Cobb, 1958).

This species has almost certainly contributed to some breeds of domesticated goats e.g. the Circassian (Zeuner, 1963), and according to Couturier (1962:652) has been itself introduced and domesticated in the Caucasus.

Capra lervia Barbary sheep, Aoudad

Antilope lervia Pallas, 1777. Oran, Algeria.
C. l. lervia. Morocco to Tunisia. Incl. *tragelaphus*.
C. l. fassini (Lepri, 1930). Libya and S. Tunisia.
C. l. ornata (Audouin, 1829). Egypt.
C. l. sahariensis (Rothschild, 1913). Algerian Sahara (type-locality Oued Mya at 28°30′N, 3°E).
*C. l. blainei** (Rothschild, 1913). N. Sudan, N.E. Tchad and Uweinat Mts (on border of Egypt, Sudan and Libya).
Ovis lervia blainei Rothschild, 1913. Border of Dongola Province and Kordofan, Sudan.
*C. l. angusi** (Rothschild, 1921). S. Sahara from Tibesti and Air to Adrar (Mali).
Ammotragus lervia angusi Rothschild, 1921. Tarrouaji Mt, Asben, Niger.

RANGE (Map 103). Mountains and low hills in the desert zone throughout most of the Sahara, south to about 14° near the Niger and through N. Tchad and Sudan to the Red Sea. Decreasing rapidly on the northern edge of its range.

REMARKS. Formerly placed in *Ovis*, this species has for most of this century been generically separated in *Ammotragus*. It is generally believed to be closer to *Capra* than to *Ovis* and has recently been placed in *Capra* by Ansell (1971b) mainly on the basis of its proven interfertility with domestic *Capra* (see for example Petzsch, 1958).

Genus *PSEUDOIS*

Pseudois Hodgson, 1846. Type-species *Ovis nayaur* Hodgson. =*Pseudovis*.

A monospecific genus, very closely related to *Capra*, although formerly included in *Ovis*.

Pseudois nayaur Bharal, Blue sheep

Ovis nayaur Hodgson, 1833. Tibetan frontier of Nepal.
P. n. nayaur. Himalayas and Tibet. =*nahoor*, *nahura*. Incl. *burrhel*=*barhal*.
P. n. szechuanensis Rothschild, 1922. China, type-locality in Shensi.

RANGE (Map 104). Alpine zone, usually above 2500 m, from Kashmir through the Himalayas and most of Tibet to Szechuan and north to the Ordos Plateau, the Ala Shan and extreme S.W. Inner Mongolia.

Genus *OVIS*
Sheep

Ovis Linnaeus, 1758. Type-species '*Ovis aries* L.', the domestic sheep. =*Ammon*, *Aries*, *Musmon*. Incl. *Argali*, *Caprovis*, *Musimon*, *Pachyceros*.

A moderately well defined genus closely related to *Capra*. *Ammotragus* and *Pseudois* were formerly included in *Ovis* but are now generally considered to be closer to *Capra*. Classification within the genus has been very unstable, but it is probable that only two or three species should be recognized. *O. ammon* is confined to the Palaearctic Region, *O. canadensis* occurs in N.E. Siberia and N.W. America whilst *O. dalli*, occurring north of *O. canadensis* in Canada and Alaska, is considered specifically distinct by most

American authors, but as a subspecies of *O. canadensis* by Heptner & Naumov (1961).

Horn cores of males short and thick, shorter than their basal circumference and than the total length of the skull . . . . **O. canadensis**

Horn cores long and thin, longer than their basal circumference and than the total length of the skull . . . . **O. ammon**

Ovis canadensis Bighorn sheep

Ovis canadensis Shaw, 1804. Alberta, Canada.

O. c. nivicola Eschscholz, 1829. Asiatic part of range, type-locality in E. Kamchatka. Incl. *alleni* (Taigonos Peninsula, N.E. Siberia), *borealis* (Syverma Mts, N. Siberia), *lydekkeri* (Yana R., N. Siberia), *potanini* (Transbaikalia), *middendorfi*, *montanus*, *storcki*.

*Ovis nivicola koriakorum** Tchernyausky, 1962. Koriak Mts, upper R. Achay-Vayam, Kamchatka.

Also other subspecies in N. America.

RANGE (Map 104). Mountains of N.E. Siberia east of Lake Baikal and R. Lena; a probably isolated segment in the Putorana Mts east of the mouth of the R. Yenesei; throughout the mountains of western N. America from British Columbia to N. Mexico and, if *dalli* is conspecific, also from British Columbia to Alaska.

REMARKS. Heptner & Naumov (1961) recognized four subspecies in Siberia, but on slender grounds, and these are disregarded following Pfeffer (1967). The specific separation between this species and *O. ammon* seems well founded, in spite of indications of a high degree of interfertility in captivity (see Gray, 1954, 1966).

Ovis ammon Argali, Urial, Mouflon

Capra ammon Linnaeus, 1758. Altai Mts.

O. a. ammon. Altai and Sayan Mts. = *typica*. Incl. *altaica*, *argali*, *asiaticus*, *dauricus*, *intermedia*, *mongolica* Severtzov, *przewalskii*.

O. a. poloi Blyth, 1841. Pamirs, Karatau, Tien Shan, Alatau north to Tarbagatai (type-locality at source of Syr Daria, Pamirs). = *polii*, *typica*. Incl. *collium* (Chinghiztau, N. of L. Balkash), *heinsii*, *karelini* (Alatau), *littledalei* (Tien Shan), *nigrimontana* (Karatau), *sairensis*.

O. a. hodgsoni Blyth, 1841. Himalayas, Tibet, Sinkiang to Great Khingan Shan (Manchuria); type-locality in Tibet. Incl. ?*adametzi* (Lob Nor), *ammonoides*, *bambhera* (Nepal), *blythi*, *brookei* (Ladak), *comosa*, *dalailamae* (Sinkiang), *darwini* (S. Gobi), *heurii*, *jubata* Peters.

O. a. kozlovi Nasonov, 1913. Yabarai Mts, Inner Mongolia.

O. a. vignei Blyth, 1841. Turkestan, Afghanistan, Kashmir and Pakistan to E. Iran and the E. Elburz Mts; perhaps also in Oman (Harrison, 1968b). Incl. *arkal*, *arkar* (Ust Urt), *blanfordi* (Baluchistan), *bochariensis* (Tadzhikistan), *cycloceros* (Hazara Hills, Afghanistan), *dolgopolovi*, *montana*, *punjabiensis* (Salt Range), *varentsowi* (Kopet Dag).

O. a. orientalis† Gmelin, 1774. S. Iran to Asia Minor; type-locality in western Elburz Mts. Incl. *anatolica* (Cilician Taurus, Asia Minor), *armeniana*, *erskinei*, *gmelini*, *isphaganica*, *isphahanica*, *laristanica*† (S. Iran), *urmiana*.

O. a. musimon† (Pallas, 1811). Sardinia (type-locality), Corsica and Cyprus. Incl. *corsico-sardinensis*, *cyprius*, *matschiei*, *musmon*, *occidentalis* (Corsica), *occidento-sardinensis*, *ophion* (Cyprus), *orientalis* Branz & Retzeburg (not of Gmelin).

*Ovis musimon sinesella** Turcek, 1949. Czechoslovakia (introduced stock) (*fide* Pfeffer, 1967). See also Turcek (1956) and Szunyoghy (1961).

RANGE (Map 104). All the mountains of central Asia from the Great Khingan in Manchuria, the Sayan and Altai, W. China and the Himalayas to Iran and Asia Minor; Oman (Harrison, 1968b); the islands of Cyprus, Corsica and Sardinia; introduced (from Corsica and Sardinia) in much of central Europe, in Iberia and in Crimea. Mapped in greater detail by Nadler *et al.* (1973).

REMARKS. The above classification is based on that of Pfeffer (1967) who gave descriptions of each subspecies recognized. The inclusion of all these forms in a single species was also supported by Heptner & Naumov (1961) who however recognized more subspecies. Karyologically these races fall into three groups: *ammon* (probably including *poloi*, *hodgsoni* and *kozlovi*) with 2n = 56; *vignei* with 2n = 58; and *orientalis* (including *musimon*) with 2n = 54 (Nadler *et al.*, 1973). The single, incomplete specimen known from Oman was considered comparable to *O. a. vignei* rather than to the nearer *O. a. orientalis* by Harrison (1968b). Nadler *et al.* (1973) treated *musimon*, *orientalis*, *vignei* and *ammon* as species but suggested that there was hybridization between *orientalis* and *vignei* in the Elburz Mountains.

This is the ancestor of the domesticated sheep which was named *O. aries* L., 1758. Following the procedure for the nomenclature of domesticated animals described on p. 8, the domesticated sheep is best referred to as '*Ovis* (domestic)'. For an account of its history see Zeuner (1963).

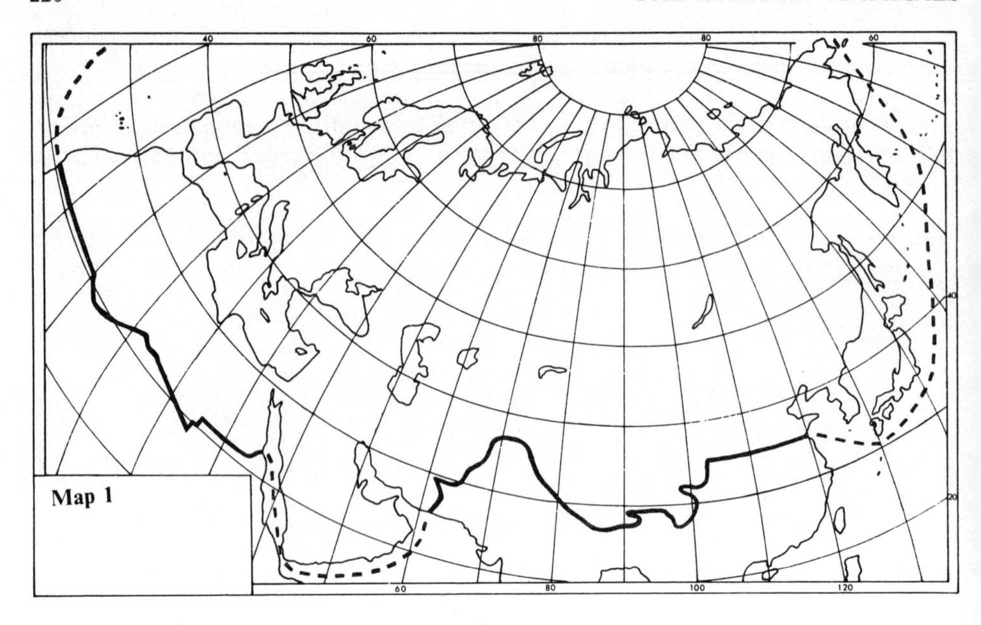

The Palaearctic Region showing the boundary used in this work.

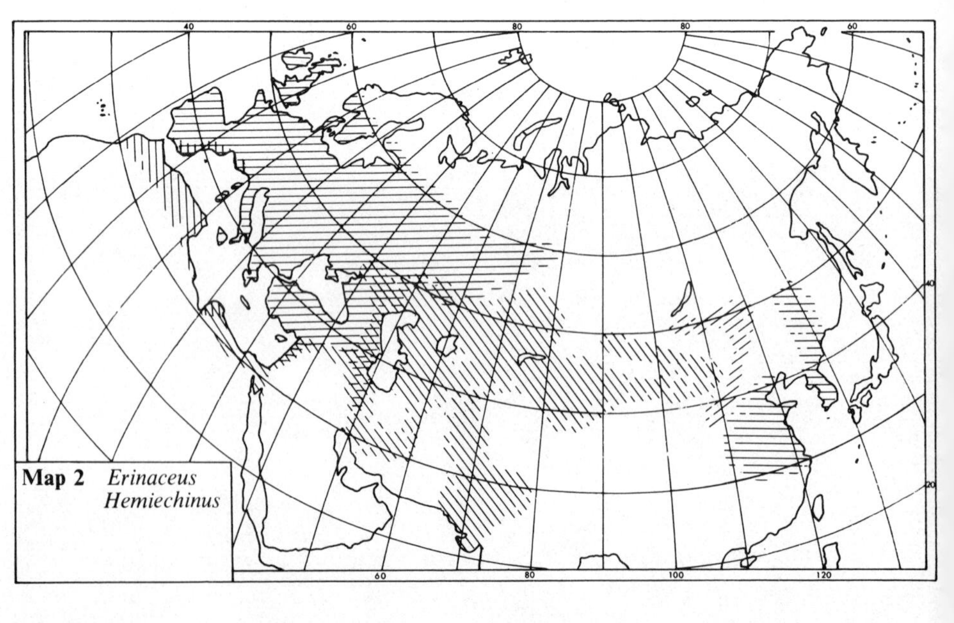

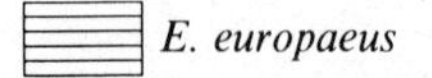

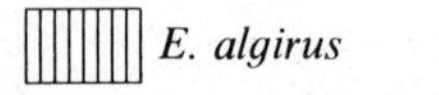

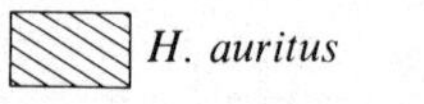

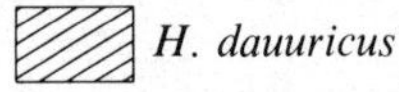

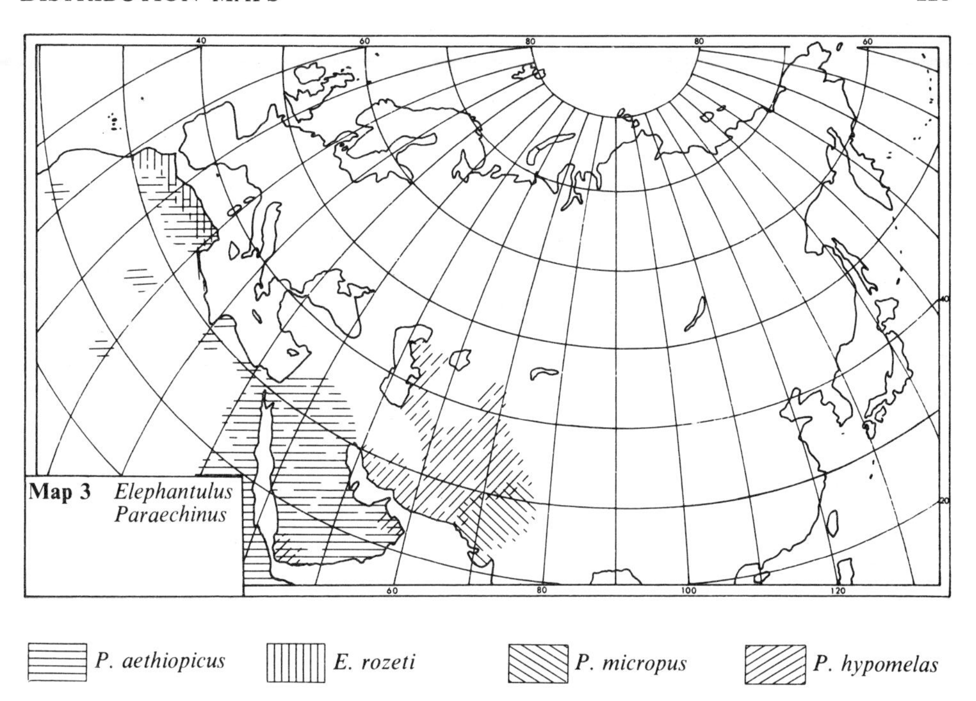
Map 3 Elephantulus
Paraechinus
P. aethiopicus
E. rozeti
P. micropus
P. hypomelas

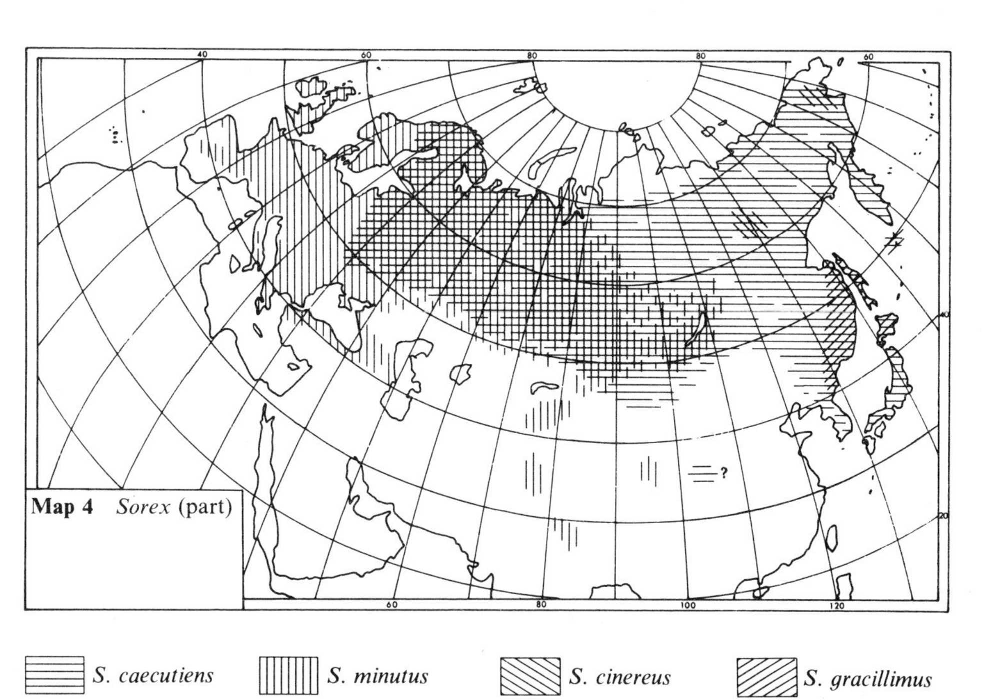
Map 4 Sorex (part)
?
S. caecutiens
S. minutus
S. cinereus
S. gracillimus

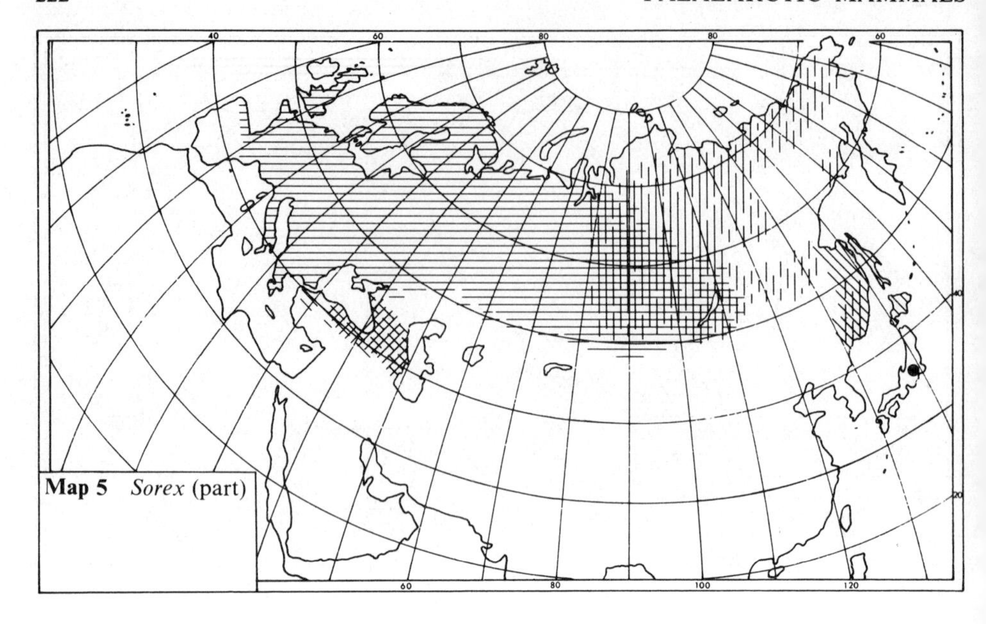

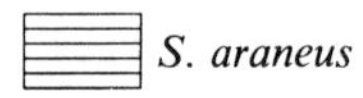

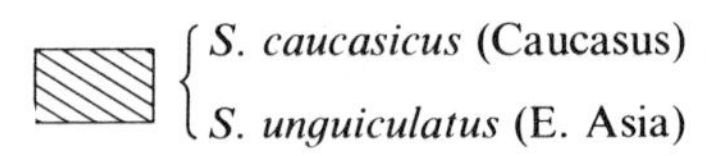

S. araneus *S. vir* { *S. caucasicus* (Caucasus) *S. unguiculatus* (E. Asia)

S. raddei ● *S. hosonoi*

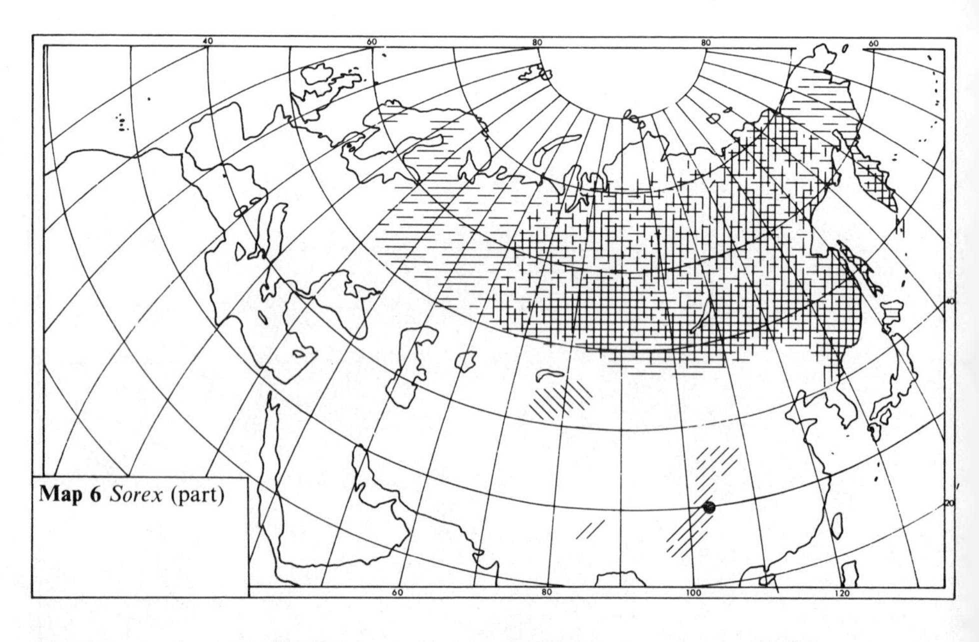

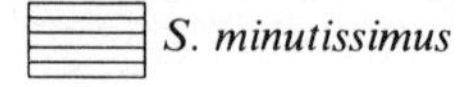
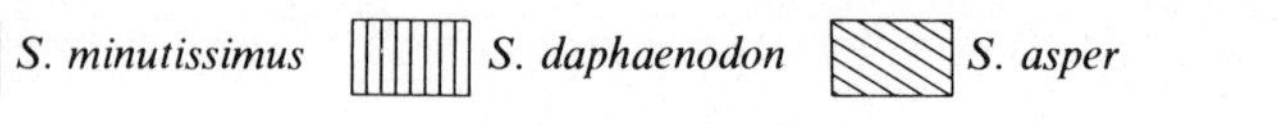
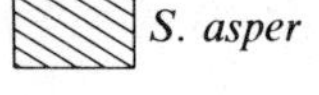
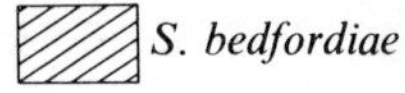

S. minutissimus *S. daphaenodon* *S. asper* *S. bedfordiae*

● *S. cylindricauda*

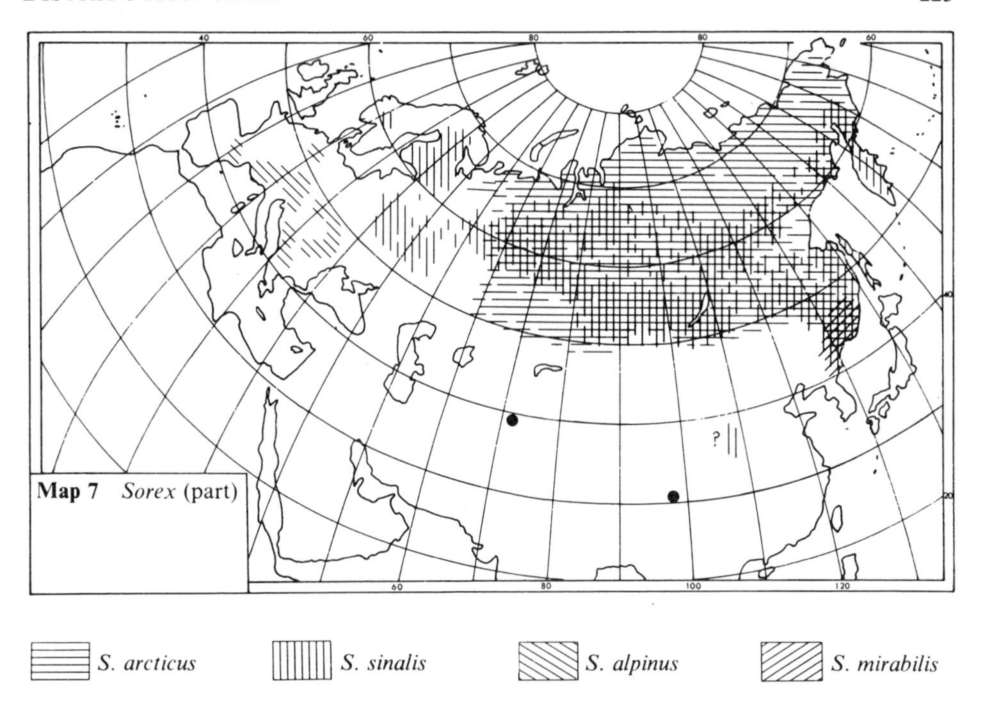

Map 7 *Sorex* (part)

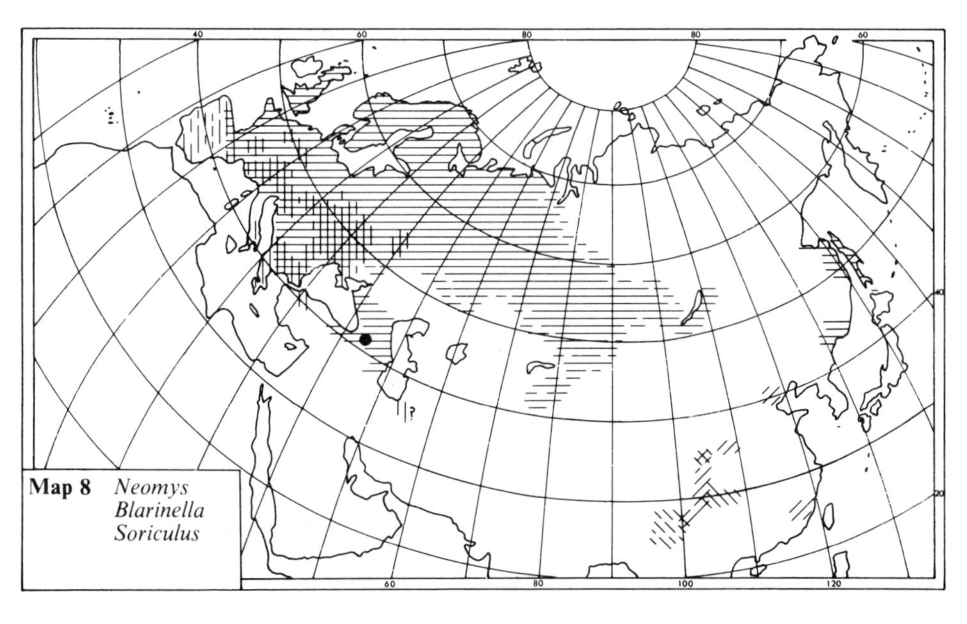

Map 8 *Neomys Blarinella Soriculus*

● *N. schelkovnikovi*

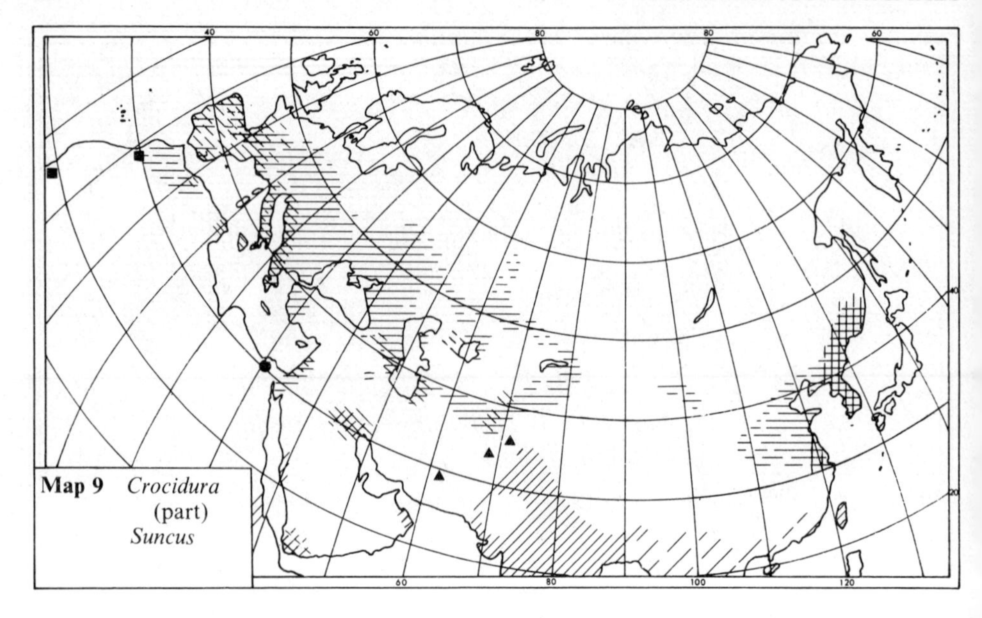

Map 9 *Crocidura* (part) *Suncus*

C. suaveolens | *C. lasiura* | *S. etruscus* | *S. murinus*

● *C. floweri* | ▲ *C. zarudnyi* | ■ *C. lusitania*

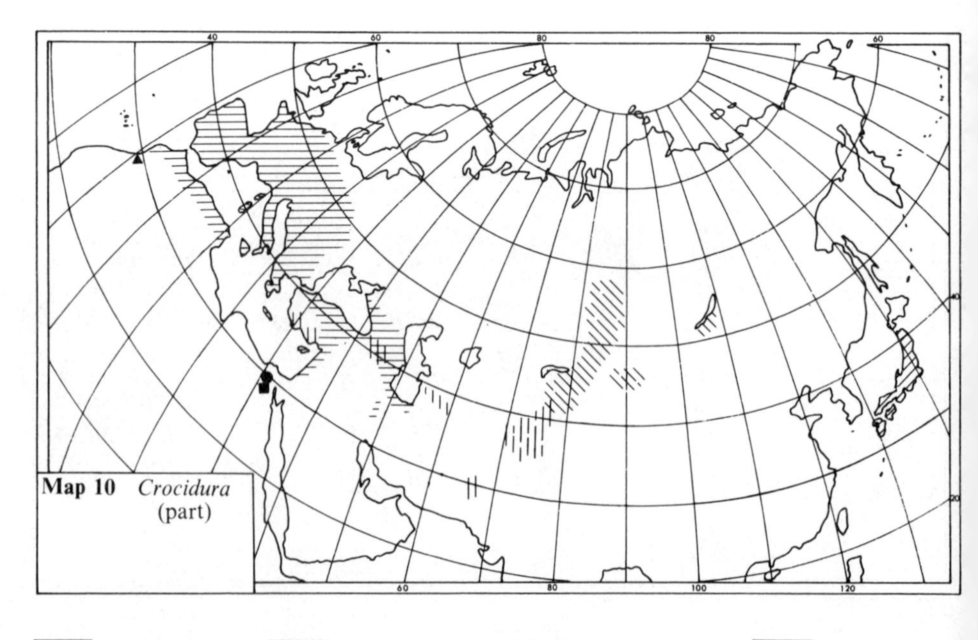

Map 10 *Crocidura* (part)

C. russula | *C. pergrisea* | *C. sibirica* | *C. dsinezumi*

● *C. religiosa* | ▲ *C. sericea* | ■ *C. flavescens*

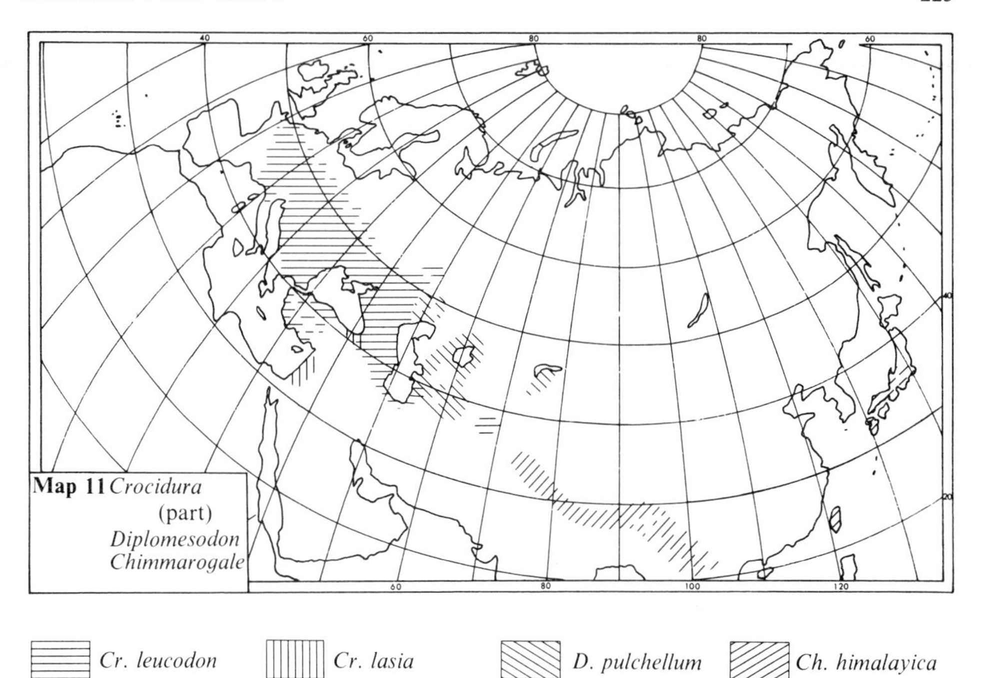

Map 11 *Crocidura* (part) *Diplomesodon Chimmarogale*

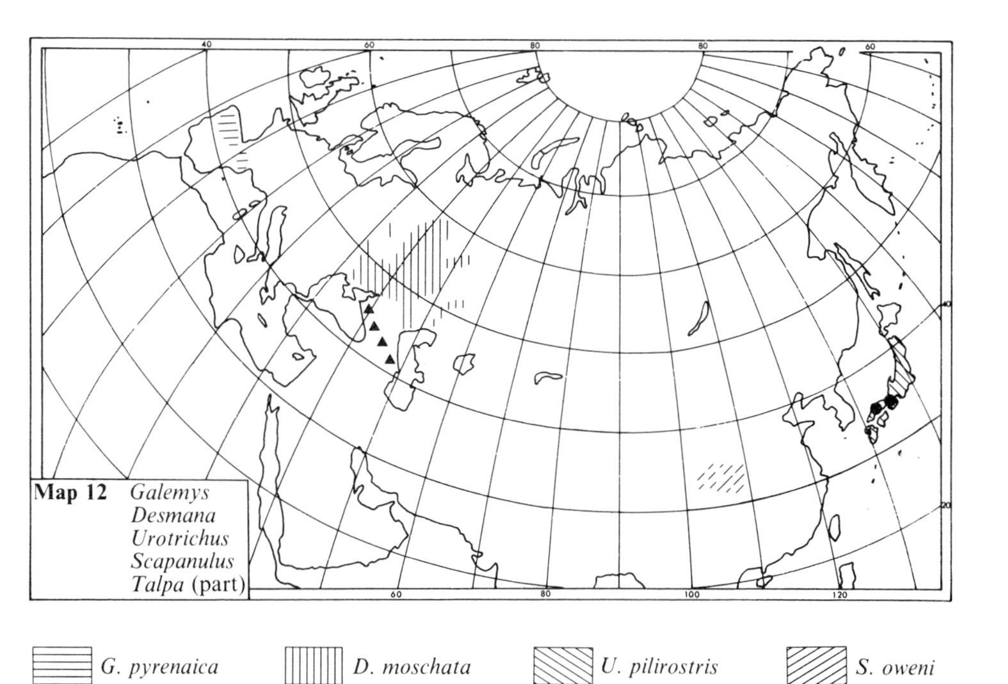

Map 12 *Galemys Desmana Urotrichus Scapanulus Talpa* (part)

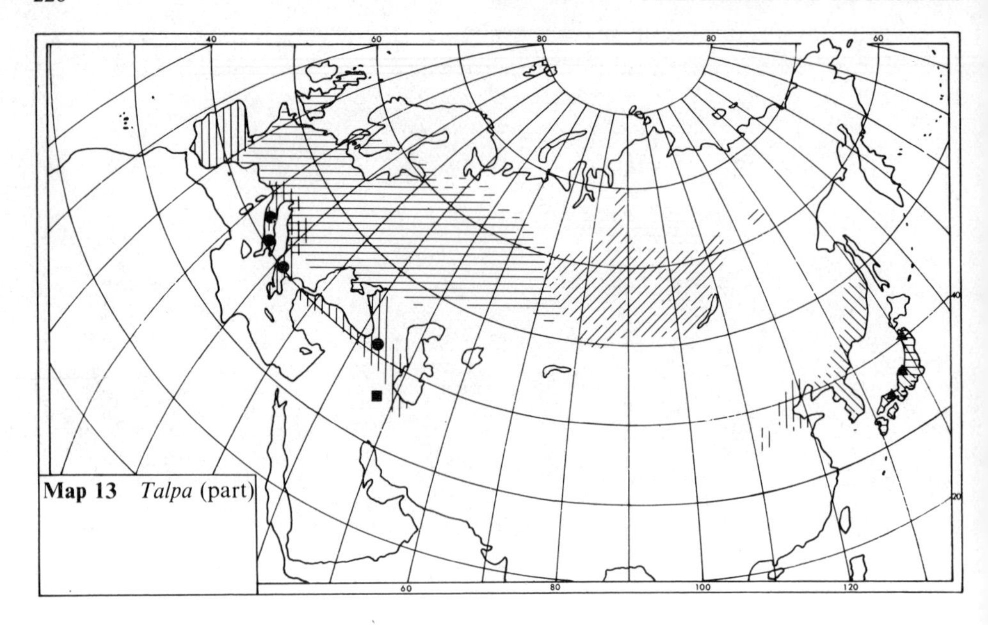

Map 13 *Talpa* (part)

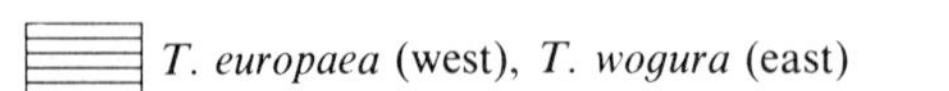 *T. europaea* (west), *T. wogura* (east)

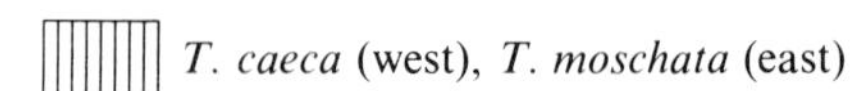 *T. caeca* (west), *T. moschata* (east)

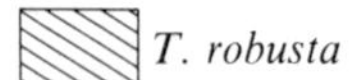 *T. robusta*

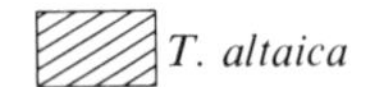 *T. altaica*

● *T. romana*

▲ *T. mizura*

■ *T. streeti*

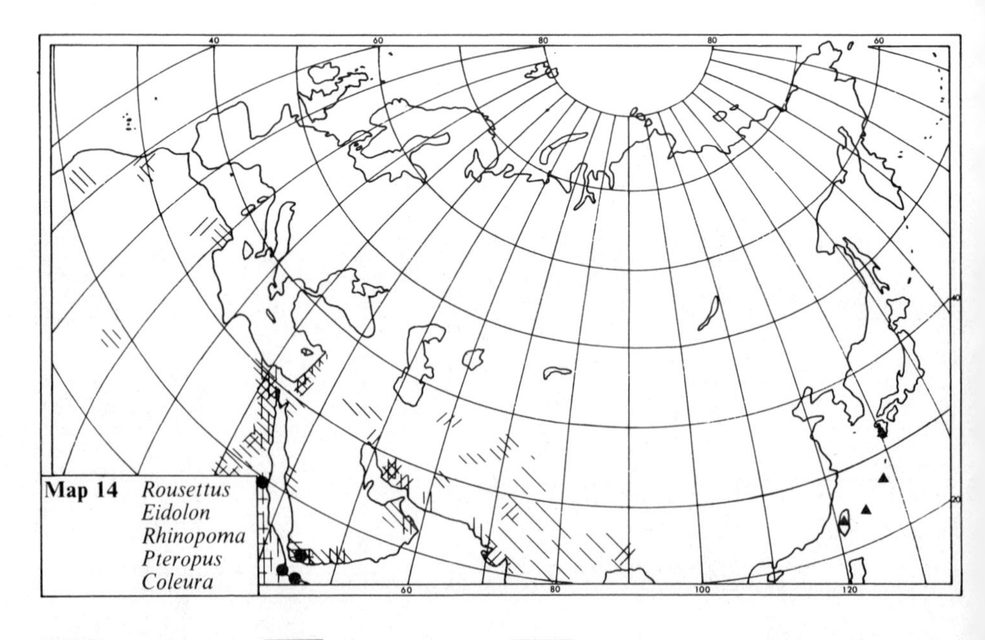

Map 14 *Rousettus*
Eidolon
Rhinopoma
Pteropus
Coleura

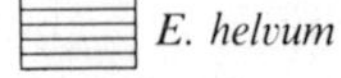 *E. helvum*

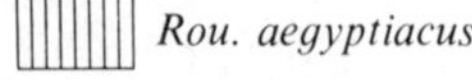 *Rou. aegyptiacus*

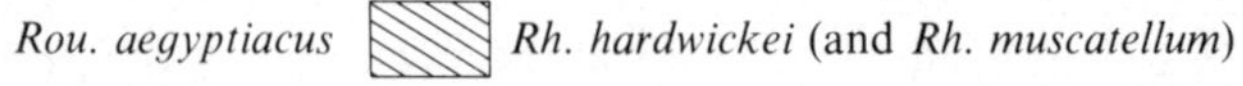 *Rh. hardwickei* (and *Rh. muscatellum*)

 *Rh. microphyllum*

● *C. afra*

▲ *P. dasymallus*

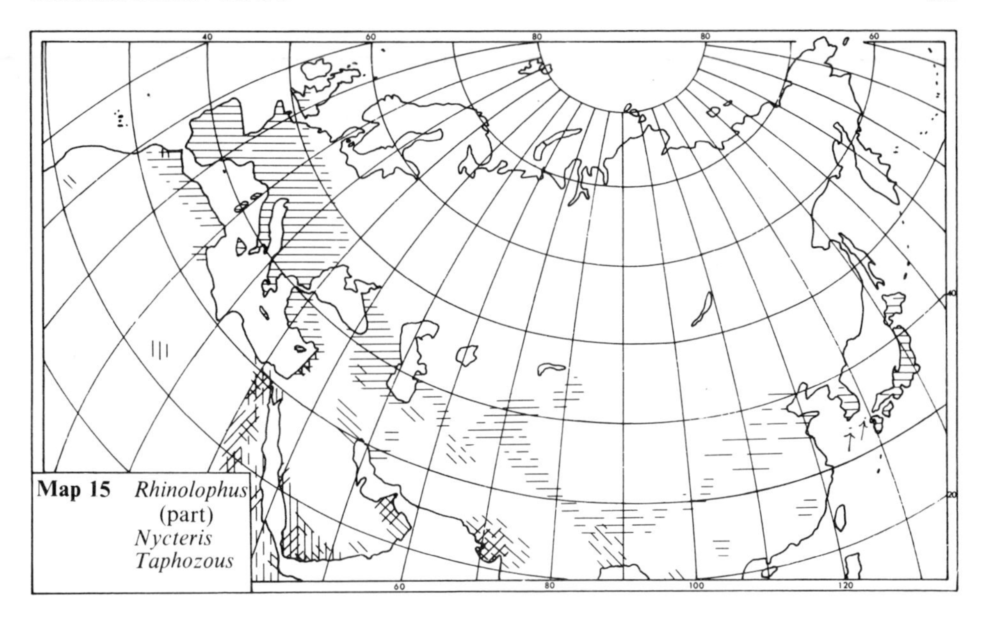

Map 15 *Rhinolophus* (part) *Nycteris* *Taphozous*

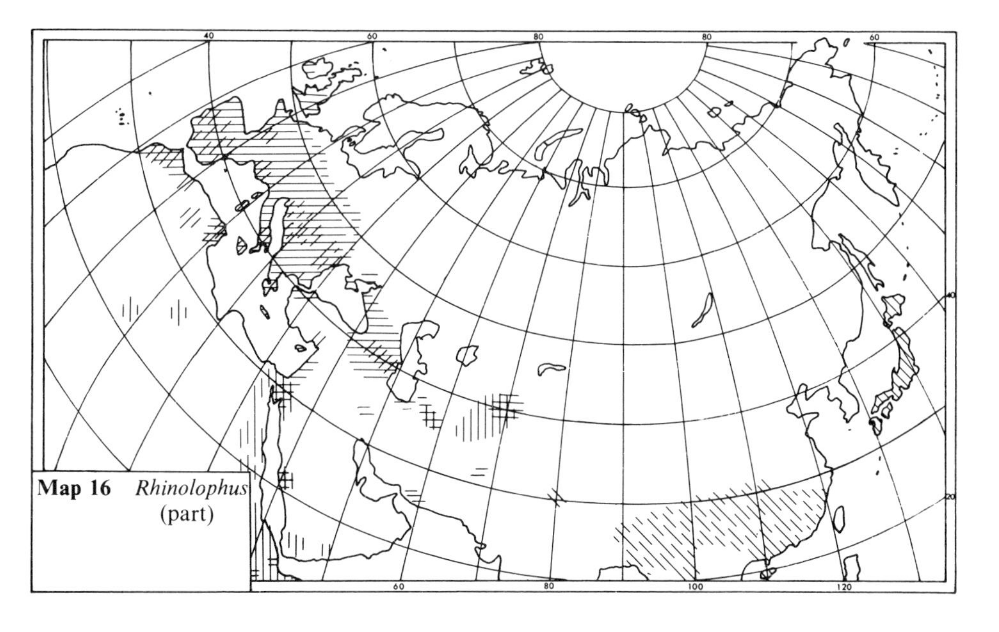

Map 16 *Rhinolophus* (part)

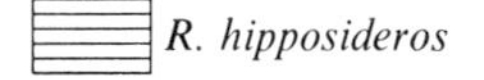

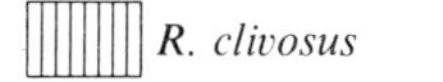

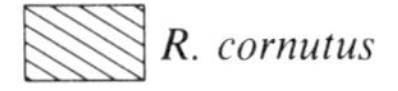

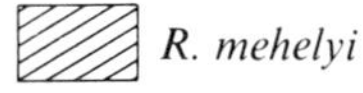

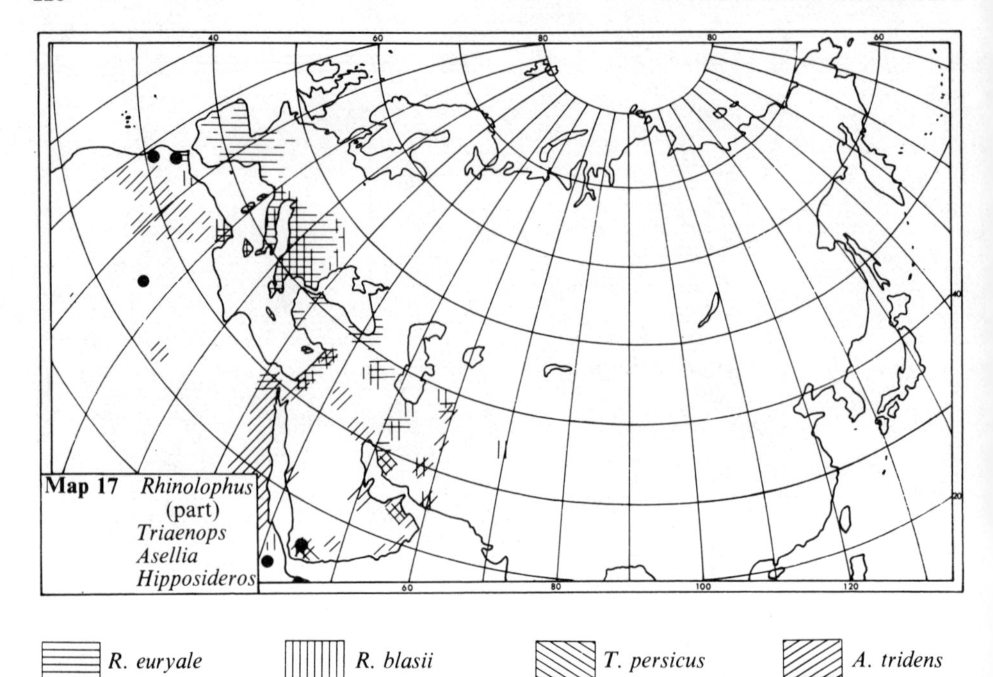

R. euryale *R. blasii* *T. persicus* *A. tridens*

● *H. caffer*

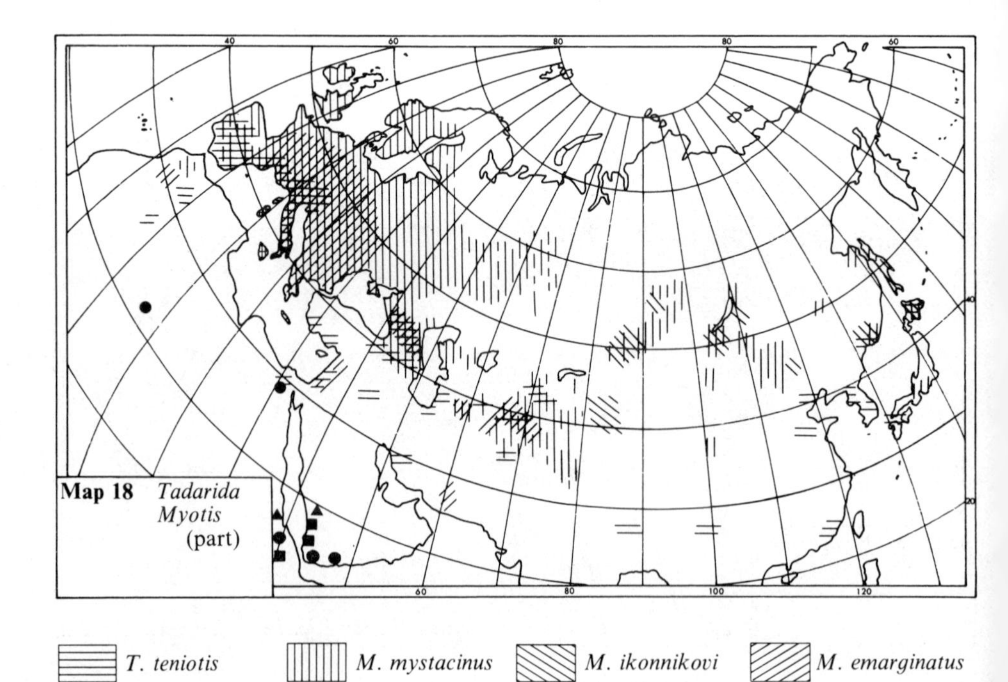

T. teniotis *M. mystacinus* *M. ikonnikovi* *M. emarginatus*

● *T. aegyptaica* ▲ *T. midas* ■ *T. pumila*

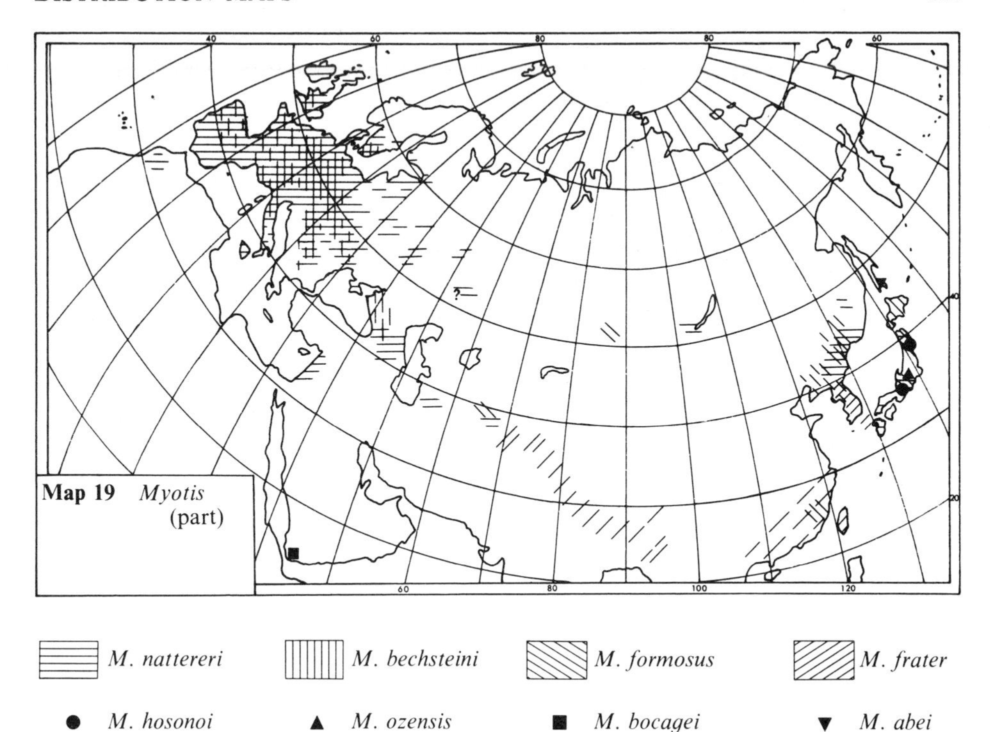

M. nattereri — M. bechsteini — M. formosus — M. frater

● M. hosonoi ▲ M. ozensis ■ M. bocagei ▼ M. abei

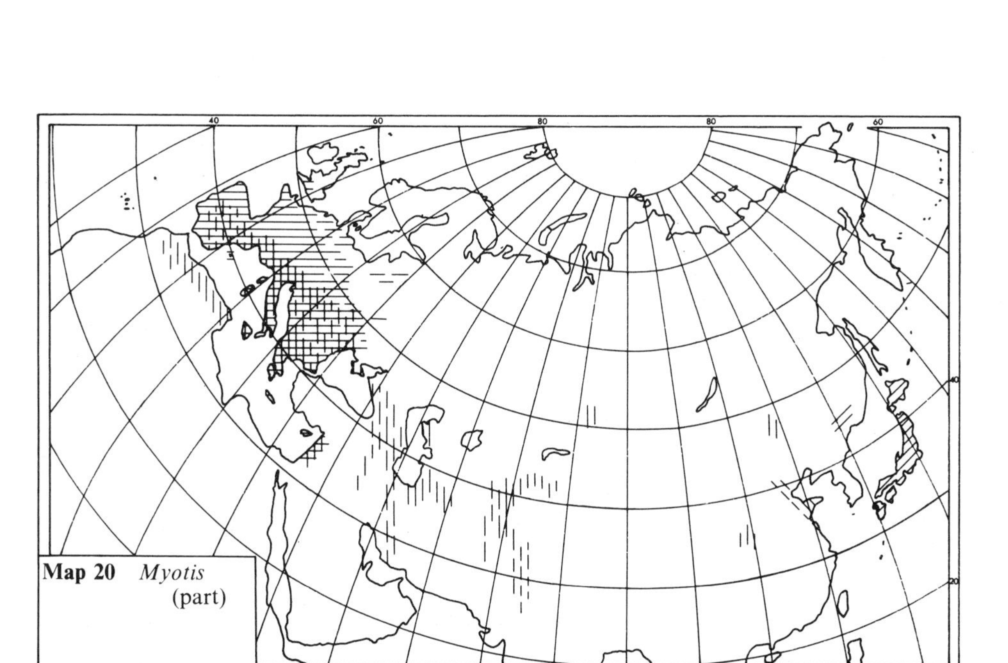

M. myotis — M. blythi — M. pequinius — M. macrodactylus

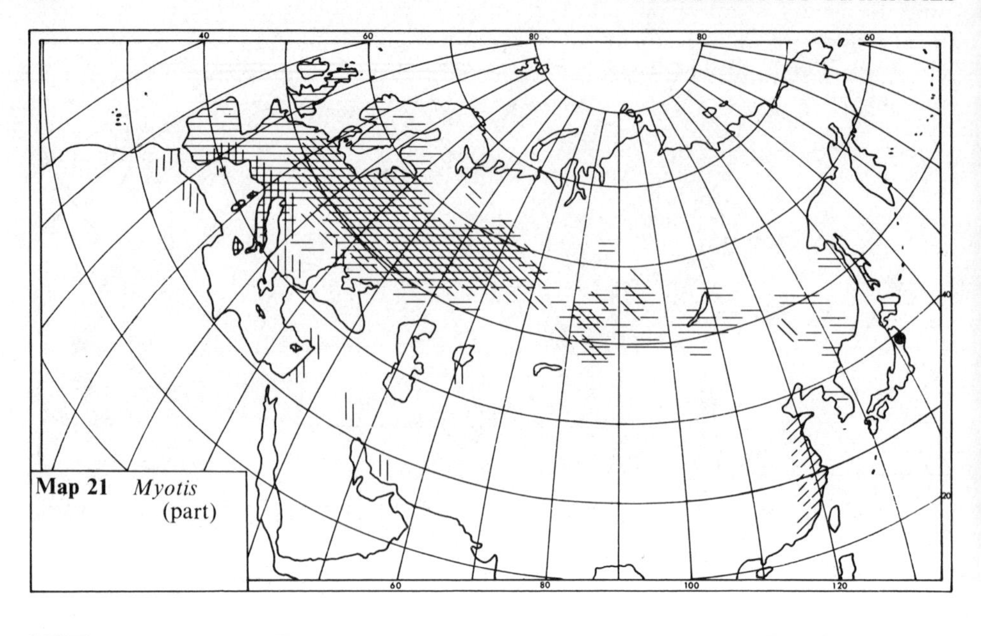

Map 21 *Myotis* (part)

M. daubentoni | *M. capaccinii* | *M. dasycneme* | *M. ricketti*

● *M. pruinosus*

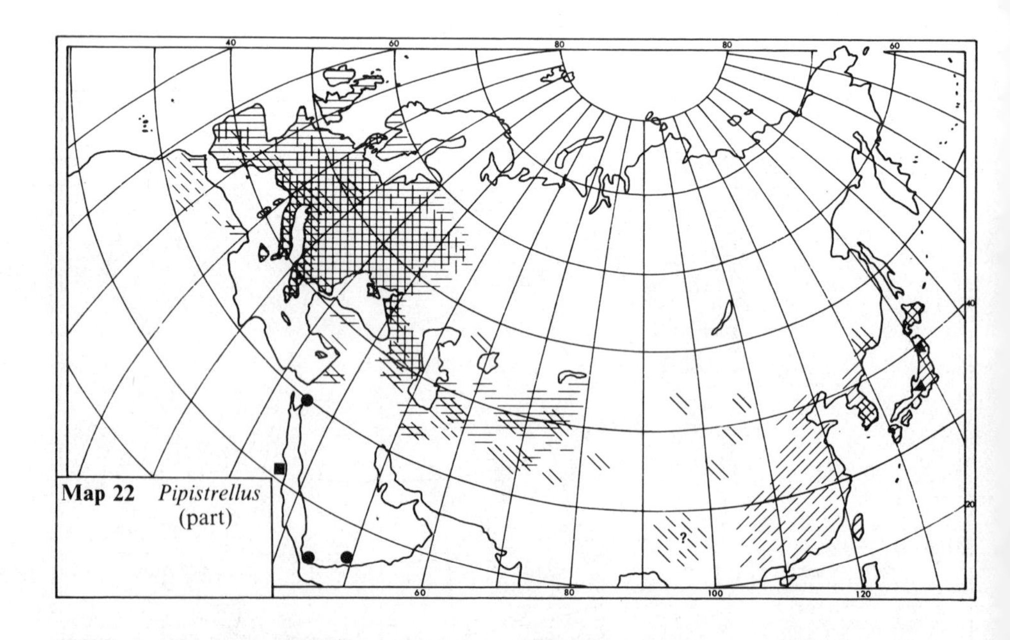

Map 22 *Pipistrellus* (part)

P. pipistrellus | *P. nathusii* | *P. savii* | *P. javanicus*

● *P. bodenheimeri* ▲ *P. endoi* ■ *P. ariel*

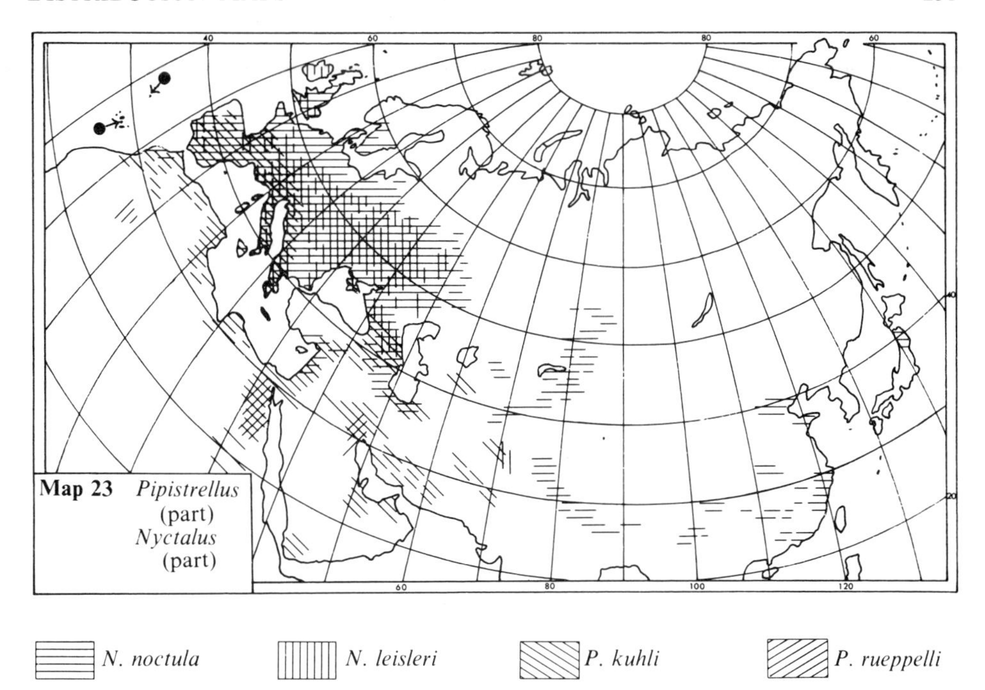

N. noctula *N. leisleri* *P. kuhli* *P. rueppelli*

● *P. maderensis*

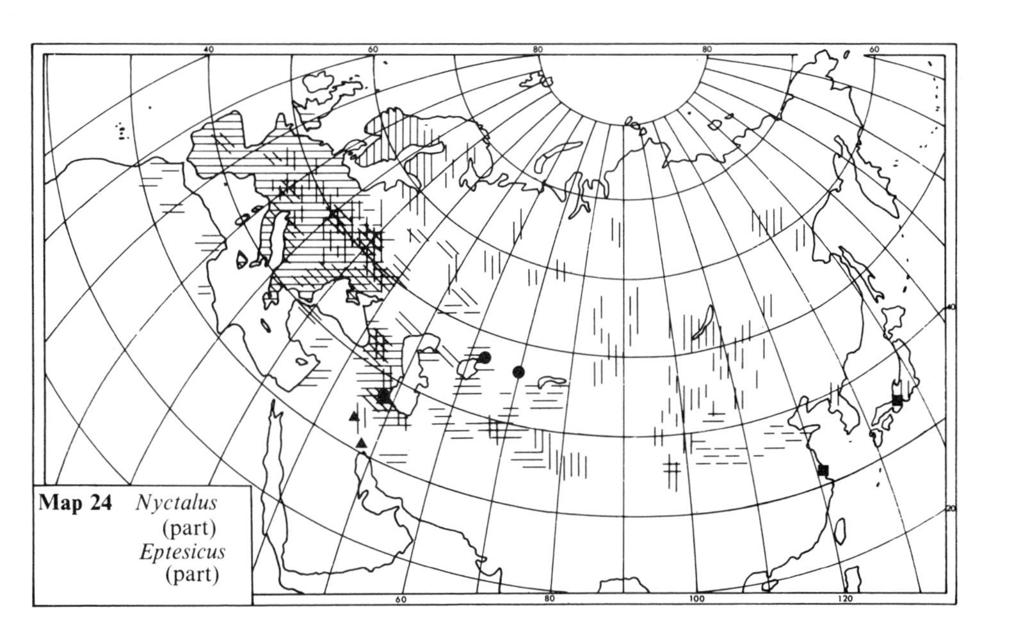

E. serotinus *E. nillsoni*

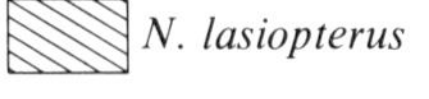
N. lasiopterus

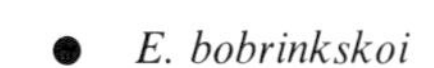
● *E. bobrinkskoi*

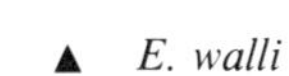
▲ *E. walli*

■ *N. aviator*

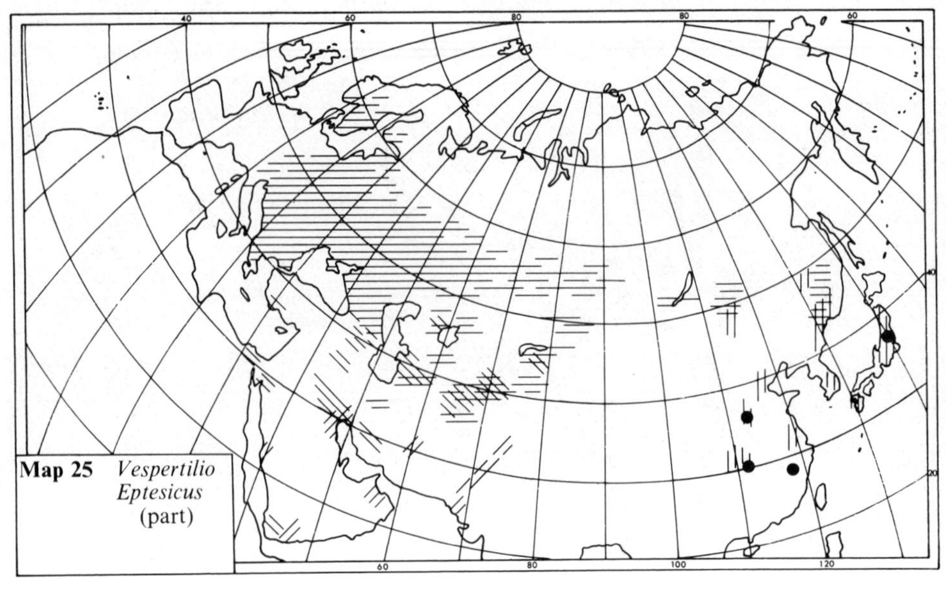

V. murinus *V. superans* *E. bottae* *E. nasutus*

● *V. orientalis*

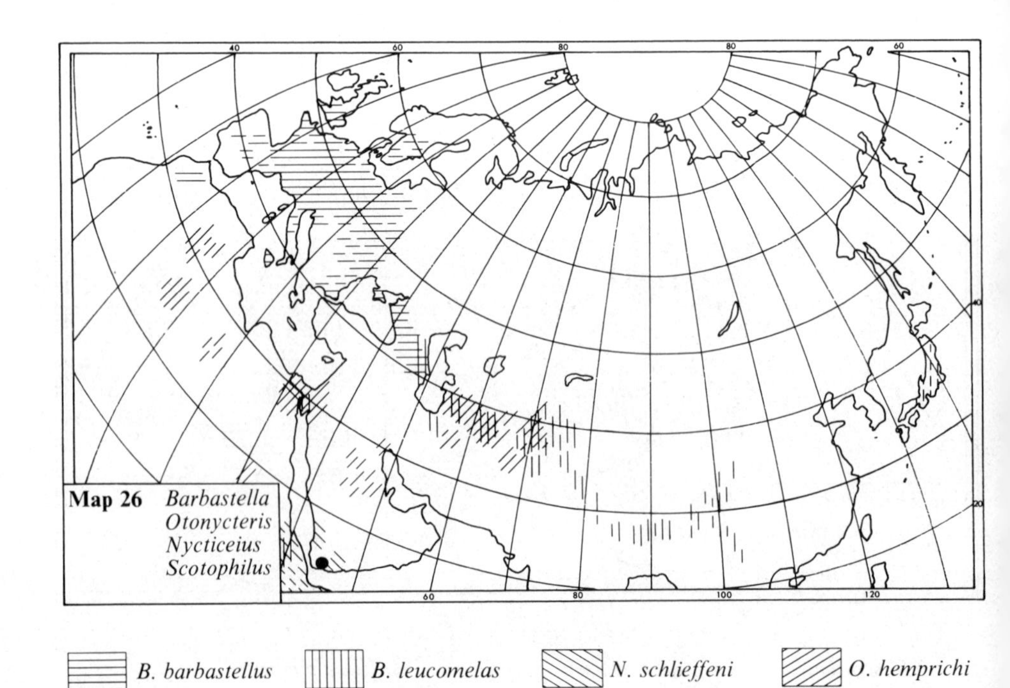

B. barbastellus *B. leucomelas* *N. schlieffeni* *O. hemprichi*

● *S. leucogaster*

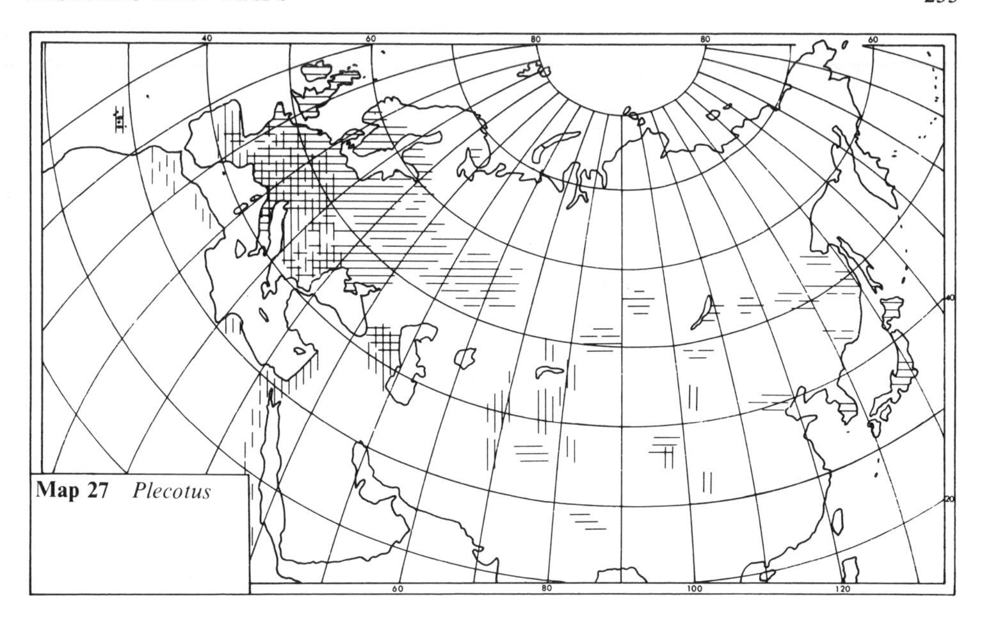

Map 27 *Plecotus*

P. auritus *P. austriacus*

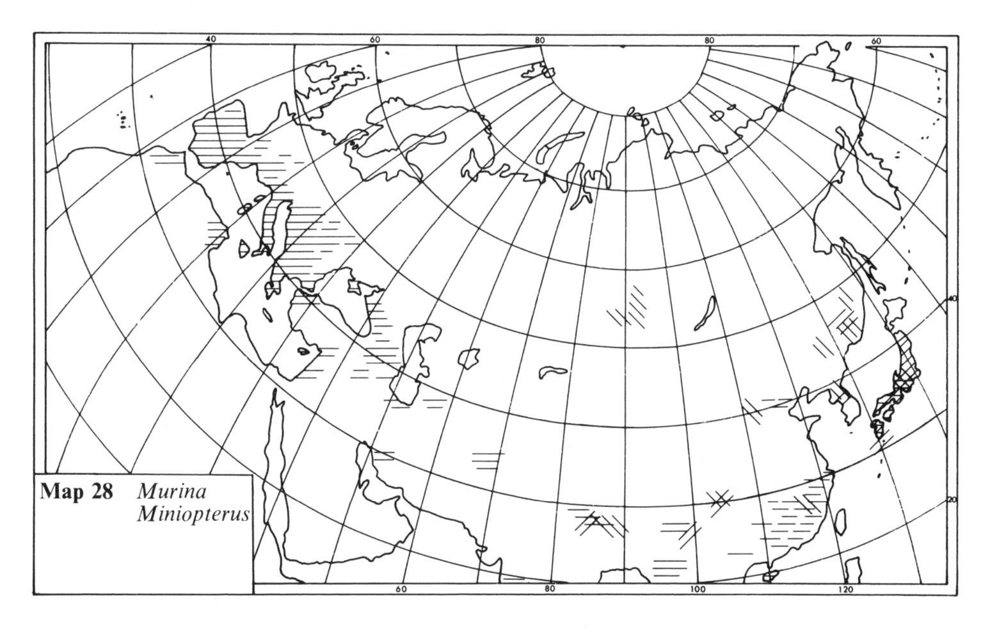

Map 28 *Murina Miniopterus*

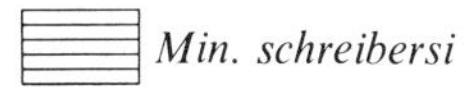 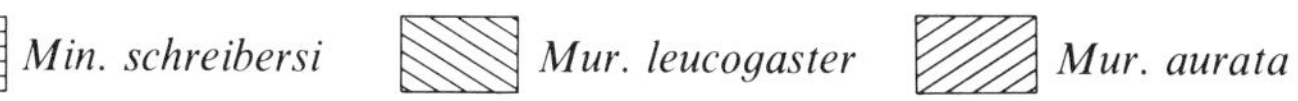 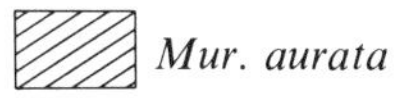

Min. schreibersi *Mur. leucogaster* *Mur. aurata*

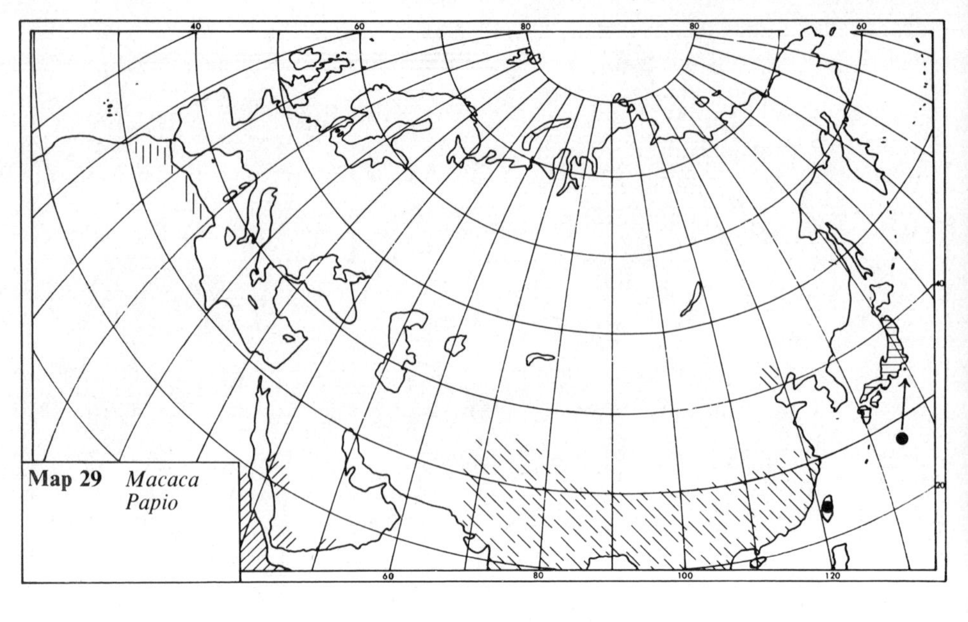

M. fuscata — *M. sylvana* — *M. mulatta* — *P. hamadryas*

● *M. cyclopus*

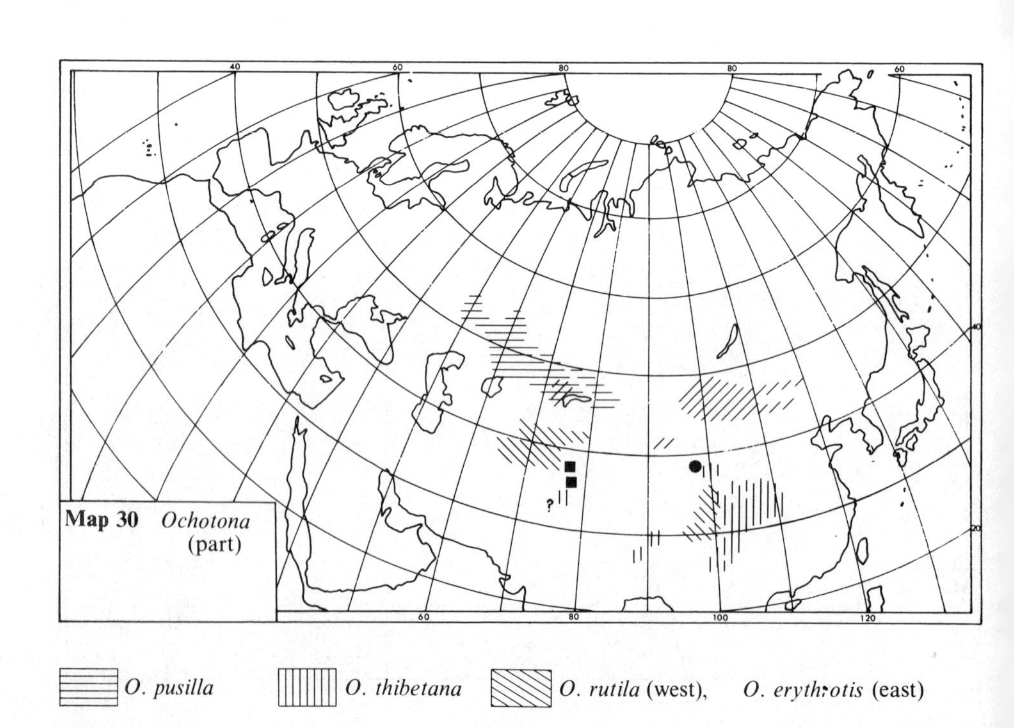

O. pusilla — *O. thibetana* — *O. rutila* (west), *O. erythrotis* (east)

O. pallasi — ● *O. thomasi* — ■ *O. ladacensis*

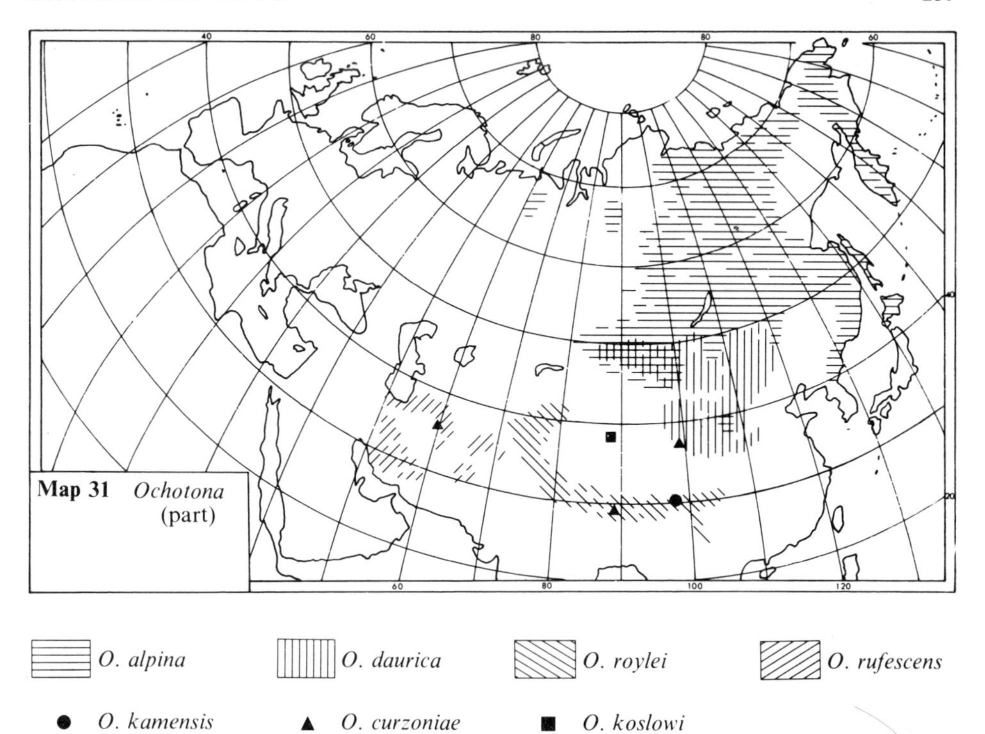

Map 31 *Ochotona* (part)

O. alpina — *O. daurica* — *O. roylei* — *O. rufescens*

● *O. kamensis* ▲ *O. curzoniae* ■ *O. koslowi*

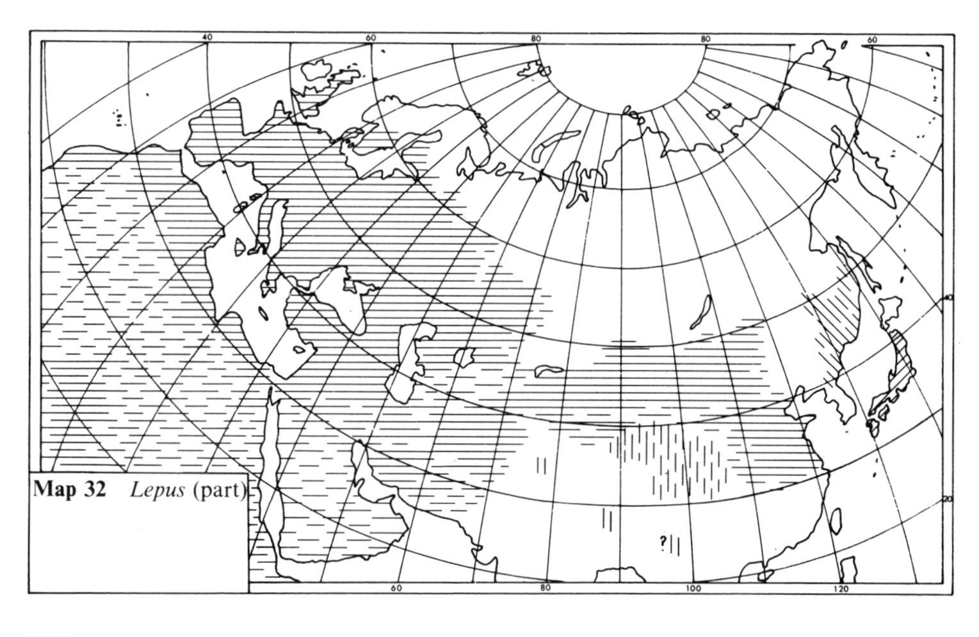

Map 32 *Lepus* (part)

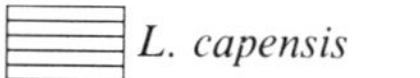 *L. capensis* 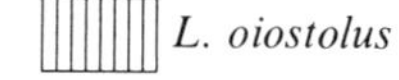 *L. oiostolus* 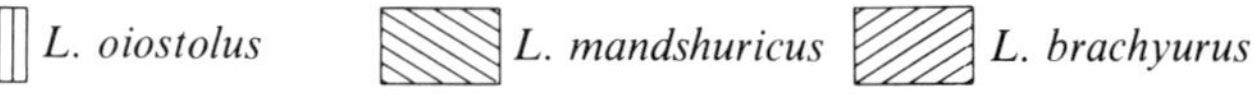 *L. mandshuricus* 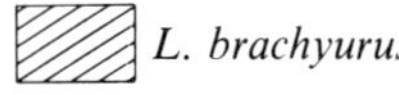 *L. brachyurus*

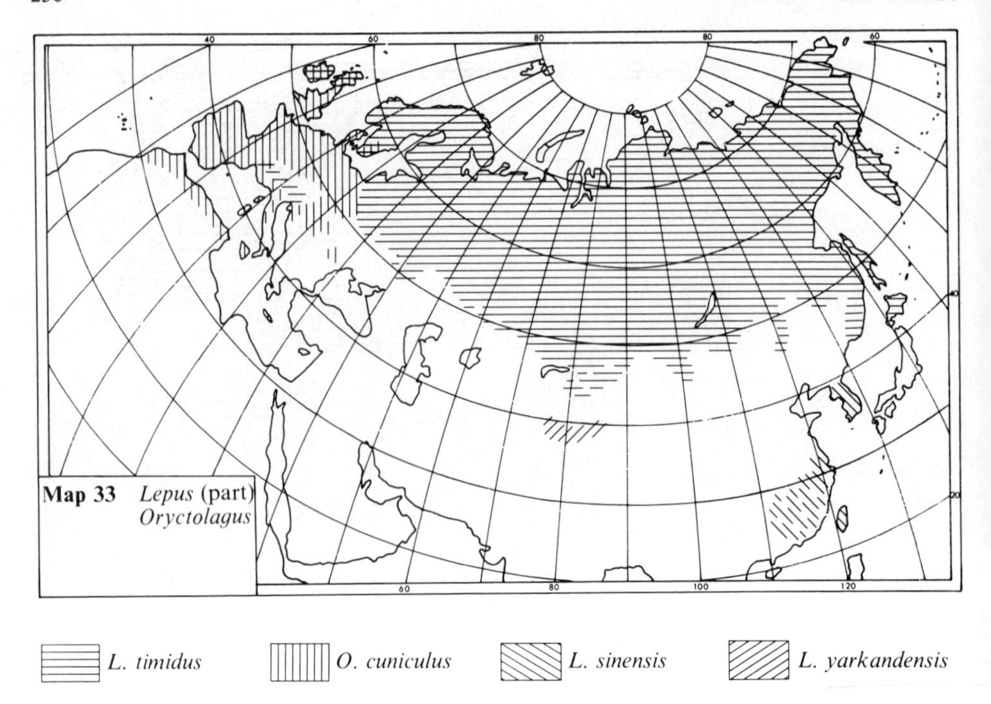
Map 33 Lepus (part)
Oryctolagus
L. timidus
O. cuniculus
L. sinensis
L. yarkandensis

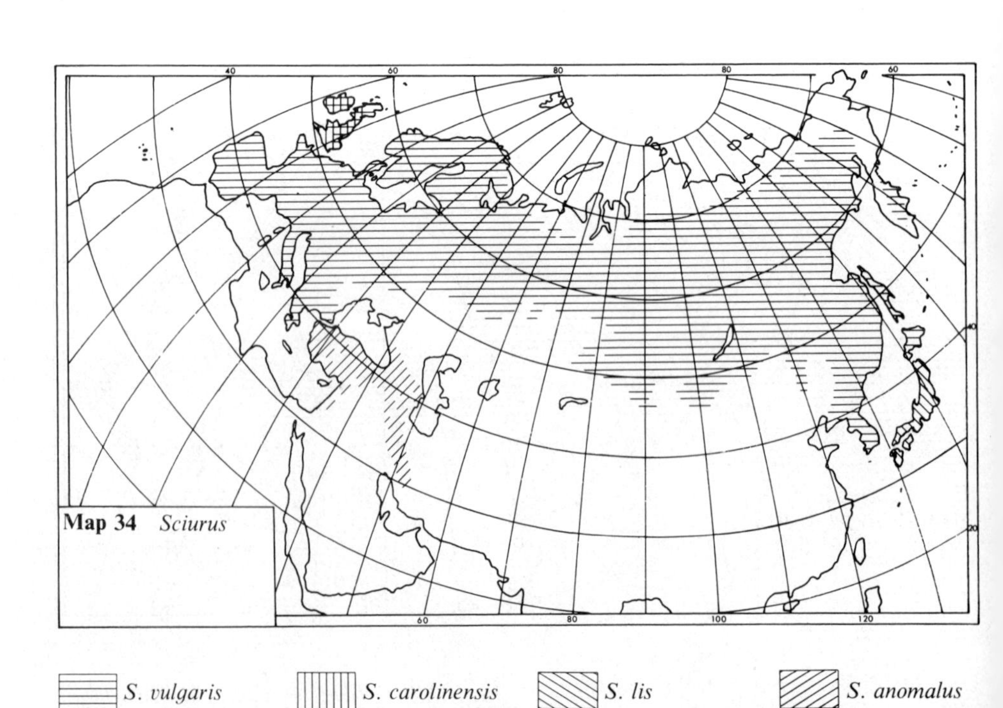
Map 34 Sciurus
S. vulgaris
S. carolinensis
S. lis
S. anomalus

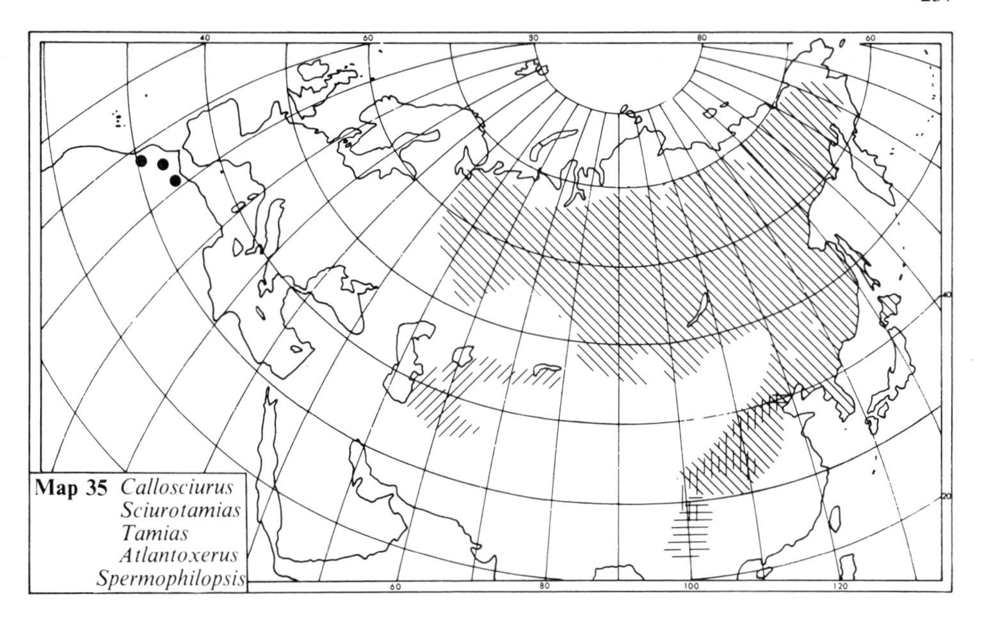

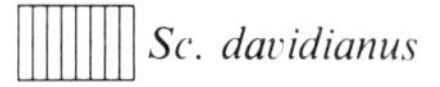

C. swinhoei *Sc. davidianus* *T. sibiricus* *Sp. leptodactylus*

● *A. getulus*

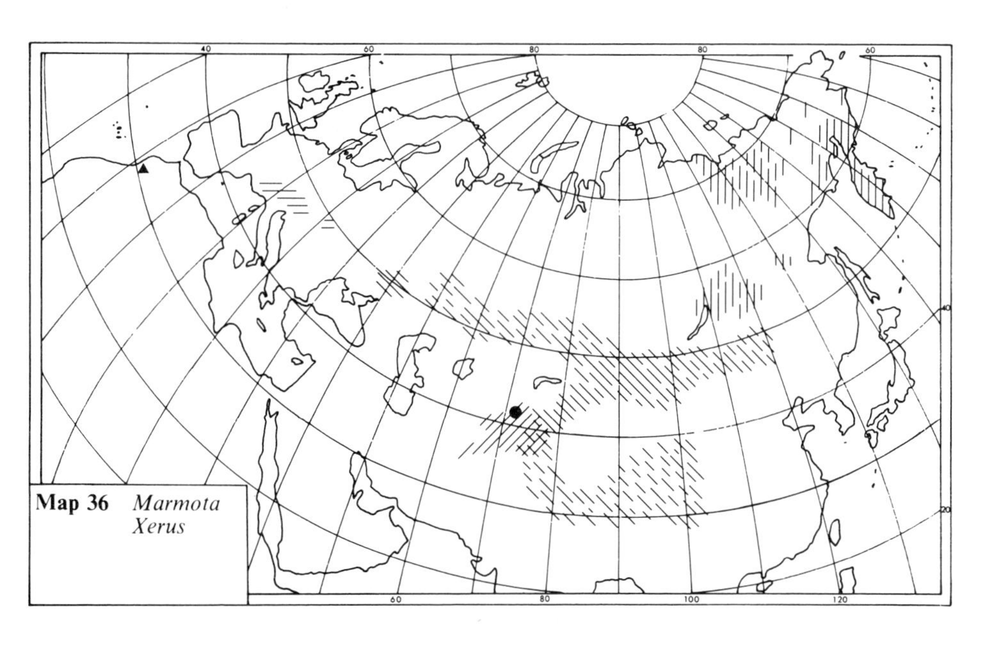

M. marmota *M. camtschatica* *M. bobak* *M. caudata*

● *M. menzbieri* ▲ *X. erythropus*

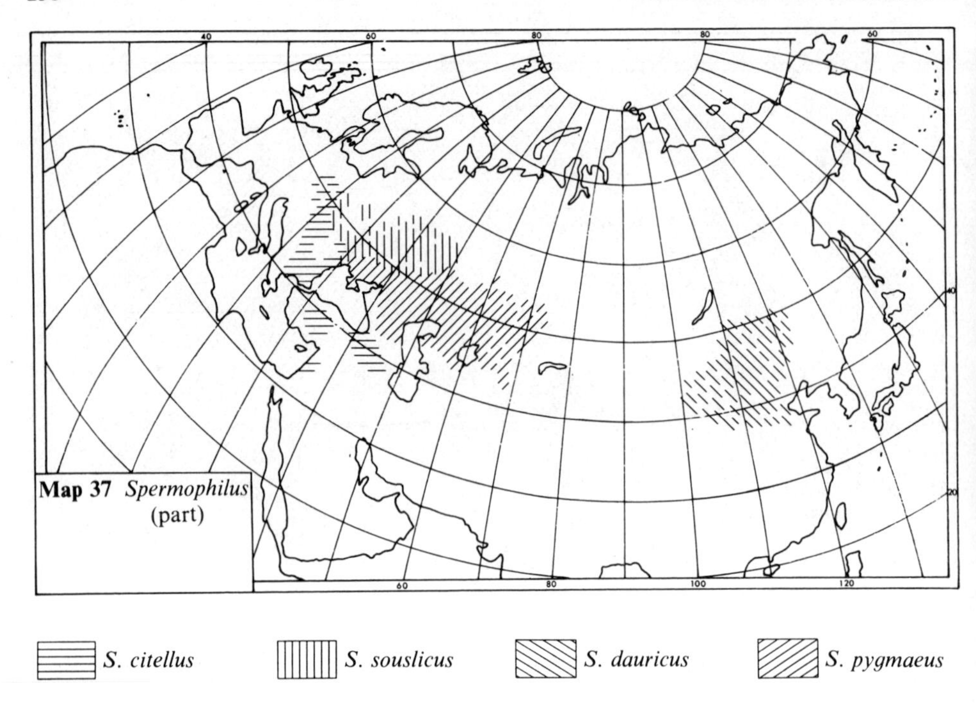
Map 37 Spermophilus (part)
S. citellus
S. souslicus
S. dauricus
S. pygmaeus

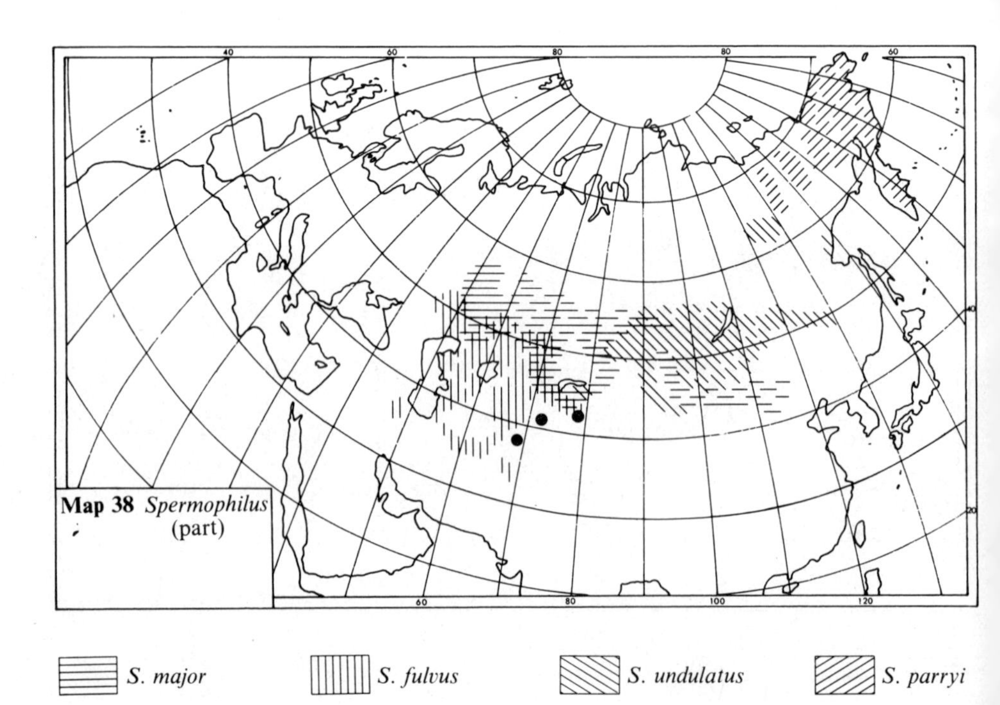
Map 38 Spermophilus (part)
S. major
S. fulvus
S. undulatus
S. parryi

● *S. relictus*

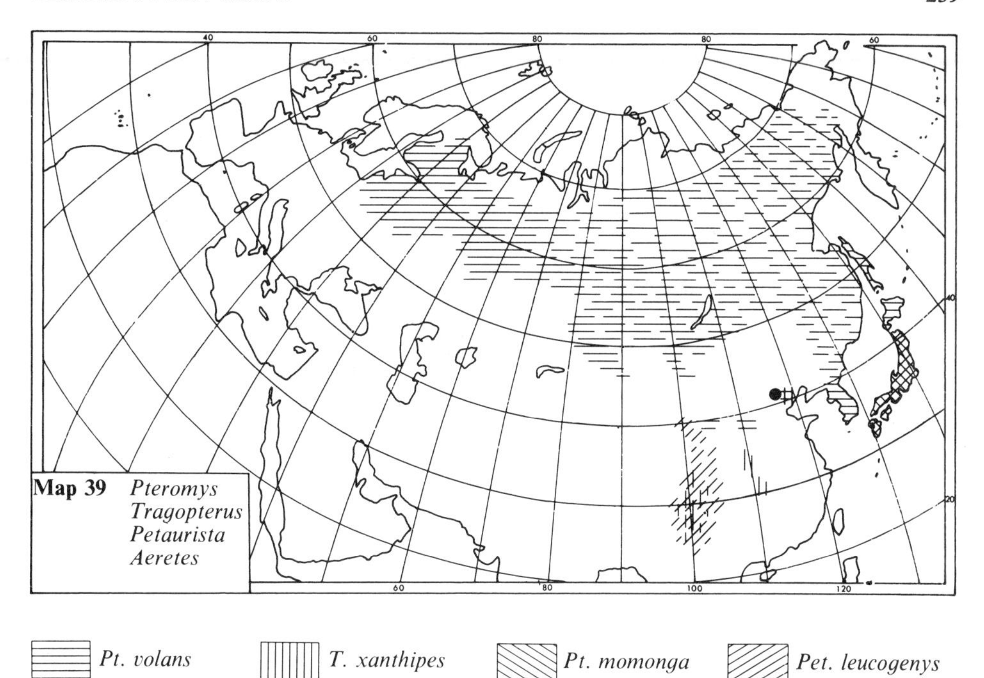

Map 39 *Pteromys Tragopterus Petaurista Aeretes*

Pt. volans — *T. xanthipes* — *Pt. momonga* — *Pet. leucogenys*

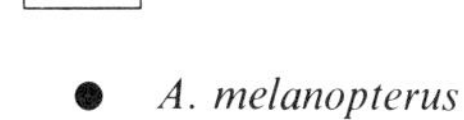

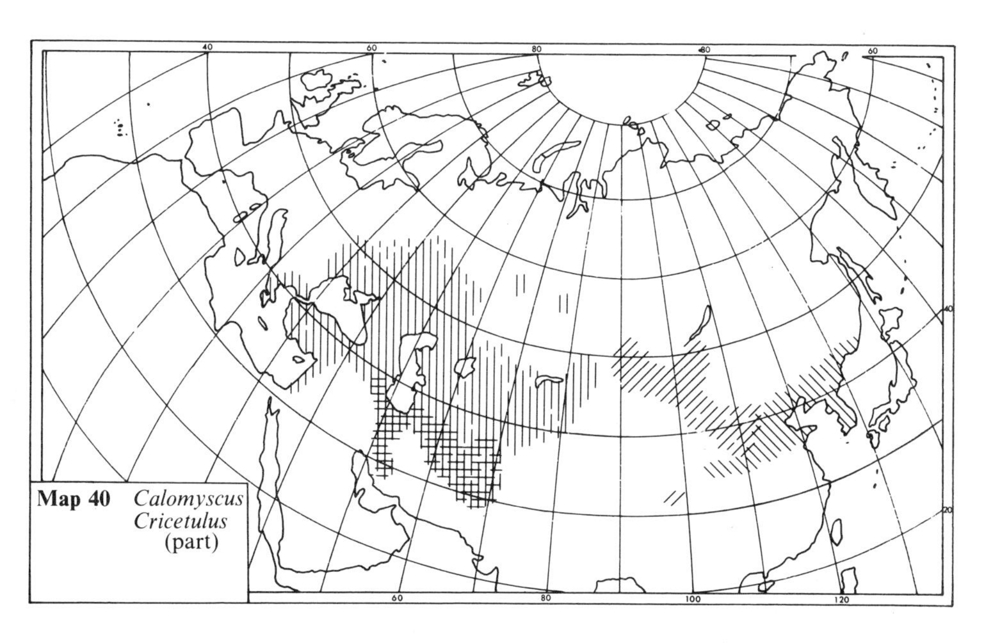

Map 40 *Calomyscus Cricetulus* (part)

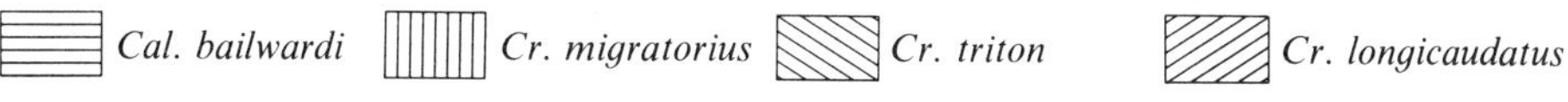

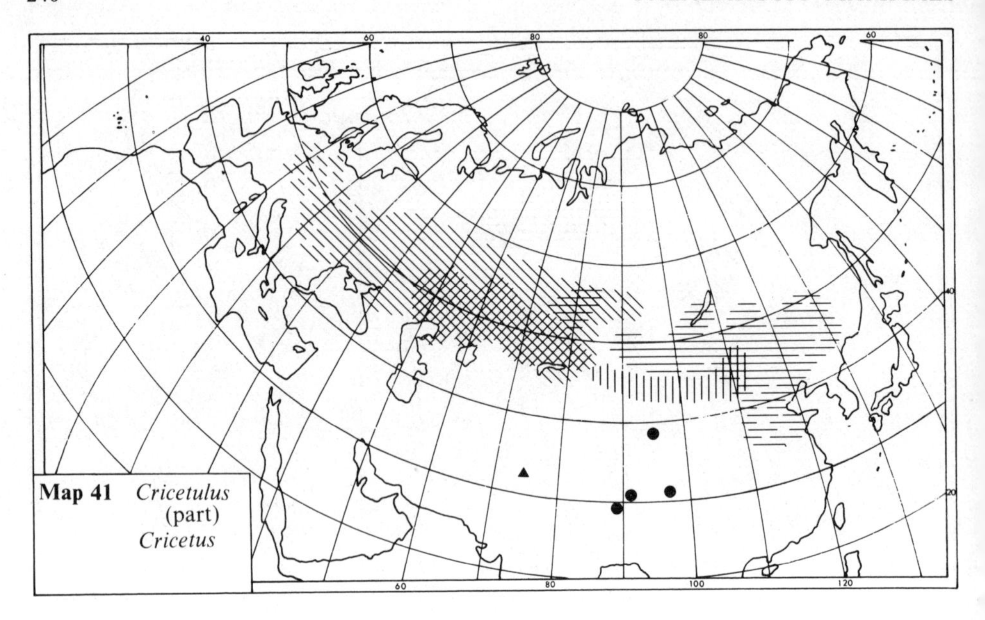

Cricetulus barabensis *Cricetulus curtatus* *Cricetus cricetus*

Cricetulus eversmanni ● *Cricetulus kamensis* ▲ *Cricetulus alticola*

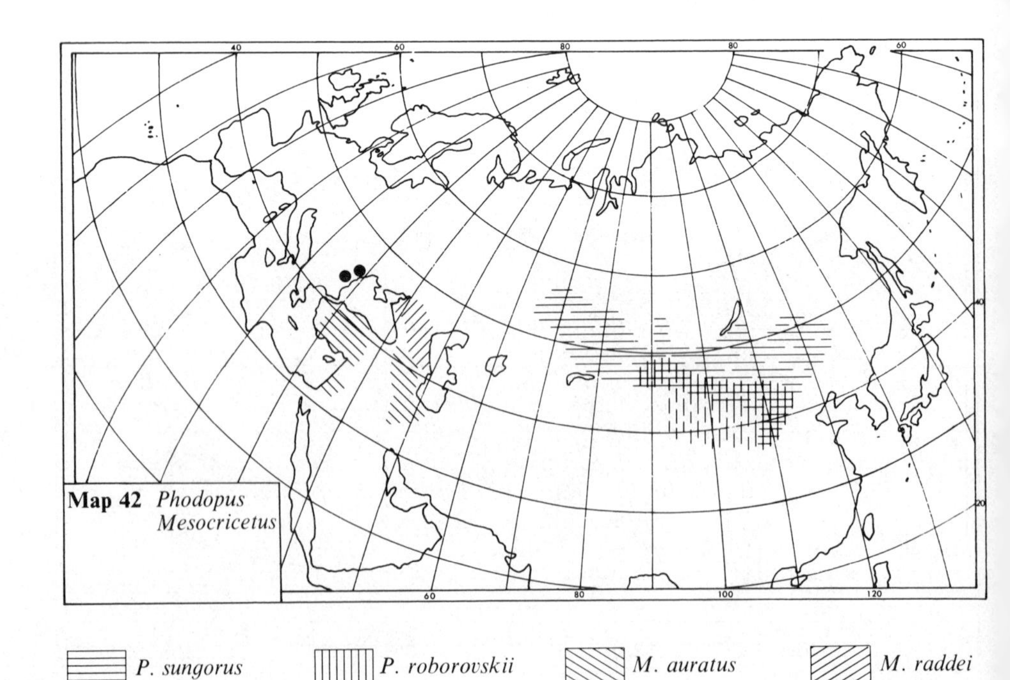

P. sungorus *P. roborovskii* *M. auratus* *M. raddei*

● *M. newtoni*

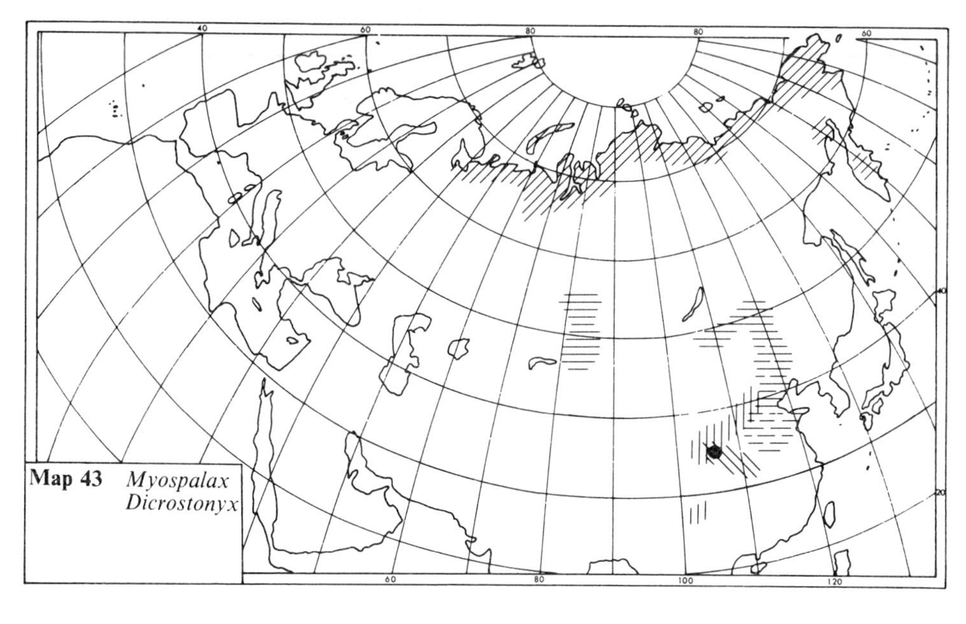

M. myospalax M. fontanieri M. rothschildi D. torquatus

● M. smithi

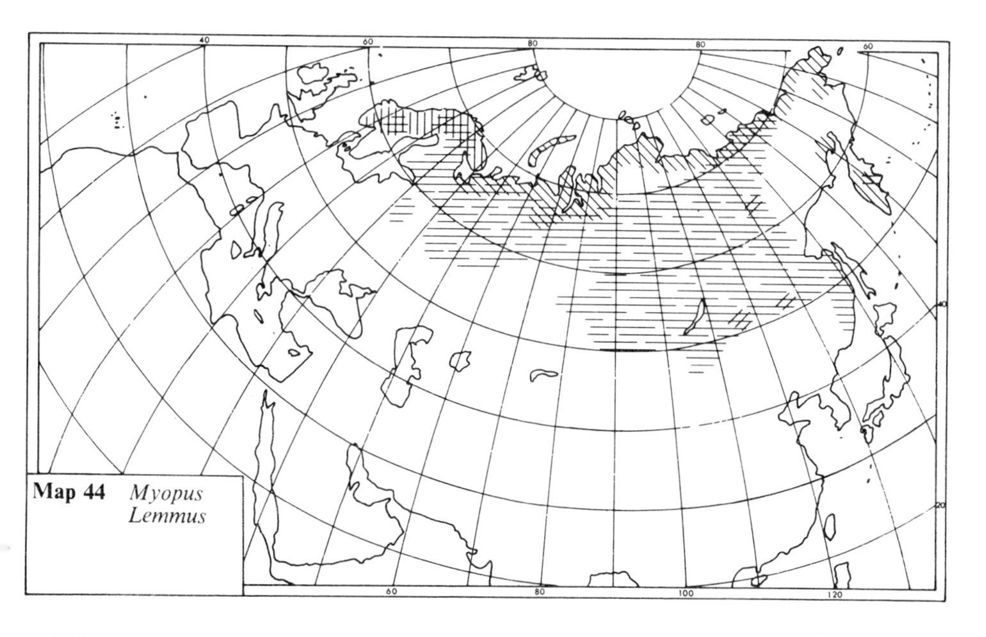

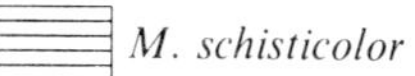

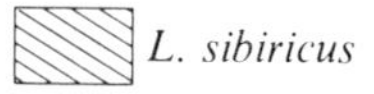
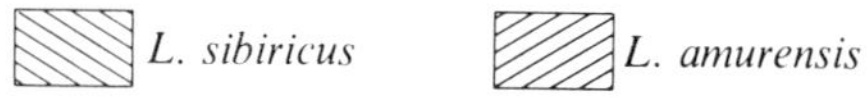

M. schisticolor L. lemmus L. sibiricus L. amurensis

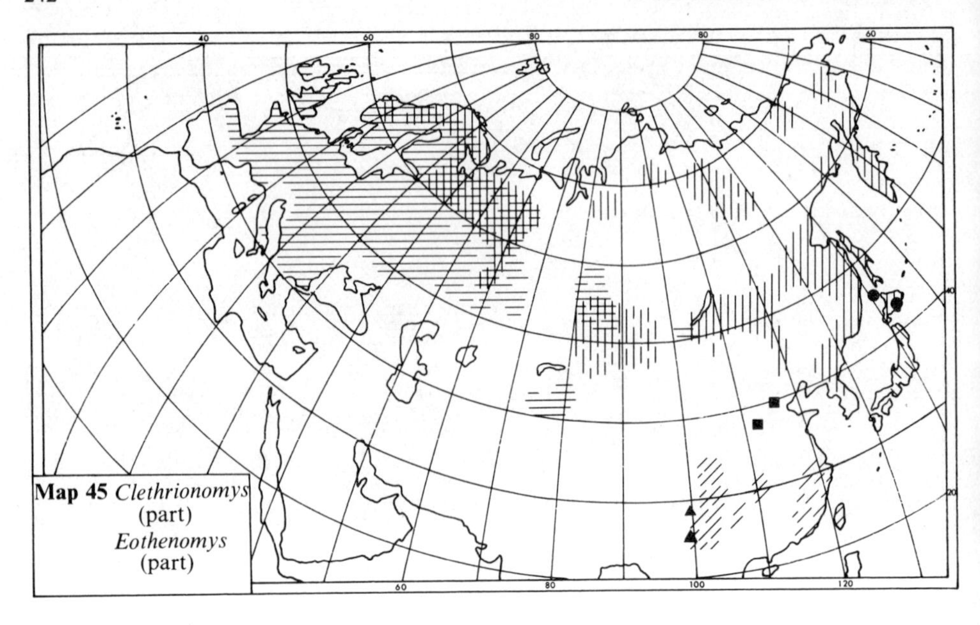

Map 45 *Clethrionomys* (part) *Eothenomys* (part)

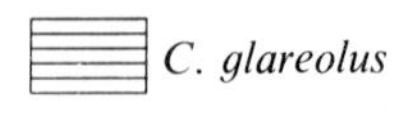 *C. glareolus* 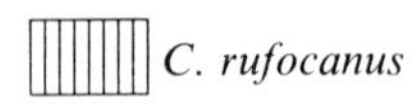 *C. rufocanus* 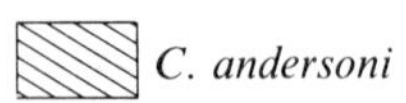 *C. andersoni* 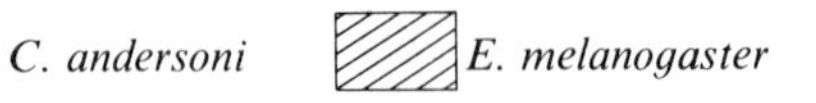 *E. melanogaster*

● *C. rex* ▲ *E. proditor* ■ *E. shanseius*

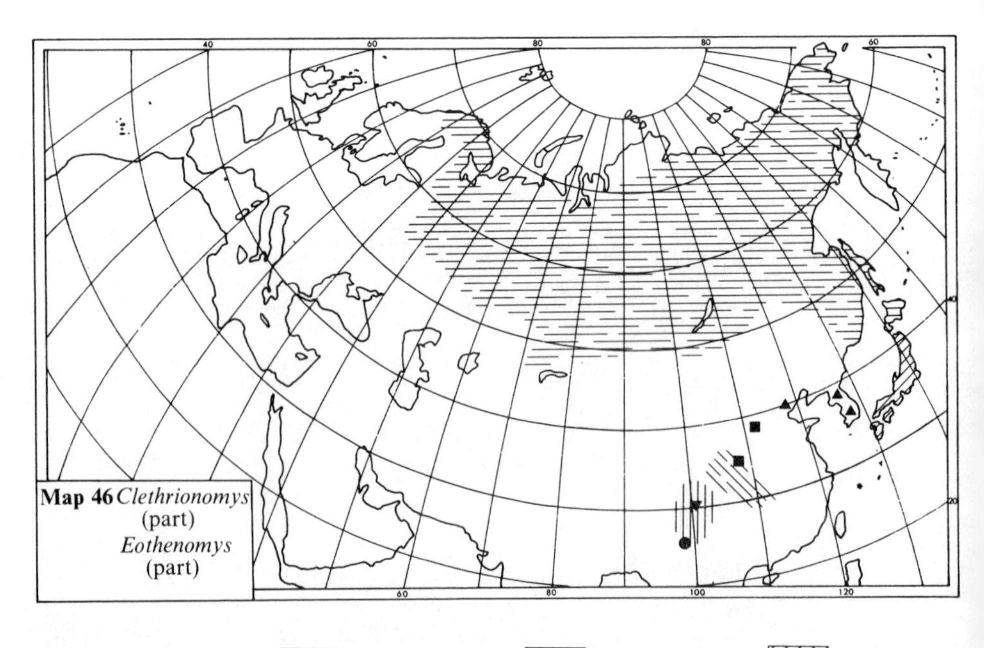

Map 46 *Clethrionomys* (part) *Eothenomys* (part)

C. rutilus 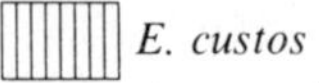 *E. custos* *E. eva* *E. smithi*

● *E. olitor* ▲ *E. regulus* ■ *E. inez* ▼ *E. chinensis*

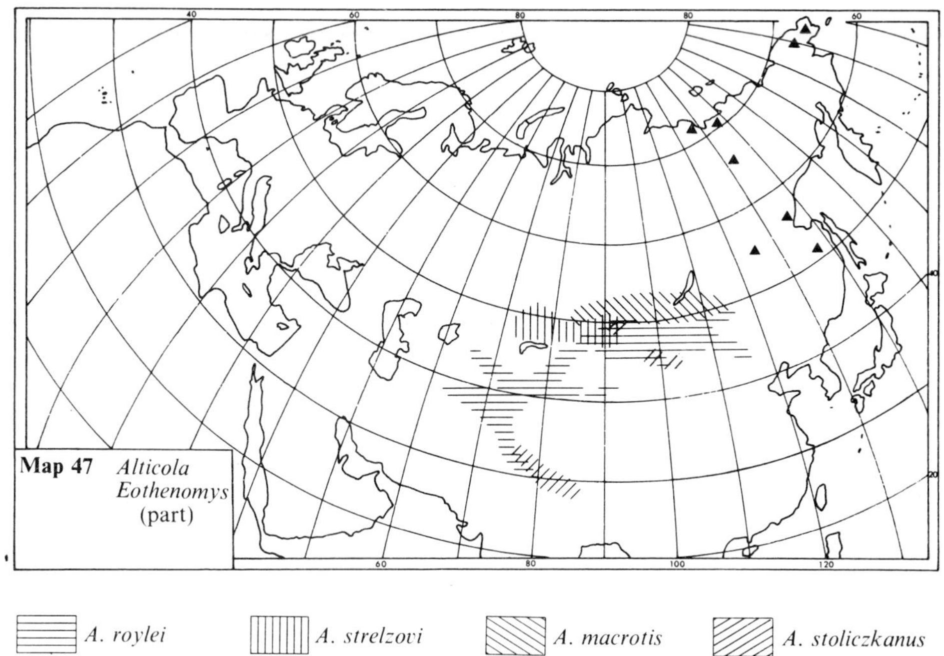

A. roylei *A. strelzovi* *A. macrotis* *A. stoliczkanus*

▲ *E. lemminus*

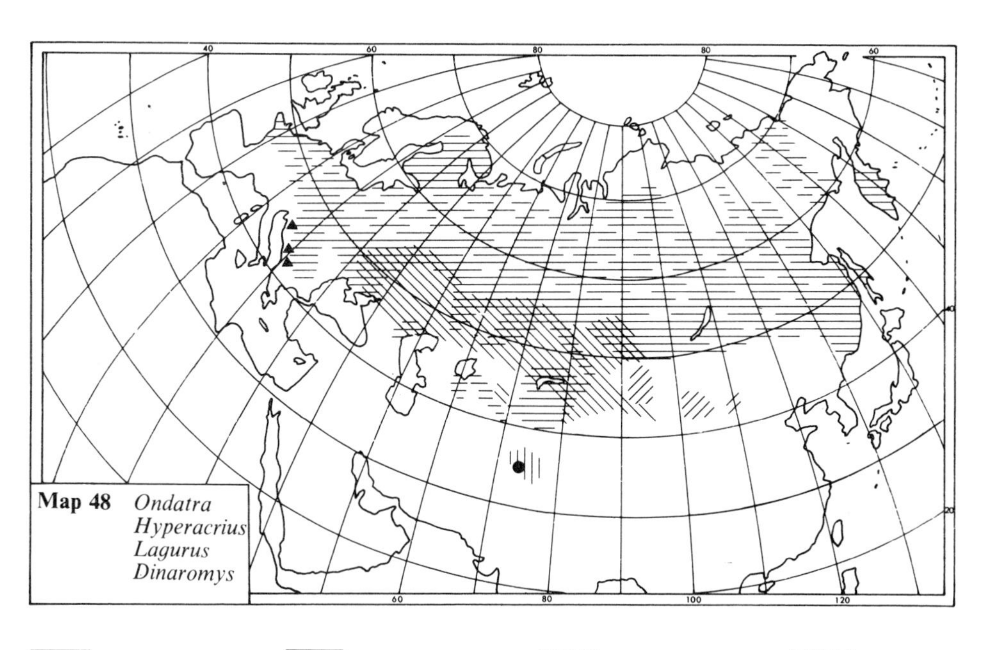

O. zibethicus *H. fertilis* *L. lagurus* *L. luteus*

● *H. wynnei* ▲ *D. bogdanovi*

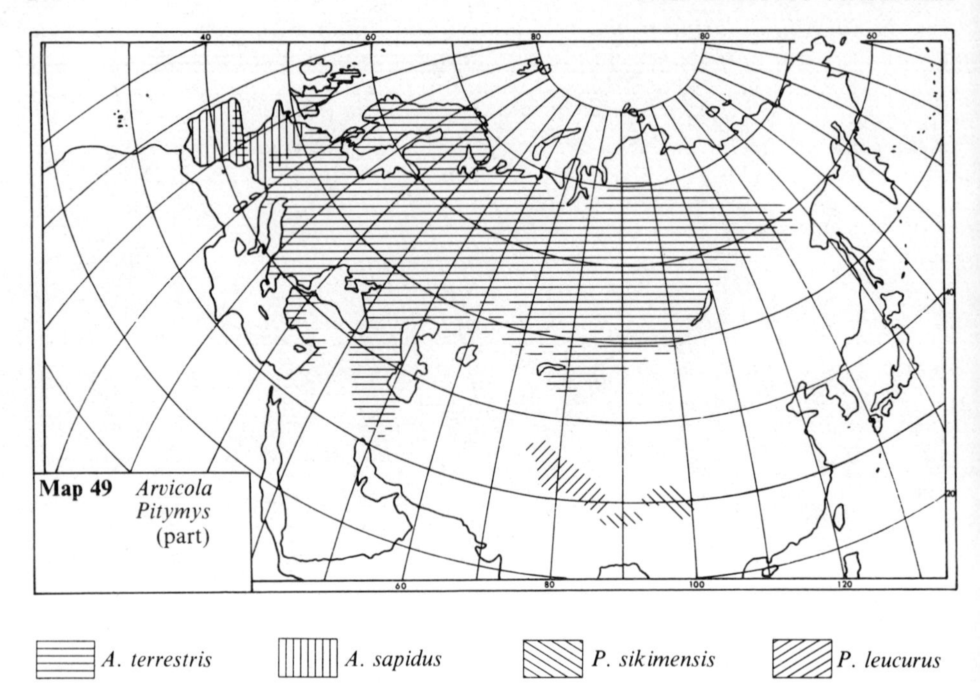
Map 49 Arvicola
Pitymys
(part)
A. terrestris
A. sapidus
P. sikimensis
P. leucurus

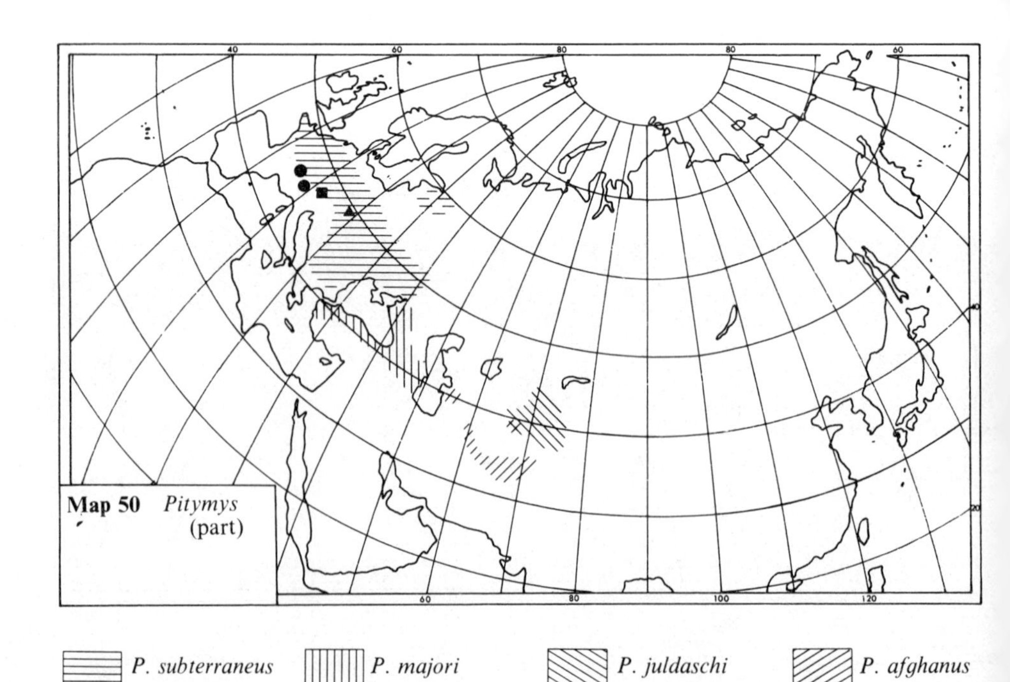
Map 50 Pitymys
(part)
P. subterraneus
P. majori
P. juldaschi
P. afghanus
P. multiplex
P. tatricus
P. bavaricus

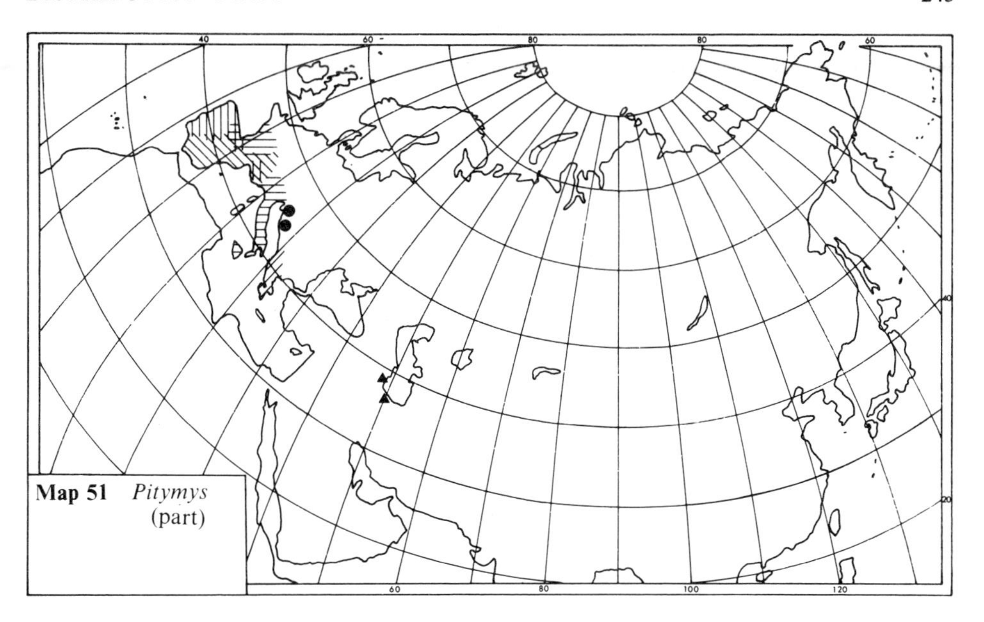
Map 51 *Pitymys* (part)

P. savii
P. lusitanicus
P. duodecimcostatus
P. thomasi
P. liechtensteini
P. schelkovnikovi

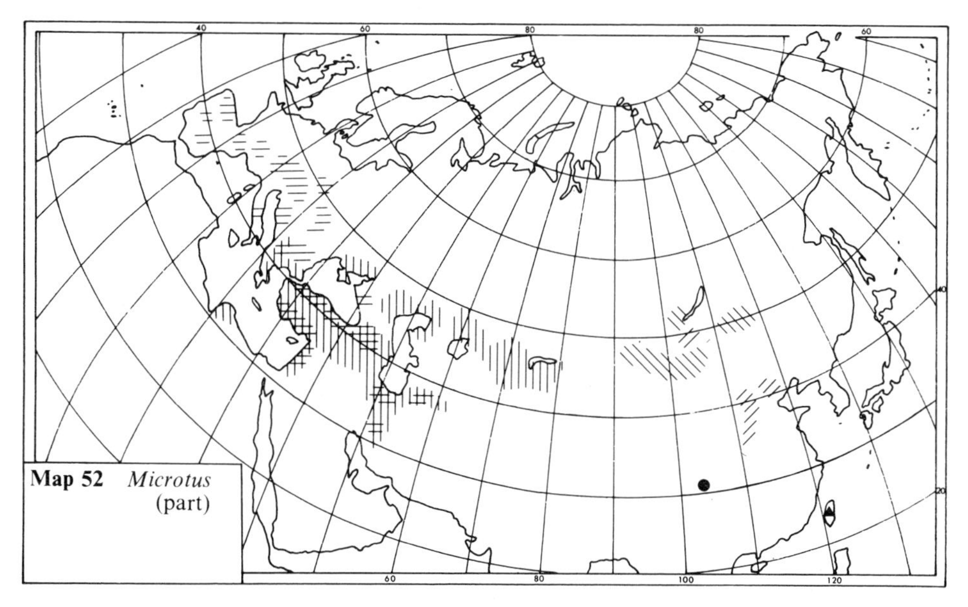
Map 52 *Microtus* (part)

M. nivalis
M. socialis
M. brandti
M. mandarinus
M. millicens
M. kikuchii

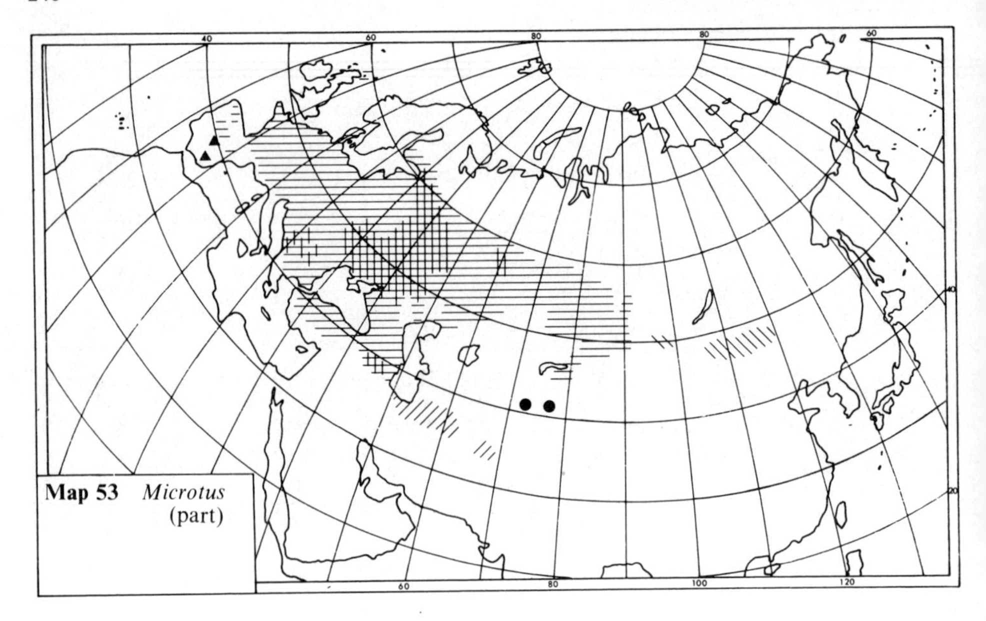

Map 53 *Microtus* (part)

M. arvalis *M. subarvalis* 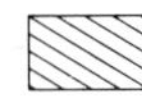 *M. mongolicus* 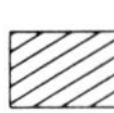 *M. transcaspicus*

● *M. ilaeus* ▲ *M. cabrerae*

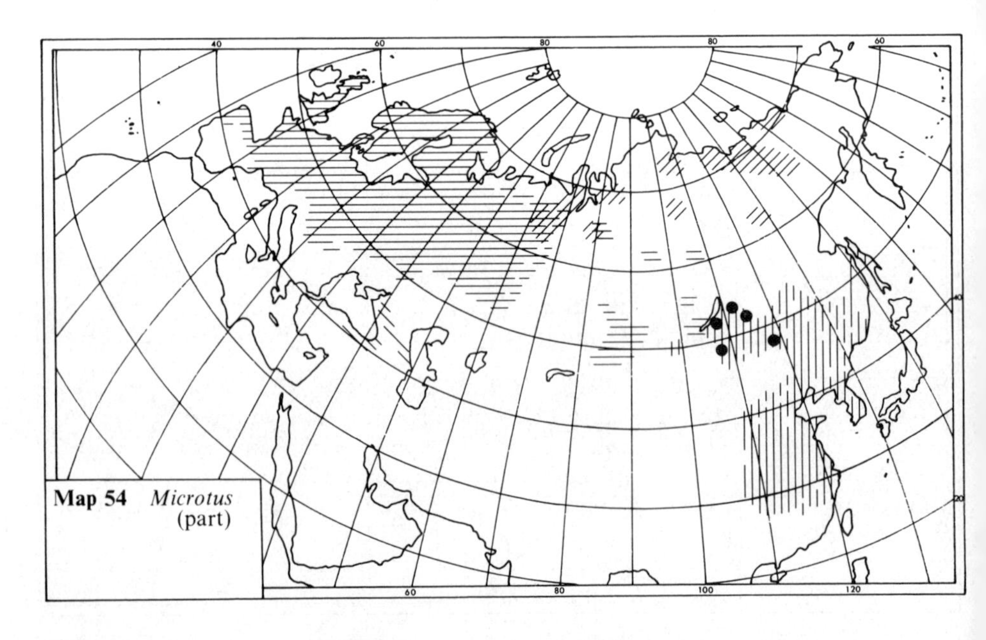

Map 54 *Microtus* (part)

M. agrestis  *M. fortis*  *M. roberti, M. gud*

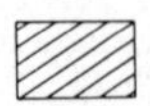 *M. middendorffi* ● *M. maximowiczii*

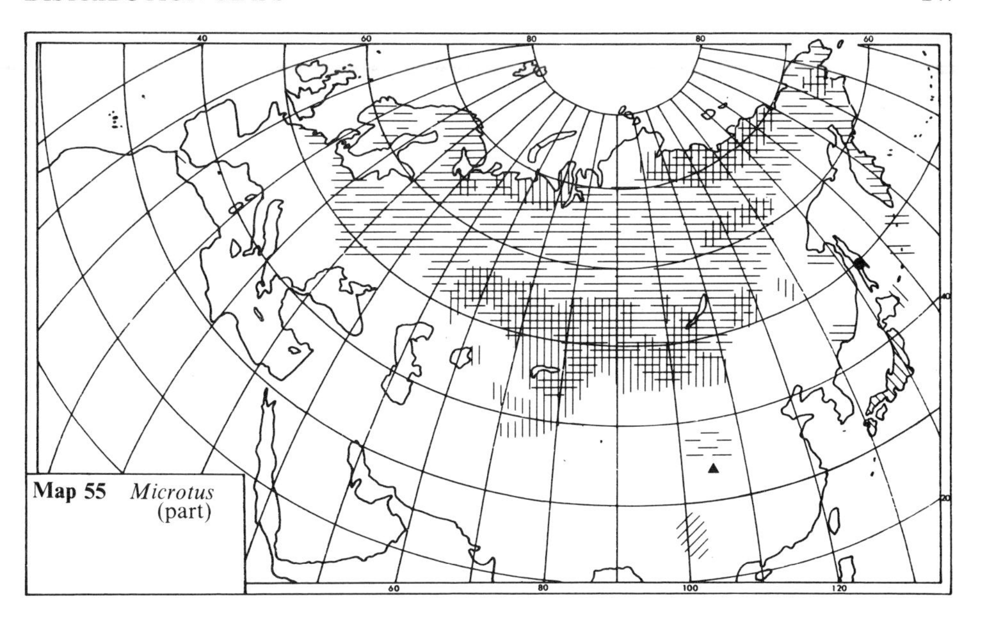

Map 55 *Microtus* (part)

M. oeconomus *M. gregalis* *M. montebelli* *M. clarkei*

● *M. sachalinensis* ▲ *M. bedfordi*

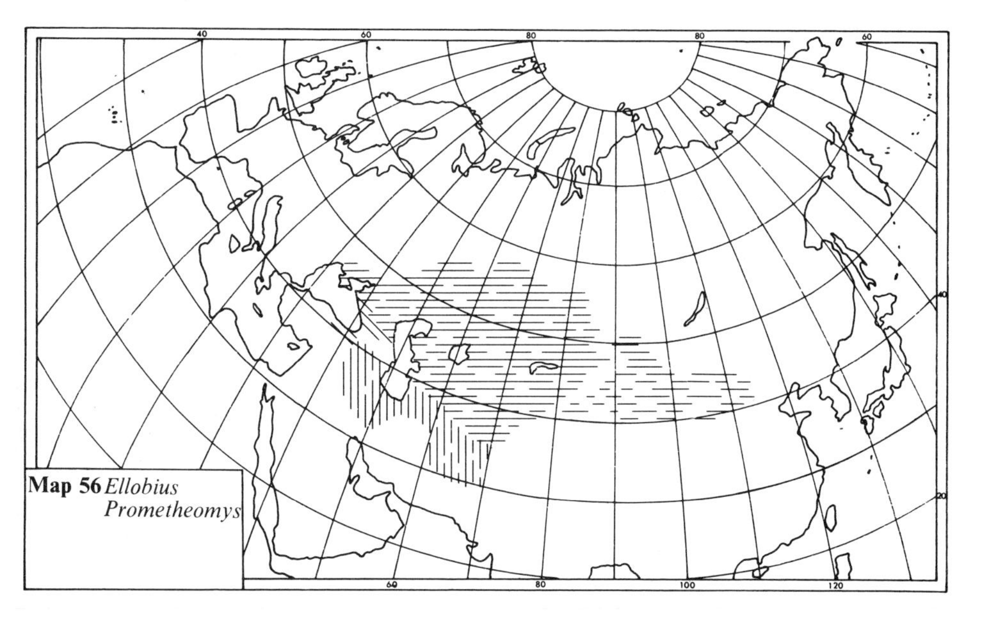

Map 56 *Ellobius* *Prometheomys*

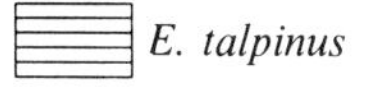 *E. talpinus* 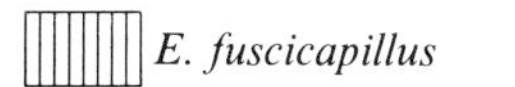 *E. fuscicapillus* 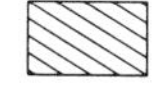 *P. schaposchnikovi*

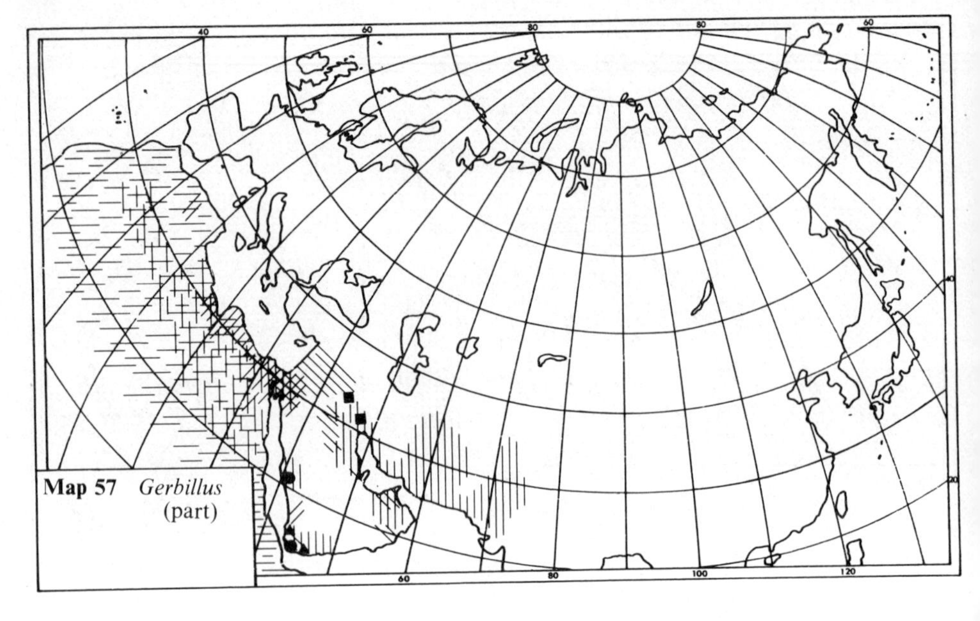

Map 57 *Gerbillus* (part)

G. campestris — *G. nanus* — *G. dasyurus* — *G. henleyi*

● *G. poecilops* ▲ *G. famulus* ■ *G. mesopotamiae*

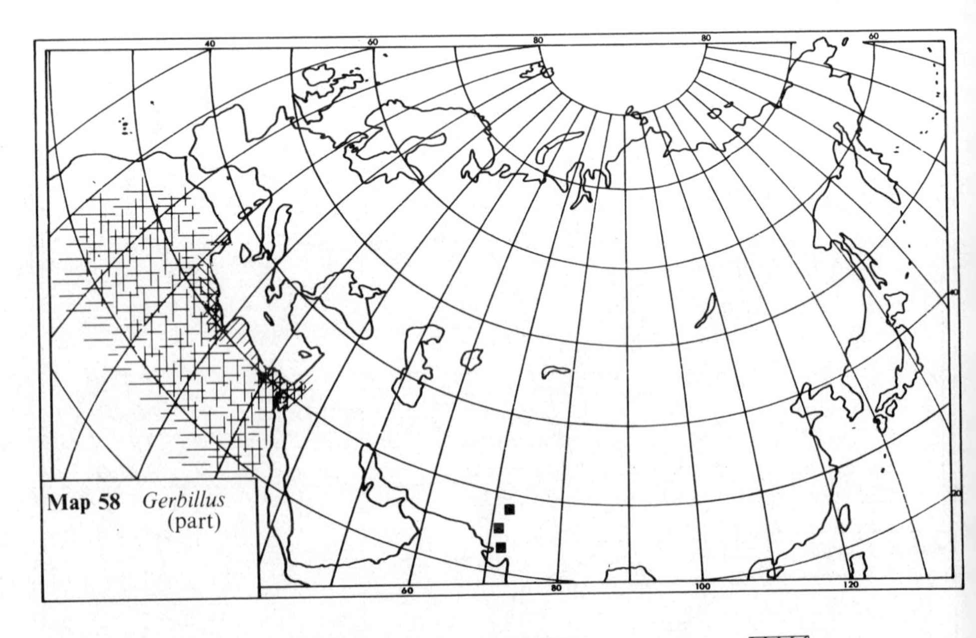

Map 58 *Gerbillus* (part)

G. gerbillus — *G. pyramidum* — *G. aureus* — *G. andersoni*

■ *G. gleadowi* ▼ *G. perpallidus*

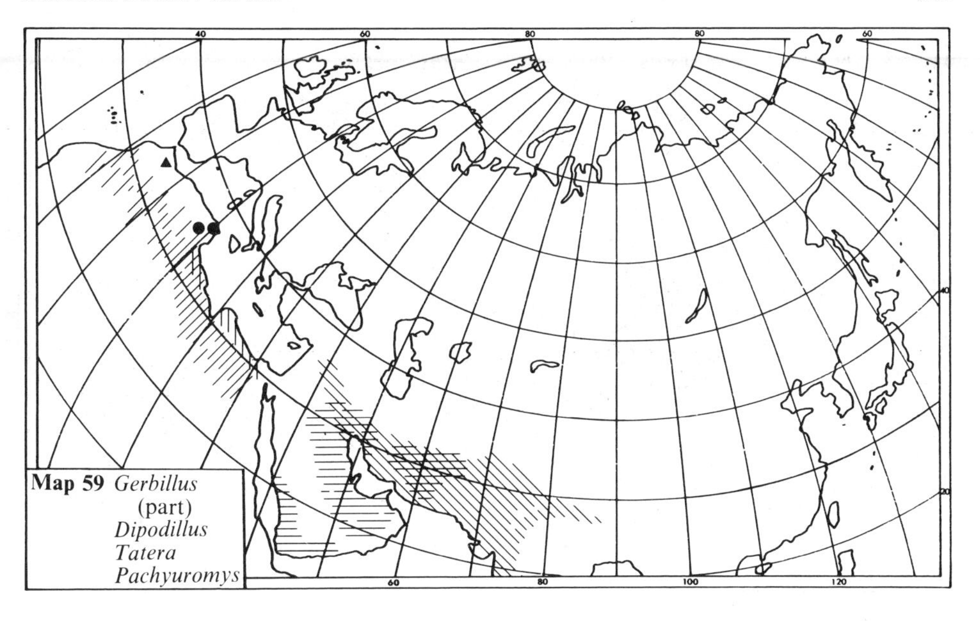

Map 59 *Gerbillus* (part) *Dipodillus* *Tatera* *Pachyuromys*

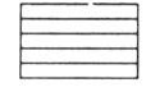 *G. cheesmani*  *D. kaiseri* 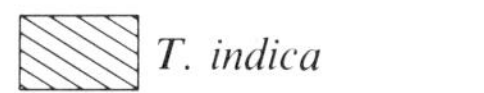 *T. indica* 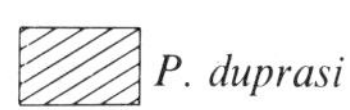 *P. duprasi*

● *D. simoni* ▲ *D. maghrebi*

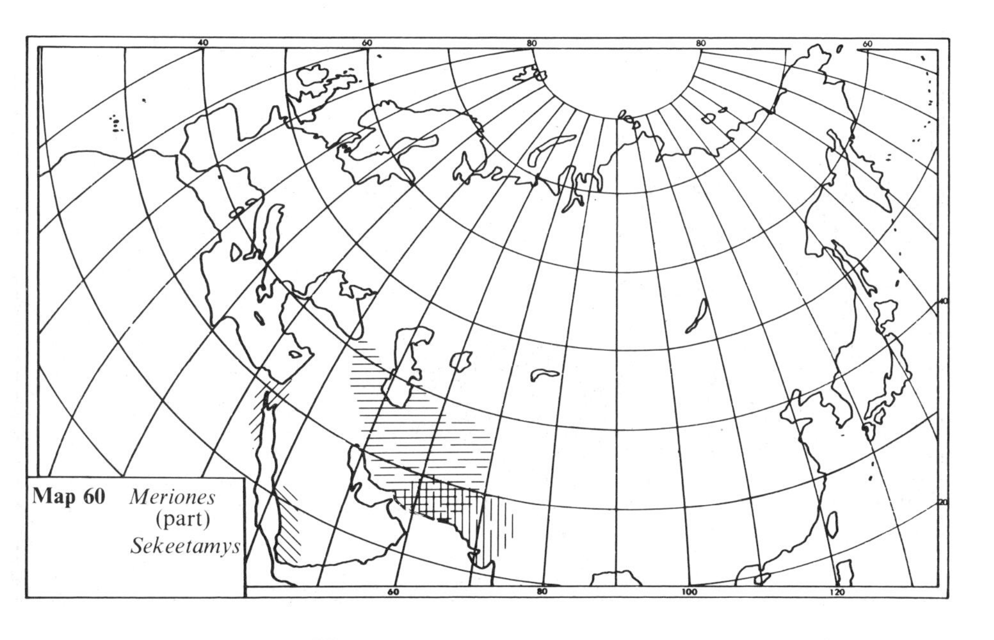

Map 60 *Meriones* (part) *Sekeetamys*

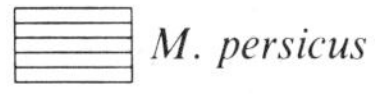 *M. persicus* 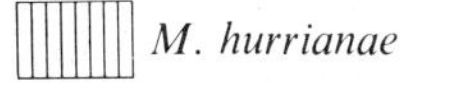 *M. hurrianae* 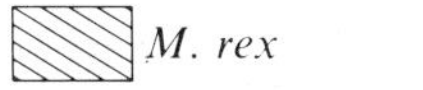 *M. rex*  *S. calurus*

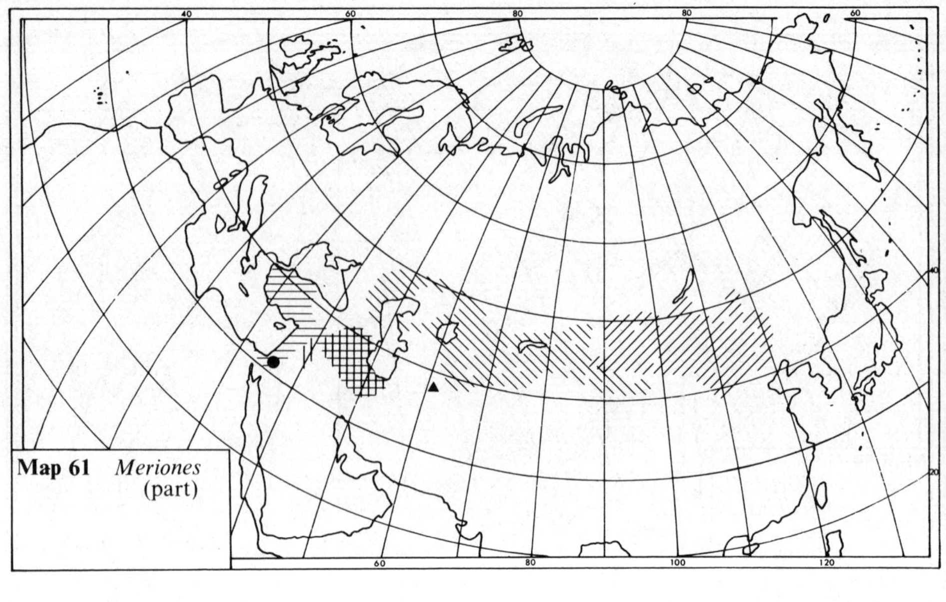

Map 61 *Meriones* (part)

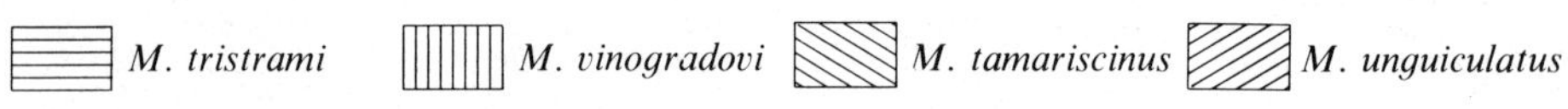

● *M. sacramenti* ▲ *M. zarudnyi*

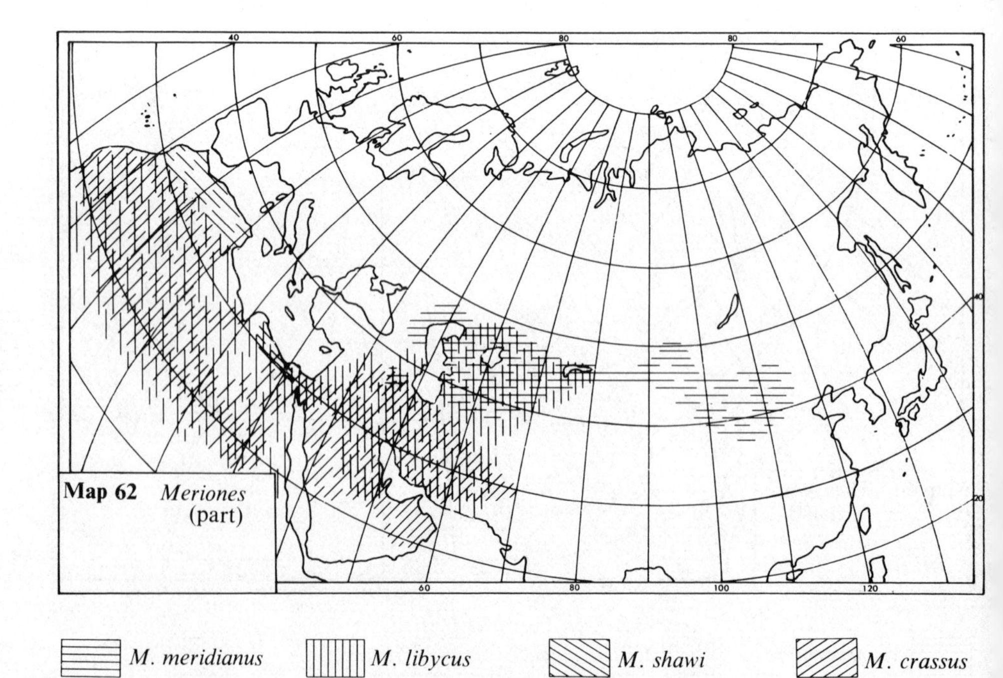

Map 62 *Meriones* (part)

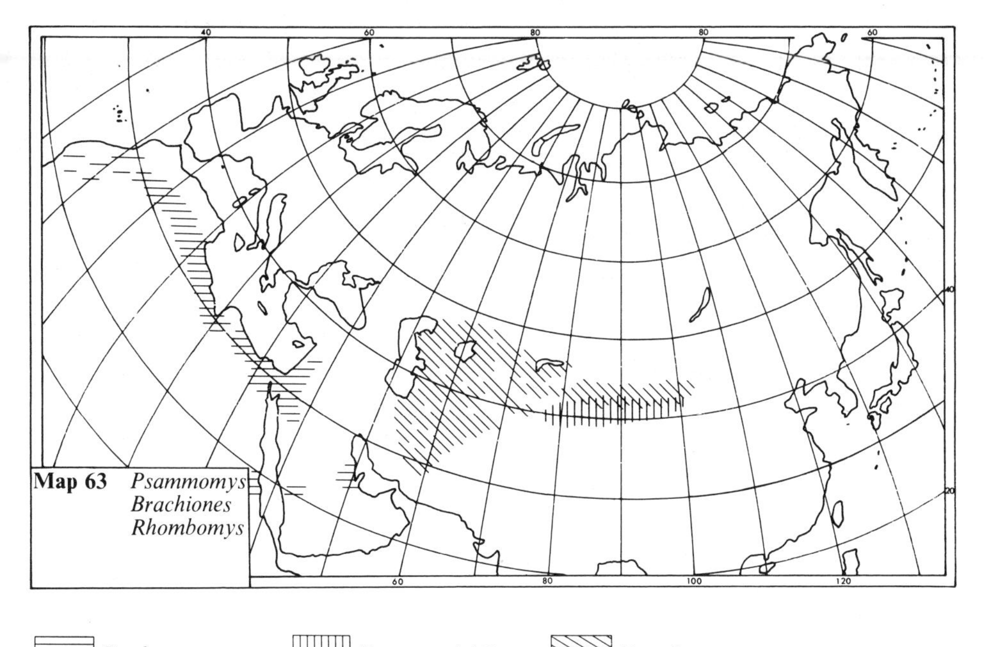
Map 63 Psammomys
Brachiones
Rhombomys
40
60
80
80
60
40
20
60
80
100
120

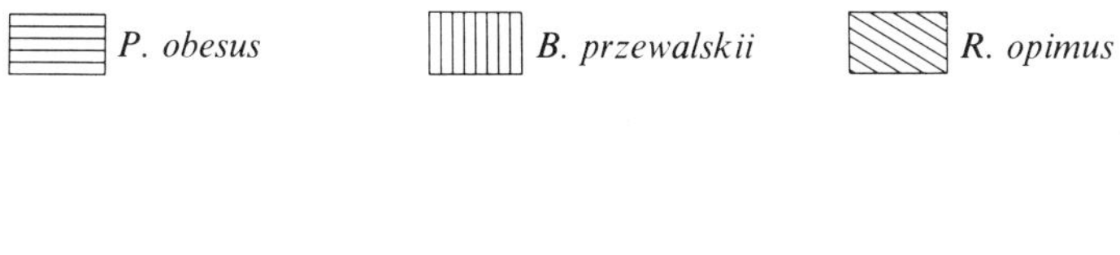
P. obesus
B. przewalskii
R. opimus

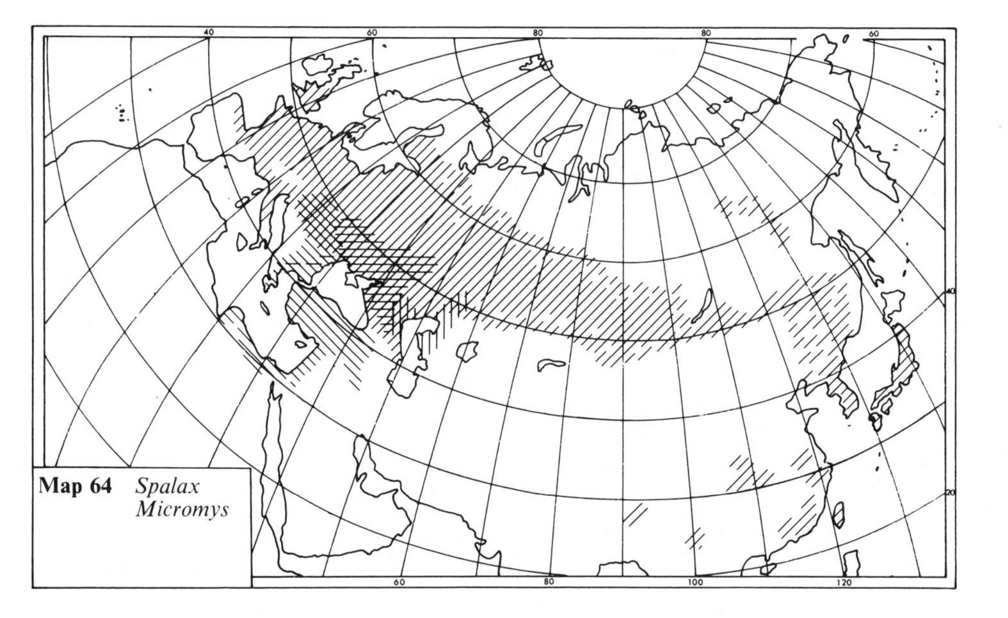
Map 64 Spalax
Micromys
40
60
80
80
60
40
20
60
80
100
120

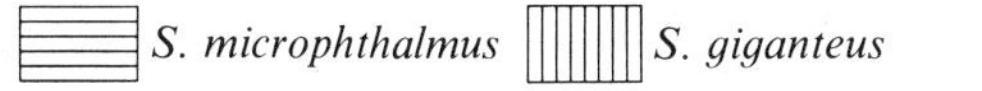
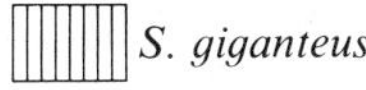
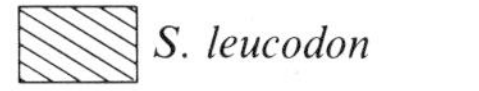
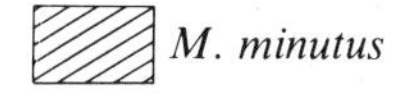
S. microphthalmus
S. giganteus
S. leucodon
M. minutus

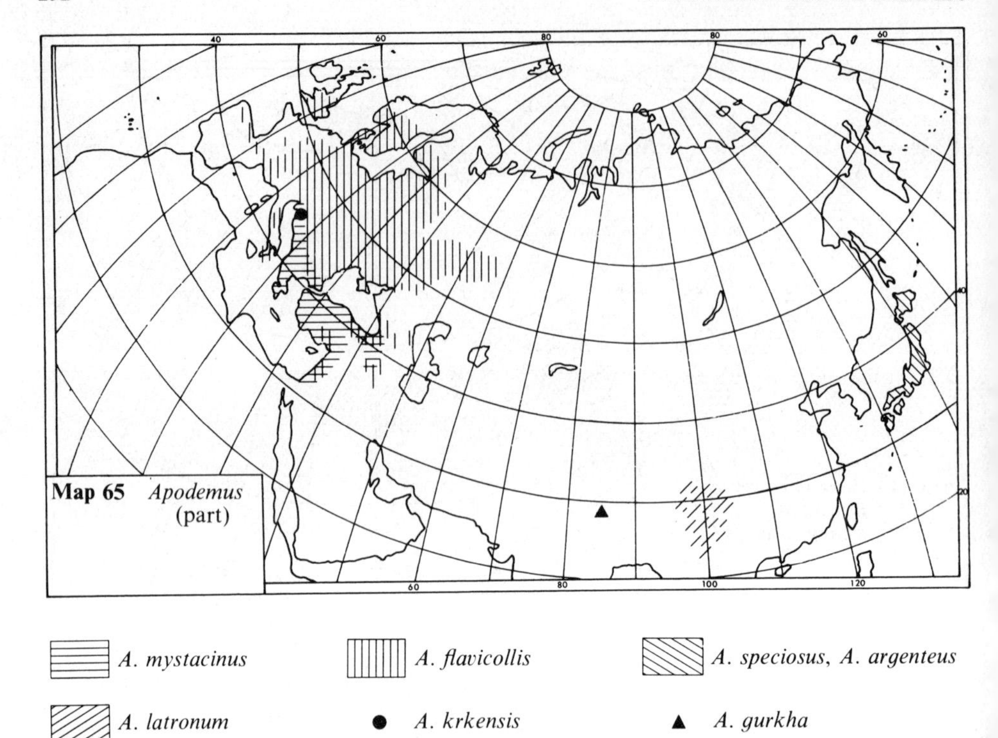

Map 65 *Apodemus* (part)

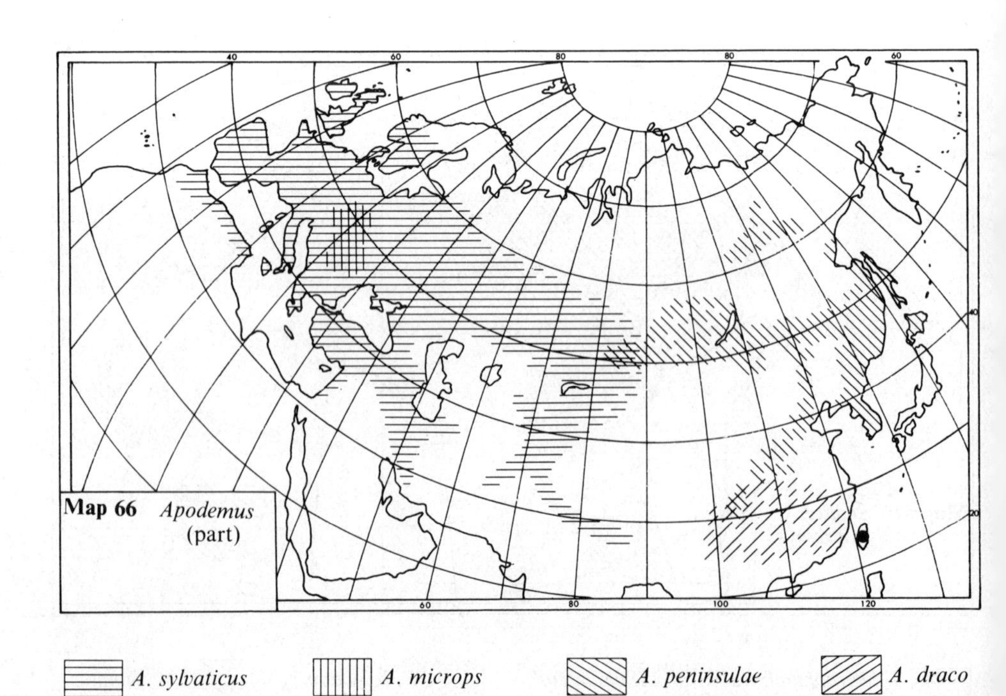

Map 66 *Apodemus* (part)

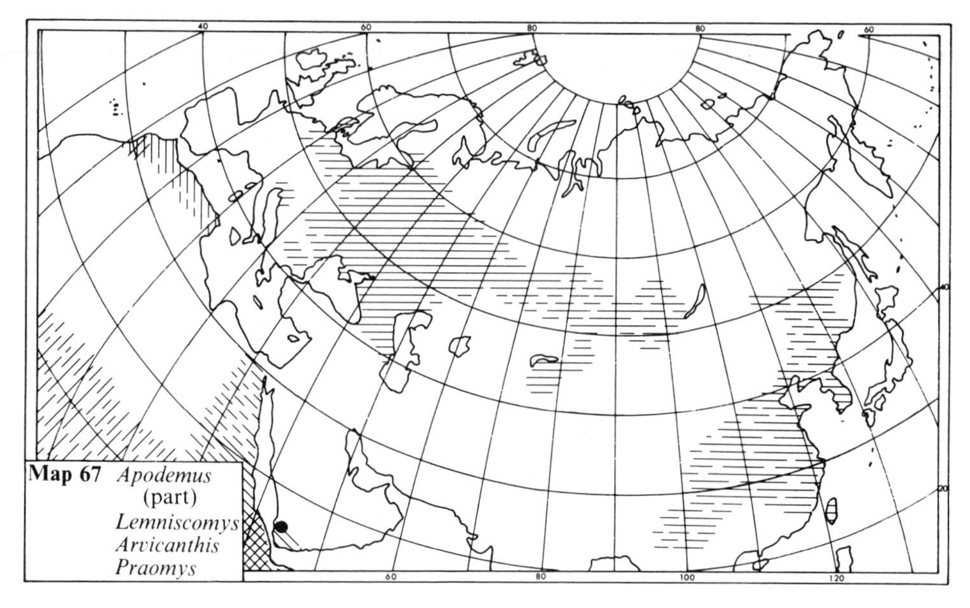

Map 67 *Apodemus* (part) *Lemniscomys* *Arvicanthis* *Praomys*

● *P. fumatus*

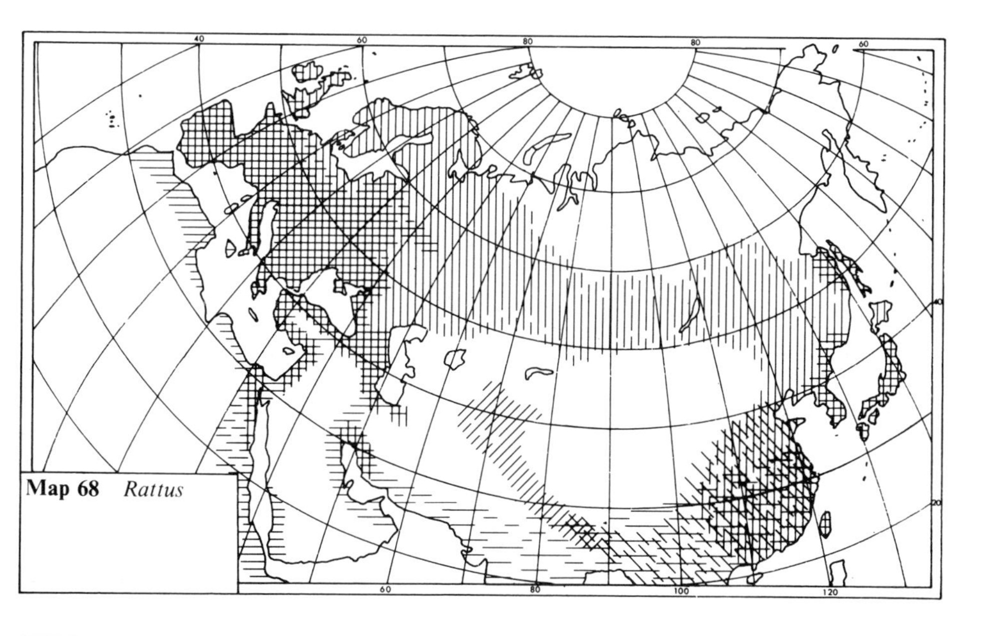

Map 68 *Rattus*

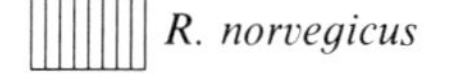

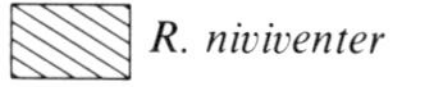

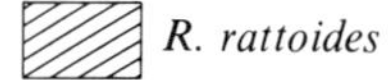

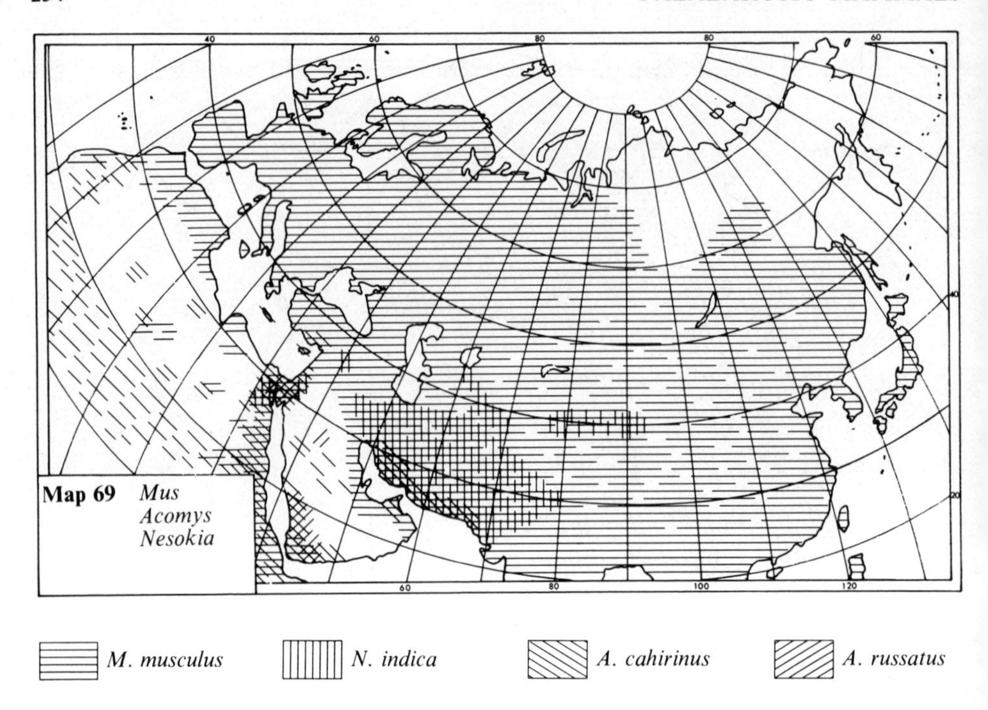

M. musculus | N. indica | A. cahirinus | A. russatus

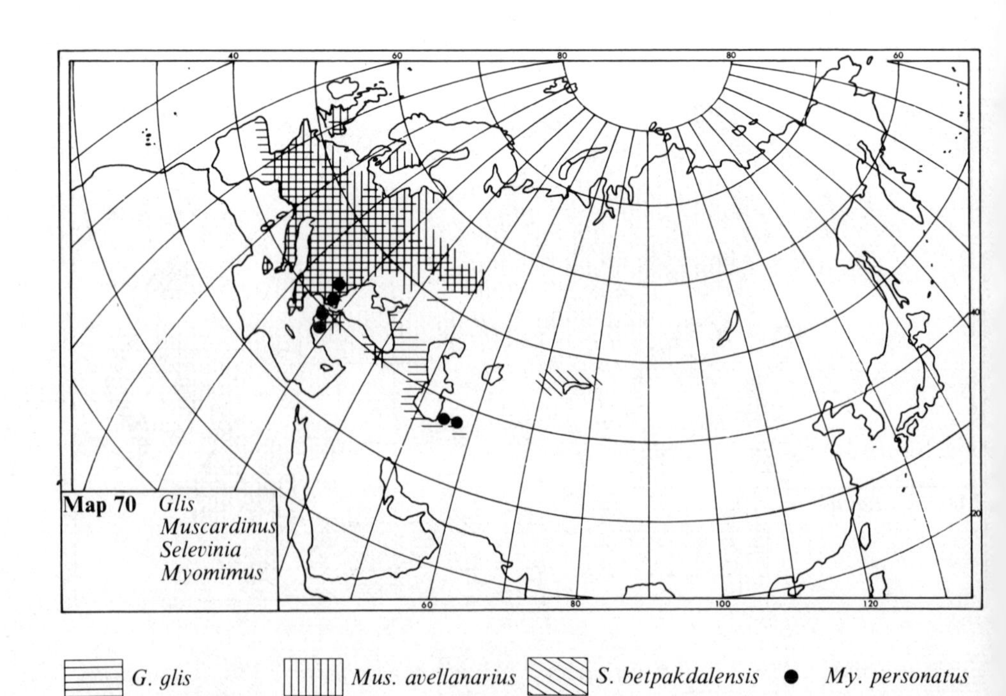

G. glis | Mus. avellanarius | S. betpakdalensis | ● My. personatus

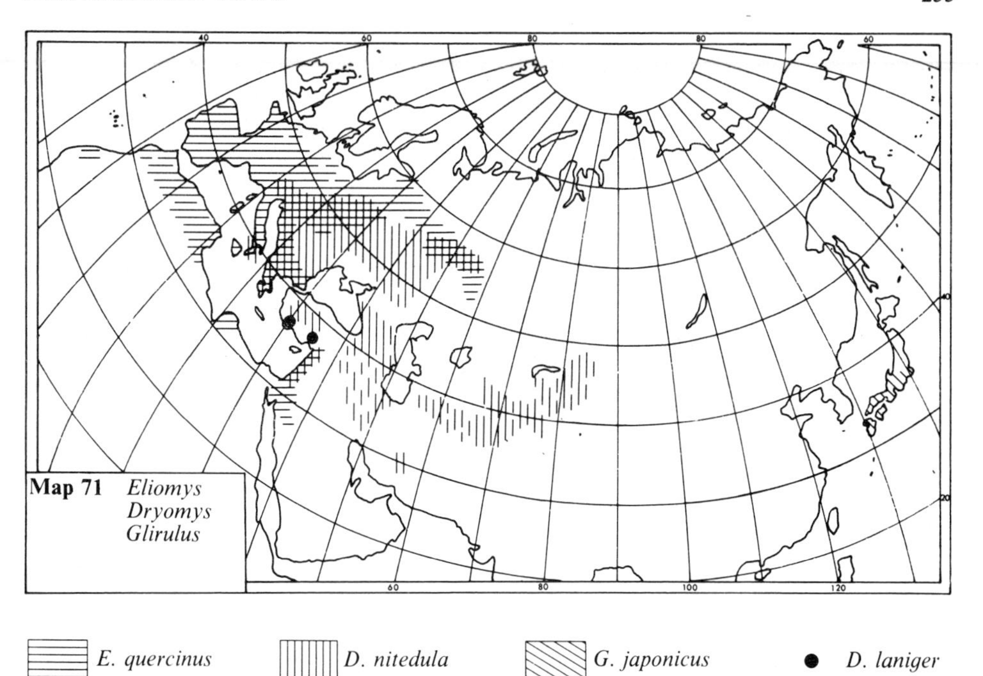
40
60
80
80
60
40
20
60
80
100
120
Map 71 *Eliomys Dryomys Glirulus*
E. quercinus
D. nitedula
G. japonicus
D. laniger

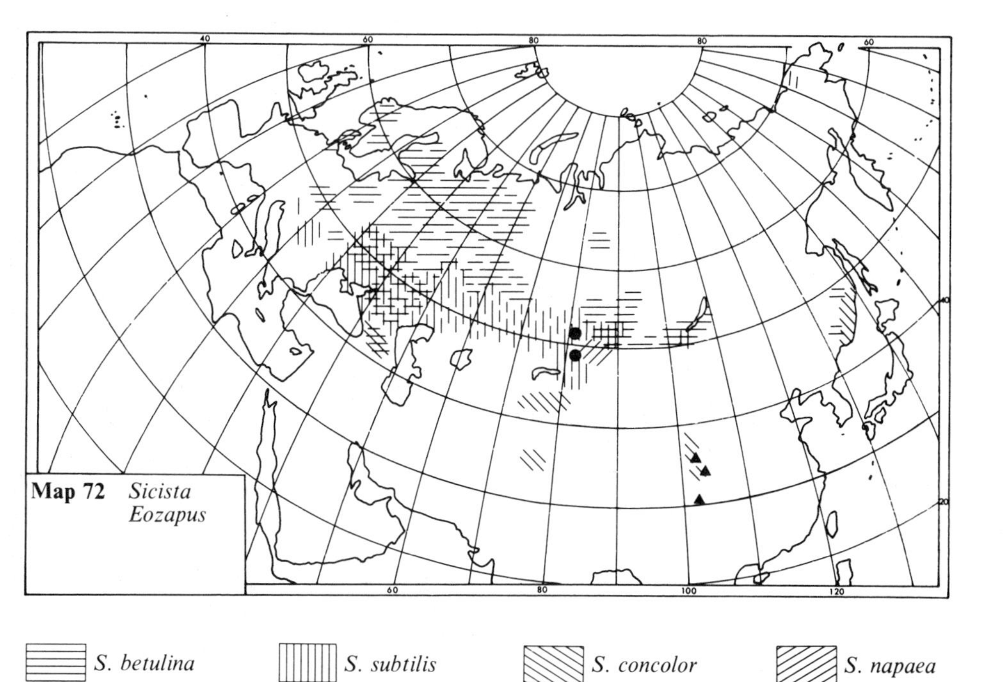
40
60
80
80
60
40
20
60
80
100
120
Map 72 *Sicista Eozapus*
S. betulina
S. subtilis
S. concolor
S. napaea
S. pseudonapaea
E. setchuanus

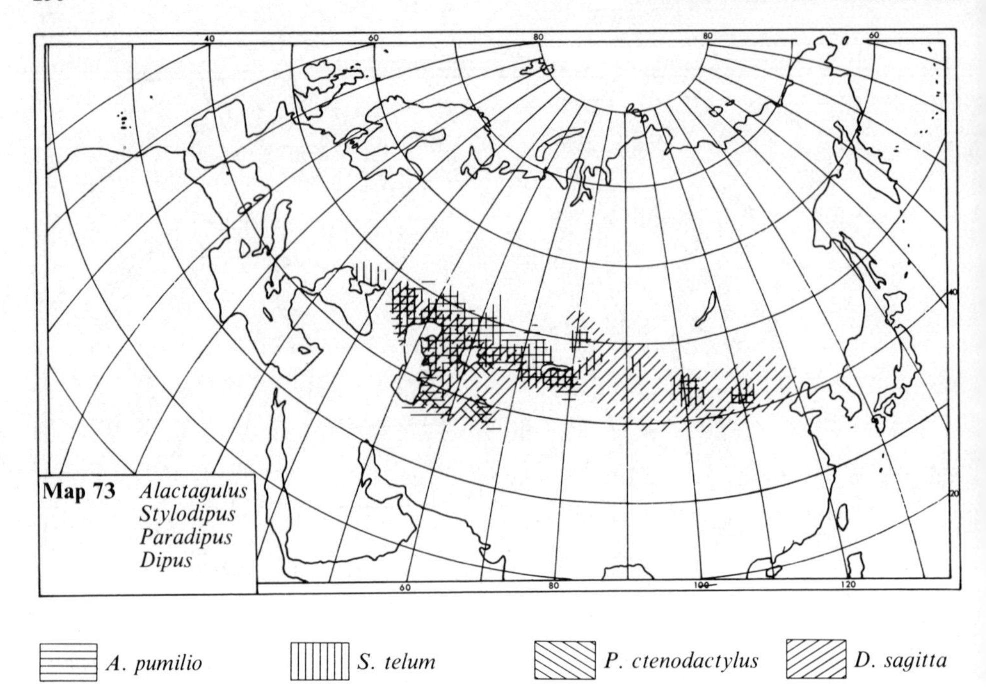
Map 73 Alactagulus Stylodipus Paradipus Dipus
A. pumilio
S. telum
P. ctenodactylus
D. sagitta

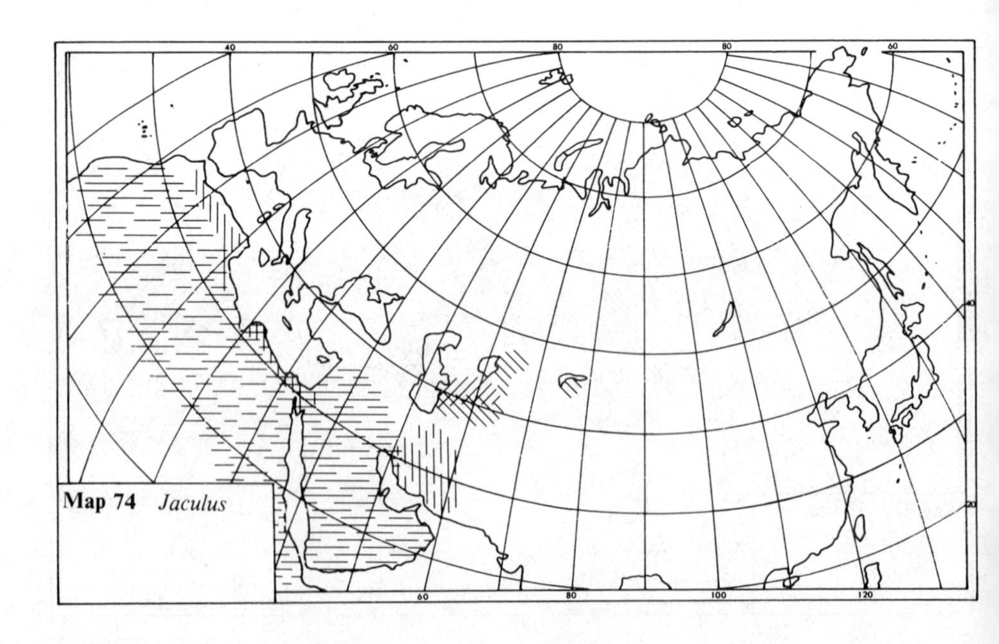
Map 74 Jaculus
J. jaculus
J. orientalis (N. Africa), J. blanfordi (Iran)
J. lichtensteini
J. turcmenicus

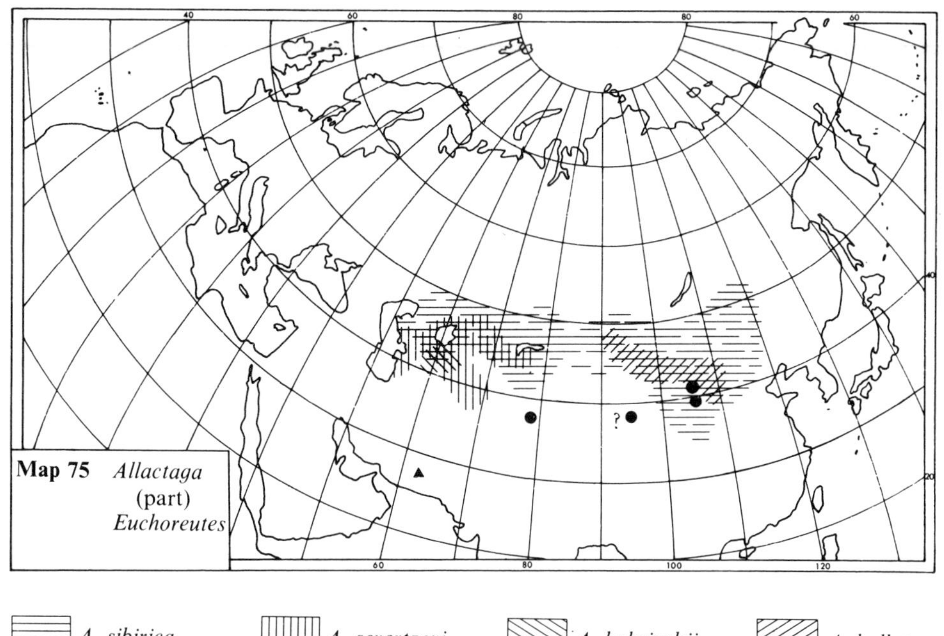
40
60
80
80
60
40
20
60
80
100
120
Map 75 Allactaga
(part)
Euchoreutes

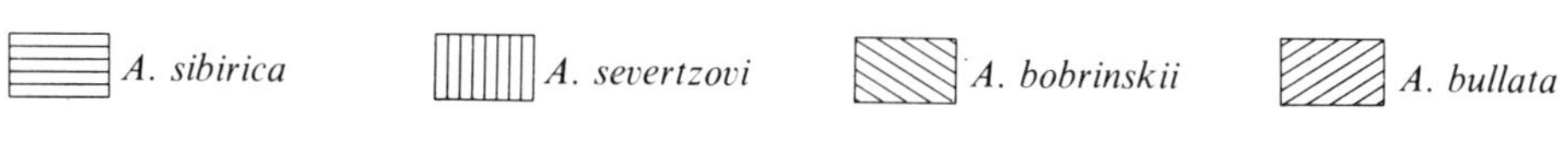
A. sibirica
A. severtzovi
A. bobrinskii
A. bullata
E. naso
A. hotsoni

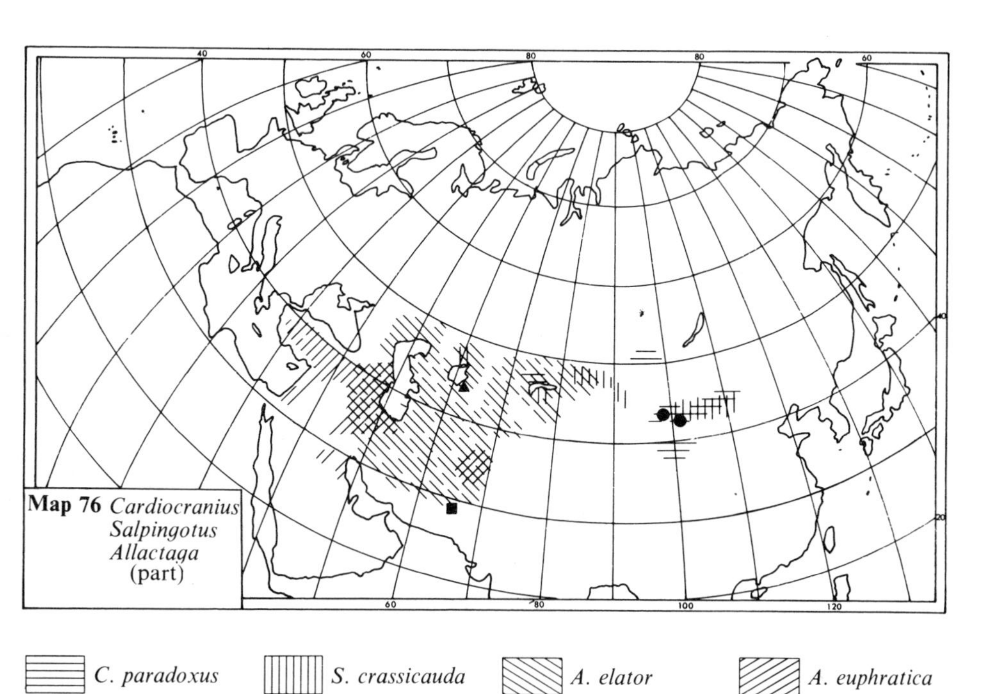
40
60
80
80
60
40
20
60
80
100
120
Map 76 Cardiocranius
Salpingotus
Allactaga
(part)
C. paradoxus
S. crassicauda
A. elator
A. euphratica
S. kozlovi
S. heptneri
S. michaelis

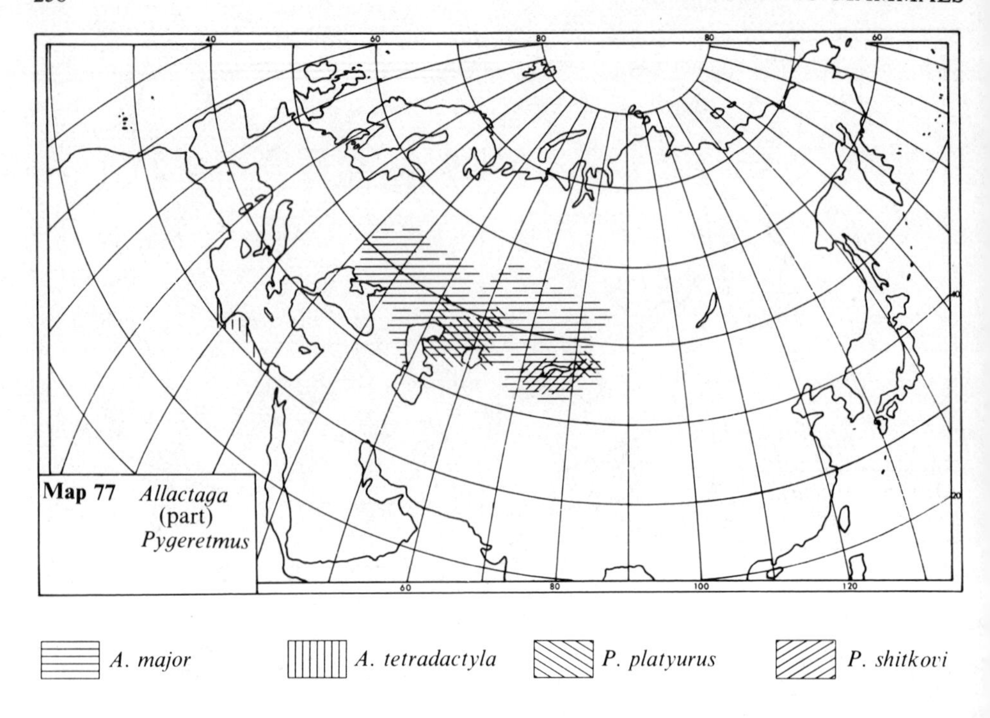

Map 77 *Allactaga* (part) *Pygeretmus*

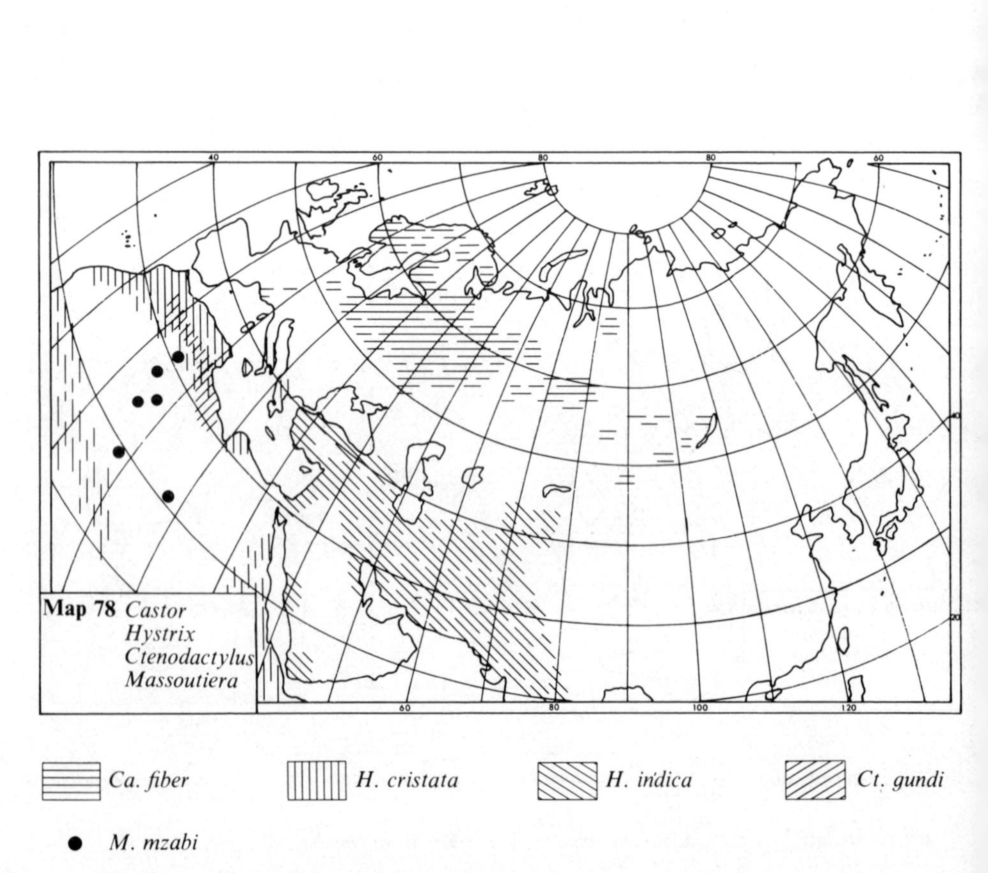

Map 78 *Castor* *Hystrix* *Ctenodactylus* *Massoutiera*

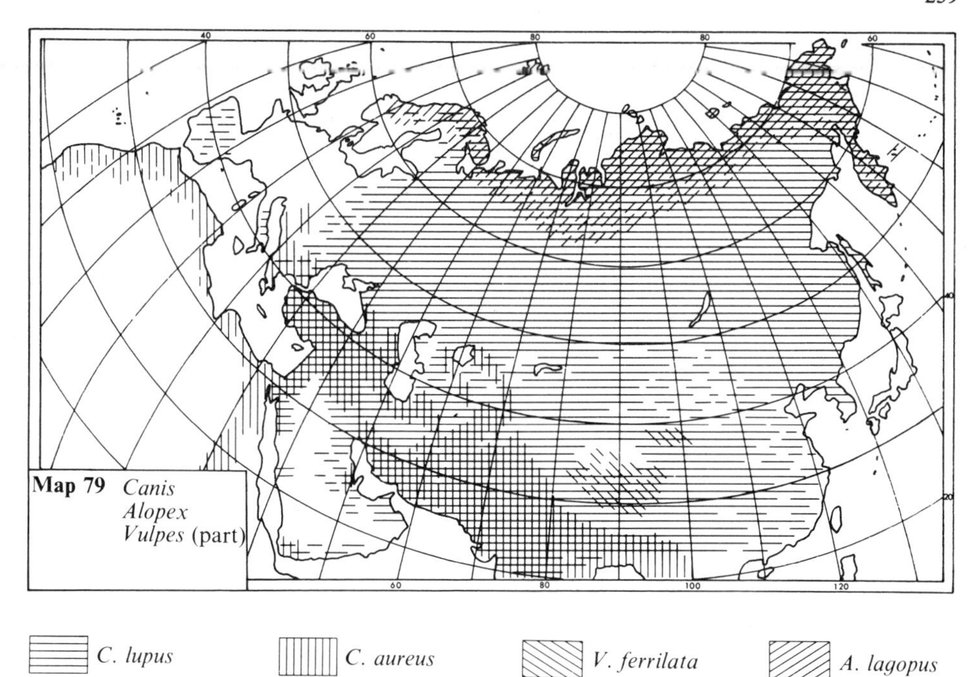

C. lupus — *C. aureus* — *V. ferrilata* — *A. lagopus*

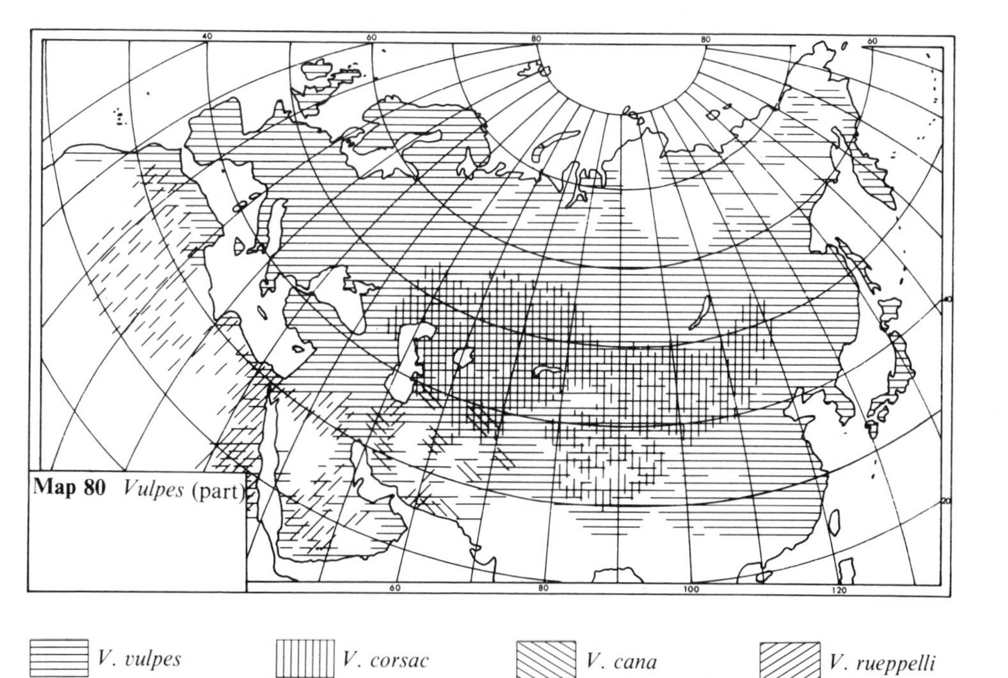

V. vulpes — *V. corsac* — *V. cana* — *V. rueppelli*

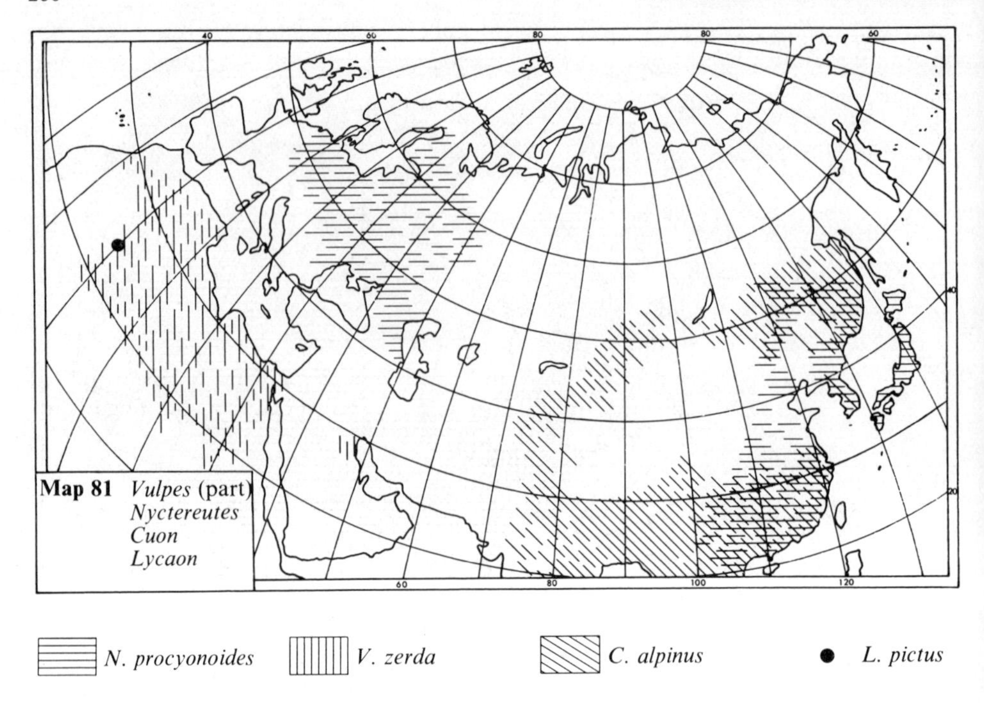
40
60
80
80
60
40
20
60
80
100
120
Map 81 *Vulpes* (part)
Nyctereutes
Cuon
Lycaon
N. procyonoides
V. zerda
C. alpinus
L. pictus

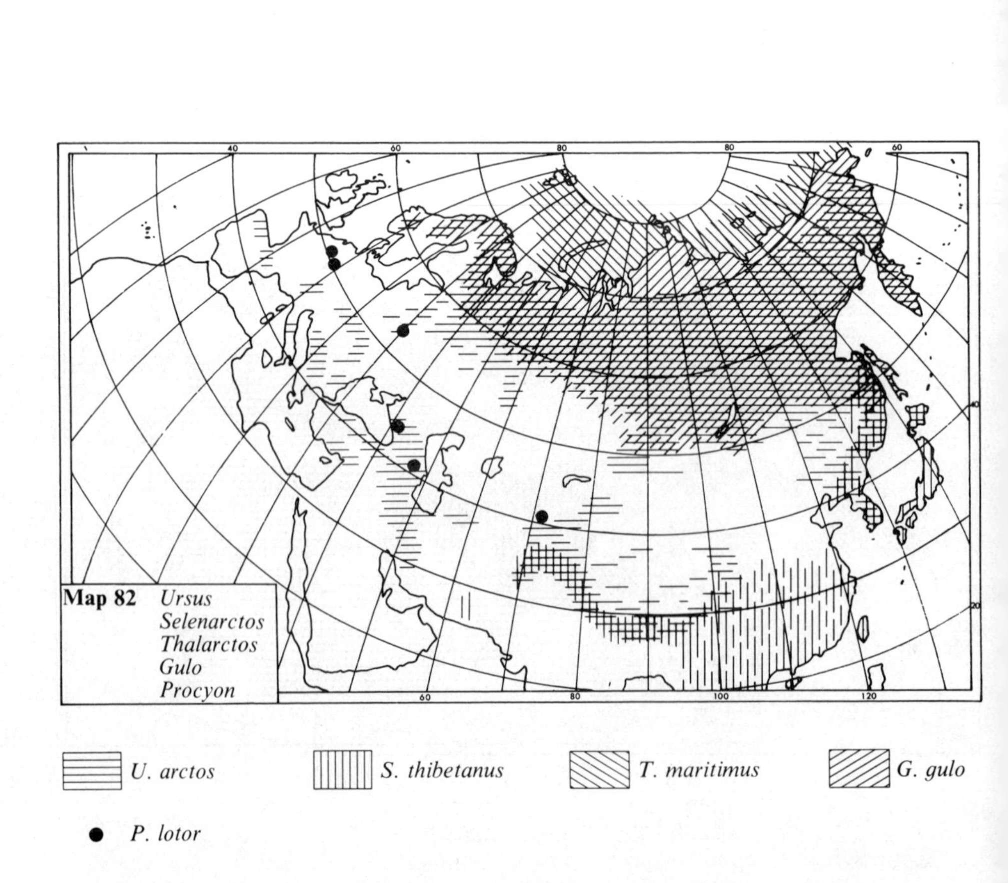
40
60
80
80
60
40
20
60
80
100
120
Map 82 *Ursus*
Selenarctos
Thalarctos
Gulo
Procyon
U. arctos
S. thibetanus
T. maritimus
G. gulo
P. lotor

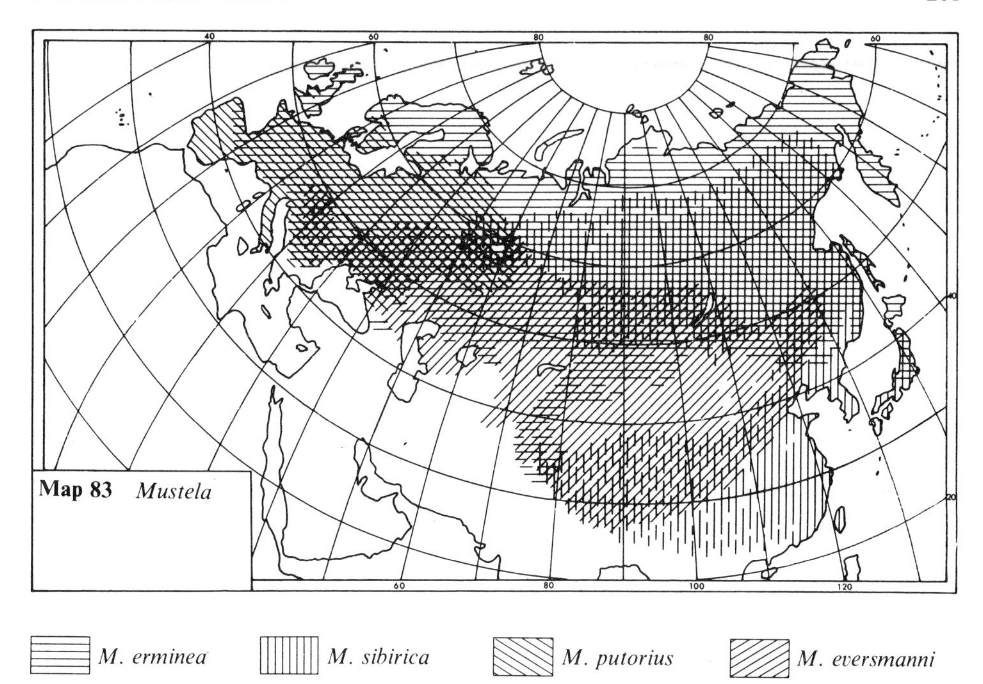
Map 83 Mustela
M. erminea
M. sibirica
M. putorius
M. eversmanni

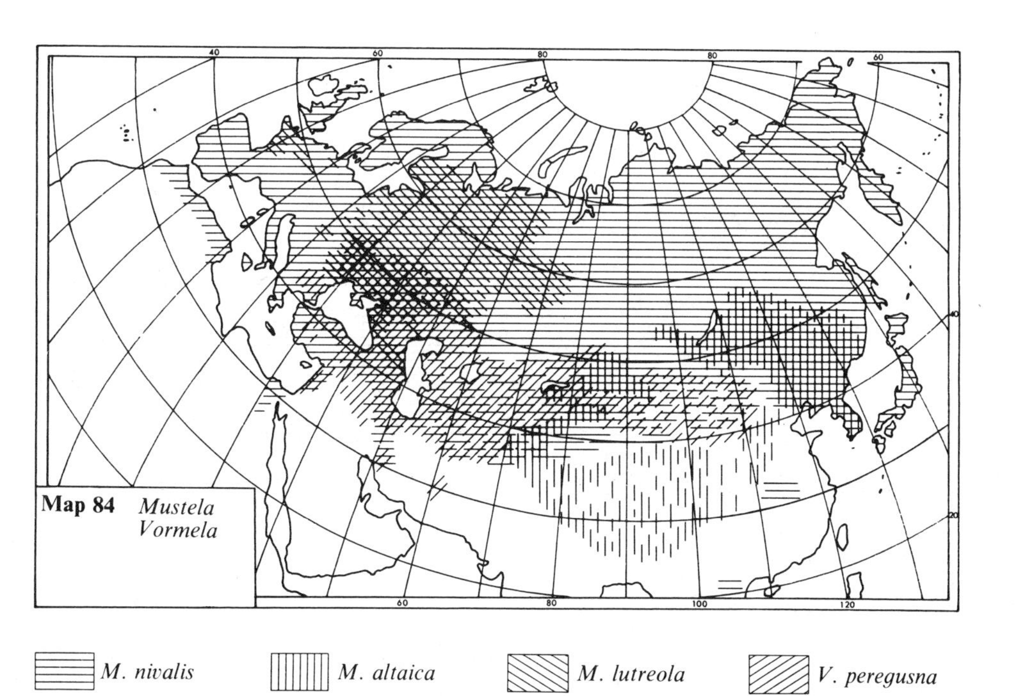
Map 84 Mustela Vormela
M. nivalis
M. altaica
M. lutreola
V. peregusna

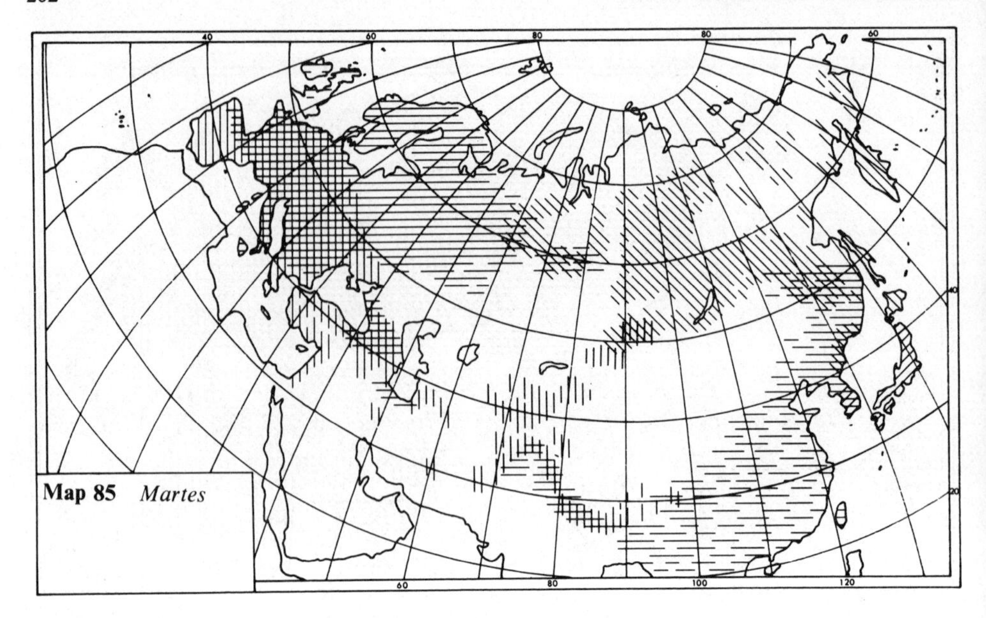

M. martes (Europe and W. Asia), *M. flavigula* (E. Asia)

M. foina *M. zibellina* *M. melampus*

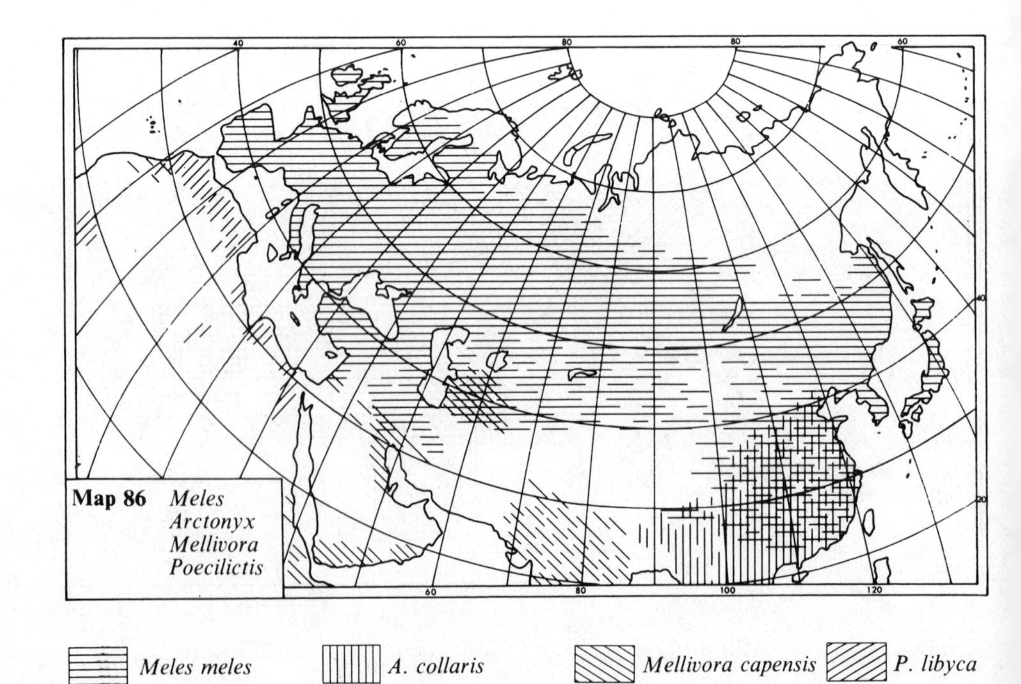

Meles meles *A. collaris* *Mellivora capensis* *P. libyca*

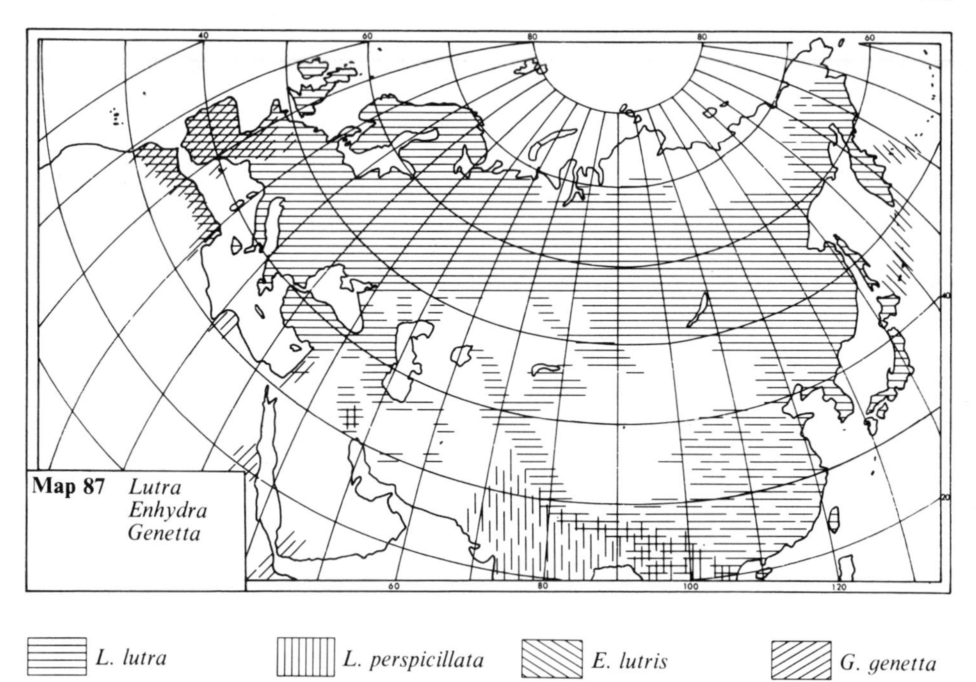
40
60
80
80
60
40
20
60
80
100
120
Map 87 Lutra
Enhydra
Genetta
L. lutra
L. perspicillata
E. lutris
G. genetta

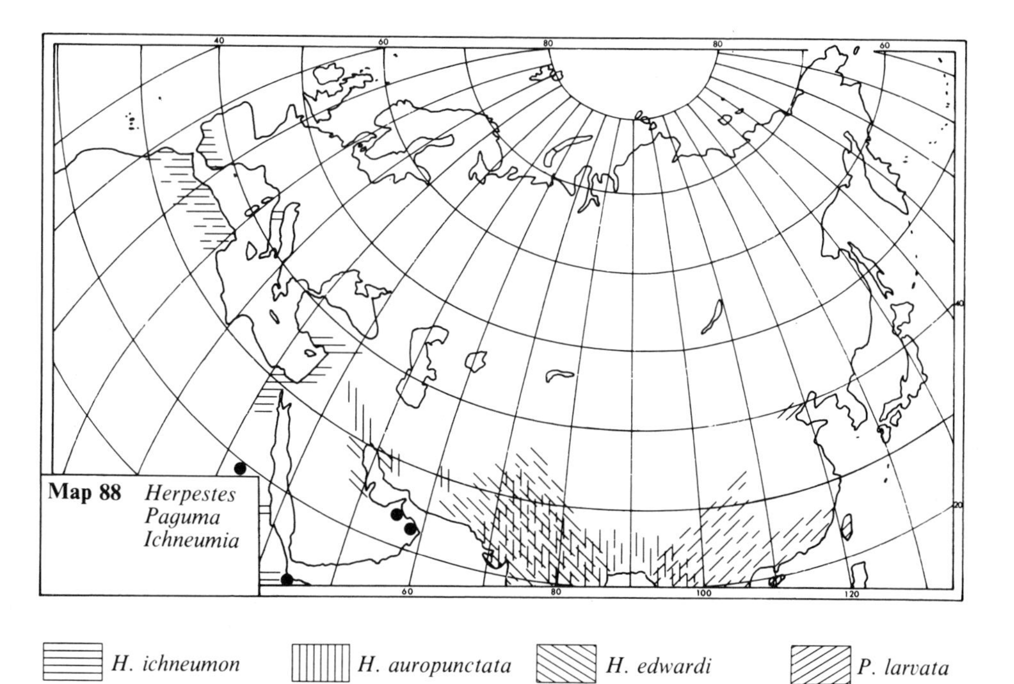
40
60
80
80
60
40
20
60
80
100
120
Map 88 Herpestes
Paguma
Ichneumia
H. ichneumon
H. auropunctata
H. edwardi
P. larvata
I. albicauda

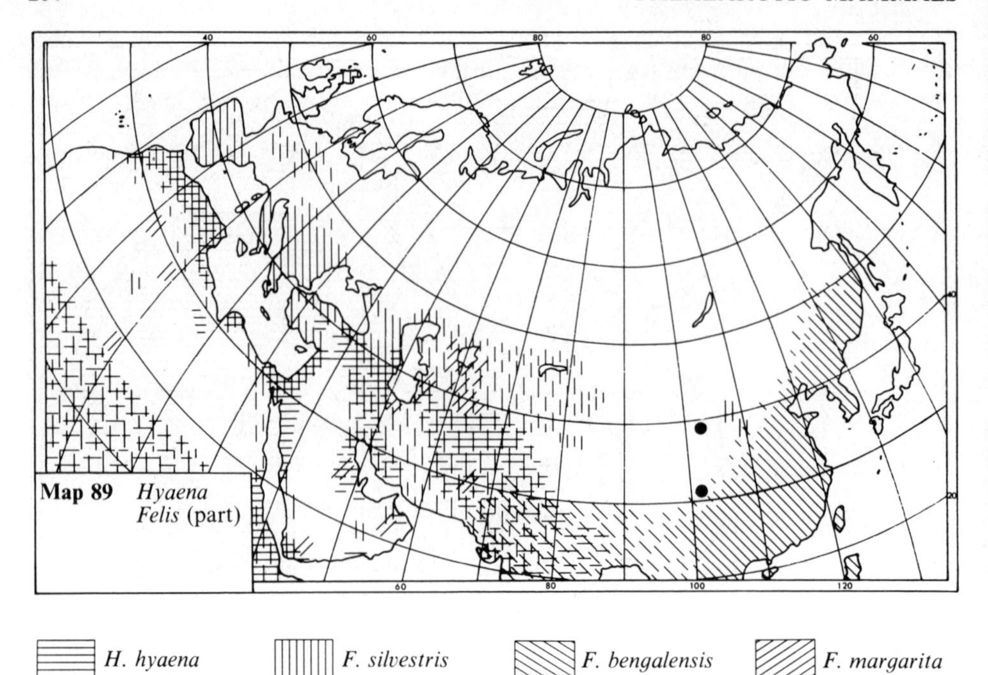

● *F. bieti*

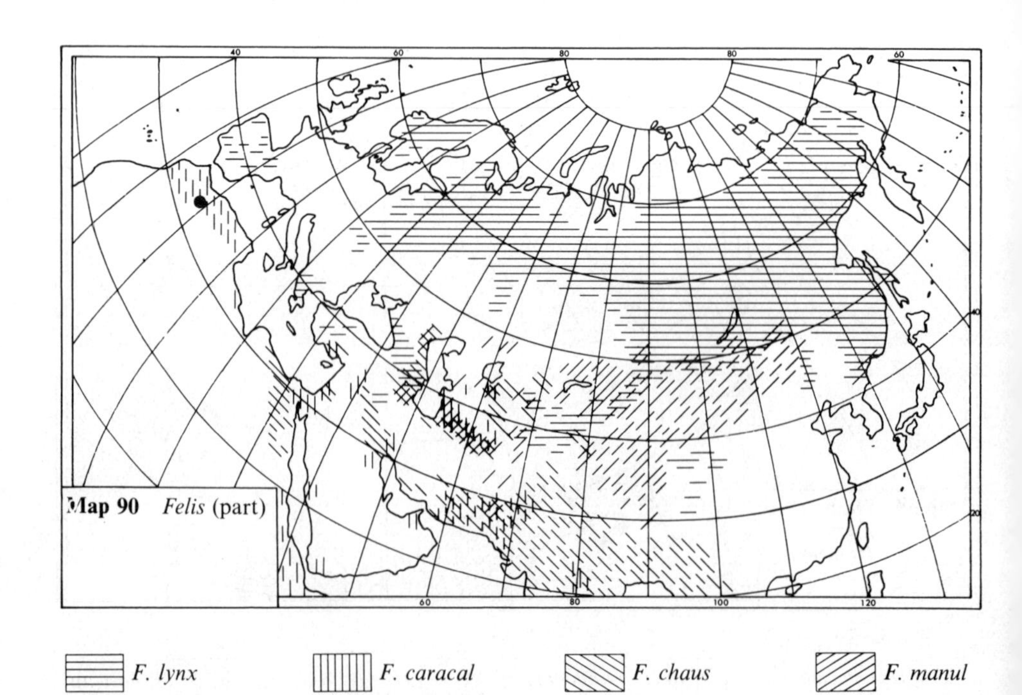

● *F. serval*

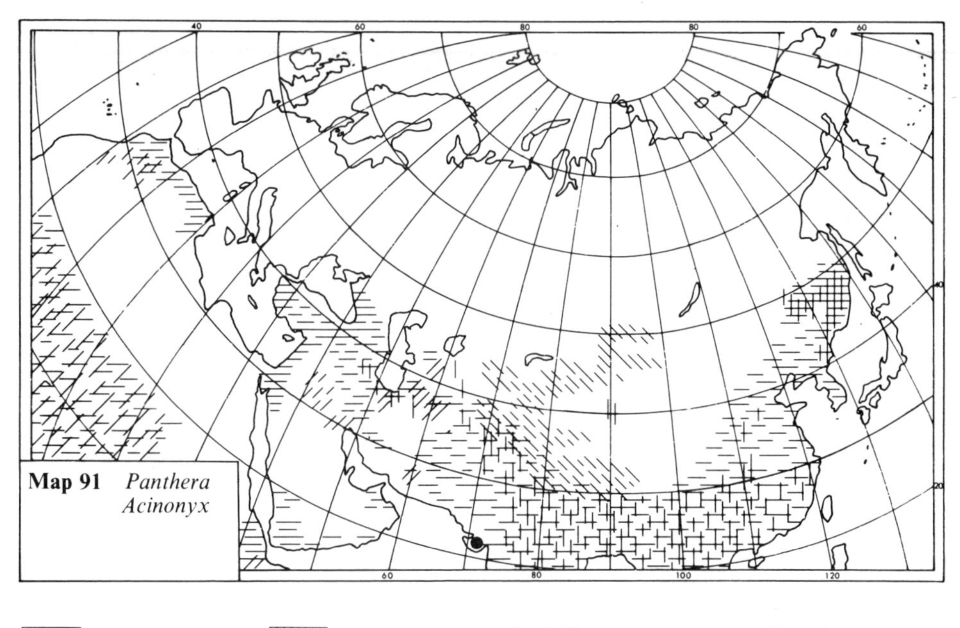

P. pardus — *P. pardus*; P. tigris — *P. tigris*; P. uncia — *P. uncia*; A. jubatus — *A. jubatus*

● *P. leo*

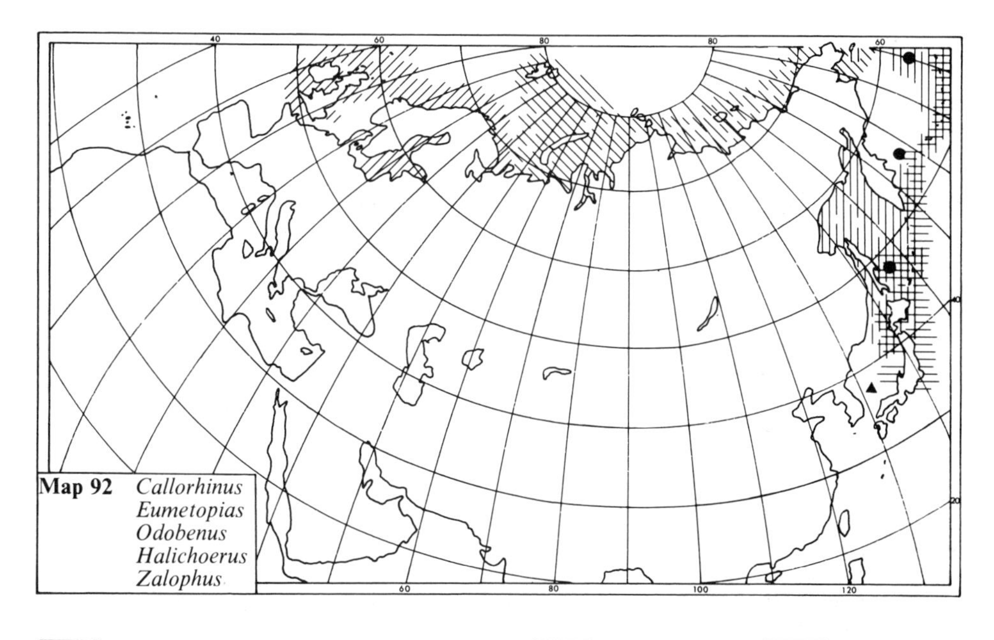

C. ursinus (● = breeding colony) — *E. jubatus* — *O. rosmarus*

H. grypus — ▲ *Z. californianus* (extinct – last Asiatic colony)

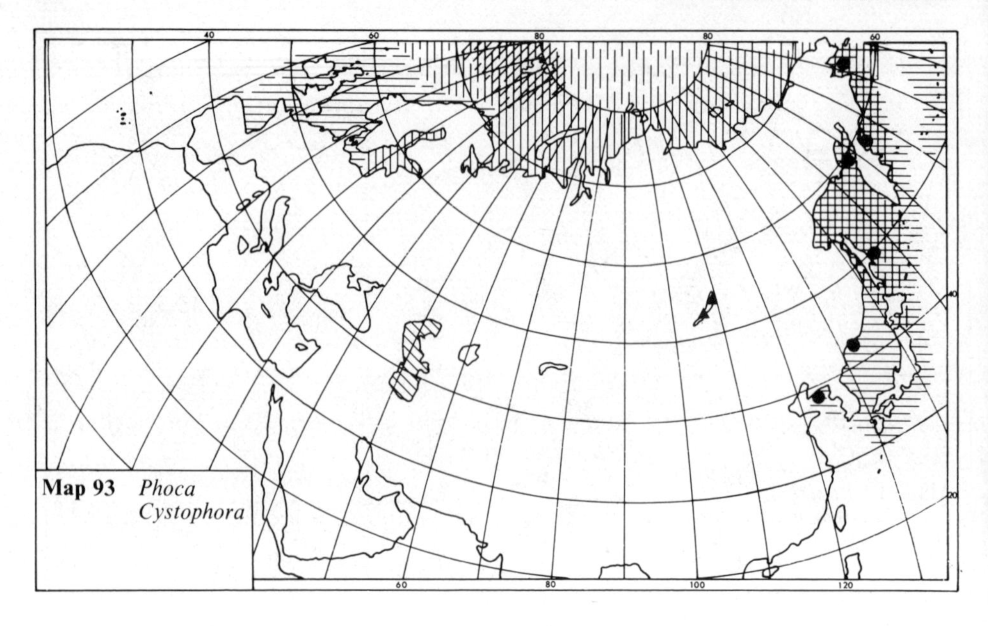

Map 93 *Phoca Cystophora*

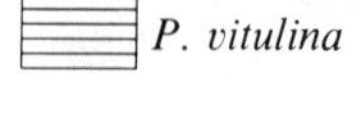
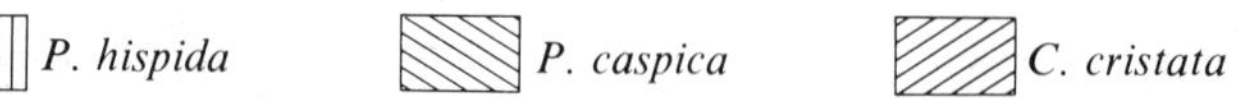
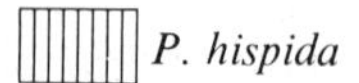
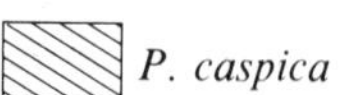

P. vitulina *P. hispida* *P. caspica* *C. cristata*

● *P. largha* ▲ *P. sibirica*

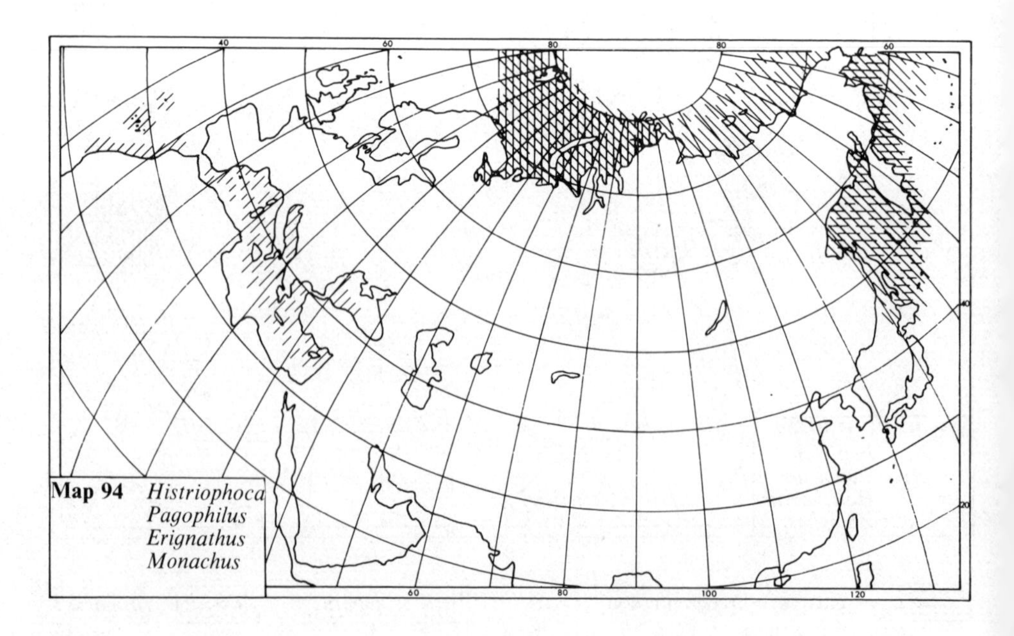

Map 94 *Histriophoca Pagophilus Erignathus Monachus*

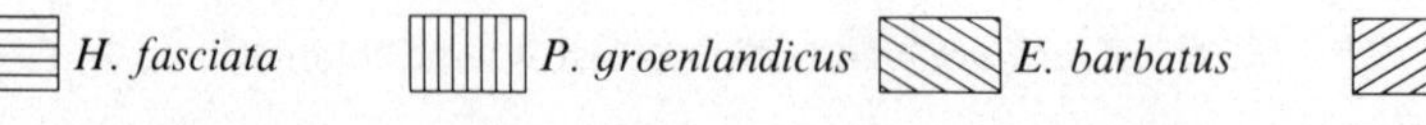

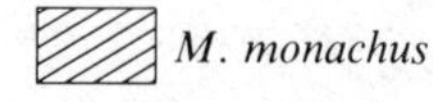

H. fasciata *P. groenlandicus* *E. barbatus* *M. monachus*

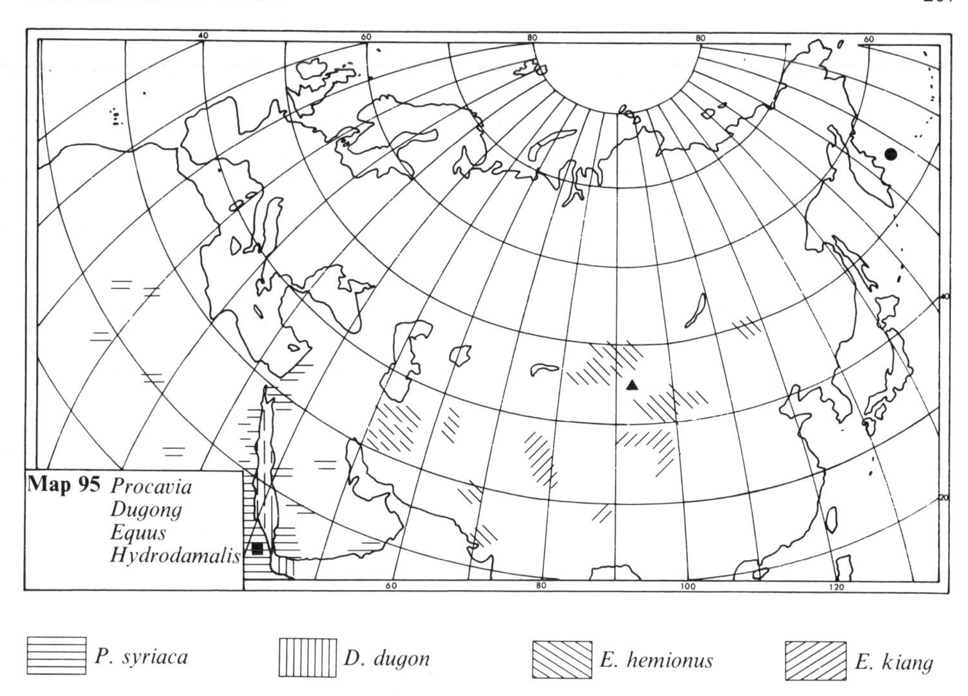

P. syriaca — *D. dugon* — *E. hemionus* — *E. kiang*

● *H. gigas* (extinct) ▲ *E. ferus* ■ *E. africanus*

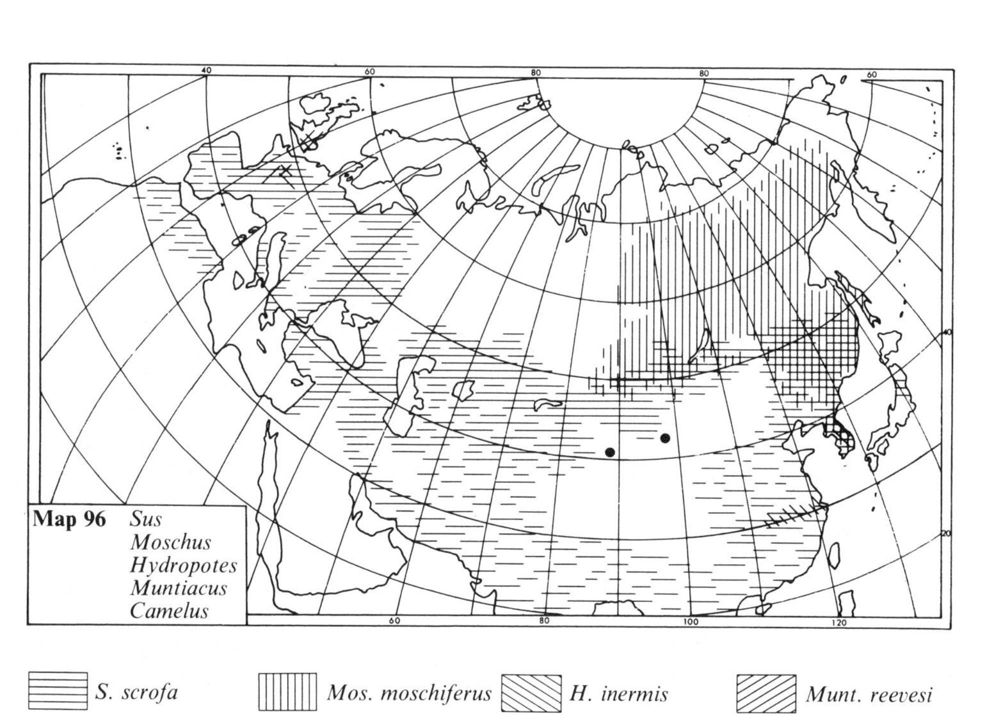

S. scrofa — *Mos. moschiferus* — *H. inermis* — *Munt. reevesi*

● *C. ferus*

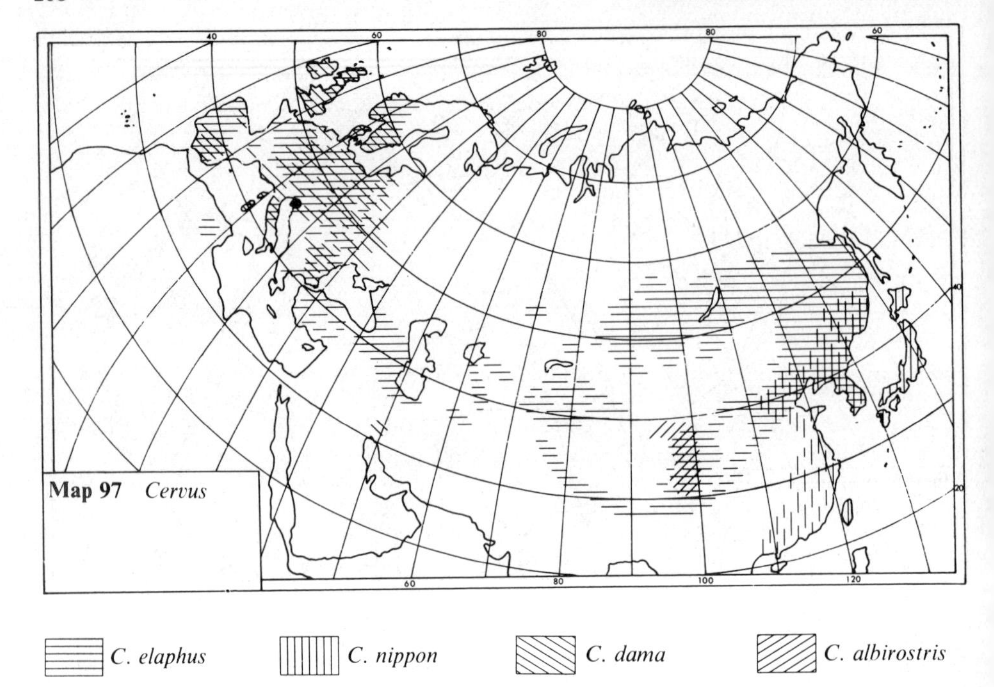

Map 97 *Cervus*

C. elaphus *C. nippon* *C. dama* *C. albirostris*

● *C. axis* (introduced)

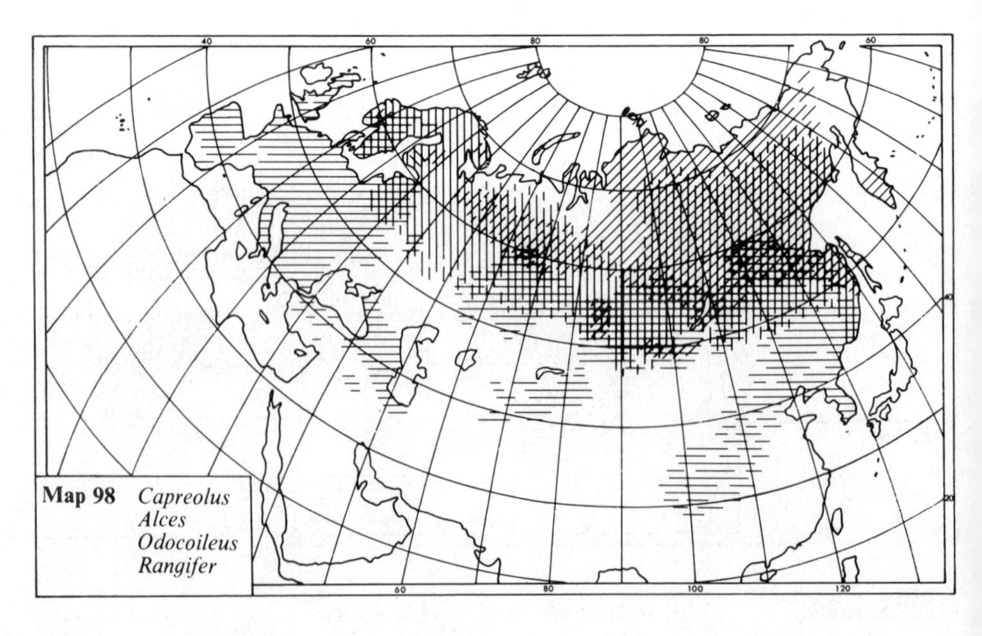

Map 98 *Capreolus* *Alces* *Odocoileus* *Rangifer*

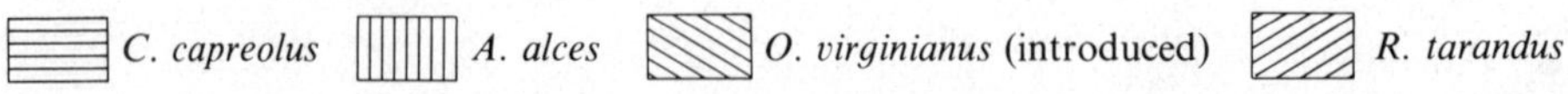

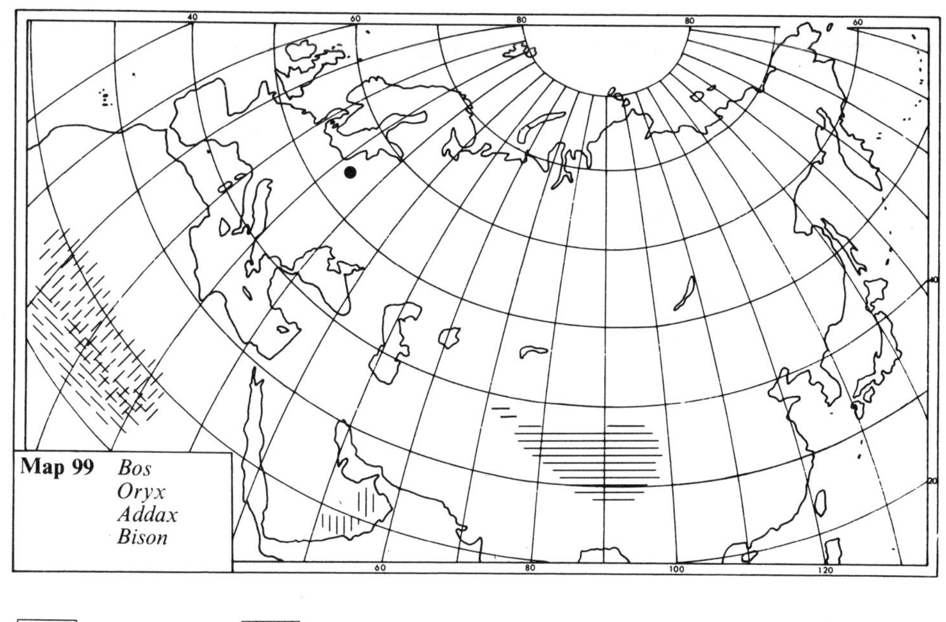

Bos mutus | *O. leucoryx* | *O. dammah* | *A. nasomaculatus*

● *Bison bison*

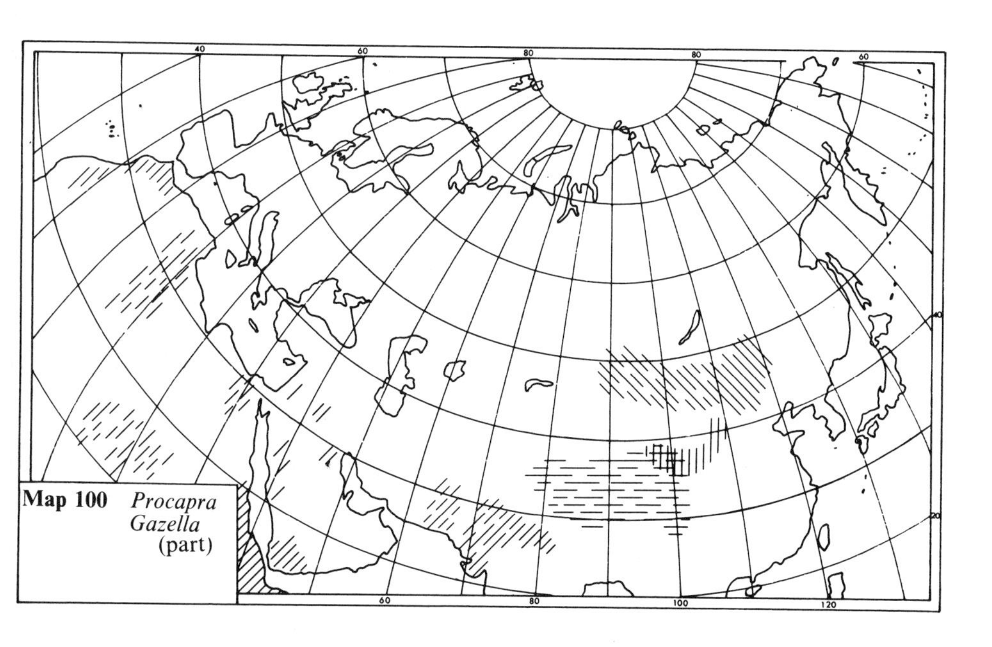

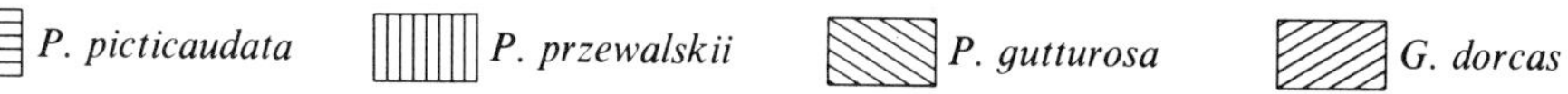

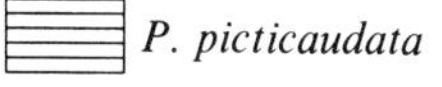
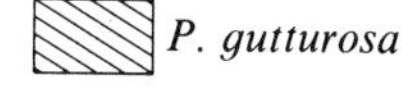
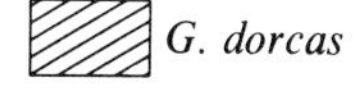

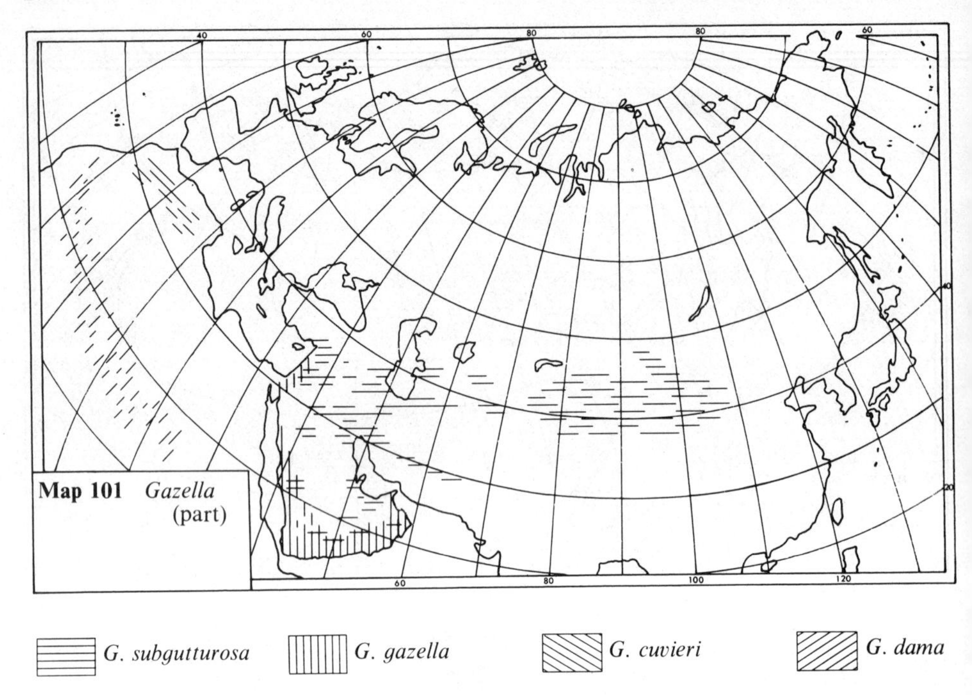

Map 101 Gazella (part)
G. subgutturosa
G. gazella
G. cuvieri
G. dama
G. leptoceros

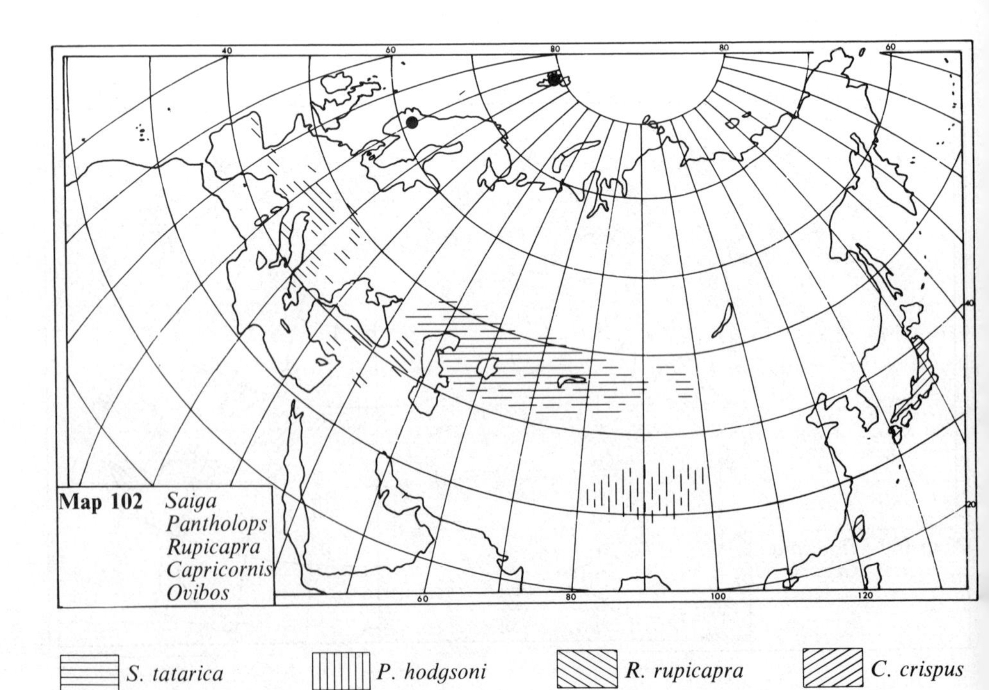

Map 102 Saiga Pantholops Rupicapra Capricornis Ovibos
S. tatarica
P. hodgsoni
R. rupicapra
C. crispus
O. moschatus (introduced)

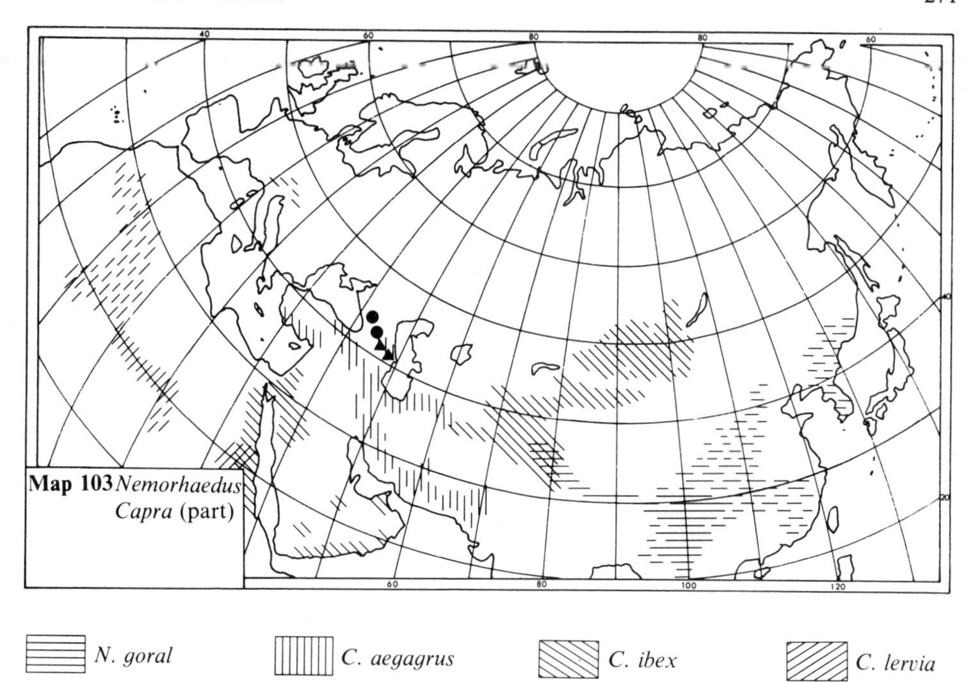

Map 103 *Nemorhaedus*
Capra (part)

N. goral | *C. aegagrus* | *C. ibex* | *C. lervia*

● *C. caucasica* ▲ *C. cylindricornis*

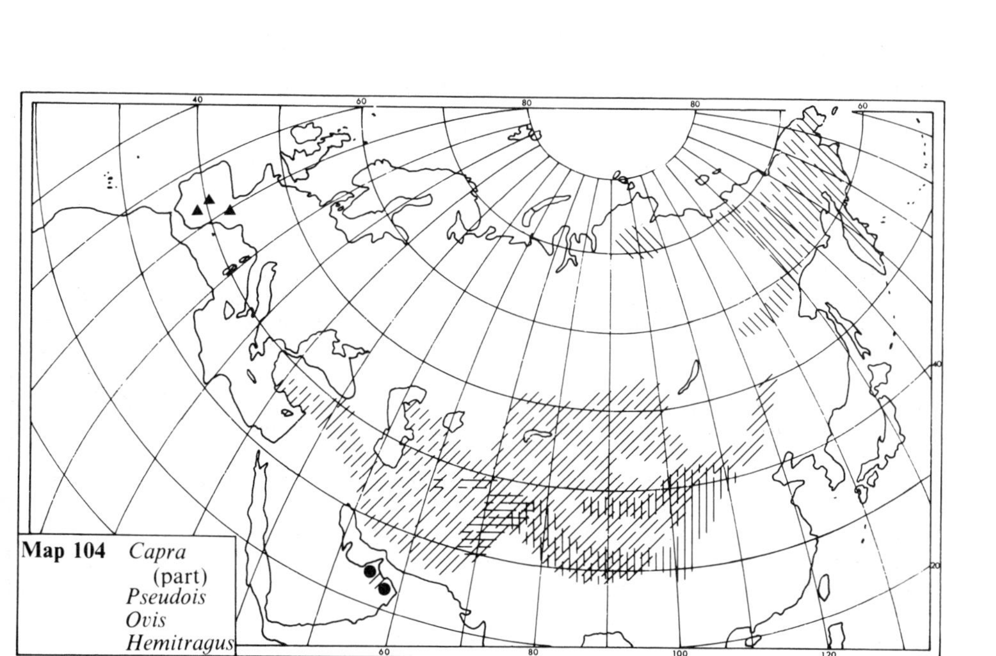

Map 104 *Capra*
(part)
Pseudois
Ovis
Hemitragus

C. falconeri | *P. nayaur* | *O. canadensis* | *O. ammon*

● *H. jayakari* ▲ *C. pyrenaica*

References

Titles in square brackets have been translated – in the case of items published in the USSR it can be assumed that the original language was Russian unless otherwise stated. An English title in a Russian journal generally means that the original paper was in Russian with an English summary. Some titles have been abbreviated, especially by omitting the names of higher taxa, and authors and dates of scientific names.

Abe, H. 1967. Classification and biology of Japanese Insectivora. I. Studies on variation and classification. *J. Fac. Agric. Hokkaido Univ.* **55**:191–265, pls 1–2.

—— 1968. Classification and biology of Japanese Insectivora. II. Biological aspects. *J. Fac. Agric. Hokkaido Univ.* **55**:429–458.

Abelentsev, V. I. 1968. [*Fauna of Ukraine*, vol. 1, part 3: *weasel family*] Kiev (in Ukrainian).

—— **& Pidoplichko, I. G.** 1956. [*Fauna of Ukraine*, vol. 1, part 1: *insectivores and bats*] Kiev (in Ukrainian).

Aellen, V. 1955. *Rhinolophus blasii*, chauve-souris nouvelle pour l'Afrique du Nord. *Mammalia* **19**:361–366.

—— 1959. Contribution à l'étude de la faune d'Afghanistan. Chiroptères. *Rev. suisse Zool.* **66**:353–386.

—— 1966. Notes sur *Tadarida teniotis*. *Rev. suisse Zool.* **73**:119–159, pl. 1.

Afanasiev, A. V. 1953. *In* Afanasiev, A. V. *et al.* [*Animals of Kazakhstan*] Alma-Ata.

—— **& Bazhanov, V. S.** 1948. *Izv. Akad. Nauk. Kazakh. SSR zool.* **51** (7):41–48.

Agrawal, V. C. & Chakraborty, S. 1971. Notes on a collection of small mammals from Nepal, with the description of a new mouse-hare (Lagomorpha: Ochotonidae). *Proc. zool. Soc. (Calcutta)* **24**:41–46.

Aimi, M. 1967. Similarity between the voles of Kii Peninsula and of northern part of Honshu. *Zool. Mag. Tokyo* **76**:44–49.

Aksenova, T. G. & Tarasov, S. A. 1974. Structural patterns of os penis in some species of the genus *Microtus*. *Zool. Zh.* **53**:609–615.

Allen, G. M. 1938–40. *The mammals of China and Mongolia*. New York.

—— 1939. A checklist of African mammals. *Bull. Mus. comp. Zool. Harvard* **83**:1–763.

—— 1942. *Extinct and vanishing mammals of the Western Hemisphere*. New York.

Almaça, C. 1973. Sur la structure des populations des *Pitymys* Ibériques. *Revta Fac. Cienc. Univ. Lisb.* **C17**:383–426.

Amtmann, E. 1965. Biometrisches Untersuchungen zur introgressiven Hybridisation der Waldmaus (*Apodemus sylvaticus*) und der Gelbhalsmaus (*Apodemus tauricus*). *Z. zool. Syst. Evolutionsf.* **3**:103–156.

Andersen, K. 1912. *Catalogue of Chiroptera in the collection of the British Museum*. Vol. 1: *Megachiroptera*. London.

Anderson, E. 1970. Quaternary evolution of the genus *Martes*. *Acta zool. fenn.* **130**:1–132.

Anderson, S. & Jones, J. K. 1967. *Recent mammals of the world: a synopsis of families*. New York.

Angermann, R. 1966a. Beiträge zur Kenntnis der Gattung *Lepus*. I. Abgrenzung der Gattung *Lepus*. *Mitt. zool. Mus. Berl.* **42**:127–144.

—— 1966b. Der taxonomische Status von *Lepus brachyurus* und *Lepus mandschuricus*. *Mitt. zool. Mus. Berl.* **42**:321–335.

—— 1967a. Beitrage zur Kenntnis der Gattung *Lepus*. III Zur Variabilität palaearktischen Schneehasen. *Mitt. zool. Mus. Berl.* **43**:161–178.

—— 1967b. Beitrage zur Kenntnis der Gattung *Lepus*. IV *Lepus yarkandensis* und *Lepus oiostolus* – zwei endemische Hasenarten Zentralasiens. *Mitt. zool. Mus. Berl.* **43**:189–203.

Anon. 1947 [*Sea otter (Kalan)*] Moscow (Central Administration of Reserves of the RSFSR). English translation: Israel Program for Scientific Translations.
— 1968. Ground squirrel nomenclature. *J. Mammal.* **49**:605.
Ansell, W. F. H. 1971a. Order Perissodactyla. *In* Meester & Setzer, 1971, part 14.
— 1971b. Order Artiodactyla. *In* Meester & Setzer, 1971, part 15.
Antonius, O. 1912. Was ist der 'Tarpan'. *Naturwiss. Wochenschr.* **27**:516.
Argyropulo, A. I. 1946. [New date on taxonomy of genus *Lagurus*] *Vestn. Akad. Nauk. kazakh. SSR.* **7/8**(16–17):44–46.
— 1948. [Survey of recent species of the family Lagomyidae] *Trudy zool. Inst. Leningr.* **7**:124–128.
— **& Vinogradov, B. S.** 1939. [Remarkable new rodent of the USSR fauna] *Priroda Mosk.* **1**:81–83.
Atallah, S. I. 1967. A new species of spiny mouse (*Acomys*) from Jordan. *J. Mammal.* **48**:258–261.
— 1970. A new subspecies of the Golden spiny mouse, *Acomys russatus* (Wagner), from Jordan. *Occas. Pap. Univ. Conn. (Biol. Sci. Ser.)* **1**:201–204.
— **& Harrison, D. L.** 1968. On the conspecificity of *Allactaga euphratica* and *Allactaga williamsi* with a complete list of subspecies. *Mammalia* **32**:628–638.
Baird, S. F. 1857. *Mammals of North America, part 1.* Philadelphia.
Banfield, A. W. F. 1961. A revision of the reindeer and caribou, genus *Rangifer*. *Bull. nat. Mus. Canada* **177**:i–vi, 1–137.
— 1974. *The mammals of Canada.* Toronto.
Bannikov, A. G. 1947a. Materials towards the study of mammals in Mongolia. 1. Jerboa. *Byull. Mosk. Obshch. Ispyt. Prir. Otd. Biol.* **52**:35–43.
— 1947b. [New species of high-mountain vole from Mongolia] *Dokl. Akad. Nauk SSSR* **56**:217–220.
— 1948. [Contribution towards a knowledge of Mongolian mammals. II] *Byull. Mosk. Obshch. Ispyt. Prir. Otd. Biol.* **53** (2):19–36.
— 1951. *Uchen. zap. gar. pedagog. Inst.* **18**:56.
— 1952. Contributions to the knowledge of Mongolian mammals. VI Mustelidae. *Byull. Mosk. Obshch. Ispyt Prir. Otd. Biol.* **57**:30–44.
— 1953. [*Identification of mammals of the Mongolian People's Republic*] Moscow (in Russian).
— 1954. [*Mammals of Mongolian People's Republic*] Moscow (in Russian).
— 1960. Notes on the mammals in Nienshan and South Gobi area (China). *Byull. Mosk. Obshch. Ispyt. Prir. Otd. Biol.* **65** (3):5–12.
— (ed.) 1961. [*Biology of the saiga.*] Moscow (English translation: Jerusalem, 1967).
— **& Skalon, V. N.** 1949. [A new subspecies of *Marmota* from Mongolia] *Dokl. Akad. Nauk SSSR* **65**:377–379.
Baranov, O. K. & Vorontsov, N. N. 1973. On serological differentiation of five Palaearctic species of *Marmota*. *Zool. Zh.* **52**:577–583.
Bashanov, B. S. & Belosludov, B. A. 1941. A remarkable family of rodents from Kasakhstan, USSR. *J. Mammal.* **22**:311–315.
Bauer, K. 1954. Die Streifenmaus (*Sicista subtilis trizona*) in Österreich. *Zool. Anz.* **152**:206–213.
— 1957. Zur Kenntnis der Fledermausfauna Spaniens. *Bonn. zool. Beitr.* **7**:296–319.
— 1960. Die Säugetiere des Neusiedlersee-Gebietes (Österreich). *Bonn. zool. Beitr.* **11**:141–344.
— **& Festetics, A.** 1958. Zur Kenntnis der Kleinsäugerfauna der Provence. *Bonn. zool. Beitr.* **9**:103–119.
Baumann, F. 1949. *Die freilebenden Säugetiere der Schweiz.* Bern.
Bazhanov, V. S. 1943. *Izv. kazakh. Fil. Akad. Nauk SSSR.* **2**:53 (NV – from Bannikov, 1954).
Baziev, Z. K. 1962. The ermine (*Mustela erminea* L.) from the Great Caucasus. *Zool. Zh.* **41**:121–124.
Belkin, A. N. 1964. A new species of seal – *Phoca insularis* sp.n. – from the Kuril Islands. *Dokl. (Proc.) Acad. Sci. URSS Biol. Sci. Sect.* **158**:756–759.
Beljaev, 1945. *Izv. Akad. Nauk. kazakh. SSR* **5**:95–96.
— 1954. *Trudy respub. Sta. Zashch. Rast. Alma-Ata* **2**:3–102.
Belyaeva, N. S. 1953. [A new wood vole, *Clethrionomys rjabovi* sp. nov. from eastern Siberia.] *Byull. Mosk. Ispyt. Prir. Otd. Biol.* **58** (6):17–20.

Berry, R. J., Evans, I. M. & Sennitt, B. F. C. 1967. The relationships and ecology of *Apodemus sylvaticus* from the Small Isles of the Inner Hebrides, Scotland. *J. Zool. Lond.* **152**:333–346.

Bibikov, D. I. 1968. *Die Murmeltiere (Gattung* Marmota). Wittenberg Lutherstadt.

Birula, A. B. 1916. *Otch. zool. Mus. ross. Akad. Nauk.*:12.

Biswas, B. & Ghose, R. K. 1970. Taxonomic notes on the Indian pale hedgehogs of the genus *Paraechinus. Mammalia* **34**:467–477.

—— **& Khajuria, H.** 1955. Zoological results of the 'Daily Mail' Himalayan Expedition 1954. Four new mammals from Khumbu, eastern Nepal. *Proc. zool. Soc. Bengal* **8**: 25–30.

Blahout, M. 1972. Zur Taxonomie der Population von *Rupicapra rupicapra* in der Hohen Tatra. *Zool. Listy* **21**:115–132.

Blaine, G. 1913.On the relationship of *Gazella isabella* to *Gazella dorcas. Ann. Mag. nat. Hist.* **11**:291–296.

Blanville, H. M. D. 1816. *Bull. Soc. philom. Paris*:75–76.

Blanc, G. & Petter, F. 1959. Présence au Maroc de l'écureuil terrestre du Sénégal *Xerus erythropus. Mammalia* **23**:239–240.

Blanchard, J. 1957. Le lièvre à oreilles bleues et à dos brun, des Alpes. *Cah. Chasse Nature* **30**:7–38.

Blanford, W. T. 1888–91. *The fauna of British India including Ceylon and Burma.* London.

Blasius, J. H. 1857. *Naturgeschichte der Säugethiere Deutschlands.* Braunschweig.

Bobrinskii, N. A. 1965. [Orders Carnivora, Pinnipedia, Perissodactyla and Artiodactyla] *In* Bobrinskii *et al.*, 1965.

—— **Kuznetsov, B. A. & Kuzyakin, A. P.** 1965. [*Key to the mammals of the USSR.*] Moscow.

Boddaert, P. 1785. *Elenchus animalium*, vol. 1. Rotterdam.

Bohlken, H. 1958. Vergleichende Untersuchungen an Wildrindern (tribus Bovini Simpson 1945). *Zool. Jb. (Allg.)* **68**:113–202.

—— 1961. Haustier und zoologische Systematik. *Z. Tierzücht. Züchtbiol.* **75**:107–113.

—— 1967. Beitrag zur Systematik der rezenten Formen der Gattung *Bison. Z. zool. Syst. EvolForsch.* **5**:54–110.

Böhme, W. 1969. Beitrag zur Kenntnis der Zwergmaus, *Micromys minutus. Faun.-Ökol. Mitt.* **3**:247–254.

Bojanus, L. H. 1827. De uro nostrate eiusque sceleto commentatio. *Nova Acta physico-med.* **13**:413–478, pls 20–24.

Bolshakov, V. N., Rossolimo, O. L. & Pokrovsky, A. V. 1969. The systematic status of Pamir-Alai high-mountain voles of the group *Microtus juldaschi. Zool. Zh.* **48**:1079–1089.

—— **& Shvartz, S. S.** 1965. New subspecies of field mouse *Clethrionomys rutilus tundrensis* subsp. nov. *Trudy Inst. Biol. Sverdlovsk* **38**:63–64.

Bonhote, J. L. 1910. On a small collection of mammals from Egypt. *Proc. zool. Soc. Lond.* **1909**:788–798.

Borovsky, S. G. & Vorontzov, N. N. 1970. *Nyctalus lasiopterus* in the West Aral Territory. *Zool. Zh.* **49**:942.

Bothma, J. D. 1971. Order Hyracoidea. *In* Meester & Setzer, 1971, part 12.

Bourdelle, E. & Dezilière, M. (1951). Sur quelques caractères ostéologiques et ostéométriques de la genette de Pyrénées. *Bull. Soc. Hist. nat. Toulouse* **86**:122–124.

Brass, E. 1911. *Aus dem Reiche der Pelze.* Berlin.

Brauner, A. A. 1914. [Mammals of Bessarabia, Khersonsk and Tavrichesk districts, pt. 1: foxes] *Zap. novoross. Obshch. Estest.* **40**.

—— 1929. [On Ukrainian polecats] *Ukrainsk. Mislivets ta Rubalka* **2–3**: 9 (NV).

Bree, P. van 1974. *Musculus dichrurus* Rafinesque 1814, a junior synonym of *Eliomys quercinus* (Linnaeus, 1776). *Säugetierk. Mitt.* **22**:258.

Brentjes, B. 1969. Der syrische Elefants als Sudform des Mammuts? *Säugetierk. Mitt.* **17**:211–212.

Brincken, 1828. *Mémoire déscriptif sur la forêt Imperiale de Bialovieza en Lithuanie* (NV).

Brink, F. H. van den 1952. Une nouvelle musaraigne dans les Pays-Bas. *Proc. K. ned. Akad. Wet.* **55c**:370–374.

—— 1953. La musaraigne masquée, espèce circumboréale. *Mammalia* **17**:96–125.

—— 1956. *Die Säugetiere Europas.* Hamburg & Berlin.

—— 1967. *A field guide to the mammals of Britain and Europe.* London.

Brosset, A. 1960. Les mammifères du Maroc oriental. Leur répartition – leur statut actuel. *Bull. Soc Sci. Maroc* **40**:243–263.
—— 1962. The bats of central and western India. Part II. *J. Bombay nat. Hist. Soc.* **59**:583–624.
—— 1963. *Myotis nattereri*, chiroptère nouveau pour l'Afrique du Nord. *Mammalia* **27**:440–443.
Bubenik, A. 1956. *Österreichischen Arbeitskreis für Wildtierforschung*:42–44.
Bühler, P. 1963. *Neomys fodiens niethammeri* ssp.n., eine neue Wasserspitzmausform aus Nord-Spanien. *Bonn. zool. Beitr.* **14**:165–170.
Burns, J. J. & Fay, F. H. 1970. Comparative morphology of the skull of the Ribbon seal, *Histriophoca fasciata*, with remarks on systematics of Phocidae. *J. Zool. Lond.* **161**:363–394.
Butler, P. M. 1956. The skull of *Ictops* and the classification of the Insectivora. *Proc. zool. Soc. Lond.* **126**:453–481.
—— *et al.* 1967. Mammalia, In *The fossil record*, London (Geological Society of London):763–787.
Cabrera, A. 1925. *Genera mammalium. Insectivora, Galeopithecia.* Madrid.
—— 1932. Los mamiferos de Marruecos. *Trab. Mus. nac. Cienc. nat. Madr. Ser. zool.* **57**:1–361.
Caslick, J. W. 1956. Colour phases of the roof rat. *J. Mammal.* **37**:255–257.
Chang, C. & Wang, T. 1963. Faunistic studies of mammals of the Chinghai Province. *Acta zool. Sinica* **15**:125–138.
Chapski, K. K. 1940. [Distribution of the walrus in the Laptev Sea in Eastern Siberia] *Problemy Arkt.* no. 6.
Charlemagne, N. V. 1920. [*Animals in the Ukraine*] Kiev (in Ukrainian).
Chasen, F. N. 1940. A handlist of Malaysian mammals. *Bull. Raffles Mus.* no. 15:i–xx, 1–209.
Chaturvedi, Y. 1964. Taxonomic status of *Tadarida tragata* (Dobson). *J. Bombay nat. Hist. Soc.* **61**:432–437.
Chaworth-Musters, J. L. & Ellerman, J. R. 1947. A revision of the genus *Meriones*. *Proc. zool. Soc. Lond.* **117**:478–504.
Churcher, C. S. 1959. The specific status of the New World red fox. *J. Mammal.* **40**:513–520.
Clinton, 1822. Letters on the natural history... of New York:193.
Clutton-Brock, J. & Corbet, G. B. 1975. *Vulpes* Frisch, 1775: proposed conservation under the plenary powers. *Bull. zool. Nomencl.* **32**:110–112.
—— —— **& Hills, M.** 1976. A review of the family Canidae, with a classification by numerical methods. *Bull. Br. Mus. nat. Hist.* (Zool.) **29**:117–199.
[Cobb, E. H.] 1958. The markhor. *Oryx* **4**:381–382. (An anonymous report of a lecture by Cobb.)
Cockrum, E. L., Vaughan, T. C. & Vaughan, P. J. 1976. A review of North African short-tailed gerbils (*Dipodillus*) with description of a new taxon from Tunisia. *Mammalia* **40**:313–326.
Corbet, G. B. 1967. The pygmy moles of Europe and Japan. *J. Zool. Lond.* **153**:567–568.
—— 1970. Patterns of subspecific variation. *Symp. zool. Soc. Lond.* **26**:105–116.
—— **Cummins, J., Hedges, S. R. & Krzanowski, W.** 1970. The taxonomic status of British water voles, genus *Arvicola*. *J. Zool. Lond.* **161**:301–316.
—— **& Hanks, J.** 1968. A revision of the elephant-shrews, family Macroscelididae. *Bull. Br. Mus. nat. Hist.* (Zool.) **16**:47–111.
—— **Hill, J. E., Ingles, J. M. & Napier, P. H.** 1974. Re-submission of *Pan* Oken, 1816 and *Panthera* Oken 1816, proposed conservation under the plenary powers. *Bull. zool. Nomencl.* **31**:29–43.
—— **& Jones, L. E.** 1965. The specific characters of the crested porcupines, subgenus *Hystrix*. *Proc. zool. Soc. Lond.* **144**:285–300.
—— **& Morris, P. A.** 1967. A collection of recent and subfossil mammals from southern Turkey (Asia Minor), including the dormouse, *Myomimus personatus*. *J. nat. Hist.* **1**:561–569.
—— **& Southern, H. N.** (eds) 1977. *The handbook of British mammals.* 2nd ed. Oxford.
—— **& Yalden, D. W.** 1972. Recent records of mammals from Ethiopia. *Bull. Br. Mus. nat. Hist.* (Zool.) **22**:213–252.
Couturier, M. A. J. 1938. *Le chamois,* Rupicapra rupicapra (*L.*) Grenoble (published by author).
—— 1954. *L'Ours brun,* Ursus arctos *L.* Grenoble (published by author).
—— 1955. Acclimatations et acclimatement de la marmotte des Alpes, dans les Pyrénées françaises. *Säugetierk. Mitt.* **3**:105–108.
—— 1962. *Le bouquetin des Alpes.* Capra aegagrus ibex ibex *L.* Grenoble (published by author).

Cranbrook, The Earl of 1957. Long-tailed field mice (*Apodemus* sp.) from the Channel Islands. *Proc. zool. Soc. Lond.* **128**:597–600.

Cretzschmar, J. 1826–1831. *In* Ruppell, E. *Atlas zu der Reise im nordlichen Afrika*, pt. 1, *Saugethiere*. Frankfurt am Main.

Cuvier, F. 1827. *Zool. J. Lond.* **3**: 140.

Dahl, S. K. 1947. [A new subspecies of Natterer's bat from the Daragalezki Mountains] *Dokl. Akad. Nauk. armyan. SSR* **7**:173–178.

Davis, D. H. S. 1965. Classification problems of African Muridae. *Zool. Africana* **1**:121–145.

Davis, W. B. 1965. Review of the *Eptesicus brasiliensis* complex in Middle America. *J. Mammal.* **46**:229–240.

De Beaux, O. 1926. Gli *Apodemus* delle Tre Venezie. *Atti Soc. liguist. Sci. nat. geogr.* **5**:52–65.

—— 1939. Una nuova sottospeci de *Vulpes Ruppelli* in Libia. *Annali Mus. libico Stor. nat.* **1**:393–396.

DeBlase, A. F. 1971. New distribution records of bats from Iran. *Fieldiana Zool.* **58**:9–14.

—— 1972. *Rhinolophus euryale* and *R. mehelyi* in Egypt and southwest Asia. *Israel J. Zool.* **21**:1–12.

—— **Schlitter, D. & Neuhauser, H.** 1973. Taxonomic status of *Rhinopoma muscatellum* from southwest Asia. *J. Mammal.* **54**:831–841.

Dehnel, A. 1949. Studies on the genus *Sorex* L. *Annls Univ. Mariae Curie-Sklodowska* **4c**:17–102.

Deleuil, R. & Labbe, A. 1955. Contributions à l'étude des chauves-souris de Tunisie. *Bull. Soc. Sci. nat. Tunisie* **8**:39–55, pls 18–22.

Delost, P. 1955. Recherches sur la biologie generale du campagnol des champs (*Microtus arvalis*). *Bull. Soc. zool. Fr.* **80**:149–162.

Denisov, V. P. 1961. Relationship of *Citellus pygmaeus* and *C. suslica* on the junction of their ranges. *Zool. Zh.* **40**:1079–1085.

—— 1963. [Hybridization of the species belonging to *Citellus* Oken.] *Zool. Zh.* **42**:1887–1889.

Dennler de la Tour, G. 1968. Zur Frage der Haustier-Nomenklatur. *Säugetierk. Mitt.* **16**:1–20.

Deparma, N. K. 1959. A new mole from N.-W. Caucasus. *Byull. Mosk. Obshch. Ispyt. Prir. Otd. Biol.* **64**, 6:31–36.

Desmarest, A. G. 1817. *Nouveau dictionaire d'histoire naturelle* vol. 17.

—— 1822. *Mammalogie* part 2. Paris.

De Winton, W. E. 1902. Soricidae. *In* Anderson, J. *Zoology of Egypt*: *Mammalia*. London:166–170.

Didier, R. & Petter, F. 1960. L'os pénien de *Jaculus blanfordi*. Étude comparée de *J. blanfordi*, *J. jaculus* et *J. orientalis*. *Mammalia* **24**:171–176.

Dinnik, N. J. 1910. [*Mammals of the Caucasus*, vol. 1.] Tiflis.

Dobroruka, L. J. 1959. *Meriones vinogradovi* in Palästine. *Säugetierk. Mitt.* **7**:173.

Dolan, J. M. 1963. Beitrag zur systematischen Gliederung des Tribus Rupicaprini. *Z. Zool. Syst. EvolForsch.* **1**:311–407.

—— 1969. The shrinking range of the Scimitar-horned oryx. *Zoonooz* **42** (10):13–15.

Dolgov, V. A. 1964. *Sorex centralis* Thomas, 1911, in the fauna of USSR. *Zool. Zh.* **43**:898–903.

—— 1966. Some patterns in the geographical variability of mammals. *Dokl. Akad. Nauk. SSSR* **171**:1230–1233.

—— 1967. Distribution and number of Palaearctic shrews (Insectivora, Soricidae). *Zool. Zh.* **46**:1701–1712.

—— 1968. [Appearance and variation of odontological characters of Palaearctic red-toothed shrews (Mammalia, *Sorex*).] *Sb. Trud. zool. Mus. MGU* **10**:179–199, pls 1–10.

—— 1974. Diagnosis of *Crocidura suaveolens* and *C. leucodon*. *Zool. Zh.* **53**:912–918.

—— **& Lukyanova, I. U.** 1966. On the structure of the genitalia of Palaearctic *Sorex* sp. *Zool. Zh.* **45**:1852–1861.

—— **& Yudin, B. S.** 1975. [Progress and problems in the investigation of the insectivorous mammals of the USSR.] *Trudy biol. Inst. Novosibirsk* **23**:5–40.

Dollman, G. 1916. On the African shrews belonging to the genus *Crocidura*. VII. *Ann. Mag. nat. Hist.* **8**, 17:188–209.

Dor, M. 1966. Restes subfossiles de *Lophiomys* trouvés en Israel. *Mammalia* **30**:199–200.

Dorogostaiski, V. Ch. 1935. [A new subspecies of corsak from southern Zabaykalya.] *Izv. Irkutsk. Gos. protivochumn. Inst. Siberia D.V.K.* **1**.

Dorst, J. & Dandelot, P. 1970. *A field guide to the larger mammals of Africa*. London.

—— & **Maurois, R. de** 1966. Presence de l'oreillard (*Plecotus*) dans l'archipel du Cap-Vert. *Mammalia* **30**: 292–301.

Dottrens, E. 1962. *Arvicola incertus* de Sélys-Longchamps était un *Pitymys*. *Archs Sci. Genève* **14**: 353–364.

Dulic, B. & Felten, H. 1962. Säugetiere (Mammalia) aus Dalmatien. *Senckenberg. biol.* **43**: 417–423.

—— & **Miric, D.** 1967. *Catalogus faunae Jugoslaviae* IV/4: *Mammalia*. Ljubljana.

—— & **Vidinic, Z.** 1967. A contribution to the study of mammalian fauna on the Dinara and Sator Mountains (S.W. Yugoslavia). *Let. slov. Akad. Znan. Umetn.*: 139–180.

Ehrenberg, C. G. 1833. *Symbolae physicae seu icones et descriptiones mammalium*. Decas 1.

Ehrstrom, K. E. 1914. Eine abweichende Form von *Apodemus agrarius* aus Finland. *Meddn. Soc. Fauna Flora fenn.* **40**: 16–18.

Elizaryeva, M. V. 1949 [Occurrence of a Jerboa (*Salpingotus*), in the boundary of USSR.] *Dokl. Akad. Nauk SSSR* **66**: 495–498.

Ellerman, J. R. 1940, 1941. *The families and genera of living rodents*. Vols 1 and 2. London.

—— 1948. Key to the rodents of South-West Asia. *Proc. zool. Soc. Lond.* **118**: 765–816.

—— 1949. *The families and genera of living rodents*. Vol. 3, part 1. London.

—— 1961. *The fauna of India. Mammalia* (2nd ed.) vol. 3 *Rodentia*. Delhi.

—— & **Morrison-Scott, T. C. S.** 1951. *Checklist of Palaearctic and Indian mammals 1758 to 1946*. London.

—— —— 1966. Ibid 2nd ed. London.

—— —— & **Hayman, R. W.** 1953. *Southern African mammals 1758 to 1951: a reclassification*. London.

Erxleben, J. C. B. 1777. *Systema regni animalis*.

Etemad, E. 1970. A note on the occurrence of giant noctule, *Nyctalus lasiopterus*. *Mammalia* **34**: 547.

—— 1973. Occurrence of a rare bat *Eptesicus walli* in Iran. *Period biol.* (Zagreb): 75.

Felten, H. 1962. Bemerkungen zu Fledermäusen der Gattungen *Rhinopoma* und *Taphozous*. *Senckenberg. biol.* **43**: 171–176.

—— 1971. Eine neue Art der Fledermaus-Gattung *Eptesicus* aus Kleinasien. *Senckenberg. biol.* **52**: 371–376.

—— **Spitzenberger, F. & Storch, G.** 1971a. Zum Mittelmeer-Program der Säugetier-Sektionen. *Natur Mus. Frankf.* **101**: 408.

—— —— —— 1971b. Zur Kleinsäugerfauna West-Anatoliens Teil I. *Senckenberg. biol.* **52**: 393–424.

—— —— —— 1973. Zur Kleinsäugerfauna West-Anatoliens Teil II. *Senckenberg. biol.* **54**: 227–290.

—— & **Storch, G.** 1965. Insektenfresser und Nagetiere aus N-Griechenland und Jugoslavien. *Senckenberg. biol.* **46**: 341–367.

—— —— 1968. Eine neue Schläfer-Art, *Dryomys laniger* n.sp. aus Kleinasien. *Senckenberg. biol.* **49**: 429–435.

—— —— 1970. Kleinsäuger von den italienischen Mittelmeer-Inseln Pantelleria und Lampedusa. *Senckenberg. biol.* **51**: 159–173.

Feng, Tso-Chien, 1973. A new species of *Ochotona* from Mount Jolmo-Lungma area. *Acta zool. Sinica* **19**: 70–75.

—— & **Kao, Yüeh-Ting,** 1974. Taxonomic notes on the Tibetan pika and allied species – including a new subspecies. *Acta zool. Sinica* **20**: 76–87.

Fetisov, A. S. 1950a. New form of shrew (*Sorex jenissejensis margarita* subsp.n.) from Eastern Siberia. *Izv. biologo-geogr. nauchno-issled. Inst. Irkutsk* **10** (4): 9–10.

—— 1950b. [A new subspecies of lynx (*Lynx lynx kozlovi* subsp.n.) from eastern Siberia.] *Izv. biologo-geogr. nauchno.-issled. Inst. Irkutsk* **12** (1): 21–22.

—— 1956. [Systematics and distribution of the mole (*Asioscalops altaica*) in Zabaikal.] *Izv. biologo-geogr. nauchno-issled. Inst. Irkutsk* **16**: 195–198.

Findley, J. S. 1972. Phenetic relationships among bats of the genus *Myotis*. *Syst. Zool.* **21**: 31–52.

Fitzgibbon, J. 1966. On a new species of *Salpingotus* from north-western Baluchistan. *Mammalia* **30**: 431–440, pl. 24.

Fitzinger, L. J. 1868. Kritische Untersuchungen uber die der natürlichen Familie der Spitzmäuse (Sorices) angehörigen Arten. *S.B. Akad. Wiss. Wien.* **57**: 121–180.

Flerov, K. K. 1952. *Fauna of the USSR. Mammals* Vol. 1, *Musk deer and deer*. Moscow & Leningrad (English translation: Jerusalem, 1960).

Flint, W. E. 1966. *Die Zwerghamster der paläarktischen Fauna*. (Neue Brehm-Bucherei No. 366) Stuttgart.

Flower, S. S. 1932. Notes on the recent mammals of Egypt. *Proc. zool. Soc. Lond.*: 369–450.

Flower W. H. & Lydekker, R. 1891. *An introduction to the study of mammals living and extinct*. London.

Fokanov, V. A. 1966. A new subspecies of *Marmota bobac* and some notes on the geographical variability of the species. *Zool. Zh.* **45**: 1862–1866.

Fooden, J. 1964. Rhesus and crab-eating macaques: integradation in Thailand. *Science, N.Y.* **143**: 363–365.

—— 1967. *Macaca fuscata* (Blyth, 1875): proposed conservation as the name for the Japanese macaque. *Bull. zool. Nomencl.* **24**: 250–251.

—— 1976. Provisional classification and key to living species of macaques (Primates: *Macaca*). *Folia primatol.* **25**: 225–236.

Foromozov, A. N. (ed.) 1965. [*Mammals of the USSR*.] Moscow.

Fortunatov, B. K. 1925. [Hybrid herds last year.] *Ascania Nova*, Kiev.

Frechkop, S. 1959. De la position systématique du genre *Nyctereutes*. *Bull. Inst. r. Sci. nat. Belg.* **35** (19): 1–20.

—— 1963. Notes sur les mammifères L – De la boccamele de Sardaigne. *Bull. Inst. r. Sci. nat. Belg.* **39** (8): 1–21.

Freye, H. A. 1960. Zur systematik der Castoridae. *Mitt. zool. Mus. Berl.* **36**: 105–122.

Gaisler, J. 1970. The bats (Chiroptera) collected in Afghanistan by Czechoslovak expeditions of 1965–1967. *Acta Sci. Nat. (Brno)* **4** (6): 1–56.

—— 1971. Systematic review and distinguishing characters of the bats (Chiroptera) hitherto recorded in Afghanistan. *Zool. Listy* **20**: 97–110.

Gambaryan, P. P. & Papanyan, S. B. 1964. [The systematics of *Meriones meridianus dahli*.] *Izv. Akad. Nauk. armyan. SSR biol.* **17** (7): 91–96.

Garzon, J. H. 1973. Primera captura de un ejemplar de *Micromys minutus* en España. *Bol. R. Soc. Espanola Hist. Natur. (Biol.)* **71**: 307.

Gauckler, A. & Kraus, M. 1970. Kennzeichnen und Verbreitung von *Myotis brandti*. *Z. Säugetierk.* **35**: 113–124.

Gautier, A. 1966. *Camelus thomasi* from the northern Sudan and its bearing on the relationship *C. thomasi – C. bactrianus*. *J. Palaeont.* **40**: 1368–1372.

Gentry, A. W. 1971. Genus *Gazella*. In Meester & Setzer, 1971, part 15. 1.

Geoffroy, I. 1826. In Bory de Saint-Vincent, *Dictionnaire classique d'histoire naturelle* **9.** Paris.

Gervais, F. L. P. 1848–52. *Zoologie et paléontologie Françaises*.

Gileva, E. A. 1972. Chromosome polymorphism in two akin forms of subarctic Microtinae (*Microtus hyperboreus* and *Microtus middendorffi*. *Dokl. (Proc.) Acad. Sci. USSR (Biol.)* **203**: 198–201.

Glagić, M. O. 1959. Contribution à la connaissance de *Dolomys* des montagnes centrales de Bosnie et Herzegovine. *Bull. Mus. Hist. nat. Belgr.* **14B**: 265–297.

Glover, R. 1956. Notes on the sika deer. *J. Mammal.* **37**: 99–105.

Gmelin, J. F. 1788. *Systema naturae* [of Linnaeus] 13th ed., vol. 1. Lipsiae.

—— 1793. *Systema de la naturae* (of Linnaeus: 13th edition, augmented and corrected).

Gohar, H. A. F. 1957. The Red Sea dugong. *Publs mar. biol. Stn. Ghardaga* **9**: 3–49.

Goldfuss, G. A. 1820. *Handbuch der Zoologie* vol. 2.

Gorbunov, S. & Kulik, I. 1974. Cadastre-information map of the *Myopus schisticolor* range. *Zool. Zh.* **53**: 144–146.

Gottlieb, G. O. 1951. Zur Kenntnis der Birkenmaus (*Sicista betulina*). *Zool. Jahrb. Jena* (Syst.) **79**: 93–113.

Grassé, P. 1955. *Traité de Zoologie: anatomie, systématique, biologie*. Tome XVII: *mammifères*. Paris.

Gray, A. P. 1954. *Mammalian hybrids: a check-list with bibliography*. Farnham Royal (Commonwealth Agricultural Bureau).

—— 1966. *Mammalian hybrids: supplementary bibliography*. Edinburgh (Commonwealth Bureau of Animal Breeding and Genetics).

Gray, J. E. 1831. *The zoological miscellany* pt. 1. London.

—— 1838. A revision of the genera of bats (Vespertilionidae). *Mag. Zool. Bot.* **2**:483–505.

—— 1863. *Catalogue of mammalia and birds of Nepal and Thibet.* 2nd ed. London.

—— 1873. On the skulls of Japanese seals, with the description of a new species, *Eumetopias elongatus. Proc. zool. Soc. Lond.*:776–779.

Gromov, I. M. 1972. [A review of the systematic categories in the vole subfamily (Microtinae) and their probable relationships.] *Sb. Trud. zool. Muz. MGU* **13**:8–33.

—— **Bibikov, D. I., Kalabukhov, N. I. & Meier, M. N.** 1965. [*Fauna of the USSR* vol. 3, pt 2: *ground squirrels (Marmotinae).*] Moscow.

—— **Gureev, A. A., Novikov, G. A., Sokolov, I. I., Strelkov, P. P. & Chaiskii, K. K.** 1963. [Mammals of the fauna of the USSR.] Moscow (2 vols).

—— **& Yanushchevich, A. I.** 1972 [*Mammals of Kirgizia*] Frunze.

Groves, C. P. 1967. On the gazelles of the genus *Procapra. Z. Säugetierk.* **32**:144–149.

—— 1969. On the smaller gazelles of the genus *Gazella. Z. Säugetierk.* **34**:38–60.

—— 1971. Request for a declaration modifying article 1 so as to exclude names proposed for domestic animals from zoological nomenclature. *Bull. zool. Nomencl.* **27**:269–272.

—— 1974. *Horses, asses and zebras in the wild.* Newton Abbot.

—— **& Mazak, V.** 1967. On some taxonomic problems of Asiatic wild asses; with the description of a new subspecies. *Z. Säugetierk.* **32**:321–355.

—— **Ziccardi, F. & Toschi, A.** 1966. Sull'Asino Selvatico Africano. *Ric. Zool. appl. Caccia Suppl.* **5**:1–30.

Gruber, U. F. 1969. Tiergeographische, ökologische und bionomische Untersuchungen an kleinen Säugetieren in Ost-Nepal. *Khumbu Himal* **3**:197–312.

Grulich, I. 1969. Kritische Populationsanalyse von *Talpa europaea* aus den West-Karpaten. *Acta Sc. nat. Brno* **3** (4):1–54.

—— 1971a. *Talpa caeca dobyi* subsp. nova in den Alpes maritimes, Frankreich. *Zool. Listy* **20**:111–129.

—— 1971b. Zur Variabilität von *Talpa caeca* aus Jugoslavien. *Prirodov. Pr. Cesk. Akad. Ved.* **5** (9):1–47.

Gubareff, N. 1941. On the fauna of the Chiroptera of the Karabakh region (Azerbaijan SSR). *Acta Mus. zool. Kiev* **1**:287–291.

Gulotta, E. F. 1971. *Meriones unguiculatus. Mammalian species* no. 3:1–5.

Gureev, A. A. 1963. *Insectivora. In* Gromov *et al.*, 1963.

—— 1964. [*Fauna of the USSR, mammals,* vol. 3, pt. 10, *Lagomorpha*] Moscow.

—— *1971.* [*Shrews (Soricidae): world fauna.*] Leningrad.

Hagen, B. 1958. Die Rötelmaus und die Gelbhalsmaus vom Monte Gargano, Apulien. *Z. Säugetierk.* **23**:50–65.

Hagmeier, E. M. 1961. Variation and relationships in North American marten. *Canad. Fld Nat.* **75**:122–138.

Hahn, H. 1934. Die Familie der Procaviidae. *Z. Säugetierk.* **9**:207–358.

Hahn, W. L. 1905. *Myotis lucifugus* in Kamchatka. *Proc. biol. Soc. Wash.* **18**:254.

Hall, E. R. 1944. Classification of the ermines of eastern Siberia. *Proc. Calif. Acad. Sci.* **23**:555–560.

—— **& Kelson, K. R.** 1959. *The mammals of North America.* New York.

Halternorth, T. 1953. *Die Wildkatzen der alten Welt: eine Übersicht über die Untergattung* Felis. Leipzig.

—— 1955. *In* Doderlein. *Bestimmruhgsbuch fur deutsche Land-und Susswassertiere.*

—— 1957. *Die Wildkatze.* Wittenberg Lutherstadt.

—— 1959. Beitrag zur Kenntnis des Mesopotamischen Damhirsches – *Cervus (Dama) mesopotamicus. Säugetierk. Mitt.* **7** Sonderheft:1–89.

—— 1963. Klassifikation der Säugetiere: Artiodactyla. *Handbuch der Zoologie (8)* **32**:1–167.

Hamar, M. 1963. A new subspecies of the field mouse (*Microtus arvalis heptneri* subsp. nova) of the Carpathian Mountains, Rumania. *Comun. Zool.* **2**:151–158.

—— **& Schutowa, M.** 1966. Neue Daten über die geographische Veränderlichkeit und die Entwicklung der Gattung *Mesocricetus. Z. Säugetierk.* **31**:237–251.

Hanak, V. 1965. Zur Systematik der Bartfledermaus *Myotis mystacinus* und über das Vorkommen von *Myotis ikonnikovi* in Europa. *Věst čsl. zeměd. Mus.* **29**:353–367.

—— 1966. Zur Systematik und Verbreitung der Gattung *Plecotus. Lynx* **6**:57–66.

—— 1967. Verzeichnis der Säugetiere der Tschecko-Slovakei. *Säugetierk. Mitt.* **15**:193–221.

—— 1969. Zur Kenntnis von *Rhinolophus bocharicus*. *Vest. Cs. spol. zool.* **33**:315–327.
—— 1970. Notes on the distribution and systematics of *Myotis mystacinus*. *Bijdr. Dierk.* **40**:40–44.
—— **& Gaisler, J.** 1969. Notes on the taxonomy and ecology of *Myotis longipes*. *Zool. Listy* **18**:195–206.
—— —— 1971. The status of *Eptesicus ognevi* and remarks on some other species of this genus. *Vest. csl. spol. zool.* **35**:11–24.
Harrison, D. L. 1948. A new central Mediterranean subspecies of field mouse (*Apodemus sylvaticus*). *Proc. zool. Soc. Lond.* **117**:650–652.
—— 1952. A new subspecies of the rabbit (*Oryctolagus cuniculus*) from Borkum Island in North-West Germany. *Ann. Mag. nat. Hist.* (12) **5**:676–678.
—— 1955. On a collection of mammals from Oman, Arabia, with the description of two new bats. *Ann. Mag. nat. Hist.* (12) **8**:897.
—— 1956a. Mammals from Kurdistan, Iraq, with description of a new bat. *J. Mammal.* **37**:257–263.
—— 1956b. Gerbils from Iraq, with description of a new gerbil. *J. Mammal.* **37**:417–422.
—— 1960. A new species of pipistrelle bat (*Pipistrellus*) from south Israel. *Durban Mus. Novit.* **5**:261–265.
—— 1962. A new subspecies of the noctule bat (*Nyctalus noctula*) from Lebanon. *Proc. zool. Soc. Lond.* **139**:337–339.
—— 1963a. Report on a collection of bats from N.W. Iran. *Z. Säugetierk.* **28**:301–308.
—— 1963b. Observations on the North African serotine bat, *Eptesicus serotinus isabellinus*. *Zool. Meded* **38**:207–212.
—— 1964a. *The mammals of Arabia*, vol. 1: *Insectivora, Chiroptera, Primates*. London.
—— 1964b. A new subspecies of Natterer's bat, *Myotis nattereri* from Israel. *Z. Säugetierk.* **29**:179–181.
—— 1967. Observations on some rodents from Tunisia, with the description of a new gerbil. *Mammalia* **31**:381–389.
—— 1968a. *The mammals of Arabia* vol. 2: *Carnivora, Artiodactyla, Hyracoidea*. London.
—— 1968b. On three mammals new to the fauna of Oman, Arabia, with the description of a new subspecies of bat. *Mammalia* **32**:317–325.
—— 1968c. A new race of Baluchistan gerbil, *Gerbillus nanus* from Oman. *Mammalia* **32**:60–63.
—— 1971. Observations on some notable Arabian mammals, with the description of a new gerbil (*Gerbillus*). *Mammalia* **35**:111–125
—— 1972. *The mammals of Arabia*, vol. 3: *Lagomorpha, Rodentia*. London.
—— 1976. Description of a new subspecies of Botta's serotine (*Eptesicus bottae*) from Oman. *Mammalia* **39**:415–418.
—— **& Lewis, R. E.** 1961. The large mouse-eared bats of the Middle East, with description of a new subspecies. *J. Mammal.* **43**:372–380.
—— **& Seton-Browne, C. J.** 1969. The influence of soil colour on subspeciation of mammals in eastern Arabia. *J. Linn. Soc.* (Zool.) **48**:467–470.
Harrison, J. L. 1948. *Chimarrogale hantu*, a new water shrew from the Malay Peninsula. *Ann. Mag. nat. Hist.* (13) **1**:282–290.
Hassinger, J. D. 1970. Shrews of the *Crocidura zarudnyi-pergrisea* group, with description of a new subspecies. *Fieldiana Zool.* **58** (2):5–8.
—— 1973. A survey of the mammals of Afghanistan. *Fieldiana Zool.* **60**:1–195.
Hausser, J., Graf, J.-D. & Meylan, A. 1975. Données nouvelles sur les *Sorex* d'Espagne et des Pyrénées. *Revue suisse Zool.* **82**:688–689.
—— **& Jammot, D.** 1974. Étude biometrique des machoires chez les *Sorex* du groupe *araneus* en Europe continentale. *Mammalia* **38**:324–343.
Hayman, R. W. 1957. A new race of the Indian smooth-coated otter from Iraq. *Ann. Mag. nat. Hist.* (12) **9**:710–712.
—— 1961. The red goral of the North-east Frontier Region. *Proc. zool. Soc. Lond.* **136**:317–324.
—— **& Hill, J. E.** 1971. Order Chiroptera, *In* Meester & Setzer, 1971: part 2.
Heaney, G. F. 1944. Occurrence of the lion in Persia. *J. Bombay nat. Hist. Soc.* **44**:467.
Heim de Balsac, H. 1940. Peuplement mammalien d'iles atlantiques françaises. *C.R. Acad. Sci. Paris* **211**:296–298.

—— 1951. Peuplement mammalien des iles atlantiques françaises: Ouessant. *C.R. Acad. Sci. Paris* **233**: 1678–1680.

—— 1968. Les Soricidae dans le milieu désertique Saharien. *Bonn. zool. Beitr.* **19**:181–188.

—— **& Barloy, J.** 1966. Revision des crocidures du groupe *flavescens-occidentalis-manni. Mammalia* **30**:601–633.

—— **& Beaufort, F.** 1966a. La crocidure de l'Ile de Sein. Sa position parmi les populations françaises de *Crocidura suaveolens. Mammalia* **30**:634–636.

—— —— 1966b. Formes geographiques de *Microtus agrestis* L. en France: Bretagne et Ile de Groix. *Mammalia* **30**:637–639.

—— & Mein, P. 1971. Les musaraignes momifiées des hypogées de Thebes. *Mammalia* **35**:220–240.

Heinrich, G. 1951. Die deutschen Waldmäuse. *Zool. Jb. (Syst.)* **80**:99–122.

—— 1952. *Apodemus flavicollis alpicola* n.n. *J. Mammal.* **33**:260.

Hemmer, H. 1972. *Uncia uncia. Mammalian Species* **20**:1–5.

—— 1974a. Studien zur Systematik und Biologie der Sandkatze *Felis margarita. Z. Köln. Zoo* **17**:11–20.

—— 1974b. *Felis margarita scheffeli*, eine neue Sandkatzen-Unterart aus der Nushki-Wuste, Pakistan. *Senckenberg. biol.* **55**:29–34.

—— **Grubb, P. & Groves, C. P.** 1976. Notes on the sand cat, *Felis margarita. Z. Säugetierk.* **41**:286–303.

Heptner, V. G. 1948a. On the nomenclature of the wood mice (*Apodemus 'flavicollis'-sylvaticus*). *Dokl. Akad. Nauk SSSR* **60**:177–178.

—— 1948b. [On the nomenclature of some mammals.] *Dokl. Akad. Nauk SSSR* **60**:709–712.

—— 1949. [New data on the distribution and geographical variation of the Mongolian gerbil.] *Byull. Mosk. Obshch. Ispyt. Prir. Otd. biol.* **54** (4):27–29.

—— 1952 [Systematics of the purcupine (*Hystrix* L.) inhabiting the USSR.] *Dokl. Akad. Nauk SSSR* **84**:1085–1088.

—— 1961. On geographical variability of pikas *(Ochotona)* inhabiting Turkmenia. *Zool. Zh.* **40**:621–622.

—— 1964. Über die morphologischen und geographischen Beziehungen zwischen *Mustela putorius* und *M. eversmanni. Z. Säugetierk.* **29**:321–330.

—— 1966. Uber die geographische Variabilität und die Nomenklatur der Iltisse. *Zool. Anz.* **176**:1–3.

—— 1969. On systematics and nomenclature of Palaearctic cats. *Zool. Zh.* **48**:1258–1260.

—— 1971. Systematic status of the Amur wildcat and some other East-Asian cats referred to *Felis bengalensis. Zool. Zh.* **50**:1720–1727.

—— **& Dolgov, V. A.** 1967. Systematic position of *Sorex mirabilis. Zool. Zh.* **46**: 1419–1421.

—— **& Ismagilov, M. I.** 1952. [A new form of suslik (*Spermophilopsis leptodactylus*).] *Dokl. Akad. Nauk SSSR* **84**:1255–1256.

—— **& Nasimovitsh, A. A.** 1967. *Der Elch* (Alces alces L). Wittenberg Lutherstadt.

—— **& Naumov, N. P.** (eds.) 1961. *Mammals of the Soviet Union*, vol. 1: *even-toed and odd-toed mammals.*] Moscow. German translation: Jena, 1966.

—— —— 1967. *Ibid* [vol. 2, part 1: *Sirenia and Carnivora*] Moscow. German translation: Jena, 1974.

—— —— 1972. *Ibid* [vol. 2, part 2: *carnivores (hyaenas and cats)*.] Moscow.

—— **& Shukurov, G. S.** 1950. [The systematics and distribution of *Microtus afghanus.*] *Dokl. Akad. Nauk SSSR* **74**:149–152.

Herold, W. & Niethammer, J. 1963. Zur systematischen Stellung des sudafrikanischen *Gerbillus paeba. Säugetierk. Mitt.* **11**:49–58.

Herre, W. 1955. *Das Ren als Haustier: eine zoologische monographie.* Leipzig.

Hershkovitz, P. 1966. Catalog of living whales. *Bull U.S. natn. Mus.* **246**:1–259.

Herter, K. 1972. Der Igel von Gran Canaria. *Zool. Beitr.* **18**:311–313.

Heuglin, M. T. von 1877. *Reise in Nordöst-Afrika* 2. Braunschweig.

Hill, J. E. 1962. Notes on some insectivores and bats from Upper Burma. *Proc. zool. Soc. Lond.* **139**:119–137.

—— 1963. A revision of the genus *Hipposideros. Bull. Br. Mus. nat. Hist.* (Zool.) **11**:1–129.

Hilzheimer, M. 1936. Uber drei neue Formen des Rentieres. *Z. Säugetierk.* **11**:154–158.

Hinton, M. A. C. 1926. *Monograph of the voles and lemmings.* London.

Hinze, G. 1950. *Der Biber*. Berlin.
Hochstrasser, G. 1969. Zur Frage der Hauskaninchen-Nomenclature. *Säugetierk. Mitt.* **17**:106–114.
Hoffmann, M. 1958. *Die Bisamratte*. Leipzig.
Hoffmeister, D. F. 1949. A new name for the meadow mouse *Microtus roberti occidentalis* Turov. *J. Wash. Acad. Sci.* **39**:205–206.
Hoogstraal, H., Wassif, K. & Kaiser, M. N. 1955. New mammal records from the Western Desert of Egypt. *Bull. zool. Soc. Egypt* **12**:7-12.
Hsu, T. C. & Benirschke, K. 1967–. *An atlas of mammalian chromosomes*. Berlin etc.
Husson, A. M. 1962. On *Blarina pyrrhonota* and *Echimys macrourus:* two mammals incorrectly assigned to the Suriname fauna. *Urtg. natuurw. StudKring Suriname* **28**:34–41.
—— **& Neurn, W. C. van.** 1959. Kleuruerscheidenheden van de mol, *Talpa europaea* L., in Nederland waargenomen. *Zool. Bijdr.* **4**:1–16.
Illiger, [J. C. W.] 1815. *Ueberblick der Saugethiere. Abh. preuss. Akad. Wiss.* **1804–1811**:39–159.
Imaizumi, Y. 1953. A new species of *Eptesicus* from Japan. *Bull. natn. Sci. Mus. Tokyo* **33**:91–95.
—— 1954a. Taxonomic studies on Japanese *Myotis. Bull. natn. Sci. Mus. Tokyo* **1**:40–58.
—— 1954b. New species and subspecies of *Sorex* from Japan. *Bull. natn. Sci. Mus. Tokyo* **1**:94–102.
—— 1955a. Taxonomic studies on the Japanese mountain mole (*Talpa mizura*). *Bull. natn. Sci. Mus. Tokyo* **2**:26–38.
—— 1955b. Systematic notes on the Korean and Japanese bats of *Pipistrellus savii* group. *Bull. natn. Sci. Mus. Tokyo* **2**:54–63.
—— 1956. A new species of *Myotis* from Japan. *Bull. natn. Sci. Mus. Tokyo* **3**:42–46.
—— 1957. Taxonomic studies on the red-backed vole from Japan. Part I. *Bull. natn. Sci. Mus. Tokyo* **3**:195–216.
—— 1959. A new bat of the '*Pipistrellus javanicus*' group from Japan. *Bull. natn. Sci. Mus. Tokyo* **4**:363–371.
—— 1960. *Coloured illustrations of the mammals of Japan*. Osaka.
—— 1961. Taxonomic status of *Crocidura dsinezumi orii. J. mammal. Soc. Japan* **2**:17–22.
—— 1967. A classification of commensal races of *Rattus rattus. J. mammal. Soc. Japan* **3**:145–151.
—— 1968. Taxonomic status of the Japanese lesser noctule, *Nyctalus noctula motoyoshii. J. mammal. Soc. Japan* **4**:35–39.
—— 1969. A new species of *Apodemus speciosus* group from Miyake Island, Japan. *Bull. natn. Sci. Mus. Tokyo* **12**:173–178.
—— 1970a. *The handbook of Japanese land mammals*, vol. 1. Tokyo.
—— 1970b. Description of a new species of *Cervus* from the Tsuchima Islands, Japan, with a revision of the subgenus *Sika. Bull. natn. Sci. Mus. Tokyo* **13**:185–194, pls 1–2.
—— 1970c. Systematic status of the extinct Japanese wolf, *Canis hodophilax. J. mammal. Soc. Japan* **5**:27–32; 62–66.
—— 1971. A new vole of the *Clethrionomys rufocanus* group from Rishiri Island, Japan. *J. mammal. Soc. Japan* **5**:99–103.
—— 1972. Land mammals of the Hideka Mountains, Hokkaido, Japan. *Mem. natn. Sci. Mus. Tokyo* **5**:131–149.
—— **& Yoshiyuki, M.** 1968. A new species of insectivorous bat of the genus *Nyctalus* from Japan. *Bull. natn. Sci. Mus. Tokio* **11**:127–134.
—— —— 1969. Results of the speleological survey in South Korea. *Bull. natn. Sci. Mus. Tokyo* **12**:255–272.
International Commission on Zoological Nomenclature. 1961. *International code of zoological nomenclature*. London.
Inukai, T. 1942. [Hair seals in Japanese waters.] *Botany Zool. Tokyo* **10**:927–932; 1025–1030.
Iredale, T. & Troughton, E. L. 1934. A checklist of the mammals recorded from Australia. *Mem. Aust. Mus.* **6**:i–xi, 1–122.
IUCN 1966. *Red data book*. Morges, Switzerland (International Union for the Conservation of Nature).
Jameson, E. W. 1961. Relationships of the red-backed voles of Japan. *Pacif. Sci.* **15**:594–604.
Jannett, F. J. & Jannett, J. Z. 1974. Drum-marking by *Arvicola richardsoni* and its taxonomic significance. *Am. Midl. Nat.* **92**:230–234.

Jenkins, P. D. 1976. Variation in Eurasian shrews of the genus *Crocidura*. *Bull. Br. Mus. nat. Hist.* (Zool.) **30**:271–309.
Jentink, F. A. 1879. On some hitherto undescribed species of *Mus* in the Leyden Museum. *Notes Leyden Mus.* **2**:13.
—— 1887. *Catalogue osteologique des mammifères*, tome IX. Leiden.
—— 1910. Description of a shrew from Surinam. *Notes Leyden Mus.* **32**:167–168.
Johnson, D. H. & Jones, J. K. 1955a. Three new rodents of the genera *Micromys* and *Apodemus* from Korea. *Proc. biol. Soc. Wash.* **68**:167–172.
—— —— 1955b. A new chipmunk from Korea. *Proc. biol. Soc. Wash.* **68**:175–176.
Jones, F. W. 1951 (ed.) A contribution to the history and anatomy of Pere David's Deer (*Elaphurus davidianus*). *Proc. zool. Soc. Lond.* **121**:319–370, pls 1–6.
Jones, G. S. 1971. Two bats new to Taiwan. *J. Mammal.* **52**:479.
Jones, J. K. 1956. Comments on the taxonomic status of *Apodemus peninsulae*. *Univ. Kans Publs Mus. nat. Hist.* **9**:337–346.
—— 1960. The least tube-nosed bat in Korea. *J. Mammal.* **41**:265.
—— **& Johnson, D. H.** 1955. A new reed vole, genus *Microtus*, from Central Korea. *Proc. biol. Soc. Wash.* **68**:193–195.
—— —— 1960. Review of the insectivores of Korea. *Univ. Kans Publs Mus. nat. Hist.* **9**:549–578.
—— —— 1965. Synopsis of the lagomorphs and rodents of Korea. *Univ. Kans Publs Mus. nat. Hist.* **16**:357–407.
Kahmann, H. 1958. Die Fledermaus *Rhinolophus mehelyi* als Glied der Säugetierfauna in Tunesien. *Zool. Anz.* **161**:227–237.
—— 1960. Der Gartenschläfer auf der Insel Lipari. *Zool. Anz.* **164**:172–185.
—— 1964. Contribution a l'etude des mammiferes du Pelopenese. *Mammalia* **28**:109–136.
—— **& Caglar, M.** 1960. Beitrage zur Säugetierkunde der Türkei. 1. *Istanb. Univ. Fen. Fak. Mecm.* B. **25** (1–2):1–21.
—— **& Niethammer, J.** 1971. Die Waldmaus (*Apodemus*) von der Insel Elba. *Senckenberg. biol.* **52**:381–392.
—— **& Tiefenbacher, L.** 1969. Der Gartenschläfer, *Eliomys quercinus* der Baleareninsel Menorca. *Säugetierk. Mitt.* **17**:242–247.
Kalabukhov, N. I. 1970. [Some peculiarities in adaptations in the steppe and yellow lemmings (*Lagurus lagurus* and *L. luteus*).] *Ekologiya* **1970** (1):69–76. English translation: *Ecology N Y* **1970**:52–57.
Kao, Y.-t. 1963. Taxonomic notes on the Chinese musk-deer. *Acta zool. sin.* **15**:479–488.
Kaplanov, L. G. & Raevskii, V. 1928. [Mammals of central industrial region.] *Trudy. gos. Mus. tsent. prom. Obl.* 21–24.
Kaznevsky, P. F. 1956. Siberian stag dispersal in the Southern Ural. *Zool. Zh.* **35**:1554–1564.
Kerr, R. 1792. *The animal kingdom.*
King, J. E. 1961. Notes on the pinnipeds from Japan described by Temminck in 1844. *Zool. Meded. Leiden* **37**:211–224.
—— 1964. *Seals of the world.* London.
Kishida, K. 1931. Notes on three species of Japanese bats. *Lansania* **3** (22):47–48.
—— 1936. [*The plants and animals of Nikko.*] ? Tokyo (NV).: 275, pl. 1.
Kittel, R. 1969. Bibliographie über den Goldhamster. *Z. Verzuchstierk.* **11**:1–115.
Klimkiewicz, M. K. 1970. The taxonomic status of the nominal species *Microtus pennsylvanicus* and *Microtus agrestis*. *Mammalia* **34**:640–665.
Knaus, W. & Schröder, W. 1975. *Das Gamswild.* Hamburg & Berlin.
Kobayashi, T. & Hayata, I. 1971. Revision of the Genus *Apodemus* in Hokkaido. *Annotnes zool. jap.* **44**:236–240.
Kock, D. 1970a. Die Verbreitungsgeschichte des Flusspferdes, *Hippopotamus amphibius* in unteren Nilgebiet. *Säugetierk. Mitt.* **18**:12–37.
—— 1970b. Zur Verbreitung der Mendesantilope, *Addax nasomaculatus* und der Spiessbocke, *Oryx gazella* im Nilgebiet. *Säugetierk. Mitt.* **10**:25–37.
Koffler, B. R. 1972. *Meriones crassus. Mammalian species* no. 9: 1–4.
König, C. 1962. Eine neue Wühlmaus aus der Umgebung von Garmisch-Partenkirchen (Oberbayern) *Pitymys bavaricus*. *Senckenberg. biol.* **43**:1–10.
Koopman, K. F. & Gudmundsson, F. 1966. Bats in Iceland. *Am. Mus. Novits* **2262**:1–6.

Kornejev, A. 1941. Zur Kenntnis der Teriofauna des Teberda-Gebeites. *Acta Mus. zool. Kiev.* **1** (1939): 169–189.

Kostron, K. 1948. The polecat of Eversmann, a new mammal from the plains of Czechoslovakia. *Acta Acad. Sci. nat. moravo-siles.* **20**: 1–96.

Kowalski, K. 1968. [New data on the distribution of mammals in the Mongolian People's Republic.] *Acta zool. cracov.* **13** (1): 1–11.

Kozlovsky, A. I. 1973. Somatic chromosomes in two species of shrews from Caucasus. *Zool. Zh.* **52**: 571–576.

—— **Orlov, V. N. & Papko, N. S.** 1972. Systematic status of caucasian (*Talpa caucasica*) and common (*T. europaea*) moles by karyological data. *Zool. Zh.* **51**: 312–315.

Kral, B. 1967. Karyological analysis of two European species of the genus *Erinaceus*. *Zool. Listy* **16**: 239–252.

—— 1975. A species of the genus *Microtus* new for the fauna of Bulgaria. *Zool. Listy* **24**: 353–360.

Kratochvil, J. 1952a. La nourriture et les races du *Putorius putorius*. *Sb. vys. Sk. zemed. les. Fak. Brno* **1**: 1–18.

—— 1952b. The voles of the genus *Pitymys* in CSR. *Acta Acad. Sci. nat. moravo-siles.* **24**: 155–194.

—— 1956. Tatra-Schneemaus *Microtus* (*Chionomys*) *nivalis mirhanreini*. *Pr. brn. Zakl. csl. Akad. Ved.* **28**: 1–39.

—— 1961. Svist horsey tatransky, nova subspecies. *Marmota marmota latirostris* ssp. nova. *Zool. Listy* **10**: 289–304.

—— 1962. [Supplementary note on the spread of *Apodemus agrarius* and *A. microps* in Czechoslovakia.] *Zool. Listy* **11**: 15–26.

—— (ed.) 1968. History of the distribution of the lynx in Europe. *Prirodov. Pr. Cesk. Akad. Ved.* **2** (4): 1–50.

—— 1970, *Pitymys*-Arten aus der Hohen Tatra. *Prirodov. Pr. Cesk. Akad. Ved.* **4** (12): 1–63.

—— **& Kral, B.** 1974. Karyotypes and relationships of Palaearctic '54-chromosome' *Pitymys* species. *Zool. Listy* **23**: 289–302.

—— **& Obrtal, R.** 1974. *Proceedings of the international symposium on species and zoogeography of European mammals.* Prague.

—— **& Rosicky, B.** 1952a. Eine neue Rasse von *Sorex alpinus* aus CSR. *Vest. csl. Spol. zool.* **16**: 51–65.

—— —— 1952b. Zur Bionomie und Taxonomie an der Tschechoslovakia lebenden *Apodemus* Arten. *Zool. Listy* **1**: 57–70.

—— **& Zejda, J.** 1962. Ergänzende Angaben zur Taxonomie von *Apodemus microps*. *In* Kratochvil, J. & Pelikan, J. (eds.) *Symposium theriologicum*. Prague: 188–194.

Kretzoi, M. 1941. Tigeriltis, Iltis und Nerz in ungarischen Pleistozän. *Földr. Közl.* **72**: 237–255.

—— 1955. *Dolomys* and *Ondatra*. *Act. geol. hung.* **3**: 347–353.

—— 1958. New names for Arvicolid homonyms. *Ann. hist.-nat. Mus. natn. hung.* **50**: 55–58.

—— 1959. New names for Soricid and Arvicolid homonyms. *Vertebr. hung.* **1**: 247–249.

—— 1964. Uber einige homonyme und synonyme Säugetiernamen. *Vertebr. hung.* **6**: 131–138.

—— 1969. Skizze einer Arvicoliden-Phylogenie-Stand 1969. *Vertebr. hung.* **11**: 155–193.

Krivosheev, V. G. & Rossolimo, O. L. 1966. Intraspecific variability and taxonomy of Siberian lemming (*Lemmus sibiricus*) in Palaearctic. *Byull. Mosk. Obshch. Ispyt. Prir. Otd. Biol.* **71** (1): 5–17.

Kryzhanovsky, V. I. & Shcherbak, N. N. (1969). On distribution and ecology of rare rodent *Cardiocranius paradoxus* of the USSR fauna. *Zbirn. Pratz zool. Mus.* **33**: 98–103.

Kullmann, E. 1965. Die Saeugetiere Afghanistans (Teil I): Carnivora, Artiodactyla, Primates. *Science* (Fac. Science, Kabul University) August 1965: 1–17.

Kumerloeve, H. 1969. Bemerkungen zum Gazellen-Vorkommen in südöstlichen Kleinasien. *Z. Säugetierk.* **34**: 113–120.

Kuroda, N. 1940. *A monograph of the Japanese mammals.* Tokyo & Osaka.

—— 1955. The present status of the introduced mammals in Japan. *J. mammal. Soc. Japan* **1**: 13–18.

—— 1957. A new name for the lesser Japanese mole. *J. mammal. Soc. Japan* **1**: 74.

—— 1967. On the four species of bats from Korea. *J. mammal. Soc. Japan* **3**: 163–166.

—— **& Okada, Y.** 1950. On two new races of *Cervus nippon* from the Southern Islands of Kyushu, Japan. *Annotnes zool. jap.* **24**: 59–64.
—— **& Uchida, T.** 1959. A new form of *Crocidura suaveolens* from Okinoshima. *Annotnes zool. jap.* **32**: 43–46.
Kurten, B. 1968. *Pleistocene mammals of Europe*. London.
—— **& Rausch, R.** 1959. Biometric comparisons between North American and European mammals. *Acta arct.* **11**: 1–44.
Kuznetsov, B. A. 1941. [The geographical variation of sables and martens in the fauna of the USSR.] *Trud. Moscow zootech. Inst.* **1**: 113–126.
—— 1944. [Order Rodentia.] *In* Bobrinskii, N. A. (ed.) [*Key to the mammals of the USSR.*] Moscow.
—— 1948. [*Animals of Kirghizia.*]
—— 1965. Order Rodentia, *In* Bobrinskii *et al.*, 1965.
Kuzyakin, A. P. 1950. [*Bats (Systematics, life history and utility to agriculture and forestry).*] Moscow.
—— 1965. Insectivora and Chiroptera, In Bobrinskii *et al.*, 1965.
Lahti, S. & Helminen, M. 1974. The beaver *Castor fiber* and *Castor canadensis* in Finland. *Acta theriol.* **19**: 177–189.
Landry, S. O. 1957. The interrelationships of the New and Old World Hystricomorph rodents. *Univ. Calif. Publ. zool.* **56**: 1–118.
Lange, J. 1972. Studien an Gazellenschädeln. Ein Beitrag zur Systematik der Kleineren Gazellen, *Gazella. Säugetierk. Mitt.* **20**: 193–249.
Laptev, I. P. 1929. [*Description of the mammals of Central Asia.*] vol. 1.
—— 1946. The Turkmenistan marten. *Izv. Akad. Nauk turkmen. SSR* **2**: 57.
Laurent, P. 1944. Observations biometriques sur le Minioptere de Schreibers. *Bull. Mus. Hist. nat. Paris* (2) **16**: 223–229.
Laurie, E. M. O. & Hill, J. E. 1954. *List of land mammals of New Guinea, Celebes and adjacent islands, 1758–1952.* London.
Lavocat, R. 1951. Le parallelisme chez les rongeurs et la classification des porcs-epics. *Mammalia* **15**: 132–138.
Lavrov, L. S. 1969. A new subspecies of the European beaver (*Castor fiber*) from the Enissei River upper flow. *Zool. Zh.* **48**: 456–457.
—— **& Orlov, V. N.** 1973. Karyotypes and taxonomy of modern beavers *(Castor). Zool. Zh.* **52**: 734–742.
Lay, D. M. 1965. A new species of mole (genus *Talpa*) from Kurdistan Province, western Iran. *Fieldiana Zool.* **44**: 227–230.
—— 1967. A study of the mammals of Iran. *Fieldiana Zool.* **54**: 1–282.
—— 1972. The anatomy, physiology, functional significance and evolution of specialized hearing organs of Gerbilline rodents. *J. Morph.* **138**: 41–120.
—— 1975. Notes on rodents of the genus *Gerbillus* from Morocco. *Fieldiana Zool.* **65**: 86–101.
—— **Anderson, J. A. W. & Hassinger, J. D.** 1970. New records of small mammals from West Pakistan and Iran. *Mammalia* **34**: 98–106.
—— **& Nadler, C. F.** 1975. A study of *Gerbillus* east of the Euphrates River. *Mammalia* **39**: 423–445.
Lazarev, A. & Paramonov, B. 1973. Finds of the collar lemming *(Dicrostonyx torquatus)* in Kamchatka. *Byull. Mosk. Obshch. Ispyt. Prir. Otd. Biol.* **78**: 142–143.
Leeuwen, L. van 1954. On the characters of *Sorex exiguus* as compared with those of *Sorex minutus. Proc. Acad. Sci. Amst.* **57c**: 332–338.
Lehmann, E. von 1955. Die Säugetiere aus Fukien (S.O.-China) im Museum A. Koenig, Bonn. *Bonn. zool. Beitr.* **6**: 147–170.
—— 1959. Eine Kleinsäugerausbeute aus Montenegro. *Bonn. zool. Beitr.* **10**: 1–20.
—— 1961. Über die Kleinsäuger der La Sila (Kalabrien). *Zool. Anz.* **167**: 214–229.
—— 1963. Die Säugetiere des Fürstentums Liechtenstein. *Jb. Hist. Ver. Fürstentum Liechtenstein* **62**: 159–362.
—— 1966a. Anpassung und 'Lokalkolorit' bei den Soriciden zweir linksrheinischer Moore. *Säugetierk. Mitt.* **14**: 127–133.
—— 1966b. Ein Hinweis auf das Vorkommen der Ostlichen Teichfledermaus *(Myotis dasycneme major)* in der Mandschurei. *Z. Säugetierk.* **31**: 329–330.

—— 1966c. Taxonomische Bemerkungen zur Säugerausbeute der Kumerloeveshen Orientreisen 1953–1965. *Zool. Beitr.* **12**:251–317.

—— 1968. Zum *Sorex arcticus* Problem in Westeuropa. *Säugetierk. Mitt.* **16**:259–261.

—— 1969a. *Vulpes vulpes toschii* ssp. nova, der Fuchs Mittel- und Suditaliens. *Z. Jagdwiss.* (Hamburg) **15**:28–31.

—— 1969b. Uber das Vorkommen des spanischen Rothirches, *Cervus elaphus bolivari*, in Nordmorokko. *Säugetierk. Mitt.* **17**:137–141.

—— 1970. Zur Taxonomie der west-europäischen Brandmaus, *Apodemus agrarius henrici* ssp. nov. *Säugetierk. Mitt.* **18**:154–155.

Lekagul, B. & McNeely, J. A. 1977. *Mammals of Thailand.* Bangkok.

Leroy, P. 1948. Sur la presence d'une Civette (*Paguma larvata*) dans les collines de Pekin. *Int. Congr. Zool.* **1948**:404–406.

Leyhausen, P. 1973. *Verhaltensstudien an Katzen* (3rd ed.) Berlin & Hamburg (Suppl. 2 to *Z. Tierpsych.*)

Liapounova, E. A. 1969. The description of the chromosome complement and evidence of the specifical independence of the *Citellus parryi. In* Vorontsov, 1969:53–54.

—— **& Vorontsov, N. N.** 1969. New data on the chromosomes of Eurasian marmots (*Marmota*). *In* Vorontsov, 1969:36–40.

Litvinov, N. J. 1960. A new subspecies of the vole *Alticola argentatus* from the isle of Olkhon (Baikal). *Zool. Zh.* **39**:1888–1891.

Lobachev, V. S. 1971. A finding of *Salpingotus crassicauda. Zool. Zh.* **50**:305–306.

Lønø, O. 1960a. Transplantation of the musk-ox in Europe and North-America. *Medd. norsk. Polarinst.* **84**:3–25.

—— 1960b. Transplantation of hares to Svalbard. *Medd. norsk. Polarinst.* **84**:26–29.

Lowe, V. P. W. & Gardiner, A. S. 1975. Hybridization between red deer (*Cervus elephus*) and sika deer (*Cervus nippon*). *J. Zool. Lond.* **177**:553–566.

Lydekker, R. 1900. A new race of Musk-ox. *Nature, Lond.* **63**:157.

—— 1915. *Catalogue of the ungulate mammals in the British Museum* (*Natural History*), vol. iv, London.

Malec, F. & Storch, G. 1963. Kleinsäuger (Mammalia) aus Makedonien, Jugoslavien. *Senckenberg. biol* **44**:155–173.

—— —— 1972. Der Wanderigel, *Erinaceus algirus*, von Malta. *Säugetierk. Mitt.* **20**:146–151.

Malygin, V. M. & Orlov, V. N. 1974. Areas of 4 species of voles (superspecies *Microtus arvalis*) by karyological data. *Zool. Zh.* **53**:616–622.

Manning, T. H. 1971. *Geographical variation in the polar bear* Ursus maritimus. Ottawa (Canad. Wildlife Service Rep. Ser. 13).

Markov, G. 1951. Species of mammals new to the Bulgarian fauna. *Izv. zool. Sofiya* **1**:344–346.

—— 1957. Untersuchungen über die Systematik von *Citellus citellus* L. *Bull. Inst. zool. Acad. bulg. Sci.* **6**:453–490.

—— 1960. *Microtus guentheri strandzensis.* subsp. nov., eine neue Wühlmaus unterart im Strandza-Gebirge, Ostbulgarien. *C. r. Acad. bulg. Sci.* **13**:615–617.

Marshall, J. 1972. Keys to Eurasian species of the genus *Mus. Mammal. Chromosome Newsl.* **13**:124–130.

—— 1977. A synopsis of Asian species of *Mus. Bull. Am. Mus. nat. Hist.* **158** (3):173–220.

Martens, J. & Niethammer, J. 1972. Die Waldmäuse (*Apodemus*) Nepals. *Z. Säugetierk.* **37**:144–154.

Martino, V. 1939. [Material on the ecology and zoogeography of southern Serbia.] *Zap. russk. nauch. Inst. Belgr.* **14**:85–106.

—— 1945. *Glasn. zemalj. Mus. Bosni Hercag.* **1**:69.

—— **& Martino, E.** 1941. Materialen zur Ökologie und Systematik der Gattung der Schläfer, *Glis. Zap. russk. nauch. Inst. Belgr.* **17**:1–10.

—— —— 1948. Nouveau compagnol de la faune de la Bosnie et de la Hercegovine. *Godisnjak biol. Inst. Saraj.* **1**:87–88.

Matschie, P. 1907. *Wiss. Ergebn. Filchner Exped. nach China.* **10,** 1:164.

—— 1913. *Veröff Inst. Jagdkunde Neudamm.* **2**:156.

Matthey, R. 1960. Chromosomes, heterochromosomes et citologie comparée des Cricetinae Palearctiques. *Caryologia* **13**:199–223.

—— 1961. Cytologie comparée des Cricetinae palearctiques et americains. *Rev. suisse Zool.* **68**:41–61.

—— 1966. Cytogenetique et taxonomie des rats appartenant au sous-genre *Mastomys*. *Mammalia* **30**:105–119.

—— **& Baccar, H.** 1967. La formule chromosomique d'*Acomys seurati* et la cytogenetique des *Acomys* palearctiques. *Rev. suisse Zool.* **74**:546.

Mazak, V. 1967. Notes on Siberian long-haired tiger, *Panthera tigris altaica* (Temminck, 1844), with a remark on Temminck's mammal volume of the 'Fauna Japonica'. *Mammalia* **31**:537–573.

—— 1970. Comments on the problem of *Mustela minuta*. *Lynx* **11**:40–44.

McKenna, M. C. 1975. Towards a phylogenetic classification of the Mammalia. *In* Luckett, W. P. & Szalay, F. S. (eds) *Phylogeny of the primates*. New York.

McLaren, I. A. 1966. Taxonomy of harbor seals of the western North Pacific and evolution of certain other hair seals. *J. Mammal.* **47**:466–473.

Medway, Lord 1965. *Mammals of Borneo*. Singapore.

Meester, J. & Setzer, H. W. (eds) 1971–. *The mammals of Africa: an identification manual*. Washington.

Mehely, L. 1913. Species generis *Spalax*. *Math.-naturw. Ber. Ung.* **28**:1–390; **29**:pls 1–33.

Mein, P. & Freudenthal, M. 1971. Une nouvelle classification des Cricetidae du Tertiaire de l'Europe. *Scripta Geol.* **2**:1–37.

Meyer, M. N. 1976. Application of morphometric, karyologic and hybridologic techniques in the taxonomy of grey voles, genus *Microtus*. *Trans. 1st int. theriol. Cong.* **1**:400–401.

—— **Orlov, V. N. & Skholl, E. D.** 1972. On the nomenclature of 46- and 54-chromosome voles of the type *Microtus arvalis*. *Zool. Zh.* **51**:157–161.

Meyer-Öhme, D. 1965. Die Saeugetiere Afghanistans (Teil III): Chiroptera. *Science* (Fac. Science, Kabul University) August 1965:42–58.

Meylan, A. 1970. Caryotypes et distribution de quelques *Pitymys* europiens. *Rev. suisse Zool.* **77**:562–575.

—— 1972. Caryotypes de quelques hybrides interspecifiques de *Pitymys*. *Experientia* **28**:1507–1510.

Migulin, A. A. 1928. [A new subspecies of the steppe polecat (*Putorius eversmanni satunini*) from the Nogaiski steppes of the Dagestan Republic.] *Ukrainsk Mislivets ta Ribalka* **9**:30 (N.V.)

Miller, G. S. 1907. The families and genera of bats. *Bull. U.S. natn. Mus.* **57**:i–xvii, 1–181, pls i–xiv.

—— 1912. *Catalogue of the mammals of western Europe*. London.

Mirić, D. 1960. Verzeichnis von Säugetieren Jugoslawiens, die nicht in der 'Checklist of Palaearctic and Indian mammals' von Ellerman and Morrison-Scott (1951) enthalten sind. *Z. Säugetierk.* **25**:35–46.

—— 1966. Bemerkungen zur Validität der Zwergmaus-unterart *Micromys minutus mehelyi* Bolkey, 1925. *Z. Säugetierk.* **31**:61–65.

—— 1968. Eine neue *Apodemus*-Art von der Insel Krk, Jugoslawien. *Z. Säugetierk.* **33**:368–376.

—— **& Dulic, B.** 1962. Neues Verbreitungsareal der Gattung *Dolomys* in Jugoslawien. *Bull. scient. Cons. Acad. RSF Yougosl.* **7**:60.

Mirza, Z. B. 1965. Four new mammal records for West Pakistan. *Mammalia* **29**:205–210.

Misonne, X. 1957. Mammifères de la Turque sud-orientale et du nord de la Syrie. *Mammalia* **21**:53–68.

—— 1960. Analyse zoogeographique des mammifères de l'Iran. *Mem. Inst. R. Sci. nat. Belg.* 2, **59**:1–157.

—— 1969. African and Indo-Australian Muridae: evolutionary trends. *Ann. Mus. r. Af. cent. Sc. Zool.* **172**:1–219, pls 1–27.

—— 1971. Order Rodentia. *In* Meester & Setzer, 1971, part. 6.

Mitchell, E. 1968. The Mio-Pliocene pinniped *Imagotaria*. *J. Fish. Res. Bd Can.* **25**:1843.

Mitchell, R. M. 1975. A checklist of Nepalese mammals. *Säugetierk. Mitt.* **23**:152–157.

—— **& Punzo, F.** 1975. *Ochotona lama* sp. nov.: a new pika from the Tibetan highlands of Nepal. *Mammalia* **39**:419–422.

Mitina, I. P. 1959. Geographical variability of the hamster *Cricetulus eversmanni*. *Zool. Zh.* **38**:1869–1875.

Mivart, S. 1890. *A monograph of the Canidae*. London.

Mohr, E. 1957. *Sirenen oder Seekühe.* Wittenberg-Lutherstadt (Neue Brehm-Bücherei No. 197).
—— 1965. *Altweltliche Stachelschweine.* Wittenberg-Lutherstadt (Neue Brehm-Bucherei No. 350).
Moore, J. C. 1959. Relationships among living squirrels of the Sciurinae. *Bull. Amer. Mus. nat. Hist.* **118**:153–206.
—— **& Tate, G. H. H.** 1965. A study of the diurnal squirrels, Sciurinae, of the Indian and Indo-Chinese subregions. *Fieldiana zool.* **48**:1–351.
Morozova-Turova, L. S. 1953. New subspecies of weasel from Middle Asia. *Zool. Zh.* **32**:1267–1269.
Morrison-Scott, T. C. S. 1946. *Suncus etruscus* (Savi) in Africa. *Mammalia* **10**:145.
Mursaloğlu, B. 1964. Statistical significance of secondary sexual variations in *Citellus citellus*, and a new subspecies of *Citellus* from Turkey. *Communs Fac. Sci. Univ. Ankara* **C 9**:252–273.
—— 1973. New records for Turkish rodents. *Communs Fac. Sci. Univ. Ankara* **C 17**:213–219, map.
Nader, I. A. 1971. Noteworthy records of bats from Iraq. *Mammalia* **35**:644–647.
—— 1974. A new record of the bushy-tailed jird, *Sekeetamys calurus calurus* from Saudi Arabia. *Mammalia* **38**:347–348.
Nadler, C. F., Hoffman, R. S. & Lay, D. M. 1969. Chromosomes of the Asian Chipmunk *Eutamias sibiricus. Experientia* **25**:868–869.
—— **Korobitsina, K. V., Hoffmann, R. S. & Vorontsov, N. N.** 1973. Cytogenetic differentiation, geographic distribution and domestication in Palaearctic sheep (*Ovis*). *Z. Säugetierk.* **38**:109–125.
Namie, M. 1889. [Descriptions of bats from Japan (VI)] *Zool. Mag. Tokyo* **1**:256–257, pl. 23.
Napier, J. R. & Napier, P. H. 1967. *A handbook of living primates.* London.
Nasonov, N. V. 1923. [*Geographical distribution of wild sheep of the Old World.*] St Petersburg.
Neuhauser, H. N. 1970. First positive record of *Pipistrellus savii* from India. *J. Bombay nat. Hist. Soc.* **67**:319–320.
—— **& DeBlase, A.** 1971. The status of *Pipistrellus aladdin* from Central Asia. *Mammalia* **35**: 273–282.
—— —— 1974. Notes on bats new to the faunal lists of Afghanistan and Iran. *Fieldiana Zool.* **62** (5):85–96.
Niethammer, G. 1963. *Die Einbürgerung von Säugetieren und Vögeln in Europa.* Hamburg & Berlin.
—— 1970. Beobachtungen am Pyrenäen-Desman, *Galemys pyrenaicus. Bonn. zool. Beitr.* **21**:157–182.
—— **& Bohmann, L.** 1950. Bemerkungen zu einigen Säugetieren Bulgariens. *Zool. Anz.* **145** (suppl.):655–671.
—— **& Niethammer, J.** 1964. Der Zwergmaulwurf (*Talpa mizura*), ein neues Relikt aus Spanien. *Naturwissenschaften* **51**:148–149.
Niethammer, J. 1959. Die nordafrikanischen Unterarten des Gartenschläfers (*Eliomys quercinus*). *Z. Säugetierk.* **24**:35–46.
—— 1965. Die Saeugetiere Afghanistans (Teil II): Insectivora, Lagomorpha, Rodentia. *Science* (Fac. Science, Kabul University) August 1965: 18–42.
—— 1969a. Zur Taxonomie der Ohrenigel in Afghanistan (Gattung *Hemiechinus*). *Z. Säugetierk.* **34**:257–274.
—— 1969b. Zur Frage der Introgression bei den Waldmäusen *Apodemus sylvaticus* und *A. flavicollis. Z. zool. Syst. EvolForsch* **7**:77–127.
—— 1969c. Zur Taxonomie europäischer Zwergmaulwürfe (*Talpa* '*mizura*'). *Bonn. zool. Beitr.* **20**:360–372.
—— 1970. Die Wühlmäuse Afghanistans. *Bonn. zool. Beitr.* **21**:1–24.
—— 1972a. Der Igel von Teneriffa. *Zool. Beitr.* **18**:307–309.
—— 1972b. Zur Taxonomie und Biologie der Kurzohrmaus. *Bonn. zool. Beitr.* **23**:290–309.
—— 1973. Zur Kenntnis der Igel (Erinacidae) Afghanistans. *Z. Säugetierk.* **38**:271–276.
—— **& Krapp, F.** In press *Handbuch der Säugetiere Europas.* (5 volumes). Wiesbaden.
—— **& Martens, J.** 1975. Die Gattungen *Rattus* und *Maxomys* in Afghanistan und Nepal. *Z. Säugetierk.* **40**:325–355.
—— **Niethammer, G. & Abs, M.** 1964. Ein Beitrag zur Kenntnis der Cabreramaus (*Microtus cabrerae*). *Bonn. zool. Beitr.* **15**:127–148.

Nikolsky, A. A. 1969. The phonotypes of Palaearctic Marmotinae. *In* Vorontsov, 1969: 32–35.
Nilsson, A. 1971. *Sorex isodon*, a shrew species new to Scandinavia. *Fauna Flora Upps.* **66**: 253–258.
Novikov, G. A. 1956. [*Carnivorous mammals of the fauna of the USSR.*] Moscow. English translation: Jerusalem, 1962.
Nyholm, E. S. 1966. Observations on some birds and mammals of Spitzbergen. *Ann. zool. fenn.* 3: 173–175.
Obaba, I. 1967. *Mustela sibirica itatsi*, introduced to Iriomote-jima, the Ryukyu Islands. *J. mammal. Soc. Japan* **3**: 127–128.
Ogilby, W. 1836. *Proc. zool. Soc. Lond.*: 131–139.
—— 1838. *Ann. Mag. nat. Hist.* **1**: 219.
Ognev, S. I. 1928. *Mammals of Eastern Europe and Northern Asia*, vol. 1, *Insectivora and Chiroptera.* Moscow. English translation: Jerusalem, 1962.
—— 1931. *Ibid*, vol. 2, *Carnivora (Fissipedia).* Moscow. English translation: Jerusalem, 1962.
—— 1935. *Mammals of USSR and adjacent countries.* vol. 3, *Carnivora.* Moscow. English translation: Jerusalem, 1962.
—— 1940. *Ibid*, vol. 4, *Rodents.* Moscow. English translation: Jerusalem, 1960.
—— 1947. *Ibid*, vol. 5, *Rodents.* Moscow. English translation: Jerusalem, 1963.
—— 1948a. *Ibid*, vol. 6, *Rodents.* Moscow. English translation: Jerusalem, 1963.
—— 1948b. Les souris domestiques à Sarajevo. *Godisnjak biol. Inst. Sarej.* **1**: 85–86.
—— 1950. *Mammals of USSR and adjacent countries*, vol. 7, *Rodents.* Moscow. English translation: Jerusalem, 1964.
—— 1951. [Erinaceidae of the Far East.] *Byull. mosk. Obshch. Ispyt. Prir. Otd. Biol.* **56**: 8–14.
Olivier, G. A. 1804. *Voyage dans l'Empire Othoman, l'Egypte et la Perse*, tome 3, Atlas. Paris.
Ondrias, J. C. 1966. The taxonomy and geographical distribution of the rodents of Greece. *Säugetierk. Mitt.* **14** (Suppl.): 1–136.
—— 1969. Die Ussuri Gross-Spitzmaus, *Crocidura lasiura*, der Agäischen Insel Lesbos. *Z. Säugetierk.* **34**: 353–358.
—— 1970. Contribution to the knowledge of *Crocidura suaveolens* from Greece. *Z. Säugetierk.* **35**: 371–381.
Orlov, V. N. 1969. *Spalax nehringi* as a distinct species. *In* Vorontsov, 1969: 94–95.
—— **& Iskhakova, E.** 1975. Taxonomy of a superspecies of *Cricetulus barabensis. Zool. Zh.* **54**: 597–604.
—— **Schvetzov, J. G., Kovalskey, J. M., Kutascheva, T. S. & Stupina, A. G.** 1974. Identification and distribution of *Microtus maximoviczii* and *M. fortis* from Transbaicalia. *Zool. Zh.* **53**: 1391–1396.
Osborn, D. J. 1965. Hedgehogs and shrews of Turkey. *Proc. U.S. natn. Mus.* **117**: 553–566.
Østbye, E. *et al.* 1974. Least shrew, *Sorex minutissimus* in Trysil, Hadmark, South Norway, *Fauna, Oslo* **27**: 225–228.
Ott, J. 1968. Nachweis naturlichen reproduktiver Isolation zwischen *Sorex gemellus* sp. n. und *Sorex araneus* in der Schweiz. *Rev. suisse Zool.* **75**: 53–75.
Palisot de Beauvois, A. M. F. J. 1796. *Catalogue raisonne du museum de Mr C. W. Peale, Philadelphia.*
Pallas, P. S. 1766. *Miscellanea Zool.*: 7.
—— 1777. *Spicilegia Zoologica.* pt. 1: 12; pt. 12: 16.
—— 1778–1779. *Novae species quadrupedum e glirum ordine.* Erlangen.
—— 1811. *Zoographia Rosso-Asiatica* vol. 1. Petropoli.
Panouse, J. B. 1951. Les chauve-souris du Maroc. *Trav. Inst. scient. cherif.* **1**: 1–120.
—— 1956. Contribution a l'etude des chauves souris du Maroc: *Pipistrellus savii* et *Barbastella barbastellus. Bull. Soc. Sci. nat. Maroc.* **35**: 259–263.
—— 1959. Presence au Maroc de *Nycteris thebaica. Bull. Soc. Sci. nat. Maroc.* **38**: 91–98.
Pasa, A. 1953. Alcuni caratteri della mammalofauna pugliese. *Memorie Biogeogr. adriat.* **2**: 1–23.
Paspalev, G. V., Martino, K. V. & Pechev, T. S. 1952. Sur certain rongeur du Massif Vitocha. *God. Sof. Univ. Biol. Geol. Geogr. Fac.* **47**: 193–237.
Pavlinin, W. N. 1966. *Der Zobel* (Martes zibellina *L.*) Wittenberg.
Pedersen, A. 1958. *Der Moschusochs.* Stuttgart.

Peshev, Tz. 1955. Systematic and biological investigations on *C. citellus* in Bulgaria. *Bull. Inst. zool. Acad. bulg. Sci.* **4–5**:277–327.

—— 1964. Is *Sorex caecutiens* to be found in Bulgaria? *Acta theriol.* **9**:368–370.

—— **Anguelova, V. & Dinev, T.** 1964. Etudes sur la taxonomie du *Myomimus personatus* en Bulgarie. *Mammalia* **28**:419–428.

—— **Dinev, T. S. & Anguelova, V.** 1960. *Myomimus personatus* – a new species of rodent to the fauna of Europe. *Bull. Inst. zool. Acad. bulg. Sci.* **9**:305–313.

Peters, W. 1851. *Naturwissenschaftliche Reise nach Mossambique*, Pt. 1 *Säugethiere*. Berlin.

—— 1870. *J. Sci. Math. Phys. Nat. Lisboa* (1) **3**:125.

—— 1872. Mittheilung uber neue Flederthiere. *Mber. K. preuss. Akad. Wiss.* 256–262.

—— 1877. *Mber. K. preuss. Akad. Wiss.* 1876:912–914.

—— 1878. *Mber. K. preuss. Akad. Wiss*:194–209.

Petrov, B. M. 1939. New vole from South Serbia. *Prirodosl. Razpr. Izdaja Zaloga prirod. Sekc. Muz. Drust. Shov.* **3**:363–365.

—— 1941. [Bemerkungen über Systematik und Okologie der Säugetiere Süd-Serbians, d. h. Mazedoniens.] *Zap. russ. nauc. inst. Beograde* **16**:57–64.

—— 1949. Materials for the classification and geographical distribution of water voles *(Arvicola terrestris)* in Serbia. *Bull. Mus. Hist. nat. Pays Serbe Beograd 1949 (B)*:171–199.

—— 1963. *Marmota menzbieri* range and western distribution border of *Marmota caudata* in Northern Tien-Shan. *Zool. Zh.* **42**:743–751.

—— 1968. Korrekturen und Bemerkungen zu den Verbreitungskarten im Van der Brink'schen Buch 'Die Säugetiere Europas' fur das Territorium Jugoslawiens. *Säugetierk. Mitt.* **16**:39–52.

—— 1971a. Eine neue unterart der Blindmaus (*Spalax leucodon*) aus Jugoslawien. *Arh. biol. Nauka Beograd* **23**:13–14.

—— 1971b. Taxonomy and distribution of moles (genus *Talpa*) in Macedonia. *Acta Mus. maced. Sci. nat.* **12**:117–136.

—— 1974. Einige Fragen der Taxonomie und die Verbreitung der Vertreten der Gattung *Talpa* in Jugoslawien. *In* Kratochvil & Obrtal, 1974:117–24.

—— **& Zivkovic, S.** 1971. Zur Kenntnis der *Pitymys liechtensteini* in Jugoslawien. *Arh. biol. Nauka. Beograd* **23**:31–32.

—— —— 1972. Zur Kenntnis der Thomas-Kleinwühlmaus, *Pitymys thomasi. Säugetierk. Mitt.* **20**:249–258.

—— —— 1974. Der taxonomische Status einiger Vertreter der untergattung *Pitymys* in Jugoslawien. *In* Kratochvil & Obrtal, 1974: 283–290.

—— —— **& Rimsa, D.** 1976. Uber die Arteigenstandigkeit der Kleinwühlmaus, *Pitymys felteni. Senckenberg. biol.* **57**:1–10.

Petter, F. 1954. Remarques biologiques sur des rats épineaux du genre *Acomys. Mammalia* **18**:389–396.

—— 1956. Evolution du dessin de la surface d'usure des molaires de *Gerbillus, Meriones, Pachyuromys* et *Sekeetamys. Mammalia* **20**:419–426.

—— 1959. Evolution du dessin de la surface d'usure des molaires des gerbillidés. *Mammalia* **23**:304–315.

—— 1961a. Eléments d'une révision des lièvres européens et asiatiques du sous-genre *Lepus. Z. Säugetierk.* **26**:1–11.

—— 1961b. Repartition geographique et écologie des rongeurs desertiques (du Sahara occidental à l'Iran oriental). *Mammalia* **25,** supplt:1–222.

—— 1961c. Les lérots des Iles Baléares et de l'ouest de la région Mediterranéenne. *Colloques int. Cent. natn. Res. scient.* **94**:97–102.

—— 1961d. Affinités des genres *Spalax* et *Brachyuromys. Mammalia* **25**:485–498.

—— 1963. Nouveaux élements d'une révision des lièvres africains. *Mammalia* **27**:238–255.

—— 1971a. Order Lagomorpha. *In* Meester & Setzer, 1971, part. 5.

—— 1971b. Subfamily Gerbillinae. *In* Meester & Setzer, 1971, part 6·3.

—— **& Chippaux, A.** 1962. Description d'une musaraigne pygmée d'Afrique equatoriale, *Suncus infinitesimus ubanguiensis* subsp. nov. *Mammalia* **26**:512–516.

—— **& Saint Girons, M.** 1965. Les rongeurs du Maroc. *Trav. Inst. scient. cherif. Ser. zool* **31**:1–58, pls 1–6.

Petzsch, H. 1958 Reflexionen zur Phylogenie der Capridae. *Wiss. Z. Martin-Luther-Univ. Halle-Wittenb.* **6**:995–1019.

Pfeffer, P. 1967. Le mouflon de Corse (*Ovis ammon musimon*); position systematique, écologie et éthologie comparées. *Mammalia* **31**: 1–262.

Phillips, C. J. 1969. Review of central Asian voles of the genus *Hyperacrius*. *J. Mammal.* **50**: 457–474.

Piechocki, R. 1958. Die Zwergmaus. *Micromys minutus* Pallas. Wittenberg Lutherstadt.

Pocock, R. I. 1917. The classification of existing Felidae. *Ann. Mag. nat. Hist.* (8) **20**: 329–350.

—— 1921. On the external characters and classification of the Mustelidae. *Proc. zool. Soc. Lond.*: 803–837.

—— 1936. The polecats of the genera *Putorius* and *Vormela* in the British Museum. *Proc. zool. Soc. Lond.*: 691–723.

—— 1939. *The fauna of British India: Mammalia* vol. I, *Primates and Carnivora* (*in part*). London.

—— 1941. *Ibid*, vol. II, *Carnivora* (*continued*). London.

—— 1947. Two new local races of the Asiatic wild ass. *J. Bombay nat. Hist. Soc.* **47**: 143–144.

—— 1951. *Catalogue of the genus* Felis. London.

Polushina, N. A. 1955. *Ecology, distribution and economic importance of the weasel family in the western regions of Ukraine.* Lvov (thesis).

Polyakova, R. S. 1962. The taxonomic position of common and Caucasian squirrels. *Zool. Zh.* **41**: 1247–1254.

Pomel, N. A. 1853. *Catalogue méthodique et déscriptif des Vertébrés fossiles.* Paris.

Popov, V. A. 1949. [Material on the ecology of the mink (*Mustela vison*) and its acclimatization in the Tatra ASSR.] *Trudy kazakhstan. Fil. Akad. Nauk SSSR Ser. biol.* **2**: 135.

Portenko, L. A. 1963. *In* Portenko, L. A., Kishchinskii, A. A. & Chernyauskii, F. B. 1963. [Mammals of the Koryatsk Mountains.] *Trudy Kamchatskoi kompleks. Eksped. Sib. Sect. AN SSSR*: 73–103.

Povolny, D. 1966. The discovery of the bear *Selenarctos thibetanus* in Afghanistan. *Zool. Listy* **15**: 305–316.

Pusanov, 1958. [*Problems of terrestrial zoogeography*] Lvov: 203–209.

Radbruch, A. 1973. Cytogenetische Analyse der Farbvererbung bei der Tabakmaus (*Mus poschiavinus*). *Z. Säugetierk.* **38**: 168–172.

Raicu, P. & Bratosin, S. 1968. Interspecific reciprocal hybrids between *Mesocricetus auratus* and *M. newtoni*. *Genet. Res. Camb.* **11**: 113–114.

Ranck, G. L. 1968. The rodents of Libya. *Bull. U.S. natn. Mus.* **275**: i–viii, 1–264, pls 1–8.

Rausch, R. L. & Rausch, V. R. 1975. Taxonomy and zoogeography of *Lemmus* spp. *Z. Säugetierk.* **40**: 8–34.

Razorenova, A. P. 1952. [Age changes in red voles (*Clethrionomys*).] *Byull. Mosk. Obshch. Ispyt. Prir. Otd. Biol.* (5): 23–28.

Reichstein, H. 1963. Beitrag zur systematischen gliederung des genus *Arvicola*. *Z. zool. Syst. EvolForsch.* **1**: 155–204.

Reig, O. A. 1972. *The evolutionary history of the South American cricetid rodents.* Ph.D. thesis, Univ. London.

[Rennie, J.] 1838. *The menageries. The natural history of monkeys, opposums and lemurs*, 1. London.

Retzius, A. 1744. *K. svenska Vetensk Akad. Handl.* **15**: 292.

Rey, J. M. 1971. Contribution to the knowledge of the pigmy-shrew *Sorex minutus*, in the Iberian Peninsula. *Boln. R. Soc. esp. Hist. nat. Biol.* **69**: 153–160.

—— 1972. Sistematica y distribucion del topillo rojo *Clethrionomys glareolus* en la peninsula iberica. *Bol. Est. cent. Ecol.* **1**: 45–56.

Richardson, J. 1825. Account of the quadrupeds and birds. *In* Parry, W. E. *Journal of a second voyage for the discovery of a North-west Passage. Zoological Appendix.* London.

Ride, W. D. L. 1970. *A guide to the native mammals of Australia.* Melbourne.

Röben, P. 1975. Zur Ausbreitung des Waschbären, *Procyon lotor* und des Marderhundes, *Nyctereutes procyonoides* in der Bundesrepublik Deutschland. *Saugetierk. Mitt.* **23**: 93–101.

Roberts, T. J. 1977. *The mammals of Pakistan.* London.

Roche, J. 1972. Systematique du genre *Procavia* et des damans en general. *Mammalia* **36**: 22–49.

Rochebrune, A. T. 1883. Faune de la Senegambie. *Acta Soc. linn. Bordeaux* (4) **7**: 49–204.

Roesler, U. & Witte, G. R. 1969. Chorologische Betrachtungen zur Subspeziesbildung einiger Vertebraten im italienischen und balkanischen Raum. *Zool. Anz.* **182**: 27–51.

Rossolimo, O. L. 1971a. Variability and taxonomy of *Dryomys nitedula. Zool. Zh.* **50**:247–258.
—— 1971b. On the taxonomic position of *Clethrionomys rjabovi. Byull. Mosk. Obshch. Ispyt. Prir. Otd. Biol.* **76** (1):63–68.
—— 1976a. Taxonomic status of the mouse-like dormouse *Myomimus* from Bulgaria. *Zool. Zh.* **55** (10):1515–1525.
—— 1976b. *Myomimus setzeri*, a new species of mouse-like dormouse from Iran. *Vest. Zool.* **4**:51–53.
Rothschild, W. 1913. On *Ovis lervia* and its subspecies. *Novit. zool.* **20**:459–460.
—— 1921. Captain Angus Buchanan's Air Expedition. III Ungulate mammals. *Novit. zool.* **28**:75–77.
Rüppell, W. P. E. S. 1835. *Neue Wirbelthiere zu den Fauna von Abyssinien gehörig.* **1**:*Säugethiere.* Frankfurt.
Saint Girons, M.-C. 1969. Le Campagnol roussatre de l'Auvergne, *Clethrionomys glareolus cantueli* ssp. nov. *Mammalia* **33**:535–539.
—— 1973. *Les mammifères de France et du Benelux.* Paris.
—— **& Bree, P. J. H. van** 1962. Recherches sur la répartition et la systematique de *Apodemus sylvaticus* en Afrique du Nord. *Mammalia* **26**:478–488.
Sanborn, C. C. 1953. Remarks on a Japanese bat, *Vespertilio macrodactylus* Temminck. *Nat. Hist. Misc. Chicago Acad. Sci.* **118**:1–3.
—— **& Hoogstraal, H.** 1953. Some mammals of Yemen and their ectoparasites. *Fieldiana Zool.* **34**:229–252.
—— —— 1955. The identification of Egyptian bats. *J. Egypt. publ. Hlth Ass.* **30**:103–121.
Satunin, K. A. 1905. *Priroda i ochota* **5**:22.
Schaller, G. B. & Kahn, S. A. (1975). Distribution and status of Markhor (*Capra falconeri*). *Biol. Conserv.* **7**:185–198.
Schaub, S. 1958. Simplicidentata (= Rodentia). *In* Piveteau: *Traité de Palaeontologie* **6** (2):659–821. Paris.
Scheffer, V. B. 1956. Little-known reference to name of a harbor seal. *J. Wash. Acad. Sci.* **46**:352.
—— 1958. *Seals, sea lions and walruses: a review of the Pinnipedia.* Stanford.
—— **& Rice, D. W.** 1963. A list of the marine mammals of the world. *Spec. sci. Rep. U.S. Fish Wildlife Service* no. 431.
Schlitter, D. A. & DeBlase, A. F. 1974. Taxonomy and geographic distribution of *Rhinopoma microphyllum* in Iran. *Mammalia* **38**:657–665.
—— **& Robbins, L. W.** 1973. Presence of *Tadarida* in the central Sahara. *Mammalia* **37**:199.
—— **& Setzer, H. W.** 1972. A new species of short-tailed gerbil (*Dipodillus*) from Morocco. *Proc. biol. Soc. Wash.* **84**:385–392.
—— —— 1973. New rodents from Iran and Pakistan. *Proc. biol. Soc. Wash.* **86**:163–173.
—— **& Thonglongya, K.** 1971. *Rattus turkestanicus* (Satunin, 1903), the valid name for *Rattus rattoides* Hodgson, 1845. *Proc. biol. Soc. Wash.* **84**:171–174.
Schreber, J. C. D. von 1777. *Die Saugetiere.* Suppl. 2. Leipzig.
Schultz-Westrum, T. 1963. Die Wildziegen der ägäischen Inseln. *Säugetierk. Mitt.* **11**:145–182.
Schwarz, E. 1948. Revision of the old-world moles of the genus *Talpa. Proc. zool. Soc. Lond.* **118**:36–48.
—— **& Schwarz, H. K.** 1943. The wild and commensal stocks of the house mouse, *Mus musculus. J. Mammal.* **24**:59–72.
—— —— 1967. A monograph of the *Rattus rattus* group. *An. Esc. nac. Cienc. biol. Mex.* **14**:79–178.
Sclater, P. L. 1858. On the general geographical distribution of the members of the class Aves. *J. Proc. Linn. Soc.* **2**:130–145.
Seba, A. 1734. *Locupletissimi rerum naturalium thesauri.* Amstelaedami.
Selander, R. K., Hunt, W. G. & Yang, S. Y. 1969. Protein polymorphism and generic heterozygosity in two European subspecies of the House mouse. *Evolution, Lancaster, Pa* **23**:379–390.
Setzer, H. W. 1955. Two new jerboas from Egypt. *Proc. biol. Soc. Wash* **68**:183–184.
—— 1956a. Two new gerbils from Libya. *Proc. biol. Soc. Wash* **69**:179–182.
—— 1956b. A new jird from Libya. *Proc. biol. Soc. Wash* **69**:205–206.
—— 1956c. Mammals of the Anglo-Egyptian Sudan. *Proc. U.S. natn. Mus.* **106**:447–587.

—— 1957a. The hedgehogs and shrews of Egypt. *J. Egypt. publ. Hlth Ass.* **32**:1–17.
—— 1957b. A review of Libyan mammals. *J. Egypt. publ. Hlth Ass.* **32**:41–82.
—— 1958a. The mustelids of Egypt. *J. Egypt. publ. Hlth Ass.* **33**:199–204.
—— 1958b. The jerboas of Egypt. *J. Egypt. publ. Hlth Ass.* **33**:87–94.
—— 1958c. The gerbils of Egypt. *J. Egypt. publ. Hlth Ass.* **33**:205–227.
—— 1959. The spiny mice (*Acomys*) of Egypt. *J. Egypt. publ. Hlth Ass.* **34**:93–101.
—— 1960. Two new mammals from Egypt. *J. Egypt. publ. Hlth Ass.* **35**:1–5.
—— 1961. The jirds of Egypt. *J. Egypt. publ. Hlth Ass.* **36**:81–90.
—— 1971. Genus *Acomys*. *In* Meester & Setzer, 1971, part 6·5.
Severtzov, N. A. 1873. [*Vertical and horizontal distribution of Turkestan animals.*]
Shen, S. 1963. [Faunal characteristics of Tibetan mammals and the history of their organization.] *Acta zool. sinica* **15**:139–150.
Shenbrot, G. 1974. Systematic status of *Allactodipus bobrinskii*. *Zool. Zh.* **53**:1697–1702.
Shidlovskyi, M. V. 1962. [*Key to the rodents of Zacaucasia*] Tbilisi.
Shortridge, G. C. 1942. Field notes on the Cape Museum's mammal survey of the Cape Province. *Ann. S. Afr. Mus.* **36**:27–100.
Sidorowicz, J. 1964. Comparison of the morphology of representatives of the genus *Lemmus* from Alaska and Palaearctic. *Acta theriol.* **8**:217–226.
—— 1971. Subspecific taxonomy of the squirrel (*Sciurus vulgaris* L.) in Palaearctic. *Zool. Anz.* **187**:123–142.
Siivonen, L. 1965. *Sorex isodon* and *S. unguiculatus* as independent shrew species. *Aquilo Ser. Zool.* **4**:1–34.
—— 1968. *Nordeuropas däggdjur*. Stockholm. (Finnish edition: *Pohjolan nisäkkäät. Helsinki*, 1967.)
—— 1969. *Sorex isodon* Turov is not synonymous with *S. centralis* Thomas. *Aquilo Ser. Zool.* **7**:42–49.
—— 1972. [*Finnish mammals.*] 2 vols. Helsinki (in Finnish).
Silverton, E. 1954. A survey of the eared seals (family Otariidae). *Scient. Results Norw. Antarct. Exped. 1927–1928* **36**:1–76.
Simonescu, V. 1971. Étude concernant le sistematique et la variabilité du genre *Micromys*. *Studii Comun. Sect. Stiint. nat. Muz. Jedetean Bacau:* 365–392.
Simpson, G. G. 1945. The principles of classification and a classification of mammals. *Bull. Amer. Mus. nat. Hist.* **85**:v–xvi, 1–350.
Sinha, Y. P. 1970. Taxonomic notes on some Indian bats. *Mammalia* **34**:81–92.
Skalon, V. N. 1936. *Izv. Gos. protivochumn. Inst. Siberia DVK* **4**:193.
—— **& Rajevsky, V. V.** 1940. [New forms of mammals from the Kondo Sosvinski Reserve.] *Nauch.-metodich zap. Gl. upr. zapoved.* **7**:193–200.
Sludskii, A. A. 1969. *Mammals of Kazakhstan*, vol. 1, pt. 1: *rodents (marmots and ground squirrels)*. Alma Ata.
Smirnov, V. S. 1960. New squirrel subspecies from the forest steppe of the Transural Territory. *Zool. Zh.* **39**:309–310.
—— 1971. *Cardiocranius paradoxus* in Kazakhstan. *Zool. Zh.* **50**:1266–1268.
Smith, C. H. 1827. *In* Cuvier: *The animal kingdom*, vol. 4. London.
Sokolov, I. I. (ed.) [*Mammals of the fauna of the USSR.*] Moscow.
Sokolov, V. E. 1954. [New species of field vole in the USSR.] *Zool. Zh.* **33**:947–950.
Soldatović, B., Zivković, S., Savic, I. & Milošević, M. 1967. Vergleichene Analyse der Morphologie und der Anzahl der Chromosomen zwischen verschiedeners Populationen von *Spalax leucodon*. *Z. Säugetierk.* **32**:238–245.
Sorokin, M. G. 1958. [On the systematic position of the racoon-dogs acclimatized in the Kalinin District.] *Byull. mosk. Obshch. Ispyt. Prir. Kalinskoe Otdelenie* **1** (NV).
Sowerby, A. de C. 1943. Mammals recorded from or known to occur in the Shangai area. *Notes Mammal. Mus. Heude* **2**:1–15.
Spitzenberger, F. 1968. Zur Verbreitung und Systematik türkischen Soricinae. *Annln naturh. Mus. Wien* **12**:273–289.
—— 1970. Erstnachweise der Wimperspitzmaus (*Suncus etruscus*) für Kreta und Kleinasian. *Z. Säugetierk.* **35**:107–113.
—— 1971a. Eine neue, tiergeographisch bemerkenswerte *Crocidura* aus der Türkie. *Annln naturh. Mus. Wien* **75**:539–562.

—— 1971b. Zur Systematik und Tiergeographie von *Microtus nivalis* und *M. gud. Z. Säugetierk.* **36**:370–380.

—— **& Steiner, H.** 1962. Uber Insektenfresser (Insectivora) und Wühlmäuse (Microtinae) der nordosttürkischen Feuchtwälder. *Bonn. zool. Beitr.* **4**:284–310.

—— —— 1964. *Prometheomys schaposchnikovi* in Nordost-Kleinasien. *Z. Säugetierk.* **29**:116–124.

Staffe, A. 1922. Uber den Schädel und das Haarkleid von *Sus leucomystax sibiricus. Arb. Lehrkanzel Tierzucht Hochschule Bodenkultur Wien* **1**:51–98.

Stalmakova, V. A. 1957. On the occurrence of the jerboa *Jaculus turcmenicus* in the Northern Kara-Kum. *Zool. Zh.* **36**:275–279.

Stebbings, R. E. 1970. A bat new to Britain, *Pipistrellus nathusii. J. Zool. Lond.* **161**:282–286, pls 1–2.

Stein, G. H. W. 1958. *Die Feldmaus.* Wittenberg Lutherstad.

—— 1960. Schädelallometrien und Systematik bei altweltlichen Maulwürfen (Talpinae). *Mitt. zool. Mus. Berlin* **36**:1–48.

Steiner, H. M. 1972. Systematik und Okologie von Wühlmäusen (Microtinae) der Vorderasiatischen Gebirge Ostpontus, Talysch und Elburz. *Sber. ost. Akad. Wiss.* **180**:99–193.

Stewart, D. R. M. 1963. The Arabian oryx (*Oryx leucoryx*). *E. Af. Wildl. J.* **1**:1–16.

—— 1964. The Arabian oryx (*Oryx leucoryx*). 2. *E. Af. Wildl. J.* **2**:168–169.

Stollmann, A. 1963. Beitrag zur Kenntnis der Luches. *Lynx lynx* in den tschechoslowakischen Karpaten. *Zool. Listy* **12**:301–316.

Strand, E. 1928. Miscellanea nomenclatoria zoologica et palaeontologica. I–II. *Arch. Naturgesch.* **92** 1926 A (8):30–75.

Strautman, E. I. 1949. *Vest. Akad. Nauk. kazakh. SSR* **5**:109–110.

Strelkov, P. P. 1972. *Myotis blythi:* distribution, geographical variability and differences from *Myotis myotis. Acta theriol.* **17**:355–380.

Strinati, P. & Aellen, V. 1958. Confirmation de la presence de *Rhinolophus mehelyi* dans le sud de la France. *Mammalia* **22**:527–536.

Stroganov, S. U. 1936. [The mammal fauna of the Valdai Hills.] *Zool. Zh.* **15**:132.

—— 1948. [New data on the systematics of the marbled polecat.] *Trav. zool. Inst. Acad. Sci. USSR* **7**, 3:129.

—— 1949a. Review of status and geographical distribution of the pygmy shrew (*Sorex tscherskii*), *Uchen. zap. tomsk. Gros. Univ.* **12**:187–188.

—— 1949b [*Key to the mammals of Karelia.*] Petrozavdsk (NV).

—— 1949c. [Systematics and distribution of some antelopes of Central Asia.] *Byull. Mosk. Obshch. Ispyt. Prir. Otd. Biol.* **54** (4):15–26.

—— 1952. [Systematics and distribution of two little known species of *Sorex* in Middle and Central Asia.] *Byull. Mosk. Obshch. Ispyt. Prir. Otd. Biol.* **57** (5):21–22.

—— 1956a. [Materials on the systematics of Siberian mammals.] *Trudy biol. nauchno-issled. Inst. perm. gos. Univ. 1 zool.*:3–10.

—— 1956b. [New species for the Siberian fauna.] *Trudy biol. nauchno-issled. Inst. perm. gos. Univ. 1 zool.*:11–14.

—— 1956c. *Trudy biol. nauchno-issled. Inst. perm. gos. Univ. 1 zool.*:17.

—— 1957. [*Animals of Siberia: insectivores.*] Moscow.

—— 1958. [Review of the subspecies of the steppe polecat (*Putorius eversmanni*) of Siberia.] *Izv. sib. Otdel. Akad. Nauk SSSR* **11**:149–155.

—— 1960. *Trudy biol. Inst. sib. Otd. Akad. Nauk SSSR* **6**.

—— 1962. [*Animals of Siberia: Carnivora.*] Moscow. English translation, Jerusalem, 1969.

—— **& Potapkina,** 1950. *Uchen. Zap. tomsk. gos. Univ.* **14**:101–139.

—— **& Turyeva, V. V.** 1948 [New subspecies of *Clethrionomys.*] *Byull. Mosk Obshch. Ispyt. Prir. Otd. Biol.* **53** (6):51–52.

Sundevall, C. J. 1842. Ofversigt af slägtet *Erinaceus. K. svenska Vetensk Akad. Handl.* **1841**:215–239.

—— 1843. Om Professor J. Hedenborgs insamlingar af Däggdjur i Nordöstra Africa och Arabien. *K. svenska Vetensk Akad. Handl.* **1842**:189–282, pls 2–4.

Suvorov, E. K. 1912. [*The Commander Islands and their fur trade.*] St Petersburg.

Swinhoe, R. 1865 (Letter to Dr J. E. Gray). *Proc. zool. Soc. Lond.*:1–2.

Szunyoghy, J. 1958. A preliminary report on the seasonal changes of the hair-color of the harvest mice and its taxonomical importance. *Annls hist.-nat. Mus. natn. hung.* **50**: S.N 9.
—— 1961. Kritische Bemerkungen zur Beschreibung des *Ovis musimon sinesella. Säugetierk. Mitt.* **9**: 6–8.
Tanaka, R. 1971. A research into variation in molar and external features among a population of the Smith's red-backed vole. *Jap. J. Zool.* **16**: 163–176.
Tatarinov, K. A. 1956. [*Mammals of the western region of the Ukraine.*] Kiev.
Tate, G. H. H. 1941. A review of the genus *Myotis* of Eurasia. *Bull. Am. Mus. nat. Hist.* **78**: 537–565.
—— 1942. Review of the vespertilionine bats. *Bull. Am. Mus. nat. Hist.* **80**: 221–297.
Tavrovskii, V. A. 1971. [*Mammals of Yakutia.*] Moscow.
Tchernyavsky, F. B. 1962. [A new form of snow sheep from the Koryaksky mountains.] *Dokl. Akad. Nauk SSSR* **145**: 1174–1176.
—— 1967. New data on the geographic variability of the Siberian lemming, *Lemmus sibiricus. Zool. Zh.* **46**: 1865–1867.
Ternovsky, D. V. 1958. *Biology and acclimatization of the American Mink* (Lutreola vison) *in the Altai.* Novosibirsk.
Thomas, O. 1905. On some new Japanese mammals. *Ann. Mag. nat. Hist.* (7) **15**: 487–495.
—— 1915. List of mammals collected on the Upper Congo. *Ann. Mag. nat. Hist.* **16**: 465–481.
Timofeev, V. V. & Nadeev, V. N. 1955. [*Sable.*] Moscow.
Timofeeva, A. A. 1962. [The finding of *Microtus fortis* in the Sakhalin Islands.] *Dokl. Irkutsk Protivoch. Inst.* **4**: 135–137.
Toktosunov, A. 1958. [*Rodents of Kirghizia.*] Frunze.
Topačevski, V. A. 1969. [Molerats (Spalacidae). *In Fauna SSSR, Mammals.*] **3** (3) (new ser. no. 99). Leningrad.
Topal, G. 1970. On the systematic status of *Pipistrellus annectens* and *Myotis primula. Annls hist.-nat. Mus. natn. hung.* **62**: 373–379.
Toschi, A. 1965. *Fauna d'Italia, Mammalia: Lagomorpha, Rodentia, Carnivora, Artiodactyla, Cetacea.* Bologna.
—— **& Lanza, B.** 1959. *Fauna d'Italia, Mammalia: generalita, Insectivora, Chiroptera.* Bologna.
Tour, G. D. de la 1975. Eine ökozoogeographische Revision der mittelasiatischen Halbesel, *Equus hemionus. Säugetierk. Mitt.* **23**: 108–111.
Trannier, M. 1974. Parenté des *Mastomys* du Maroc et du Sénégal. *Mammalia* **38**: 558–560.
Trouessart, E. L. 1898. *Catalogus Mammalium.* Berolini.
Trumler, E. 1959. Die Unterarten des Kiang, *Hemionus kiang. Säugetierk. Mitt.* **7**: 17–24.
—— 1961. Entwurf einer Systematik der rezenten Equiden und ihrer fossilen Verwandten. *Säugetierk. Mitt.* **9**: 109–125.
Tsuchiya, K. 1974. Cytological and biochemical studies of *Apodemus speciosus* group in Japan. *J. mammal. Soc. Japan* **6**: 67–87.
Turcek, F. J. 1949. *Muflon,* Ovis musimon *Schr., na Slovensku.* Bratislava (*fide* Pfeffer, 1967).
—— 1956. Uber den Mufflon, *Ovis musimon* in der Slowakie (C.S.R.). *Säugetierk. Mitt.* **4**: 167–171.
Turkin, V. N. & Satunin, K. A. 1904. [*Animals of Russia,* vol. 4.] Moscow.
Tzalkin, V. I. 1949. *Byull. Mosk. Obshch. Ispyt. Prir. Otd. Biol.* **4** (2): 20.
—— 1950. [On *Capra aegagrus turcmenicus* subsp. nov. in Turkmenia.] *Dokl. Acad. Nauk. SSSR.* **20**: 323–326.
Udagawa, T. 1954. The gem-faced civet occurring in the central part of Honshu. *Misc. Rep. Yamashina Inst. Orn. Zool.* **4**: 174–175.
Valverde, J. A. 1968. Nueva ardilla del S.E. Espanol. *Boln. R. Soc. esp. Hist. nat. (biol).* **65**: 225–248.
Van Peenen, P. F. D., Ryan, P. F. & Light, R. H. 1969. *Preliminary identification manual for mammals of South Vietnam.* Washington.
Vasilieva, M. V. 1964a. On taxonomic relationships of mountainous susliks (*Citellus*) of the Tien-Shan. *Zool. Zh.* **43**: 904–909.
—— 1964b. In [*First annual report of the conference of the soil-biology faculty, Moscow State Univ.*] Moscow: 125–127.
Vasiliu, G. D. & Decei, P. 1964. Uber den Luchs (*Lynx lynx*) der rumänischen Karpaten. *Säugetierk. Mitt.* **12**: 155–183.

Vasin, B. N. 1955. [New species of grey field vole from Sakhalin (*Microtus sachalinensis* sp.nov.)] *Zool. Zh.* **34**:427–431.

Verbeek, N. A. M. 1974. Two sightings of the Pine Marten (*Martes martes*) on Corsica. *Mammalia* **38**:751–752.

Vereshchagin, N. K. 1945. [New fauna findings in Talysh.] *Priroda, Mosk.* **6**:67–68.

—— 1955. Caucasian elk (*Alces alces caucasicus* subsp.nova) and data on elk history in Caucasus. *Zool. Zh.* **34**:460–463.

Vericad, J. R. & Balcells, R. E. 1965. Fauna mastozoologica de las Pituisas. *Boln. R. Soc. esp. Hist. nat. Secc. Biol.* **63**:233–264.

Vesmanis, I. 1975. Morphometrische Untersuchungen an algerischen Wimperspitzmäusen, 1: die *Crocidura russula*-Gruppe. *Senckenberg. Biol.* **56**:1–19.

Vetulani, T. 1927. Weitere Studien über den polnischen Konik (polnisches Landpferd). *Bull. int. Acad. pol. Sci. Lett.* **1927B**:835–949.

Vietinghoff-Riesch, A. F. von 1960. Der Siebenschläfer (*Glis glis*). *Monogr. Wildsäugetiere* **14**:1–196.

Vinogradov, B. S. 1958. On the structure of the external genitalia in white-toothed shrews (genus *Crocidura*). *Zool. Zh.* **37**:1236–1243.

—— **& Bondar, E. P.** 1949. A new species of Jerboa belonging to the genus *Jaculus*. *Dokl. Akad. Nauk SSSR* **65**:559–562.

—— **& Gromov, I. M.** 1952. [Rodents of the USSR.] *Opred. Faune SSSR* no. 48.

Volobuev, V. T., Graphodatsky, A. S. & Ternovsky, D. V. 1974. Comparative karyotype studies in European and American minks. *Mamm. Chrom. Newsl.* **15**:6.

—— **Ternovsky, D. & Graphodatsky, A.** 1974. Taxonomic status of the ferret (*Putorius putorius furo*) by karyological data. *Zool. Zh.* **53**:1738–1740.

Vorontsov, N. N. 1958. A new species of fat-tailed gerbil (*Pygerethmus vinogradovi*) from Zaissan's Valley. *Zool. Zh.* **37**:96–104.

——1966. New data on the biology and taxonomic position of *Prometheomys schaposchnikovi*. *Zool. Zh.* **45**:619–623.

—— 1969 (ed.). *The mammals: evolution, karyology, taxonomy, fauna.* Novosibirsk.

—— **& Ivanitskaya, E. Y.** 1973. Comparative karyology of North Palaearctic pikas (*Ochotona*). *Zool. Zh.* **53**:584–588.

—— **& Kriukova, E. P.** 1969. *Phodopus przhewalskii* species nova, a new species of desert hamsters from the Zaisson Basin. *In* Vorontsov, 1969:102–104.

—— **& Liapounova, E. A.** 1969. The chromosomes of Palaearctic ground squirrels. *In* Vorontsov, 1969:41–47.

—— —— 1970. Chromosome numbers and speciation in the Sciuridae. *Byull. Mosk. Obshch. Ispyt. Prir. Otd. Biol.* **75** (3):112–126.

—— —— **Zakarjan, G. G. & Ivanov, V. G.** 1969. The karyology and taxonomy of the genus *Ellobius*. *In* Vorontsov, 1969:127–129.

—— **Orlov, O. J. & Malygina, N. A.** 1969. Biology and taxonomy of fat-tailed jerboas (*Pygerethmus*) and comparative karyology of the genera *Pygerethmus* and *Alactagulus*. *In* Vorontsov, 1969:74–84.

—— **& Smirnov, V. M. 1969.** *Salpingotus heptneri* sp.nov. – a new species of three-toed dwarf jerboas from Kiril-Kums Desert. *In* Vorontsov, 1969:60–68.

Wagner, R. 1829. Ueber den Zahnbau der Gattung *Lagomys*. *Isis, Jena* **22**:1132–1141.

Wahrman, J., Gottein, R. & Nevo, E. 1969. Geographical variation of chromosome forms in *Spalax*. *In* Benirschke, K. (ed.), *Comparative mammalian cytogenetics*. New York.

Walker, E. P. 1975. *Mammals of the world.* Baltimore.

Wallin, L. 1963. Notes on *Vespertilio namiyei*. *Zool. Bidr. Upps.* **35**:397–416.

—— 1969. The Japanese bat fauna. *Zool. Bidr. Upps.* **37**:223–440.

Wang, S. 1959. Further report on the mammals of north-eastern China. *Acta zool. Sinica* **11**:344–352.

—— **& Cheng, Chang-Lin,** 1973. Notes on Chinese hamsters (Cricetinae). *Acta zool. Sinica* **19**:61–68.

Wassif, K. 1954. The bushy-tailed gerbil *Gerbillus calurus* of South Sinai. *J. Mammal.* **35**:243–248.

—— 1959. On a collection of mammals from the Egyptian oases of Bahariya and Farafra. *Ain Shams Sci. Bull.* **4**:137–147.

Waterhouse, G. R. 1838. *Proc. zool. Soc. Lond.* (1837): 103.
Weigel, I. 1969. Systematische Ubersicht über die Insectenfresser und Nager Nepals. *Khumbu Himal (Munich)* **3**: 149–196.
Wettstein, O. V. 1953. Die Insectivora von Kreta. *Z. Säugetierk.* **17**: 4–13.
—— 1954. Uber die Rötelmäuse Osterreichs. *Säugetierk. Mitt.* **2**: 118–124.
Witte, G. 1962. Zur systematik der Haselmaus (*Muscardinus avellanarius*). *Bonn. zool. Beitr.* **4**: 311–320.
Wolf, H. 1940. Zur Kenntnis der Säugetierfauna Bulgariens. *Mitt. naturw. Inst. Sofia* **13**: 153–168.
Wood, A. E. 1950. Porcupines, palaeogeography and parallelism. *Evolution, Lancaster Pa.* **4**: 87–98.
—— 1955. A revised classification of the rodents. *J. Mammal.* **36**: 165–187.
Wood, J. G. 1879. *Natural History* ?3rd ed. London.
Yom-Tov, Y. 1967. On the taxonomic status of the hares (genus *Lepus*) in Israel. *Mammalia* **31**: 246–259.
Yoshikura, M. 1956. Insectivores and bats of South Sakhalin. *Kumamoto J. Sci.* **2**: 259–280.
Yoskiyuki, M. 1970. A new species of insectivorous bat of the genus *Murina* from Japan. *Bull. natn. Sci. Mus. Tokyo* **13**: 195–198.
—— 1971. A new bat of the Leuconoe group in the genus *Myotis* from Honshu, Japan. *Bull. natn. Sci. Mus. Tokyo* **14**: 305–310.
Yudin, B. S. 1964. The geographical distribution and interspecific taxonomy of *Sorex minutissimus* in West Siberia. *Acta theriol.* **8**: 167–179.
—— 1967. [A species of shrew new to the Palaearctic from the Kurile Islands.] *Izv. Sibirsk. AN SSSR ser. biol. med. nauk.* **5**: 155–157.
—— 1971. [*Insectivorous mammals of Siberia (Key)*.] Novosibirsk.
—— 1972. Contribution to the taxonomy of the transarctic common shrew (*Sorex cinereus*) of the USSR fauna. *Theriology* (Novosibirsk) **1**: 45–50.
Yung, M. 1966. A new subspecies of the narrow-skulled vole from Inner Mongolia, China. *Acta zootax. sin.* **2**: 183–186.
Yurgenson, P. B. 1947. [Hybridization between sable and marten – a historical enquiry.] *Trud. Petchorsk-Vlilsk Gos. zapoved* **5**: 145.
Zahavi, A. & Wahrman, J. 1957. The cytotaxonomy, ecology and evolution of the gerbils and jirds of Israel. *Mammalia* **21**: 341–380.
Zeuner, F. E. 1963. *A history of domesticated animals.* London.
Zimmermann, E. A. W. von 1780. *Geographische Geschichte des Menschen, und der allgemein verbreiteten vierfussigen Thiere.* Leipzig.
Zimmermann, K. 1953a. Die Wildsäuger von Kreta. 4. Die Rodentia Kretas. *Z. Säugetierk.* **17**: 21–51.
—— 1953b. Die Hausmaus von Helgoland *Mus musculus helgolandicus* spec. nov. *Z. Säugetierk.* **17**: 163–166.
—— 1962. Die Untergattungen der Gattung *Apodemus. Bonn. zool. Beitr.* **13**: 198–208.
—— 1964. Zur Säugetier-Fauna Chinas. Ergebnisse der Chinesisch-Deutschen Sammelreise durch Nord- und Nordost-China 1956. No. 15. *Mitt. zool. Mus. Berl.* **40**: 87–140.
Zivkovic, S. & Petrov, B. 1974. Record of a vole of the *Microtus arvalis* group with 54 chromosomes in Yugoslavia. *Genetika* **6**: 283–288.
Zukowsky, L. 1959. Persische Panther. *Zool. Gart. Leipzig.* **24**: 329–344.
—— 1964. Weitere Mitteilungen über Persische Panther. *Zool. Gart. Leipzig* **28**: 151–182.

Index

All generic and subgeneric names are given, including junior synonyms. Species-group names are included only if (1) they are used as the accepted name of a species; or (2) they are additional to the *Checklist* (i.e. in the main those proposed subsequent to 1946). Scientific names that are used as the accepted names of genera and species are in bold. Only the principal reference is given.

abei, Myotis 48
aberrans, Meles 175
abnormis, Sorex 19
Abra 66
Abrana 66
Acanthion 159
Acanthomys 142
Achlis 203
Acinonyx 185
Acomys 142
Acosminthus 142
Addax 207
aegagrus, Capra 214
Aegoceros 213
Aegoryx 206
aegyptiaca, Tadarida 63
aegyptiacus, Rousettus 38
Aeretes 87
aeruginosus, Gerbillus 121
Aethechinus 14
aethiopicus, Paraechinus 16
afer, Triaenops 45
afghanus, Pitymys 108
afra, Coleura 40
africana, Loxodonta 191
africanus, Equus 195
agrarius, Apodemus 137
agrestis, Microtus 115
Agricola 110
airensis, Felis 182
Alactagulus 156
alaicus, Ellobius 117
albicauda, Ichneumia 179
albirostris, Cervus 200
Alcelaphus 207
Alcelaphus 202
Alces 202
alces, Alces 202
alexandra, Poecilictis 174
Alexandromys 110
algirus, Erinaceus 15
Allactaga 153
Allactodipus 153
Allocricetulus 90
Allolagus 70
Alobus 51
Alopedon 161
Alopex 162
alpicola, Apodemus 134
alpina, Ochotona 69
alpinus, Apodemus 134
alpinus, Cuon 165
alpinus, Sorex 23
Alsomys 132
altaica, Martes 173
altaica, Mustela 169
altaica, Mustela 170
altaica, Talpa 34
altaicus, Panthera 184
Alticola 103
alticola, Cricetulus 91
Ambliodon 178
Amblyotus 56
americanus, Alces 202
Ammon 217
ammon, Ovis 218
Ammotragus 213
amphibius, Hippopotamus 196
Amphisorex 17, 24
amplus, Meriones 127
amurensis, Lemmus 97
anatolicus, Eptesicus 57
andersoni, Clethrionomys 100
anderssoni, Vespertilio 58
angarensis, Martes 173
angdawai, Ochotona 68
angusi, Capra 217
Anisonyx 82
anomalus, Neomys 25
anomalus, Sciurus 78
Anotis 129
Anteliomys 100
Antilopinae 208
Anurocyon 164
Aoudad 217
aphanasievi, Marmota 81

Aphrontis 76
Apodemus 132
aquilus, Gerbillus 123
aralensis, Alactagulus 156
araneus, Sorex 20
araxenus, Myotis 49
Arbusticola 106
Arcticonus 165
arcticus, Sorex 22
Arctocephalus 186
Arctogale 168
Arctomys 80
Arctonyx 175
arctos, Ursus 166
arenaceous, Jaculus 152
Argali 217
argentea, Talpa 33
argenteus, Apodemus 136
argenteus, Glis 144
argyropuloi, Apodemus 134
argyropuloi, Spermophilus 84
ariel, Pipistrellus 54
Aries 213, 217
arispa, Crocidura 29
Aristippe 58
armenica, Crocidura 29
armoricanicus, Microtus 115
Artiodactyla 195
arvalis, Microtus 113
Arvicanthis 138
Arvicola 105
Aschizomys 100
Asellia 44
Asinohippus 194
Asinus 194
Asiocricetus 90
Asioscalops 32
Asioscaptor 32
Aspalax 129
asper, Sorex 21
aspromontis, Dryomys 146
Ass 195
asyutensis, Gerbillus 121
asyutensis, Meriones 128
Atalapha 60
atallahi, Lepus 71
Atelerix 14
Atlantoxerus 79
atrimaculata, Talpa 33
aurata, Murina 62
auratus, Meriones 127
auratus, Mesocricetus 92
aureus, Canis 162
aureus, Gerbillus 122
auritus, Plecotus 61
auropunctatus, Herpestes 178
austriacus, Plecotus 61
Austritragus 212
avellanarius, Muscardinus 144
averini, Martes 173
aviator, Nyctalus 56
Axis 199
axis, Cervus 199
azizi, Meriones 127

Baboon 65
bactrianus, Canis 161
Badgers 175
bahram, Equus 194
bailwardi, Calomyscus 89
balcanica, Crocidura 27
balcanicus, Spermophilus 83
balchanensis, Pitymys 108
balkarica, Mustela 168
barabensis, Cricetulus 91
barabensis, Sorex 19
barakshin, Alticola 104
barbarus, Lemniscomys 138
Barbastella 60
barbastellus, Barbastella 60
barbatus, Erignathus 190
barberi, Tamias 86
batinensis, Eptesicus 56
Bats 37
baussencis, Miniopterus 62
bavaricus, Pitymys 108
Bears 165
beaucournui, Talpa 34
Beavers 87
bechsteini, Myotis 49
becki, Sorex 19
bedfordi, Microtus 116
bedfordiae, Sorex 24
bengalensis, Felis 183
bennetti, Macropus 13
beringianus, Alopex 162
beringianus, Sorex 22
bernisi, Clethrionomys 99
betpakdalensis, Selevinia 147
betulina, Sicista 148
Bezoar 214
Bharal 217
bieti, Felis 181
Bifa 145
Bison 206
Bison 206
bison, Bison 206
blainei, Capra 217
blanci, Gerbillus 121
blanfordi, Equus 194
blanfordi, Jaculus 153
Blanfordimys 106
Blarinella 25
blasii, Rhinolophus 44
blythi, Myotis 50
Boar 196
bobak, Marmota 81
bobrinskii, Allactaga 155

bobrinskoi, Eptesicus 56
bocagei, Myotis 49
bodenheimeri, Pipistrellus 54
bogdanovi, Dinaromys 104
Borioikon 96
boristhenicus, Spermophilus 83
borkumensis, Oryctolagus 74
Bos 205
bosniensis, Clethrionomys 99
Bosovis 213
bottae, Eptesicus 57
Bovidae 204
Bovinae 205
Brachiones 128
brachycercus, Pitymys 109
Brachyotis 46
brachyurus, Lepus 73
brandti, Microtus 116
brandti, Myotis 48
brunnescens, Gerbillus 120
brunnescens, Spermophilus 84
bucharensis, Myotis 49
buchariensis, Sorex 23
bulgaricus, Myomimus 147
bullata, Allactaga 155
buselaphus, Alcelaphus 208

cabrerae, Microtus 114
caeca, Talpa 34
caecutienoides, Sorex 22
caecutiens, Sorex 20
caffer, Hipposideros 44
cahirinus, Acomys 142
californianus, Zalophus 186
caliginosus, Marmota 81
Callocephalus 188
Callorhinus 186
Callosciurus 78
Callotaria 186
Calogale 178
Calomyscus 89
calurus, Sekeetamys 124
Camelidae 197
Camels 197
Camelus 197
campestris, Gerbillus 120
Campicola 110
Campsiurus 167
camtschatica, Marmota 81
camtschatica, Sorex 22
cana, Vulpes 164
canadensis, Castor 88
canadensis, Cervus 201
canadensis, Ovis 218
Canidae 161
Canis 161
Cansumys 90
cantueli, Clethrionomys 99
capaccinii, Myotis 50
Capaccinius 46
Capella 212
capensis, Lepus 71
capensis, Mellivora 175
Capra 213
Caprea 203
Capreolus 203
capreolus, Capreolus 204
Capricornis 212
Capricornulus 212
caprimulga, Allactaga 155
Caprina 212
Caprinae 211
Caprios 32
Capromyidae 159
Caprovis 217
Caracal 183
Caracal 180
caracal, Felis 183
Cardiocraniinae 157
Cardiocranius 157
Caribou 203
carinthiacus, Microtus 115
Carnivora 160
carolinensis, Sciurus 78
carpatanus, Sorex 19
carpathicus, Cervus 201
Caryomys 100
caspica, Phoca 189
caspicus, Microtus 113
Castor 87
Castoridae 87
Cateorus 56
Catolynx 180
Cats 180
Catus 180
caucasica, Capra 216
caucasica, Capreolus 204
caucasica, Talpa 34
caucasicus, Alces 202
caucasicus, Bison 206
caucasicus, Sorex 21
caudata, Marmota 82
caudata, Sorex 19
caudatus, Meriones 127
cedrorum, Microtus 112
Cemas 212
centralis, Cervus 200
Centuriosus 196
Cercopithecidae 64
Cercopteropus 38
cernjavskii, Arvicola 105
Cervaria 180
Cervidae 198
Cervulus 199
Cervus 199
Chaerephon 63
Chamois 212
Charronia 172

Chaus 180
chaus, Felis 182
cheesmani, Gerbillus 123
Cheetah 185
chejuensis, Apodemus 137
Cheliones 125
Chimmarogale 31
chinensis, Eothenomys 101
Chionobates 70
Chionomys 110
Chipmunk 85
Chiroptera 37
Chiroscaptor 32
Chiru 211
Chital 199
chitralensis, Nesokia 143
chiumalaiensis, Cricetulus 91
Chodsigoa 24
chouei, Sorex 20
Chrysaeus 164
Chrysonycteris 44
Chrysopteron 46
cilanica, Ochotona 67
cinerea, Tadarida 63
cinereus, Lasiurus 60
cinereus, Sorex 22
Citellus 82
citellus, Spermophilus 83
clanceyi, Apodemus 134
clarkei, Microtus 115
Clethrionomys 97
clivosus, Rhinolophus 43
Cnephaeus 56
coeruleus, Dinaromys 105
Coleura 40
collaris, Arctonyx 176
collinsi, Jaculus 151
Colobotis 82
Colus 211
Comastes 46
Comopithecus 65
concolor, Sicista 149
Conothoa 66
coreensis, Pipistrellus 54
coriakorum, Spermophilus 85
cornutus, Rhinolophus 43
corsac, Vulpes 163
Corsira 17
Coypu 159
coypus, Myocastor 159
Craseomys 97
crassicauda, Salpingotus 158
crassus, Meriones 127
cremea, Talpa 33
cretensis, Felis 181
Cricetidae 88
Cricetinae 88
Cricetiscus 89
Cricetulus 90
Cricetus 89
cricetus, Cricetus 90
crispus, Capricornis 212
cristata, Cystophora 191
cristata, Hystrix 159
Crocidura 26
Crossogale 31
Crossopus 24
crymensis, Vulpes 163
Ctenodactylidae 159
Ctenodactylus 160
ctenodactylus, Paradipus 151
cufrana, Vulpes 164
cufrensis, Jaculus 151
Cuniculus 74, 96
cuniculus, Oryctolagus 74
Cuon 164
curcio, Clethrionomys 99
curtatus, Cricetulus 92
curzoniae, Ochotona 69
custos, Eothenomys 101
cuvieri, Gazella 210
cyanotis, Lepus 71
cyclopis, Macaca 64
cylindricauda, Sorex 24
cylindricornis, Capra 216
Cynaelurus 185
Cynailurus 185
Cynalopex 162
Cyon 164
Cystophoca 191
Cystophora 191

Dactylocerus 199
dahli, Meriones 126
dalmaticus, Eliomys 145
Dama 199, 202
dama, Cervus 199
dama, Gazella 210
dammah, Oryx 207
danubialis, Micromys 131
daphaenodon, Sorex 23
Dassie 192
dasycneme, Myotis 51
dasymallus, Pteropus 38
dasyurus, Gerbillus 121
dathei, Panthera 184
daubentoni, Myotis 50
daurica, Ochotona 68
dauricus, Mustela 171
dauricus, Spermophilus 83
dauuricus, Hemiechinus 15
davidianus, Elaphurus 202
davidianus, Sciurotamias 79
Deer 198
deltae, Crocidura 30
dementievi, Allactaga 154

denticulatus, Eliomys 145
Desman 32
Desmana 32
Desmaninae 32
Desmans 32
devius, Clethrionomys 99
Dhole 165
diamesus, Dryomys 146
diardi, Rattus 139
dichotomus, Rangifer 203
dichruroides, Apodemus 134
Dicrostonyx 96
Dieba 161
dietzi, Apodemus 134
dinaricus, Pitymys 108
Dinaromys 104
Dinops 63
Diplomesodon 31
Dipodidae 149
Dipodillus 123
Dipodinae 150
Dipodipus 150
Dipus 150
discolor, Gerbillus 121
diversicornis, Procapra 211
dobyi, Talpa 34
Dogs 161, 165
Dolomys 104
domestica, Mustela 170
Dorcas 208
dorcas, Gazella 209
Dormice 143
dorogostaiskii, Canis 161
dorothea, Pitymys 109
draco, Apodemus 137
dromedarius, Camelus 197
Dromedary 197
Dryomys 146
dsinezumi, Crocidura 28
dugon, Dugong 193
Dugong 193
Dugong 193
Dugongidae 193
Dugongidus 193
Dugungus 193
dukelskiae, Microtus 116
duodecimcostatus, Pitymys 110
duprasi, Pachyuromys 124
Dymecodon 36
Dysopes 63
dzigguetai, Equus 194

edwardsi, Herpestes 178
ehiki, Mustela 170
Eidolon 38
Elaphoceros 199
Elaphurus 201
Elaphus 199
elaphus, Cervus 200
elater, Allactaga 154
elbaensis, Gerbillus 122
elbaensis, Jaculus 151
Elephantidae 191
Elephants 191
Elephant-shrews 37
Elephantulus 37
Elephas 191
Eliomys 145
elisarjewi, Cricetulus 90
Elius 144
Elk 202
Ellobius 117
elongatus, Zalophus 186
emarginatus, Myotis 48
Emballonuridae 40
Endecapleura 118
endoi, Pipistrellus 53
enez-groezi, Microtus 115
enezsizunensis, Crocidura 27
Enhydra 176
Enhydrinae 176
Enhydris 176
ensicornis, Oryx 207
Eolagurus 116
Eoscalops 32
Eospalax 93
Eothenomys 100
Eozapus 149
Epimys 138
epiroticus, Microtus 113
Episoriculus 24
Eptesicus 56
Equidae 193
Equus 194
Eremaelurus 180
Eremiomys 116
Eremodipus 151
Ericius 15
Erignathus 190
Erinaceidae 13
Erinaceolus 15
Erinaceus 14
erminea, Mustela 168
erythroleucus, Praomys 140
erythropus, Xerus 79
erythrotis, Ochotona 68
etruscus, Suncus 31
Euarctos 165
Euarvicola 110
Eucapra 213
Eucervaria 180
Eucervus 199
Euchoreutes 158
Euchoreutinae 158
Eudorcas 208
Euhyaena 179
Euhyrax 192
Euibex 213

Eulagus 70
Eulepus 70
Eumetopias 187
Eumustela 168
Euotomys 97
euphratica, Allactaga 155
europaea, Talpa 33
europaeus, Erinaceus 14
Euroscaptor 32
Eurosorex 17
euryale, Rhinolophus 44
Euryalus 42
Eutamias 85
Euvespertilio 46
Euvesperugo 51
Euxerus 79
eva, Eothenomys 102
eversmanni, Cricetulus 91
eversmanni, Mustela 171
Evotomys 97
exiguus, Sorex 19

falconeri, Capra 216
famulus, Gerbillus 120
farsi, Meriones 127
fasciata, Histriophoca 189
favillus, Gerbillus 122
favillus, Jaculus 151
Felidae 179
Felis 180
felteni, Pitymys 109
Fennecus 162
ferrilata, Vulpes 164
ferrumequinum, Rhinolophus 42
fertilis, Hyperacrius 104
ferus, Camelus 197
ferus, Equus 194
Fiber 87, 106
fiber, Castor 87
flavescens, Crocidura 30
flavicollis, Apodemus 134
flavigula, Martes 174
flavimanus, Callosciurus 78
floweri, Crocidura 26
fodiens, Neomys 25
Foetorius 168
foina, Martes 173
fontanieri, Myospalax 93
formosovi, Capra 215
formosus, Myotis 50
fortis, Microtus 114
Foxes 162
Fox, flying 38
frater, Myotis 49
fruticus, Macropus 13
fudisanus, Rhinolophus 43
fujiensis, Myotis 47
fulvus, Spermophilus 85
fumatus, Praomys 141
furvus, Nyctalus 55
fuscata, Macaca 64
fuscipes, Jaculus 152
fuscocapillus, Ellobius 117

Gale 168
Galemys 32
Galeopardus 180
gallagheri, Gerbillus 121
galliae, Castor 87
Galomys 32
garganicus, Clethrionomys 99
garganicus, Sorex 20
Gazella 208
gazella, Gazella 209
gazella, Oryx 206
Gazelles 209
gemellus, Sorex 20
geminae, Apodemus 134
Genet 177
Genetta 177
genetta, Genetta 177
Gerbillinae 118
Gerbillus 118
gerbillus, Gerbillus 121
Gerbils 118
gerritmilleri, Pitymys 110
getulus, Atlantoxerus 79
giganteus, Spalax 130
gigas, Hydrodamalis 193
glareolus, Clethrionomys 98
Glareomys 97
gleadowi, Gerbillus 122
Gliridae 143
Glirulus 146
Glis 144
Glis 129
glis, Glis 144
Glutton 174
gmelini, Equus 194
Goa 211
Goats 213
golzmajeri, Sciurus 77
Goral 212
goral, Nemorhaedus 212
gordoni, Felis 181
gracilicornis, Gazella 209
gracillimus, Sorex 19
granarius, Sorex 21
grandis, Calomyscus 89
grandis, Microtus 113
gregalis, Microtus 116
griseomaculata, Talpa 33
groenlandicus, Pagophilus 189
grypus, Halichoerus 190
gud, Microtus 112
Gulo 174
gulo, Gulo 174
gundi, Ctenodactylus 160

Gundis 159
gurkha, Apodemus 137
gusevi, Talpa 34
gutturosa, Procapra 211

Halichoerus 190
Halicore 193
Halicyon 188
Haliphilus 188
Halticus 153
Haltomys 151
hamadensis, Gerbillus 122
Hamadryas 65
hamadryas, Papio 65
Hamster 89
Hamsters 88
hantu, Chimmarogale 31
hanuma, Mus 141
Harana 199
hardwickei, Rhinopoma 40
Hares 70
harrisoni, Acomys 142
harrisoni, Felis 182
harrisoni, Rhinopoma 40
Hartebeest 208
haymani, Gerbillus 120
Hedgehogs 14
helgolandicus, Mus 141
Heliomys 89
Heliophoca 190
Heliosorex 26
heljanensis, Crocidura 28
helvum, Eidolon 39
Hemiechinus 15
Hemionus 194
hemionus, Equus 194
Hemiotomys 105
Hemitragus 213
hemprichi, Otonycteris 59
Hendecapleura 118
henleyi, Gerbillus 121
henrici, Apodemus 137
heptneri, Microtus 115
heptneri, Mustela 169
heptneri, Salpingotus 158
heptneri, Spermophilus 84
heptopotamicus, Mustela 171
heptopotamicus, Sorex 19
heptopotamicus, Spermophilopsis 79
Herinaceus 14
hermani, Apodemus 135
Herpestes 178
hertigi, Micromys 132
hessei, Apodemus 134
himalayana, Ochotona 68
himalayica, Chimmarogale 31
hippagrus, Equus 195
Hippelaphus 199
Hippopotamidae 196
Hippopotamus 196
Hippopotamus 196
Hipposiderinae 44
Hipposideros 44
hipposideros, Rhinolophus 43
Hippotraginae 206
Hircus 213
hispanicus, Plecotus 61
hispida, Phoca 189
Histriophoca 189
hiwaensis, Talpa 35
hodgsoni, Pantholops 211
hoffmanni, Sciurus 77
Homalurus 17
hoogstraali, Gerbillus 123
Horse 194
hosonoi, Myotis 49
hosonoi, Sorex 23
hotsoni, Allactaga 155
hoveli, Myotis 49
hülleri, Sorex 20
hurrianae, Meriones 126
Hyaena 179
Hyaena 179
hyaena, Hyaena 179
Hyaenidae 179
Hydrelaphus 203
Hydrodamalis 193
Hydrogale 24
Hydromustela 168
Hydropotes 203
Hydrosorex 24
Hyperacrius 104
hypomelas, Paraechinus 16
hypsibius, Soriculus 24
Hypsugo 51
Hypudeus 96
Hyracoidea 192
Hyrax 192
Hyrax 192
Hystricidae 158
Hystrix 159

Ibex 215
ibex, Capra 215
ibicensis, Crocidura 28
Ichneumia 179
ichneumon, Herpestes 178
iconicus, Apodemus 135
Ictis 168
Idmi 209
Idomeneus 125
ifranensis, Apodemus 135
ikonnikovi, Myotis 48
ilaeus, Microtus 114
iliensis, Spermophilus 84
ilimpiensis, Martes 173
ilvanus, Apodemus 135
imaizumii, Clethrionomys 100

imaizumii, Talpa 35
imberbis, Tragelaphus 205
imhausi, Lophiomys 94
indica, Hystrix 159
indica, Nesokia 143
indica, Tatera 124
inermis, Hydropotes 203
inez, Eothenomys 102
inflatus, Gerbillus 121
innae, Microtus 114
innominabilis, Talpa 33
Insectivores 13
insularis, Phoca 189
intermedius, Allactaga 156
intermedius, Glis 144
Inuus 64
Isomys 138
Isotus 46

Jackal 162
Jaculus 151
jaculus, Jaculus 151
jamesi, Gerbillus 120
japonensis, Eptesicus 57
japonicus, Glirulus 147
japonicus, Macaca 64
javanicus, Pipistrellus 53
jayakari, Hemitragus 213
Jerboas 149
Jirds 118
jubatus, Acinonyx 185
jubatus, Eumetopias 187
juldaschi, Pitymys 107

kaguyae, Myotis 49
kahmanni, Apodemus 137
kaiseri, Dipodillus 123
kalininensis, Nyctereutes 164
kamensis, Cricetulus 91
kamensis, Ochotona 68
kaneii, Mustela 168
Kangurus 13
karelicus, Apodemus 137
karelicus, Sorex 19
karpinskii, Sorex 20
Karstomys 132
Kemas 212, 213
kerulenica, Mustela 169
khumbuensis, Rattus 139
Kiang 195
kiang, Equus 195
kikuchii, Microtus 115
kilikiae, Apodemus 135
kirgisorum, Microtus 114
Kolinsky 170
Kolonokus 168
koriakorum, Ovis 218
Korin 208
kosidianus, Rhinolophus 43
koslowi, Ochotona 69
kozlovi, Felis 182
kozlovi, Marmota 81
kozlovi, Salpingotus 158
kozlovi, Sorex 23
kratochvilli, Talpa 33
krkensis, Apodemus 135
kroecki, Muscardinus 145
Kuban 216
Kudu 205
kuhli, Pipistrellus 53
Kulan 194
kulan, Equus 194
kurilensis, Phoca 188
kutscheruki, Sorex 23

labensis, Pitymys 109
ladacensis, Ochotona 69
Lagomorpha 65
Lagomys 66
lagopus, Alopex 162
Lagurus 116
lagurus, Lagurus 117
lama, Ochotona 68
Lamprogale 172
laniger, Dryomys 146
laptevi, Odobenus 187
largha, Phoca 189
larvata, Paguma 178
lasia, Crocidura 29
Lasiopodomys 110
lasiopterus, Nyctalus 55
Lasiopus 179
lasiura, Crocidura 29
Lasiurus 60
Latax 176
latirostris, Marmota 80
latronum, Apodemus 137
lebanoticus, Nyctalus 55
Leiponyx 38
leisleri, Nyctalus 55
Lemmimicrotus 110
Lemmings 96
lemminus, Eothenomys 103
Lemmiscus 116
Lemmus 96
lemmus, Lemmus 97
Lemniscomys 138
Leo 184
leo, Panthera 185
Leopard 184
leosollicitus, Gerbillus 121
Leporidae 70
Leptailurus 180
Leptoceros 208
leptoceros, Gazella 210
leptodactylus, Spermophilopsis 79
Lepus 70
lervia, Capra 217
Leucocyon 162

Leucodon 26
leucodon, Crocidura 29
leucodon, Spalax 130
leucogaster, Murina 62
leucogaster, Scotophilus 59
leucogenys, Petaurista 86
leucomelas, Barbastella 60
Leuconoe 46
Leucorrhynchus 24
leucoryx, Oryx 207
leucurus, Alticola 103
leucurus, Pitymys 107
lewisi, Acomys 142
lhasaensis, Ochotona 67
libyca, Poecilictis 174
libycus, Meriones 127
lichtensteini, Jaculus 151
liechtensteini, Pitymys 109
Lion 185
liparensis, Eliomys 145
Liponycteris 41
Liponyx 38
Lipotus 175
lis, Sciurus 78
Lithotragus 212
loginovi, Microtus 112
longepedis, Dinaromys 105
longicaudatus, Cricetulus 91
Lophiomyinae 94
Lophiomys 94
lorenzi, Capra 215
Lotor 167
lotor, Procyon 167
Loxodon 191
Loxodonta 191
lucifugus, Myotis 51
Lupulus 161
Lupus 161
lupus, Canis 161
luridus, Meriones 127
lusitania, Crocidura 27
lusitanicus, Pitymys 110
luteus, Lagurus 117
Lutra 176
lutra, Lutra 176
Lutreola 168
lutreola, Mustela 170
Lutrinae 176
lutris, Enhydra 177
Lutrogale 176
Lutronectes 176
Lycaon 165
Lynceus 180
Lynchus 180
Lynx 182
Lynx 180
lynx, Felis 182

Macaca 64
Macaques 64
macdonaldi, Triaenops 45
macedonicus, Microtus 112
Machlis 199
macrobullaris, Plecotus 61
macrocephalicus, Myotis 50
macrodactylus, Myotis 51
macrodon, Sorex 30
Macroechinus 16
Macropodidae 13
Macropus 13
Macroscelidea 37
Macroscelides 37
Macroscelididae 37
Macrospalax 129
macrotis, Alticola 104
Macrotus 61
maderensis, Pipistrellus 54
maghrebi, Dipodillus 124
Magotus 64
major, Allactaga 156
major, Spermophilus 84
majori, Pitymys 109
makedonicus, Clethrionomys 99
makrami, Gerbillus 121
makrami, Sekeetamys 124
mandarinus, Microtus 116
mandshuricus, Lepus 74
Mangusta 178
manul, Felis 182
margarita, Felis 182
margarita, Sorex 22
maritimus, Gerbillus 123
maritimus, Thalarctos 166
Markhor 216
Marmota 80
marmota, Marmota 80
Marmots 80
Marsipolaemus 58
Marsupialia 13
Marsupials 13
Martens 172
Martes 172
martes, Martes 172
martinoi, Arvicola 105
martinoi, Glis 144
martinoi, Microtus 112
martinoi, Rhinolophus 42
martinoi, Spalax 130
martinoi, Spermophilus 83
Massoutiera 160
Mastomys 140
matruhensis, Crocidura 27
Matschiea 208
maximowieczi, Microtus 114
maximus, Elephas 191
maximus, Spermophilus 85
maxwelli, Lutra 176
Mediocricetus 92

Megachiroptera 38
megalodus, Acomys 142
Megalotis 162
mehelyi, Rhinolophus 44
melampus, Martes 173
Melanarctos 165
melanogaster, Eothenomys 101
melanopterus, Aeretes 87
meldensis, Microtus 113
Meledes 175
Meles 175
meles, Meles 175
Melesium 175
Melinae 175
Melitoryx 175
Mellivora 175
Mellivorinae 175
menzbieri, Marmota 81
meridianus, Meriones 126
meridionalis, Alces 202
Meriones 125
Mesechinus 15
Mesocricetus 92
mesopotamiae, Gerbillus 121
Mesospalax 129
Meteorus 58
metwallyi, Hemiechinus 15
Mice 130
Mice, birch 148
Mice, jumping 147
michaelis, Salpingotus 158
Microchiroptera 39
Microhippus 194
Micromys 131
microphthalmus, Spalax 130
microphyllum, Rhinopoma 40
microps, Apodemus 135
micropus, Paraechinus 16
Microspalax 129
Microtinae 94
Microtus 110
Micrurus 106
midas, Tadarida 63
middendorffi, Microtus 116
migratorius, Cricetulus 90
milleri, Apodemus 134
millicens, Microtus 115
minima, Talpa 34
minimus, Sorex 19
Miniopterus 61
Mink 170
minutissimus, Sorex 19
minutus, Micromys 131
minutus, Sorex 19
mirabilis, Sorex 23
Misothermus 96
mitchelli, Ochotona 68
miyakensis, Apodemus 136
mizura, Talpa 35
Mogera 32
Mole-rat 129
Moles 32
Mole-voles 117
Molossidae 62
momonga, Pteromys 86
Monachus 190
monachus, Monachus 190
mongolicus, Microtus 114
Mongooses 178
Monkeys 64
montanus, Clethrionomys 99
montebelli, Microtus 115
Moose 202
Mops 63
moschata, Desmana 32
moschata, Talpa 36
moschatus, Ovibos 213
Moschidae 198
moschiferus, Moschus 198
Moschus 198
mosquensis, Mustela 171
Mouflon 218
mulatta, Macaca 64
multiplex, Pitymys 108
Muntiacus 199
Muntjac 199
Muntjaccus 199
Muntjacus 199
Muridae 130
Murina 62
murina, Talpa 33
murinus, Suncus 30
murinus, Vespertilio 58
mursaevi, Ochotona 68
Mus 141
Muscardinus 144
muscatellum, Rhinopoma 40
Musculus 141
musculus, Mus 141
Musimon 217
Muskrat 106
Musmon 217
Mustela 168
Mustelidae 167
Mustelina 168
Mustelinae 168
mustersi, Calomyscus 89
mutus, Bos 206
Mygale 32
Mygalina 32
Mylarctos 165
Myocastor 159
Myodes 96
Myogale 32
Myogalea 32
Myomimus 147
Myomys 140
Myomyscus 140

Myopus 96
Myospalacinae 93
Myospalax 93
myospalax, Myospalax 94
Myotalpa 93
Myotis 46
myotis, Myotis 50
Myoxus 144
Myrmarctos 165
mystacinus, Apodemus 133
mystacinus, Myotis 47
mzabi, Massoutiera 160

Naemorhaedus 212
nalutensis, Gerbillus 122
Nanger 208
Nannospalax 129
Nannugo 51
nanus, Gerbillus 120
napaea, Sicista 149
naso, Euchoreutes 158
nasomaculatus, Addax 207
nasutus, Eptesicus 56
nathusii, Pipistrellus 53
nattereri, Myotis 49
nayaur, Pseudois 217
nebulosa, Talpa 33
neglectus, Felis 182
Nemomys 132
Nemorhaedus 212
Nemorhedus 212
Nemorrhedus 212
Nemotragus 212
Neoaschizomys 97
Neodon 106
Neomys 24
nepalensis, Equus 195
nepalensis, Sorex 24
Nepus 193
Nesocia 143
Nesokia 143
newtoni, Mesocricetus 93
Nicteris 41
niethammeri, Neomys 25
nikolajevi, Microtus 112
niloticus, Arvicanthis 138
nilssoni, Eptesicus 57
nippon, Cervus 200
nitedula, Dryomys 146
nivalis, Microtus 112
nivalis, Mustela 169
niviventer, Rattus 140
nobilis, Mustela 171
Noctula 56
noctula, Nyctalus 55
Noctules 54
Noctulinia 54
norikuranus, Rhinolophus 43
norvegicus, Rattus 140
nubica, Oryx 207
nudiventris, Taphozous 41
Nyctalus 54
Nyctereutes 164
Nycteridae 41
Nycteris 41
Nycterops 41
Nycticeius 59
Nycticejus 59
Nycticeus 59
Nycticeyx 59
Nyctinoma 63
Nyctinomas 63
Nyctinomia 63
Nyctinomus 63
Nystactes 46

obensis, Talpa 33
obesus, Psammomys 128
obscura, Martes 173
obscura, Vormela 171
occidentalis, Mustela 171
occiduus, Gerbillus 123
Ochetomys 105
Ochotona 66
Ochotonidae 65
Ocypetes 62
Odmaelurus 177
Odobenidae 187
Odobenus 187
Odocoileus 202
Odontodorcus 198
oeconomus, Microtus 115
Oedocephalus 159
ogasimanus, Rhinolophus 43
ognevi, Crocidura 29
ognevi, Marmota 81
ognevi, Martes 173
ognevi, Mustela 171
Ognevia 17
Ogotoma 66
ohtai, Talpa 35
oiostolus, Lepus 73
okinoshimae, Crocidura 28
olchonensis, Alticola 103
olitor, Eothenomys 101
omanensis, Eptesicus 57
Onager 194
Onager 194
Oncoides 180
Ondatra 106
opimus, Rhombomys 129
Oreosciurus 76
orientalis, Jaculus 153
orientalis, Mustela 171
orientalis, Vespertilio 59
Orthaegoceros 213
Oryctolagus 74
Oryx 207

Oryx 206
osiris, Gazella 209
Otariidae 186
Otocolobus 180
Otonycteris 59
Otters 176
Ovibos 213
Ovis 217
oweni, Scapanulus 37
Ox 206
Ox, musk 213
oxiana, Felis 182
Oxygous 161
Oxyrhin 17
oyaensis, Microtus 113
ozensis, Myotis 49

Pachyceros 217
Pachyomus 56
Pachyotus 59
Pachyura 30
Pachyuromys 124
Pagomys 188
Pagophilus 189
Pagophoca 189
Paguma 178
pallasi, Ochotona 69
Pallasiinus 110
Pallasiomys 125
pallescens, Apodemus 137
pallidior, Vormela 171
pallidus, Mustela 171
Palmatus 199
Palm-civet 178
palmyrae, Gerbillus 121
Paludicola 105
Panthera 184
Pantholops 211
Panugo 54
Papio 65
Paradipus 151
paradoxa, Selevinia 147
Paradoxodon 30
paradoxus, Cardiocranius 157
Paraechinus 16
Paralces 202
Parameriones 125
Paramyotis 46
Parascaptor 32
Pardina 180
Pardus 184
pardus, Panthera 184
parryi, Spermophilus 85
parvidens, Alticola 103
Pasang 214
pashtomus, Eptesicus 57
patrovi, Clethrionomys 99
Paurodus 26
Pelagios 190
Pelagocyon 190
peleponnesiacus, Spalax 130
peninsulae, Apodemus 136
pequinius, Myotis 51
perforatus, Taphozous 41
peregusna, Vormela 171
pergrisea, Crocidura 29
Perissodactyla 193
Peroechinus 14
perpallidus, Gerbillus 122
perpallidus, Meriones 128
persicus, Meriones 126
persicus, Triaenops 45
personatus, Myomimus 147
perspicillata, Lutra 176
Petalia 41
Petaurista 86
Petromys 132
Phaiomys 106
Phaulomys 100
Phoca 188
Phocidae 187
Phodopus 89
picticaudata, Procapra 211
pictus, Lycaon 165
Pigs 196
Pika 66
Pikas 65
pilirostris, Urotrichus 37
Pinalea 24
pindicus, Glis 144
Pinnipedia 185
Pipistrelles 51
Pipistrellus 51
pipistrellus, Pipistrellus 52
piriformis, Mustela 171
pirinus, Clethrionomys 99
Pithes 64
Pitymys 106
Platycercomys 156
Platyceros 199
Platycranius 103
Platystomus 193
platyurus, Pygeretmus 157
Plecotus 61
Plerodus 30
Poecilictis 174
poecilops, Gerbillus 120
Poephagus 205
Polecats 171
Poliailurus 180
Porcupines 158
portenkoi, Lemmus 97
portenkoi, Sorex 22
Praomys 140
Praticola 105
Primates 63
primigenius, Bos 206
princeps, Sorex 22

Prionailurus 180
Proboscidea 191
Procapra 210
Procavia 192
Procaviidae 192
Procops 199
Procyon 167
Procyonidae 166
procyonoides, Nyctereutes 164
proditor, Eothenomys 101
Prodorcas 210
Proedromys 110
Proeulagus 70
Profelis 180
Prolagus 65
Prometheomys 117
propinquus, Eptesicus 57
Protelas 179
Protemnodon 13
Prox 199
pruinosus, Myotis 51
przewalskii, Brachiones 128
przewalskii, Procapra 211
Przewalskium 199
przhewalskii, Phodopus 89
Psammomys 128
psammophilous, Gerbillus 121
Pseudaxis 199
Pseudocervus 199
pseudogriseus, Cricetulus 91
Pseudois 217
pseudonapaea, Sicista 149
Pseudovis 217
Pterocyon 38
Pteromys 86
Pteropodidae 38
Pteropus 38
Pterygistes 54
Ptychorhina 44
pulchellum, Diplomesodon 31
pulchellus, Cervus 200
pulcher, Miniopterus 62
pumila, Tadarida 63
pumilio, Allactagulus 156
punctulata, Talpa 33
Pusa 176, 188
pusilla, Ochotona 67
Putorius 168
putorius, Mustela 171
Pygerethmus 156
Pygeretmus 156
pygmaeus, Spermophilus 83
pyramidum, Gerbillus 122
pyrenaica, Capra 216
pyrenaica, Genetta 177
pyrenaicus, Galemys 32
pyrrhonota, Sorex 20

quadraticauda, Blarinella 25
quercinus, Eliomys 145

Rabbit 74
Raccoon 166
Raccoon-dog 164
raddei, Mesocricetus 92
raddei, Sorex 21
ralli, Spermophilus 84
Rangifer 203
rarus, Jaculus 151
Rat, Crested 94
Ratel 175
Ratellus 175
Rats 138
rattoides, Rattus 139
Rattus 138
rattus, Rattus 139
ravijojla, Dryomys 146
reevesi, Muntiacus 199
regulus, Eothenomys 102
Reindeer 203
relictus, Spermophilus 84
religiosa, Crocidura 27
rex, Clethrionomys 99
rex, Meriones 126
Rhesus 64
Rhim 210
Rhinocrepis 42
Rhinolophidae 42
Rhinolophinae 42
Rhinolophus 42
Rhinopoma 39
Rhinopomatidae 39
Rhombomys 129
Rhyneptesicus 56
Rhytina 193
ricketti, Myotis 51
Rickettia 46
Rigoon 190
rjabovi, Clethrionomys 98
roberti, Microtus 112
roborovskii, Phodopus 89
robusta, Mustela 171
robusta, Talpa 35
Rodentia 74
romana, Talpa 34
Romicia 51
Rosmarus 187
rosmarus, Odobenus 187
rothschildi, Myospalax 93
Rousettus 38
roylei, Alticola 103
roylei, Ochotona 68
rozeti, Elephantulus 37
rubiginosus, Rhinolophus 43
rueppelli, Pipistrellus 54
rueppelli, Vulpes 164
rufescens, Apodemus 135
rufescens, Ochotona 69

rufina, Gazella 210
rufocanus, Clethrionomys 99
rufogriseus, Macropus 13
Ruminantia 197
Ruminants 197
Rupicapra 212
rupicapra, Rupicapra 212
russatus, Acomys 142
russula, Crocidura 27
ruthenus, Microtus 113
rutila, Ochotona 68
rutilus, Clethrionomys 98
Rytina 193

sabaneevi, Martes 172
Sable 173
Sacalius 161
Saccolaimus 41
sachalinensis, Microtus 114
sacramenti, Meriones 128
sagitta, Dipus 150
Saiga 211
Salmacis 64
Salpingotus 157
sapidus, Arvicola 105
sardus, Prolagus 66
satunini, Mustela 171
saurica, Alticola 103
savii, Pipistrellus 54
savii, Pitymys 109
scaloni, Microtus 115
Scalopinae 36
Scapanulus 37
Scaptochirus 32
Scarturus 153
schaposchnikowi, Prometheomys 117
scheffeli, Felis 182
schelkovnikovi, Neomys 25
schelkovnikovi, Pitymys 109
schisticolor, Myopus 96
schlieffeni, Nycticeius 59
schreibersi, Miniopterus 62
Scirtetes 153
Scirtomys 153
Scirtopoda 151, 153
Sciuridae 75
Sciuropterus 86
Sciurotamias 79
Sciurus 76
scorodumovi, Vulpes 163
Scotophilus 59
Scotozous 51
Scrofa 196
scrofa, Sus 196
Sea-cows 192
Seal, fur 186
Sea-lions 186
Seals 187
Sekeetamys 124
Selenarctos 165
selenginus, Meriones 126
Selevinia 147
Seleviniidae 147
Selysius 46
Semicricetus 92
semideserta, Allactaga 154
semenovi, Alopex 162
semotus, Apodemus 137
senegalensis, Oryx 207
sepiacea, Talpa 33
septentrionalis, Vulpes 163
serbicus, Pitymys 108
sericea, Crocidura 30
Serotines 56
serotinus, Eptesicus 57
Serow 212
Serval 183
Serval 180
serval, Felis 183
Servalina 180
setchuanus, Eozapus 149
setonbrownei, Gerbillus 120
setzeri, Myomimus 147
severtzovi, Allactaga 155
shanseius, Eothenomys 102
shawi, Meriones 127
Sheep 217
shikokensis, Sorex 20
shiroumanus, Sorex 23
shitkovi, Pygeretmus 157
Shrew-moles 36
Shrews 17
shukurovi, Ochotona 69
Siaga 211
sibirica, Allactaga 154
sibirica, Crocidura 29
sibirica, Mustela 170
sibirica, Phoca 189
sibiricus, Lemmus 97
sibiricus, Sus 196
sibiricus, Tamias 85
Sica 199
Sicista 148
Sicistinae 148
Sikaillus 199
sikimensis, Pitymys 107
silanus, Sorex 20
silvatica, Equus 194
silvestris, Equus 194
silvestris, Felis 180
Simia 64
simoni, Dipodillus 123
sinalis, Sorex 22
sinensis, Lepus 73
sinesella, Ovis 218
Sinisus 196
Siphneus 93
Sirenia 192

sirtalensis, Microtus 116
Sminthus 148
smithi, Eothenomys 102
smithi, Myospalax 94
socialis, Microtus 112
somaliae, Vulpes 164
sordida, Talpa 33
Sorex 17
Soricidae 17
Soricidus 17
Soriculus 24
Sousliks 83
sowerbyi, Apodemus 136
Spalacidae 129
Spalacomys 143
Spalax 129
speciosa, Macaca 64
speciosus, Apodemus 136
Spermophilopsis 79
Spermophilus 82
Squirrels 75
Squirrels, Flying 86
stankovici, Arvicola 105
steini, Talpa 34
Stellerus 193
Stemmatopus 191
Stenocranius 110
Stoat 168
stoliczkanus, Alticola 103
strandzensis, Microtus 112
streeti, Talpa 34
streetorum, Crocidura 29
streetorum, Talpa 34
strelzowi, Alticola 104
Strepsiceros 205
stroganovi, Felis 182
stroganovi, Sorex 19
Strongyloceros 199
Stylocerus 199
Stylodipus 153
suaveolens, Crocidura 27
subarvalis, Microtus 113
subgutturosa, Gazella 209
subpunctulata, Talpa 33
subsolanus, Gerbillus 123
subterraneus, Pitymys 108
subtilis, Sicista 148
Suidae 196
Suiformes 196
Sumeriomys 110
Suncus 30
sungorus, Phodopus 89
Sunkus 30
superans, Vespertilio 58
suramensis, Pitymys 109
Sus 196
suslicus, Spermophilus 83
swinhoei, Callosciurus 78
Sylvanus 64
sylvanus, Macaca 65
sylvaticus, Apodemus 134
Sylvicola 110
Synotus 60
syriaca, Procavia 192

Tadarida 63
Tahr 213
taigica, Sicista 148
talassicus, Microtus 116
Talpa 32
Talpidae 32
Talpinae 32
talpinus, Ellobius 117
Talpoides 129
talpoides, Urotrichus 36
talyschensis, Talpa 34
tamariscinus, Meriones 126
Tamias 85
tanaiticus, Alactagulus 156
tanaiticus, Ellobius 117
Taphonycteris 41
Taphozous 41
Tarandus 203
tarandus, Rangifer 203
tarimensis, Vulpes 163
Tarimolagus 70
Tarpan 194
tatarica, Mustela 170
tatarica, Saiga 212
Tatera 124
tatrica, Rupicapra 213
tatricus, Pitymys 108
tatricus, Sorex 23
tauricus, Apodemus 134
tauricus, Cervus 201
Taurus 205
Taxus 175
teberdina, Mustela 168
telum, Stylodipus 153
temmincki, Suncus 30
tenebrosa, Murina 62
Tenes 76
teniotis, Tadarida 63
terrestris, Arvicola 105
Terricola 106
tetradactyla, Allactaga 156
Thalarctos 166
Thalassarctos 166
Thalassiarchus 166
thebaica, Nycteris 41
thessalicus, Spalax 130
thibetana, Ochotona 67
thibetanus, Selenarctos 165
thomasi, Ochotona 67
thomasi, Pitymys 110
thomasi, Salpingotus 158
Thos 161
thracius, Spermophilus 83

Tibetholagus 66
Tiger 184
Tigris 184
tigris, Panthera 184
timidus, Lepus 72
tokudae, Talpa 36
tomensis, Martes 173
torquatus, Dicrostonyx 96
toschii, Vulpes 163
Trachelocele 208
Tragelaphus 205
Tragops 208
Tragopsis 208
Tragus 213
transcaspicus, Microtus 114
transcaucasica, Panthera 184
transrypheus, Sorex 22
transuralensis, Talpa 33
traubi, Hyperacrius 104
trebevicensis, Dinaromys 105
Triaenops 45
Trichaelurus 180
Trichechus 187
tridens, Asellia 45
tripolitanicus, Jaculus 152
tristrami, Meriones 126
triton, Cricetulus 92
Trogopterus 87
Tscherskia 90
tschuktschorum, Sorex 19
tundrensis, Clethrionomys 98
tuneti, Rhinolophus 44
tungussensis, Martes 173
Tur 216
turcmenica, Capra 214
turcmenicus, Jaculus 153
Turocapra 213
turovi, Microtus 112
turovi, Mustela 170
Turus 213
tuvinicus, Alticola 103
tuvinicus, Castor 87
tuvinicus, Mustela 171
Tylopoda 197

ubsanensis, Dipus 151
uenoi, Plecotus 61
Ujhelyiana 129
uliginosus, Microtus 114
Uncia 184
uncia, Panthera 185
undulatus, Spermophilus 84
unguiculatus, Meriones 126
unguiculatus, Sorex 23
uralensis, Martes 172
Urial 218
Urocitellus 82
Urocricetus 90
Urolynchus 180
Urotragus 212
Urotrichus 36
Ursarctos 165
Ursidae 165
ursinus, Callorhinus 186
Ursitaxus 175
Ursus 165
Urus 205
Urus 206
ussuriensis, Vespertilio 58
uxantisi, Crocidura 27

vagneri, Glis 144
valverdi, Eliomys 145
Vansonia 51
variscicus, Clethrionomys 99
vastus, Jaculus 151
velessiensis, Talpa 33
ventromaculata, Talpa 33
versicolor, Gerbillus 121
Vespertilio 58
Vespertilionidae 45
Vesperugo 58
Vesperus 58
vinogradovi, Dicrostonyx 96
vinogradovi, Meriones 126
vinogradovi, Pitymys 109
vinogradovi, Pygeretmus 157
vir, Sorex 21
virginianus, Odocoileus 202
vison, Mustela 170
vitimensis, Martes 173
vitulina, Phoca 188
Viverridae 177
volans, Pteromys 86
Voles 97
volgensis, Clethrionomys 98
Vormela 171
vulgaris, Sciurus 77
Vulpes 162
vulpes, Vulpes 163
Vulpicanis 161

walie, Capra 215
Wallabia 13
Wallaby 13
walli, Eptesicus 56
Walrus 187
Wapiti 200
wardi, Ovibos 213
wassifi, Gerbillus 120
wassifi, Paraechinus 16
Weasel 169
Weasel, striped 174
wettsteini, Sorex 20
whitchurchi, Jaculus 151
Wisent 206
wogura, Talpa 35
Wolf 161

wollebaeki, Zalophus 186
Wolverine 174
wynnei, Hyperacrius 104

xanthipes, Trogopterus 87
Xerus 79

Yak 206
yakushimae, Cervus 200
yarkandensis, Lepus 74
yemeni, Praomys 141

zachidovi, Marmota 81
zakariai, Dipodillus 123
Zalophus 186
Zapodidae 147
Zapodinae 149
zarudnyi, Crocidura 29
zarudnyi, Meriones 128
zayidi, Taphozous 41
zerda, Vulpes 164
Zeren 211
Zibellina 172
zibellina, Martes 173
zibethicus, Ondatra 106
zimmermanni, Crocidura 28
Zokor 93
Zokors 93
zygomaticus, Hyperacrius 104